THE NATURE OF
MODERN MATHEMATICS
THIRD EDITION

Other Titles by the Same Author

Trigonometry for College Students, Second Edition
Karl J. Smith

Introduction to Symbolic Logic
Karl J. Smith

Beginning Algebra for College Students, Second Edition
Karl J. Smith and Patrick J. Boyle

Study Guide for Beginning Algebra for College Students, Second Edition
Karl J. Smith and Patrick J. Boyle

Intermediate Algebra for College Students
Karl J. Smith and Patrick J. Boyle

Study Guide for Intermediate Algebra for College Students
Karl J. Smith and Patrick J. Boyle

Algebra and Trigonometry
Karl J. Smith and Patrick J. Boyle

College Algebra
Karl J. Smith and Patrick J. Boyle

Analytic Geometry: A Refresher
Karl J. Smith

Precalculus Mathematics: A Functional Approach
Karl J. Smith

THE NATURE OF MODERN MATHEMATICS

THIRD EDITION

KARL J. SMITH

Santa Rosa Junior College

BROOKS/COLE PUBLISHING COMPANY
Monterey, California
A Division of Wadsworth, Inc.

Printed in the United States of America

10 9 8 7 6 5 4 3 2

Library of Congress Cataloging in Publication Data

Smith, Karl J
    The nature of modern mathematics.

    Includes index.
    1. Mathematics—1961-   I. Title.
QA39.2.S6 1980   510   79-20064
ISBN 0–8185–0352–1

Acquisition Editor: *Craig Barth*
Production Editor: *Cece Munson*
Interior and Cover Design: *Katherine Minerva*
Cover Drawing: *Robert Bordow*
Typesetting: *Holmes, San Jose, California*

To my wife, Linda,
and our children, Melissa and Shannon

PREFACE

This book was written to create a positive attitude toward mathematics—a realization that mathematics is not an endless procession of dull manipulations, theorems, proofs, and irrelevant topics. Rather than simply presenting the technical details needed to proceed to the next course, the book attempts to give insight into what mathematics is, what it accomplishes, and how it is pursued as a human enterprise.

Intended primarily for a one- or two-semester course addressed to the liberal arts student, this book presupposes almost no knowledge from previous courses. The student is asked not to "do math problems" but rather to "experience math." For this reason many of the topics may seem unorthodox when compared with previous experiences the student has had in mathematics courses.

Historical notes and history charts are provided to give the student some insight into the humanness of mathematics. Rather than being strictly biographical reports, these notes focus on the people. Nearly every major mathematician will have some part of his or her life to tell on the pages of this book. I hope that a glimpse of the people of mathematics will give you a glimpse of the nature of mathematics.

I frequently encounter people who tell me about their unpleasant experiences with mathematics. I have a true sympathy for these people, and I recall one of my elementary teachers who assigned additional arithmetic problems as punishment. This can only create negative attitudes toward mathematics, which is indeed unfortunate. If elementary teachers and parents have a positive attitude toward mathematics, their children will see some of the beauty of the subject. I want the student to come away from this course with the feeling that mathematics can be pleasant, useful, and practical—and enjoyed for its own sake.

This text is designed so that professor and students have many options. Although some chapters are dependent on others, the development of the course allows for great flexibility in changing the order of the chapters and in deemphasizing or eliminating certain sections without disrupting the continuity or requiring supplementary material. For example, if you don't like to consider logic as the second chapter, you can insert it wherever you wish, if at all. A typical three-unit course could cover from five to six chapters.

One of the most frequently occurring suggestions that I received from reviewers and users of the book was to rearrange the chapters. Nearly everyone had their "own arrrangement." This simply points out one of the strongest features of this book—that the order in which the material is presented depends on the preference of the instructor and the make-up of the students taking the course. The material can be easily adapted to almost any order or arrangement. Several alternative course outlines are described in the Instructor's Manual accompanying this text. The dependence of the chapters is as follows:

Chapter	Prerequisite
1	none
2	none
3	none
4	none
5	none
6	Chapter 4
7	none
8	none
9	Chapter 8
10	none

Chapter 1 introduces the student to the spirit and style of the course by showing how patterns are used in mathematics. Patterns are applied to such diverse ideas as the golden section and the metric system. However, any of the sections in this chapter can be omitted without any loss of continuity in the development of the remaining material in this book.

Chapter 2 combines two fundamental concepts of mathematics—sets and logic—by showing how logical reasoning applies to everyday arguments as well as to relationships among sets and to the material that will follow in the book.

Calculator and computer applications from Chapter 3 are integrated throughout the rest of the book, but they are always found in the problem sets and are indicated by the words *Calculator Problem* or *Computer Problem*. The instructor has several options for handling the computer material.

1. Omit Chapter 3.
2. Treat Chapter 3 merely as an introduction to calculators and computers. In this case, only Sections 3.1 through 3.3 need be covered.
3. Treat calculators (Section 3.2) early in the course and computers as the last section in the course (time permitting).
4. Emphasize Chapter 3, including programming, and use the computer throughout the course. I have experienced success using this material with students of varied abilities. The computer problems can be discussed in class or assigned, and computer solutions can be used as part of the course's development. This is one of the features of the book: *computers can be integrated into the entire course*. The computer language used is BASIC, but the material could easily be adapted to FORTRAN or other computer languages.

Chapter 4 concerns itself with the development of our numeration system as well as with the different sets of numbers that are of interest to mathematicians and laypersons.

Chapter 5 treats arithmetic from a layperson's point of view by discussing interest as it is applied to installment buying, automobile buying, home buying, and insurance. Although important to the consumer, Sections 5.3, 5.4, and 5.5 are not typical of a usual mathematics course. Depending on the class, the instructor might wish to skip these sections or to cover them only as a reading assignment. Chapter 5 is designed to help the reader make more informed decisions regarding personal finances.

Chapter 6, on the nature of algebra, returns to the thread begun in Chapter 4 and builds the idea of a mathematical system related to the structure of algebra. Chapter 7, on the nature of geometry, takes a layperson's look at non-Euclidean geometry, topology, and the famous Königsberg Bridge Problem.

Chapter 8 develops ways of counting without actually enumerating all possibilities. Permutations and combinations lead quite naturally to the notion of probability in Chapter 9, although Chapter 9 may be studied without Chapter 8 if the instructor is selective in the problems assigned. The final Chapter, Chapter 10, deals with statistics, with emphasis on how statistics can be manipulated to support almost any argument.

A wide range of problems is offered, since students who enroll in a survey course come from a wide range of backgrounds. There are quite a number of problems of the routine or drill type, as well as a large selection of thought-provoking problems. Much interesting material appears in the problems, and the reader can learn quite a bit just by reading them. The answers to the odd-numbered problems are provided in the back of the book, and complete solutions to all of the problems, as well as sample tests, are available in the Instructor's Manual. The "Mind Bogglers" and "Problems for Individual Study" sections should be utilized as interest demands. They are not to be forced on the student but should be investigated for their own sake. Hopefully, these sections will stimulate class discussion and still other avenues of investigation.

I also thank the many people who wrote me offering suggestions for revising the book: Jan Boal, T. W. Buquoi, Chris C. Braunschweiger, T. A. Bronikowski, Eugene Callahan, Michael W. Carroll, Joseph Cavanaugh, Ralph De Marr, Samuel L. Dunn, Ernest Fandreyer, Ralph Gellar, Charles E. Johnson, Linda H. Kodama, Daniel Koral, Helen Kriegsman, C. Deborah Laughton, John Mullen, John Palumbo, Gary Peterson, O. Sassian, Clifford H. Wagner, and Barbara Williams, as well as all those who responded to my questionnaire.

As the 23rd person mentioned in the acknowledgments, I wish to thank William A. Leonard for his detailed and thoughtful review. His excellent suggestions and comments were very helpful to me. I appreciate the suggestions and comments of the reviewers of the first two editions: Jerald T. Ball, Charles M. Bundrick, William J. Eccles, James J. Jackson, Martha C. Jordan, William Leahey, James R. Smart, Arnold Villone, and Stephen S. Willoughby. I also extend my gratitude to the reviewers of the third edition: George Berzsenyi, James R. Choike, Thomas C. Craven, Charles Downey, Gregory N. Fiore, John J. Hanevy, Caroline Hollingsworth, Judy D. Kennedy, George McNulty, and James V. Rauff.

I would especially like to thank my editor, Robert J. Wisner of New Mexico State, for his countless suggestions and ideas, and Craig Barth, Joan Marsh, Cece Munson, and Jamie Brooks of Brooks/Cole, as well as Jack Thornton for his sterling leadership and inspiration.

Finally, my thanks go to my wife, Linda, whose suggestions and help were invaluable. I will never be able to repay her and my family for those hours lost while I was working on the manuscript and for those evenings she spent "proofing." Without her, this book would exist only in my dreams.

Karl J. Smith

CREDITS

Page 6. B. C. cartoon reprinted by permission of Johnny Hart and Field Enterprises, Inc.

Page 10. Radiolaria form from *Report on the Scientific Results of the Voyage of H. M. S. Challenge* (Vol. 18), 1887.

Page 11. Photo of Pascal courtesy of Smith Collection, Rare Book and Manuscript Library, Columbia University.

Page 12. From MENSA, the high I.Q. Society, 1701 W. 3rd Street, Brooklyn, N.Y. 11223. Reprinted by permission. MENSA is an organization whose members have scored at or above the 98th percentile of the general population on any standardized I.Q. test.

Page 13. Peanuts cartoon © 1967 United Feature Syndicate, Inc. Reprinted by permission.

Page 16. Melancholia by Albrecht Dürer. Courtesy of The Bettmann Archive, Inc.

Page 19. Title page of an arithmetic book of Petrus Apianus, Ingolstadt. It shows Pascal's triangle as it was first printed in 1527.

Page 26. Bathers by Georges Seurat. Courtesy of the Trustees, The National Gallery, London.

Page 27. Dynamic Symmetry of a Human Face by Leonardo da Vinci, and *Proportions of the Human Body* by Albrecht Dürer. © VEB Bibliographisches Institut Leipzig. Reprinted by permission.

Page 28. Parthenon courtesy of The Bettmann Archive, Inc.

Page 32. Photo of *David* by Michelangelo. Firenze, Galleria dell' Accademia. Courtesy of the Gallerie Fiorentine.

Page 32. B. C. cartoon reprinted by permission of Johnny Hart and Field Enterprises, Inc.

Page 36. Dennis the Menace cartoon © 1972. Reprinted by permission of Hank Ketcham.

Page 37. Graffiti by Leary. Reprinted with permission of The McNaught Syndicate, Inc.

Page 38. Peanuts cartoon © 1963 United Feature Syndicate, Inc. Reprinted by permission.

Page 38. Photo © 1973 by United Press International, Inc. Reprinted by permission.

Page 40. News article © 1971 United Press International. Reprinted by permission.

Page 44. Photo courtesy of Department of Transportation, Ashland, Ohio.

Page 45. Page reproduced from *First Book of Arithmetic for Pupils Uniting Oral and Written Exercises*, by E. E. White. American Book Co., 1890.

Page 46. Limerick and cartoon from *The Metrics Are Coming! The Metrics Are Coming!*, by R. Cardnell. Copyright © 1975 by Dorrance & Company. Reprinted by permission.

Page 48. Halve Your Meter courtesy of Dr. Charles B. Rhinehart.

Page 52. Wizard of Id reprinted by permission of Johnny Hart and Field Enterprises, Inc.

Page 56. Apple crisp recipe courtesy of Mrs. Rigmor Sovndal; cake recipe courtesy of Mrs. Linda Smith.

Page 62. Photo of Aristotle from *The Endless Quest*, by F. W. Westaway (New York: Hillman-Curl Inc., 1936).

Page 63. Photo of Leibniz courtesy of Smith Collection, Rare Book and Manuscript Library, Columbia University.

Page 64. Photo of Boole courtesy of Smith Collection, Rare Book and Manuscript Library, Columbia University.

Page 66. Cartoon by Robert Mankoff from *Saturday Review* (6-11-77). Copyright © 1977 by Robert Mankoff. Reprinted by permission of Robert Mankoff.

Page 69. Photo of Galileo from *The Endless Quest*, by F. W. Westaway (New York: Hillman-Curl Inc., 1936).

Page 70. B. C. cartoon reprinted by permission of Johnny Hart and Field Enterprises, Inc.

tion, Carrollton, Texas.

Page 160. Photos courtesy of IBM.

Page 162. "A Programmer's Lament" from *The Compleat Computer.* © 1976, Dennie Van Tassel. Reprinted by permission of the publisher, Science Research Associates, Inc.

Page 164. Lion courtesy of Honeywell Information Systems.

Page 165. Margin note reprinted with the permission of *The Wall Street Journal,* © Dow Jones & Company, Inc. (1973).

Page 166. Flow chart from General Library Information Leaflet No. 1. University of California, Davis. Reprinted by permission.

Page 169. Computer composition from *Cybernetic Serendipity,* edited by J. Reichardt. © 1968 by W. J. Mackay & Co., Ltd.; London: *Studio International*; New York: Frederick A. Prager. Reprinted by permission.

Page 171. Limerick from *Cybernetic Serendipity,* edited by J. Reichardt. © 1968 by W. J. Mackay & Co., Ltd.; London: *Studio International*; New York: Frederick A. Prager. Reprinted by permission.

Page 174. "Machine Translation" from *The Compleat Computer.* © 1976, Dennie Van Tassel. Reprinted by permission of the publisher, Science Research Associates, Inc.

Page 175. "Computer Career Opportunities" courtesy of Honeywell, Inc.

Page 177. Cartoon by Mal, © 1971 Washington Star Syndicate, Inc. Reprinted by permission.

Page 183. Flowers (left) courtesy of Computra, Inc., Computer Art for People, 2513 Moore Road, Muncie, Ind. 47302. Graphic displays courtesy of IMLAC Corporation, Needham, Mass. Bottom (center) courtesy of IEEE Computer Society.

Page 185. Illustration courtesy of Dr. E. E. Herron, Biostereometrics Laboratory of Texas Institute for Rehabilitation and Research.

Page 191. Cartoon by Sidney Harris. Reprinted with permission of DATAMATION® magazine, © Copyright by TECHNICAL PUBLISHING COMPANY, A Division of DUN-DONNELLEY PUBLISHING CORPORATION, A DUN & BRADSTREET COMPANY, 1978—all rights reserved.

Page 191. Computer poems from *Cybernetic Serendipity,* edited by J. Reichardt. © 1968 by W. J. Mackay & Co., Ltd.; London: *Studio International*; New York: Frederick A. Prager. Reprinted by permission.

Page 192. Table 3.4 adapted from *Programming Languages,* by the Software Writing Group. A publication of the Digital Equipment Corporation. Reprinted by permission.

Page 193. Cartoon by Stew Burgess. Reprinted with permission of DATAMATION® magazine, © Copyright by TECHNICAL PUBLISHING COMPANY, A Division of DUN-DONNELLEY PUBLISHING CORPORATION, A DUN & BRADSTREET COMPANY, 1971—all rights reserved.

Page 197. Floorplan developed by Saphier, Lerner, Schindler, Environetics, Inc., 600 Madison Avenue, New York, N. Y. 10017. Reprinted by permission.

Page 204. Peanuts cartoon © 1963 United Feature Syndicate, Inc. Reprinted by permission.

Page 206. Rhind Papyrus reproduced by courtesy of the Trustees of the British Museum.

Page 208. Photo from *Science Awakening,* by B. L. Van Der Waerden, 1961, Oxford University Press. Courtesy of the Berlin Museum and Wolters-Noordhoff Publishing Company.

Page 209. News article © 1972 United Press International. Reprinted by permission.

Page 210. "Eden's Fall Pippins" from *Encyclopedia of Wit, Humor, and Wisdom.* Copyright 1949 Abingdon Press. Reprinted by permission.

Page 213. Mayan numerals from *An Introduction to the Study of the Mayan Hieroglyphics,* by S. G. Morley. Reprinted by permission of the Scholarly Press.

Page 214. From *Margarita Philosophica Nova,* 1512. Courtesy of Museum of the History of Science, University of Oxford.

Page 219. Drawing from an unpublished manuscript of a Venetian monk, in *A History of Mathematics,* by C. Boyer. Copyright 1968 by John Wiley & Sons, Inc. Reprinted by permission.

Page 221. Photo of Euler courtesy of Smith Collection, Rare Book and Manuscript Library, Columbia University.

Page 222. Euler quotation from *In Mathematical Circles,* by H. Eves. Copyright 1969 by Prindle, Weber, & Schmidt, Inc. Reprinted by permission.

Page 226. Cartoon courtesy of Patrick J. Boyle.

Page 229. Ascending, Descending, by M. C. Escher. Reproduction courtesy of the Escher Foundation—Haags Gemeentemuseum—The Hague and Vorpal Galleries, San Francisco, Chicago, New York, Laguna Beach.

Page 231. The Vanishing Leprechaun puzzle © 1968 by the W. A. Elliott Company, Toronto. All rights reserved under the Universal Copyright Convention.

Page 244. B. C. cartoon reprinted by permission of Johnny Hart and Field Enterprises, Inc.

Page 252. Peanuts cartoon © 1969 United Feature Syndicate, Inc. Reprinted by permission.

Page 254. B. C. cartoon reprinted by permission of Johnny Hart and Field Enterprises, Inc.

Page 255. Illustration courtesy of the British Library Board.

Page 256. Illustration courtesy of the British Library Board.

Page 257. Table used with permission of Macmillan Publishing Co., Inc. and George Allen & Unwin (Publishers) Ltd., from *Number: The Language of Science* (4th ed.), by T. Dantzig. Copyright 1930, 1933, 1939, 1954 by Macmillan Publishing Co., Inc., renewed 1958, 1961, 1967 by Anna G. Dantzig.

Page 258. Cartoon courtesy of Patrick J. Boyle.

Page 266. Photo of Gauss courtesy of Smith Collection, Rare Book and Manuscript Library, Columbia University.

Page 271. Photo of Fermat courtesy of Smith Collection, Rare Book and Manuscript Library, Columbia University.

Page 276. Pascal's triangle from *Science and Civilization in China* (Vol. 3), by J. Needham. Reprinted by permission of the publisher, Cambridge University Press.

Page 463. "Match the Mayors" from The Santa Rosa *Press Democrat*, January 24, 1971. Reprinted by permission.

Page 463. Advertisement courtesy of US JVC CORP. Reprinted by permission.

Page 463. *Peanuts* cartoon © 1974 United Feature Syndicate, Inc. Reprinted by permission.

Page 464. Problem 25 from *The American Mathematical Monthly*, January 1970. Reprinted by permission.

Page 465. Advertisement courtesy of *Wendy's International, Inc.*

Page 467. *The Ancient of Days*, frontispiece of William Blake's *Europe, A Prophecy*, 1794. The Library of Congress, Washington, D. C. (Lessing J. Rosenwald Collection).

Page 467. Limerick by Sir Arthur Eddington from *Fantasia Mathematica*, by C. Fadiman. Copyright © 1958 by Clifton Fadiman. Reprinted by permission of Simon & Schuster, Inc.

Page 470. *Ancient Greek and Roman Dice* reproduced by courtesy of the Trustees of the British Museum.

Page 481. "The Drunkard's Walk" reproduced from *How to Take a Chance*, by Darrell Huff. Illustrated by Irving Geis. By permission of W. W. Norton & Company, Inc., and Victor Gollancz Ltd., London. Copyright © 1959 by W. W. Norton & Company, Inc.

Page 482. Photo courtesy of Juliet Lee.

Page 484. "Le Trente-et-un, ou la maison de pret sur nantissement," by Darcis. Courtesy of Bibliotheque National, Paris.

Page 486. Photo by Ron Pietersma, *The Dairyman*, January 1974. Reprinted by permission.

Page 493. *B. C.* cartoon reprinted by permission of Johnny Hart and Field Enterprises, Inc.

Page 493. *Buz Sawyer* cartoon © 1972 by King Features Syndicate, Inc. Reprinted by permission.

Page 497. Photo courtesy of WIDE WORLD PHOTOS.

Page 498. Monte Carlo example reproduced from *How to Take a Chance*, by Darrell Huff. Illustrated by Irving Geis. By permission of W. W. Norton & Company, Inc., and Victor Gollancz Ltd., London. Copyright © 1959 by W. W. Norton & Company, Inc.

Page 500. Odds chart from *The Complete Illustrated Guide to Gambling*, by A. Wykes. Copyright © 1964 Aldus Books Limited, London. Reprinted by permission.

Page 501. Photo of Cardano courtesy of The Bettman Archive, Inc.

Page 502. *B. C.* cartoon reprinted by permission of Johnny Hart and Field Enterprises, Inc.

Page 506. 1972 Miracle Whip Sweeps blank © 1972 Kraftco Corporation. Reprinted by permission.

Page 507. Mortality table courtesy of the Society of Actuaries, Chicago.

Page 510. Lottery tickets courtesy of Hospital's Trust, Dublin, Ireland; Ohio Lottery Commission; Pennsylvania State Lottery Commission; Commission on Special Revenue, Lottery Division, State of Connecticut; New Hampshire State Lottery Commission; Massachusetts State Lottery Commission; Michigan State Lottery Commission.

Page 510. Fey's slot machine from *The Complete Illustrated Guide to Gambling*, by A. Wykes. Copyright © 1964 Aldus Books Limited, London. Reprinted by permission.

Page 510. Problem 11 from *Introduction to Statistics*, by F. W. Carlborg. Copyright © 1968 Scott, Foresman, and Company, Inc. Reprinted by permission of the author.

Page 512. 1972 Reader's Digest Sweepstakes © 1972 The Reader's Digest Association, Inc. Reprinted by permission.

Page 521. *B. C.* cartoon reprinted by permission of Johnny Hart and Field Enterprises, Inc.

Page 522. Photo of Wright Brothers' plane courtesy of The Bettmann Archive, Inc.

Page 528. Photo courtesy of the Atlanta Braves.

Page 536. Advertisement © 1971 by SAAB-Scania of America, Inc. Reprinted by permission.

Page 537. *Peanuts* cartoon © 1961 United Feature Syndicate, Inc. Reprinted by permission.

Page 542. Historical note from *Mathematics and the Modern World* (2nd ed.), by M. F. Triola. (Menlo Park, Ca.: The Benjamin/Cummings Publishing Company, 1978.)

Page 542. Photo courtesy of U.S. Air Force Museum, Dayton, Ohio.

Page 548. Cartoon by David Pascal from *MacLean's*. Copyright MacLeans-Hunter Ltd., Toronto. Reprinted by permission.

Page 553. Photo courtesy of United Press International.

CONTENTS

7

THE NATURE
OF GEOMETRY
388

8

THE NATURE
OF COUNTING
430

9

THE NATURE OF
PROBABILITY
466

10

THE NATURE
OF STATISTICS
522

APPENDIX A
FINAL REVIEW
561

THE NATURE OF
MODERN MATHEMATICS
THIRD EDITION

-3000

-2800: Great Pyramid

-2580: Cheops Pyramid

-2500

-2200: Mathematical tablets of Nippur—oldest example of a magic square

-2000

-1900: Hammurabi King of Babylonia

-1850: Moscow Papyrus—25 mathematical problems

-1700: Stonehenge

-1700: Early Babylonian tablet, Plimpton 322, written
-1650: Rhind Papyrus—85 numerical problems

-1500

-1350: King Tutankhamen
-1300: Approximate beginning of iron age
-1250: Moses leads exodus from Egypt
-1200: Trojan war

-1350: Rollin Papyrus—elaborate math problems

-1000: Phoenician's invent modern alphabet

-1000

-850: Homer's *Iliad* and *Odyssey*

-753: Rome founded

Thales

Pythagoras

-538: Persians capture Babylon

-600: Thales
-540: Pythagoras—geometry

1

THE NATURE OF PATTERNS AND INDUCTIVE REASONING

The idea that aptitude for mathematics is rarer than aptitude for other subjects is merely an illusion which is caused by belated or neglected beginners.

J. F. Herbart

1.1 INDUCTIVE REASONING

A NOTE FOR THE STUDENT

There are many reasons for reading a book, but the best reason is because you want to read it. Although you are probably reading this first page because you were required to do so by your instructor, it is my hope that in a short while you will be reading this book because you want to read it. It was written for people who think they don't like mathematics, or people who are math avoiders, or people who think they can't work math problems, or people who think they are never going to use math. Don't worry if you feel your previous math background is weak—you don't need to know any high school algebra or geometry. On the other hand, if you have had these subjects you will find that the material in this book is not a repetition. As you begin your trip through this book I wish you a BON VOYAGE!

There is a common belief that mathematics is a difficult, dull subject that is to be pursued only in a clear-cut, logical fashion. This belief is perpetuated by the way mathematics is presented in most textbooks; often mathematics is reduced to a series of definitions, methods to solve various types of problems, and theorems. These theorems are justified by means of proofs and deductive reasoning. I do not mean to minimize the importance of proof in mathematics, for it is the very thing that gives mathematics its strength. But the power of the imagination is every bit as important as the power of deductive reasoning. As the mathematician Augustus De Morgan once said, "The moving power of mathematical invention is not reasoning but imagination."

It is often the artificial organization of mathematics that bewilders the beginning student. Textbooks rarely show the long history in the development of a concept or any of the blind alleys that were taken. The fact is that the mathematician seeks out relationships in simple cases, looks for patterns, and only then tries to generalize. It is usually much later that the generalization is proved and finds its way into some textbook.

Let's look at some patterns. By studying numerical relationships that exhibit patterns, we can learn much about mathematics and in the meantime have a lot of fun doing so.

MULTIPLES OF 9

A very familiar pattern is found in the ordinary "times" tables. By pointing out patterns, teachers can make it easier for children to learn some of their multiplication tables. For example, consider the multiplication table for 9s:

$$
\begin{array}{ll}
1 \times 9 = 9 & 6 \times 9 = 54 \\
2 \times 9 = 18 & 7 \times 9 = 63 \\
3 \times 9 = 27 & 8 \times 9 = 72 \\
4 \times 9 = 36 & 9 \times 9 = 81 \\
5 \times 9 = 45 & 10 \times 9 = 90
\end{array}
$$

What patterns do you notice? You should be able to see many number relationships by looking at the totals. For example, notice that the sum of the digits to the right of the equality is 9 in all the examples ($1 + 8 = 9$, $2 + 7 = 9$, $3 + 6 = 9$, and so on). Will this always be the case for multiplication by 9? (Consider $11 \times 9 = 99$. The sum of the digits here is 18. However, notice the result if you add the digits of 18.) Do you see any other patterns? Can you explain why they "work"?

The pattern from the multiplication described in the text generates a sequence of numbers: 9, 9, 9, ... Having

ANOTHER 9 PATTERN

Complicated arithmetic problems can sometimes by solved by using patterns. Given a difficult problem, a mathematician will often try to solve a simpler, but similar, problem. For example, suppose we wish to compute the following number.*

$$\frac{(999,999,999)(999,999,999)}{1 + 2 + 3 + 4 + 5 + 6 + 7 + 8 + 9 + 8 + 7 + 6 + 5 + 4 + 3 + 2 + 1}$$

Instead of doing a lot of terrible arithmetic, let's study the following pattern.

$$\frac{(1)(1)}{1} = 1$$

$$\frac{(22)(22)}{1 + 2 + 1} = \frac{484}{4} = 121$$

$$\frac{(333)(333)}{1 + 2 + 3 + 2 + 1} = \frac{110,889}{9} = 12,321$$

Continue computation like this until you see a pattern, and then predict what the number

$$\frac{(999,999,999)(999,999,999)}{1 + 2 + 3 + 4 + 5 + 6 + 7 + 8 + 9 + 8 + 7 + 6 + 5 + 4 + 3 + 2 + 1}$$

represents.   *123,456,78987   /5,-4*

INDUCTIVE REASONING

The type of reasoning just used—first observing patterns and then predicting answers for complicated problems—is an example of what is called *inductive reasoning*. It is a very important method of thought and is sometimes called the *scientific method*. It involves reasoning from particular facts or individual cases to a general *conjecture*—a statement you think may be true. That is, a generalization is made on the basis of some observed occurrences. The more individual occurrences we observe, the better able we are to make a correct generalization. Peter in the *B.C.* cartoon makes the mistake of generalizing on the basis of a single observa-

*Recall that we can use parentheses to indicate multiplication. That is, (1)(1) means 1 × 1, and (22)(22) means 22 × 22.

found a "nine pattern," let's try to find an "eight pattern."

8 · 1 = 8		
8 · 2 = 16	1 + 6 =	7
8 · 3 = 24	2 + 4 =	6
8 · 4 = 32	3 + 2 =	5
8 · 5 = 40	4 + 0 =	4
8 · 6 = 48	4 + 8 = 12	1 + 2 = 3
8 · 7 = 56	5 + 6 = 11	1 + 1 = 2
8 · 8 = 64	6 + 4 = 10	1 + 0 = 1
8 · 9 = 72	7 + 2 =	9
8 · 10 = 80	8 + 0 =	8
8 · 11 = 88	8 + 8 = 16	1 + 6 = 7

The "eight pattern" is:

8, 7, 6, 5, 4, 3, 2, 1, 9, 8, 7, 6, 5, . . .

Verify the "seven pattern":

7, 5, 3, 1, 8, 6, 4, 2, 9, 7, 5, 3, 1, 8, . . .

Can you find the "six pattern"?
    6, 3, 9

tion. We must keep in mind that an inductive conclusion is always tentative and may have to be revised on the basis of new evidence. For example, consider this pattern:

$$1 \times 9 = 9$$
$$2 \times 9 = 18$$
$$3 \times 9 = 27$$
$$\vdots$$
$$10 \times 9 = 90$$

You might have made the conjecture that the sum of the digits of the answer to any product involving a 9 is always 9. Certainly the first ten examples substantiate this speculation. But the conjecture is suspect because it is based on so few cases. It is shattered by the very next case, wherein $11 \times 9 = 99$.

On the other hand, we can make a prediction about the exact time of sunrise and sunset tomorrow. This is also an example of inductive reasoning, since our prediction is based on a large number of observed cases. Thus there is a very high probability that we will be successful in our prediction.

Although mathematicians often proceed by inductive reasoning to formulate new ideas, they are not content to stop at the "probable" stage. So they often formalize their predictions into theorems and then try to prove these theorems deductively. Proof and deductive reasoning will be discussed in Chapter 2. Right now, however, consider some additional patterns in Problem Set 1.1 to see if you can reason inductively and formulate some conjectures.

PROBLEM SET 1.1

A Problems

1. Study the following pattern:

$$(1 \times 9) + 2 = 11$$
$$(12 \times 9) + 3 = 111$$
$$(123 \times 9) + 4 = 1111$$
$$(1234 \times 9) + 5 = 11,111$$
$$(12345 \times 9) + 6 = 111,111$$
$$(123456 \times 9) + 7 = 1,111,111$$

Examine this pattern, and then guess what number $(1234567 \times 9) + 8$ represents. Don't do a lot of calculation.

Hint for Problem 2: Use patterns.

2. Simplify:

$$\frac{(999,999,999) \times (999,999,999)}{1 + 2 + 3 + 4 + 5 + 6 + 7 + 8 + 9 + 8 + 7 + 6 + 5 + 4 + 3 + 2 + 1}$$

3. Compute:

   a. 1 × 142,857   b. 2 × 142,857   c. 3 × 142,857
   d. 4 × 142,857   e. 5 × 142,857   f. 6 × 142,857

          g. Can you describe a pattern?

4. Compute:

   a. 123456789 × 9   b. 123456789 × 18
   c. 123456789 × 27   d. Can you describe a pattern?

5. Using the pattern in Problem 4, predict the following *without* direct calculation:

   a. 123456789 × 36   b. 123456789 × 45.
   c. 123456789 × 72   d. 123456789 × 81

6. Find a pattern illustrated in Figure 1.1. What do you think the next figure would look like?

7. Find a pattern illustrated in Figure 1.2. What do you think the next figure would look like?

8. a. What is the "six pattern"?
   b. What is the "five pattern"?

9. Find the multiplication patterns for:

   a. 4   b. 3   c. 2   d. 1

10. a. Write 1/9 in decimal form (divide 1 by 9).
   b. Write 2/9 in decimal form.
   c. Write 3/9 in decimal form.
   d. Using parts a.–c., predict the decimal representation of 8/9.

   Enter these answers in the table in Problem 11.

B Problems

11. Now complete the table.

   a. Write 13/9 in decimal form.
   b. Write 14/9 in decimal form.
   c. Write 15/9 in decimal form.

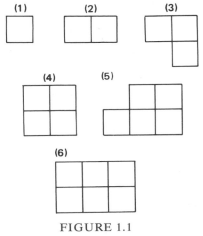

FIGURE 1.1

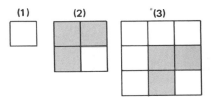

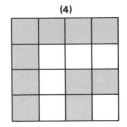

FIGURE 1.2

$\frac{1}{9}$	$\frac{2}{9}$	$\frac{3}{9}$	$\frac{4}{9}$	$\frac{5}{9}$	$\frac{6}{9}$	$\frac{7}{9}$	$\frac{8}{9}$	$\frac{9}{9}$
10a. answer	10b. answer	10c. answer	.444 . . .	.555 . . .	.666 . . .	.777 . . .	10d. answer	////
$\frac{10}{9}$	$\frac{11}{9}$	$\frac{12}{9}$	$\frac{13}{9}$	$\frac{14}{9}$	$\frac{15}{9}$	$\frac{16}{9}$	$\frac{17}{9}$	$\frac{18}{9}$
1.111 . . .	1.222 . . .	1.333 . . .	11a. answer	11b. answer	11c. answer	1.777 . . .	1.888 . . .	////

12. a. Using the table in Problem 11, how should 9/9 be written as a decimal?
     b. Since 9/9 = 1, do you think your answer for 12a. is *equal* to 1?
     c. Is there anything wrong with the following "proof"?

$$\frac{1}{3} = .3333\ldots$$ Divide 3 into 1

$$3\left(\frac{1}{3}\right) = 3(.3333\ldots)$$ Multiply both sides of the equation by 3

$$1 = .9999\ldots$$ Multiplication

     d. Does the "proof" in part c. change your answer and reasoning for part a.?

Hint for Problem 14: Consider the simpler problems

$$9 \times 1 - 1$$
$$9 \times 21 - 1$$
$$9 \times 321 - 1$$
$$\vdots$$

and try to detect a pattern.

13. Find:

     a. the sum of the first 25 consecutive odd numbers
     b. the sum of the first 50 consecutive odd numbers
     c. the sum of the first 100 consecutive odd numbers

14. Find $(9 \times 987654321) - 1$ without doing direct multiplication.

15. a. Construct a right triangle very carefully, and measure the two acute angles. What is the sum of their measures?
     b. Repeat the above experiment. Are your answers the same?
     c. Make a conjecture about the sum of the measures of the acute angles in any right triangle.

16. a. Draw a circle and any diameter in that circle. Now choose any point P on the circumference of the circle, and then draw the triangle with the endpoints of the diameter and P as the vertices. What is the measure of the angle P?
     b. Repeat the experiment above choosing a different point on the circle.
     c. Repeat the experiment above by drawing a circle with a different diameter.
     d. Can you make a conjecture about the size of angle P?

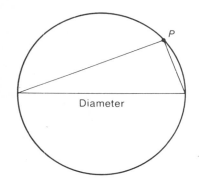

Diameter

17. How many numbers would be in each of the sequences below if all the missing terms indicated by the ellipses (the three dots) were included?

Hint for Problem 17b.: Divide each term by 3.
Hint for Problem 17c.: Subtract 3 from each term and then divide by 4.

     a. 5, 6, 7, 8, 9, . . . , 65
     b. 3, 6, 9, 12, 15, . . . , 240
     c. 11, 15, 19, 23, . . . , 399
     d. 37, 50, 63, 76, 89, . . . , 1558

Mind Bogglers

Speaking of cards, an interesting number fact concerning cards is shown below. Consider the

18. A magician divides a deck of cards into two equal piles of 26 cards each and places the piles face down on a table. He then picks up one of the piles, counts down from the top to the seventh card, and shows it to the audience

without looking at it himself. These seven cards are replaced face down in the same order on top of the pile. He then picks up the other pile and deals the top three cards face up in a row in front of him. He forms three columns of cards by making them add up to 10 (face cards count 10, and others count at face value). That is, if the card in the first column is a four, he counts out six more cards; if the card is a queen, no additional cards are needed. The remainder of this pile is placed on top of the first pile. Next the magician adds the values of the three face-up cards, and this number of cards down in the deck is the card that was originally shown to the audience. Explain why this trick works.

19. A magician begins a trick by giving someone a sealed envelope. She gives a deck of cards to another person and asks that individual to riffle-shuffle the cards twice. Now she obtains a six-digit number by going through the deck, cards face up, and removing the first six spades that have digit values. The digits turn out to be, in order,

$$1,4,2,8,5,7.$$

She arranges the cards in a row on a table and rolls a die to obtain a random digit from 1 to 6. The audience multiplies this digit by

$$142,857$$

while the magician retrieves and cuts open the sealed envelope. It contains the correctly predicted product. Explain how this trick was done.

20. A very magical mathematics teacher had a student select a two-digit number between 50 and 100 and write it on the board out of view of the instructor. Next, the student was asked to add 76 to the number, producing a three-digit sum. If the digit in the hundreds place is added to the remaining two-digit number and this result is subtracted from the original number, the answer is 23, which was predicted by the instructor. How did he know the answer would be 23?

Problems for Individual Study

21. *Recreational Mathematics.* What are some mathematical puzzles, tricks, magic stunts? What is the mathematical explanation for number tricks? What games are based on mathematical problems? What puzzles have mathematical explanations? How is mathematics related to chess, card games, athletics?
Exhibit suggestions: puzzles, tricks, games of mathematical nature; mathematical analysis of games and tricks.
References: Kasner, E., and Newman, James, *Mathematics and the Imagination,* pp. 170–180 (New York: Simon and Schuster, 1940).
Schaaf, William, *A Bibliography of Recreational Mathematics* (Washington, D.C.: National Council of Teachers of Mathematics, 1970).
See also the *Journal of Recreational Mathematics.*

---

number of letters in each of the names of the cards:

Ace	*3*
Two	*3*
Three	*5*
Four	*4*
Five	*4*
Six	*3*
Seven	*5*
Eight	*5*
Nine	*4*
Ten	*3*
Jack	*4*
Queen	*5*
King	*4*

The total is 52, the same as the number of cards in the deck!

Hint: Compare this number with the one we used in Problem 3. This is not a coincidence.

76, in honor of our Bicentenial

This problem is dedicated to Bill Leonard of Cal State Fullerton because 23 is his favorite number. (His book, No Upper Limit, Creative Teaching Associates, *1977, has 23 chapters. It is a delightful book, and I recommend it to you.)*

22. *Mathematical Forms in Nature.* What forms in nature represent mathematical patterns? What shapes in nature are geometric and/or symmetric? What activities in nature illustrate mathematical functions?

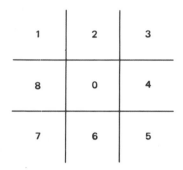

Examples of the five regular solids as they can be found in nature. These are skeletons of minute marine animals called radiolaria.

Exhibit suggestions: samples of spiral shells, crystals, ellipsoidal stones or eggs, spiraling sunflower seed pods, branch distributions illustrating Fibonacci series, snowflake patterns, body bones acting as levers, ratio of food consumed to body size.

References: Bergamini, David, *Mathematics,* Chap. 4 (New York: Time, Inc., Life Science Library, 1963).

Carson, Judithlynne, "Fibonacci Numbers and Pineapple Phyllotaxy," *Two Year College Mathematics Journal,* June 1978, pp. 132–136.

Newman, James, *The World of Mathematics* (New York: Simon and Schuster, 1956).

"Crystals and the Future of Physics," pp. 871–881.

"On Being the Right Size," pp. 952–957.

"On Magnitude," pp. 1001–1046.

"The Soap Bubble," pp. 891–900.

See also the *Fibonacci Quarterly.*

23. *Two-Dimensional Tic-Tac-Toe.* Develop a winning strategy for playing tic-tac-toe. Label the squares as shown in Figure 1.3. You must consider all possible games; look for patterns, and use inductive reasoning.

24. *Three-Dimensional Tic-Tac-Toe.* Develop a strategy for playing three-dimensional tic-tac-toe by using inductive reasoning. Label the squares as shown in Figure 1.4. An article that will help you to do this is "Three-

FIGURE 1.3

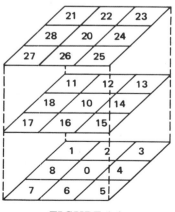

FIGURE 1.4

Dimensional Tic-Tac-Toe,'' by Gene Mercer and John Kolb, in *The Mathematics Teacher*, February 1971 (Vol. LXIV, No. 2). This article also poses five questions; see if you can answer any of them.

1.2 MATHEMATICAL PATTERNS

PASCAL'S TRIANGLE

The following pattern of numbers provides us with one of the richest sources of other number patterns, and it is certainly one of the most elegant of all number arrays.

```
                        1
                  1.        1
              1       2       1
          1       3       3       1
      1       4       6       4       1
    1     5     10      10      5     1
  1     6     15      20      15      6     1
1    7    21     35     35.    21     7     1
1   8    28     56     70     56     28    8     1
                        :
                        :
```

This pattern, which continues indefinitely, is known as *Pascal's triangle*.

Do you see any patterns associated with this triangular array? Here are some examples to get you started.

1. Each row begins and ends with a 1.
2. Do you see the numbers 1,2,3,4, . . . in the array? Look at the first diagonal (the diagonal consisting of all 1s is usually called the 0th diagonal).
3. If you are given the first several rows, do you see a pattern that will give you the next row? Each number (except the first and last) is the sum of the two numbers directly to the right and left above it.

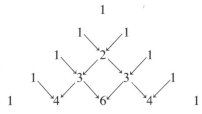

Can you write down the next row of the triangle? Other patterns will be considered in Problem Set 1.2.

Historical Note

Blaise Pascal (1623–1662) has been described as "the greatest 'might-have-been' in the history of mathematics." Pascal was a person of frail health, and, because he needed to conserve his energy, he was forbidden to study mathematics. This aroused his curiosity and forced him to acquire most of his knowledge of the subject by himself. At 16 he wrote an essay on conic sections that astounded the mathematicians of his time. At 18 he had invented one of the first calculating machines. However, at 27, because of his health, he promised God that he would abandon mathematics and spend his time in religious study. Three years later he broke this promise and wrote Traite du triangle arithmetique, *in which he investigated what we today call Pascal's triangle. The very next year he was almost killed when his runaway horse ran over a bank. He took this to be a sign of God's displeasure with him and again gave up mathematics—this time permanently.*

Patterns are sometimes used as part of an IQ test. There is an organization called Mensa that people who have scored at or above the 98th percentile on a standardized IQ test can join. The test on this page is one of their "quickie" tests. You might want to see how well you can do. Notice the use of number sequences as part of the test.

SCORING

Give yourself 1 point for each correct answer. If you completed the test in 15 minutes or less, give yourself an additional 4 points.

1-M; 2-15; 3-8; 4-6; 5-5; 6-22; 7-2; 8-3; 9-2; 10-4; 11-4; 12-4; 13-2; 14-3; 15-3.

IF YOU SCORED:

15 to 19 points: You are exceptionally intelligent—a perfect Mensa candidate!
13 or 14 points: This should put you in the upper 2% of the population.
10, 11, or 12 points: An honorable score.
Less than 10 points: Forget about joining Mensa. But you're in good company—many world-famous figures don't have exceptional IQs either! And whatever you scored, don't take this test too seriously.

For more information about Mensa, write: Mensa, 50 E. 42 St., New York, N.Y. 10017.

NUMBER SEQUENCES

ARE YOU A GENIUS?

Each problem is a sequence of some sort—that is, a succession of either letters, numbers, or drawings—with the last item in the sequence missing. Each sequence is arranged according to a different rule, and, in order to identify the missing item, you must figure out what that rule is.

Now it's your turn to play. Give yourself a maximum of 20 minutes to answer the 15 questions. If you haven't finished in that time, stop anyway. For the problems made up of drawings, it is always the top row that needs to be completed by choosing a drawing from the bottom row.

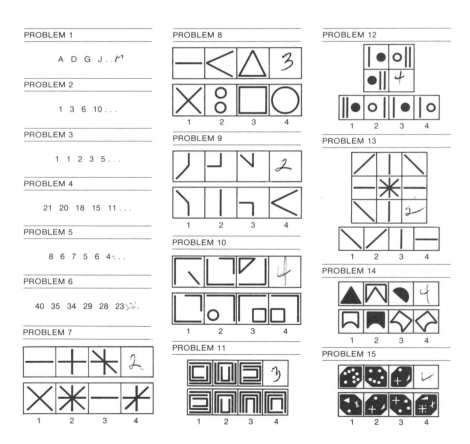

PROBLEM 1

A D G J . . .

PROBLEM 2

1 3 6 10 . . .

PROBLEM 3

1 1 2 3 5 . . .

PROBLEM 4

21 20 18 15 11 . . .

PROBLEM 5

8 6 7 5 6 4 . . .

PROBLEM 6

40 35 34 29 28 23 . . .

PROBLEM 7

PROBLEM 8

PROBLEM 9

PROBLEM 10

PROBLEM 11

PROBLEM 12

PROBLEM 13

PROBLEM 14

PROBLEM 15

Sequences of numbers occur throughout mathematics. A *sequence,* or *progression,* is a set of numbers that is arranged in order with a first *term* (or *element*), a second term, a third term, and so on. A sequence may stop at a particular place, or it may continue indefinitely. See if you can recognize a pattern in each of the following indicated sequences. Try to fill in the blank space with the result of a pattern you "see." There may be more than one answer.

A. 1. 1,4,7,10,13, ____
   2. 20,14,8,2,$^-$4,$^-$10, ____
   3. $a, a + d, a + 2d, a + 3d, a + 4d,$ ____
B. 4. 2,4,8,16,32, ____
   5. $10, 5, {}^5/_2, {}^5/_4, {}^5/_8,$ ____
   6. $a, ar, ar^2, ar^3, ar^4,$ ____
C. 7. 1,2,1,1,2,1,1, ____
   8. 1,2,4,7,11,16, ____
   9. 1,3,4,7,11,18, ____

The first three examples are called *arithmetic progressions*. There is a *common difference* between successive terms. That is, if any term is subtracted from the next term, the result is always the same, and this number is called the common difference. In Example 1 the common difference is 3, so the next term after 13 must be 16.

The next three examples are called *geometric progressions*. There is a *common ratio* between successive terms. If any term is divided into the next term, the result is always the same, and this number is called the common ratio. The common ratio in Example 4 is 2, so the term after 32 is 64. In a geometric progression, every term after the first can be obtained by multiplying the immediately preceding term by the common ratio. In Example 5, this ratio is $^1/_2$. If we multiply 10 by $^1/_2$, we get 5; if we multiply 5 by $^1/_2$, we get $^5/_2$, and so on. The other examples are left for Problem Set 1.2.

The *Peanuts* cartoon illustrates another example of a geometric progression. Charlie Brown is supposed to send a copy of the letter to six of his friends. Let's see how many people become involved in a very short time if everyone carries out the directions in the letter.

1st mailing:	6
2nd mailing:	36
3rd mailing:	216
4th mailing:	1296
5th mailing:	7776
6th mailing:	46,656
7th mailing:	279,936
8th mailing:	1,679,616
9th mailing:	10,077,696
10th mailing:	60,466,176
11th mailing:	362,797,056

⋮

© 1967 United Feature Syndicate, Inc.

Actually, the total number of letters sent is this number plus the number of previous letters sent. Thus, after the 3rd mailing, the number of letters is 6 + 36 + 216 = 258.

At the time of the 1970 census, the U.S. population was 203,235,298. The predicted population in 1980 is about 222,200,000.

After 11 mailings more letters will have been sent than there are people in the United States! The number of letters in only two more mailings would exceed the population of the world.

FIBONACCI NUMBERS

A well-known sequence in mathematics is the Fibonacci sequence:

$$1,1,2,3,5,8,13,21,34,55,89,144, \ldots$$

These numbers have interested mathematicians for centuries—at least since the 13th century, when Leonardo Fibonacci wrote *Liber Abaci,* which discussed the advantages of Hindu-Arabic numerals over Roman numerals. In this book, he had a problem that related Fibonacci numbers to the birth patterns of rabbits. Let's see what he did. The problem was to find the number of rabbits alive after a given number of generations. Suppose a pair of rabbits will produce a new pair of rabbits in their second month and thereafter will produce a new pair every month. The new rabbits will do exactly the same. Starting with one pair, how many pairs will there be in 10 months if no rabbits die?

To solve this problem, we could begin by direct counting.

Number of Months	Number of Pairs	Pairs of Rabbits (the shaded pairs are ready to reproduce in the next month)
start	1	
1	1	
2	2	
3	3	
4	5	
5	8	
⋮	⋮	Same pair (rabbits never die)

Instead of continuing in this fashion, Leonardo looked for a pattern:

$$1,1,2,3,5,8,13,?$$

Do you see a pattern for this sequence? (For each term, consider the sum of the two preceding terms.)

Using this pattern, Leonardo was able to compute the number of rabbits alive after ten months (it is the tenth term after the first 1), which is 89. (Continuing with the above sequence: . . . ,8,13,21,34,55,89, ...) He could also compute the number of rabbits after the first year or after any other interval. Without a pattern, the problem would indeed be a difficult one.

Many mathematicians, such as Verner Hoggatt and Brother Brosseau, have devoted large portions of their careers to the study of Fibonacci numbers. There is even a mathematical society called the Fibonacci

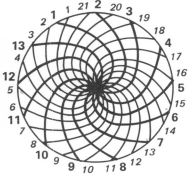

FIGURE 1.5 The arrangement of the pods (phyllotaxy) of a sunflower

Melancholia engraving by Dürer

Association, which publishes *The Fibonacci Quarterly* and operates a Fibonacci Bibliographical and Research Center.

You may wonder why there is so much interest in the Fibonacci sequence. Some of its applications are far-reaching. It has been used in botany, zoology, physical science, business, economics, statistics, operations research, archeology, fine arts, architecture, education, psychology, sociology, and poetry.* An example of Fibonacci numbers in nature is illustrated by a sunflower. The seeds are arranged in spiral curves as shown in Figure 1.5. If we count the number of clockwise spirals (13 and 21 in this example), they are successive terms in the Fibonacci sequence. This is true of all sunflowers and, indeed, of the seed head of any composite flower such as the daisy or aster.

Closely related to the Fibonacci numbers are the Lucas numbers (after the 19th-century French mathematician Lucas):

$$1,3,4,7,11,18,29, \underline{\quad\quad}$$

Can you fill in the blank? (Hint: This sequence begins differently but is generated in the same way as the Fibonacci sequence.)

What properties of Lucas or Fibonacci numbers can you discover? These numbers can also be found in Pascal's triangle. Can you find the Fibonacci numbers embedded in Pascal's triangle?

MAGIC SQUARES

A magic square is an arrangement of numbers in the shape of a square with the sums of each vertical column, each horizontal row, and each diagonal all equal. The magic square below is a rather special one and is thought to have been developed by Dürer in 1514 (notice that the date appears in the magic square). It is called *Melancholia*.

16	3	2	13
5	10	11	8
9	6	7	12
4	15	14	1

*For more information you can write to the Fibonacci Association (c/o V. E. Hoggatt, Jr., San Jose State University, San Jose, Calif. 95114) or refer to Verner Hoggatt's *Fibonacci and Lucas Numbers* (Boston: Houghton Mifflin, 1969).

All the rows, columns, and diagonals add up to 34, so it is a magic square. However, this one has some special properties. A few are listed here. See if you can find any more.

1. The four corners add up to 34.
2. The four center squares (10,11,6,7) add up to 34.
3. Opposite pairs of squares (3,2,15,14) add up to 34.
4. Slanting squares (2,8,15,9) add up to 34.
5. The sum of the first eight numbers equal the sum of the second eight numbers.
6. The sum of the squares of the first eight numbers equals the sum of the squares of the second eight numbers.

There are formal methods for finding magic squares, but we will not describe them here.

Magic squares can be constructed using the first 9, 16, 25, 36, 49, 64, and 81 consecutive numbers. You may want to try some of them. One with the first 25 numbers is shown in Figure 1.6.

Benjamin Franklin was fond of magic squares and invented the one shown in Figure 1.7. In this magic square, not only do the rows, columns, and diagonals add up to 260, but the bent diagonals (indicated by the broken lines) also add up to 260. The four corners plus the four middle boxes total 260. All four-box subsquares total 130. So do any four numbers equidistant from the center point.

23	12	1	20	9
4	18	7	21	15
10	24	13	2	16
11	5	19	8	22
17	6	25	14	3

FIGURE 1.6

52	61	4	13	20	29	36	45
14	3	62	51	46	35	30	19
53	60	5	12	21	28	37	44
11	6	59	54	43	38	27	22
55	58	7	10	23	26	39	42
9	8	57	56	41	40	25	24
50	63	2	15	18	31	34	47
16	1	64	49	48	33	32	17

FIGURE 1.7 Benjamin Franklin's magic square

One final magic square that is rather interesting is called IXOHOXI and is shown in Figure 1.8. It is a magic square whose sum is 19,998. It is

Historical Note

Benjamin Franklin was interested in magic squares, but he often complained of the amount of time they took. Howard Eves, in his delightful book In Mathematical Circles, *quotes Dr. Franklin as saying that the time consumed with these pastimes was time "which I still think I might have employed more usefully."*

WORLD'S LARGEST MAGIC SQUARE?

Leon H. Nissimov of San Antonio, TX claims to have broken the record for the largest magic square listed in the 1976 Guinness Book of World Records. His magic square adds up to 999,999,999,989, and the old record was 578,865.

A magic square is a square arrangement of numbers whose numbers added vertically, horizontally, or diagonally make the same total.

Actually, this record is not much to brag about. There is a method for building magic squares (see Problem 27), and anyone could build a larger one, if enough time was spent on it. Here is your chance to make your claim to fame by beating this record.

CATALOG OF STANDARD MAGIC SQUARES

Although there are many variations of magic squares, the standard magic square is defined as a square array of positive integers 1 through N^2, arranged so that the sum of every row, every column, and the two diagonals is the same. N is called the order of the square.

Order	Number of different standard magic squares
1	1
2	0
3	8
4	440
5	275,305,224

interesting because it is still a magic square when it is turned upside down and also when it is reflected in a mirror.

FIGURE 1.8 IXOHOXI

PROBLEM SET 1.2

A Problems

1. Given Pascal's triangle as shown on page 11, write down the next two rows.

2. Consider the powers of 11:

$$11^0 = 1$$
$$11^1 = 11$$
$$11^2 = 121$$
$$11^3 = 1331$$

Relate the powers of 11 to Pascal's Triangle.

3. a. How is the sequence of numbers 1,2,3,4,5,6, . . . related to Pascal's Triangle?
   b. How is the sequence of numbers 1,3,6,10,15,21,28, . . . related to Pascal's Triangle?

c. How is the sequence of numbers 1,4,10,20,36,56,84, . . . related to Pascal's Triangle?

4. Fill in each missing term.

   a. 1,4,7,10,13,16, ____
   b. 20,14,8,2,⁻4,⁻10, ____
   c. $a, a + d, a + 2d, a + 3d, a + 4d,$ ____
   d. 2,5,8,11,14, ____

5. Classify each of the sequences in Problem 4 as arithmetic, geometric, or neither.

6. Fill in each missing term.

   a. $a, ar, ar^2, ar^3, ar^4,$ ____
   b. 1,2,1,1,2,1,1,1,2,1,1,1,1, ____
   c. 1,2,4,7,11,16, ____
   d. 1,3,4,7,11,18,29, ____

7. Fill in each missing term.

   a. 3,6,12,24,48, ____
   b. 5,⁻15,45,⁻135,405, ____
   c. 100,99,97,94,90, ____
   d. $p, pq, pq^2, pq^3, pq^4,$ ____

8. Classify each of the sequences in Problem 6 as arithmetic, geometric, or neither.

9. Classify each of the sequences in Problem 7 as arithmetic, geometric, or neither.

10. Fill in each missing term.

    a. 2,5,2,5,5,2,5,5,5, ____
    b. 2,4,8,16,32,64, ____
    c. $10, 5, \frac{5}{2}, \frac{5}{4}, \frac{5}{8},$ ____
    d. 225, 625, 1225, 2025, ____

11. Classify each of the sequences in Problem 10 as arithmetic, geometric, or neither. If arithmetic, give the common difference; if geometric, give the common ratio.

12. Fill in each missing term.

    a. 5,⁻5,⁻15,⁻25,⁻35, ____
    b. $1, \frac{1}{2}, \frac{1}{3}, \frac{2}{3}, \frac{1}{4}, \frac{3}{4}, \frac{1}{5}, \frac{2}{5}, \frac{3}{5}, \frac{4}{5}, \frac{1}{6},$ ____
    c. $\frac{4}{3}, 2, 3, 4\frac{1}{2},$ ____
    d. 1,8,27,64,125, ____

13. Fill in the missing term: 8,5,4,9,1, ____ (This is a tricky one, so here's a hint: A secretary who does a lot of filing might solve this one more easily than a mathematician.)

14. Fill in each missing term.

    a. A,A,B, Я,C,Ɔ, D, ____
    b. A,C,E,G,I,K, ____

Historical Note

The title page of an arithmetic book by Petrus Apianus in 1527 is reproduced below. It was the first time Pascal's Triangle appeared in print.

Historical Note

1	48	31	50	33	16	63	18
30	51	46	3	62	19	14	35
47	2	49	32	15	34	17	64
52	29	4	45	20	61	36	13
5	44	25	56	9	40	21	60
28	53	8	41	24	57	12	37
43	6	55	26	39	10	59	22
54	27	42	7	58	23	38	11

THE EULER MAGIC SQUARE

The mathematician Leonhard Euler also worked with magic squares. In the square above, each vertical and horizontal row adds up to 260, with each halfway point totaling 130. Adding extra magic for chess players, a knight starting at box 1 would hit all 64 boxes in numerical order.

c. A,E,F,H,I,K,L,M,N, _____

d. dog, three, hippopotamus, twelve, lion, four, tiger, _____

15. What are Fibonacci numbers?

16. Use patterns to fill in the blanks in the following magic squares.

a.

8	1	6
3		7
4	9	2

b.

21	7		18
10		15	
14	12	11	17
9	19		

17. What is the largest number that you know? (A million? A billion?)

B Problems

18. What is the largest number you can write using only three digits? Using only three distinct (different) digits? Using only even digits? Using only distinct even digits?

19. Suppose we multiply

$$\underbrace{9 \cdot 9 \cdot 9 \cdot \ldots \cdot 9}_{1000 \text{ factors}}$$

(The three dots we've used here indicate that some of the factors have been left out. Altogether there are 1000 factors of 9.) What is the last digit?

20. Find the value of x, as shown in Figure 1.9, for a magic square consisting of nine consecutive natural numbers.

4		
	x	3
6		

FIGURE 1.9

21. Here is a problem that was proposed by Professor Richard Andree of the University of Oklahoma in 1971 and, to my knowledge, is still unsolved.

Build a sequence as follows:

a. Pick any number—7 for example.
b. If the last number in the sequence is odd, the next number is found by tripling it and adding 1.
c. If the last number is even, the following number is half of it. For example,

7,22,11,34,17,52,26,13,40,20,10,5,16,8,4,2,1.

The conjecture (still unproved) is that all such sequences eventually go to 1. Try this for several examples.

Current research on this problem indicates that all numbers less than 10^{40} go to 1.

Mind Bogglers

22. *Pick a Number*
    Take a number between 1 and 10.
    Add 16.
    Multiply by 2.
    Subtract 8.
    Divide by 4.
    Subtract one-half the original number.
    What is your answer?
    Show that the answer to this problem is always the same.

23. *More Number Magic*
    Select any four digits (not all the same).

    For example, 7,3,8, and 4

    Rearrange these four digits in descending and ascending order to form two numbers.

    Descending: 8743
    Ascending: 3478

    Subtract the smaller number from the larger.

    8743
    − 3478
    5265 difference

    Using this difference, repeat the process.

    Descending: 6552
    Ascending: − 2556
    3996

The leading digit of a number might be 0, but be sure to retain the four digits.

    Repeat this process again and again until you see a pattern. Try another number. What happens? What pattern results for any four digits?

24. *Fibonacci Magic Trick*
    a. The magician asks the audience for any two numbers (you can limit it to the counting numbers between 1 and 10 to keep the arithmetic manageable.

EXAMPLE:

Numbers		Solution hint
a.	5	(1)	n
	9	(2)	m

b.	*14*	*(3)*	$n + m$
c.	*23*	*(4)*	$n + 2m$
d.	*37*	*(5)*	$2n + 3m$
	60	*(6)*	$3n + 5m$
	97	*(7)*	$5n + 8m$
	157	*(8)*	$8n + 13m$
	254	*(9)*	$13n + 21m$
	411	*(10)*	$21n + 34m$
e.	*1067*		$55n + 88m$ $= 11(5n + 8m)$

b. Add these to obtain a third number.

c. Add the second and third numbers to obtain a fourth.

d. Continue until ten numbers are obtained.

e. Ask the audience to add the ten numbers, while you instantly give the sum.

The trick depends on the magician's ability to multiply quickly and mentally by 11. Consider the following pattern of multiplication by 11:

$11 \times 10 = 110$
$11 \times 11 = 121$
$11 \times 12 = 132$
$11 \times 13 = 1\ 3$
↑ ↑ original two digits
$= 143$
↑ sum of the two digits

$11 \times 52 = 5\ 7\ 2$
↑ sum of the two digits

$11 \times 74 = 7\ (11)\ 4$
$= 814$, doing the usual carry.

For the example in the margin,

$11 \times 97 = 9\ (16)\ 7$
$= 1067$

Explain how this trick works, and why I called it a Fibonacci magic trick.

10	7	8	11
14	11	12	15
13	10	11	14
15	12	13	16

FIGURE 1.10

	6	5	7	9
4	10	9	11	13
10	16			
2				
4				

FIGURE 1.11

25. Consider the square shown in Figure 1.10.
Circle any number.
Cross out all the numbers in the same row and column.
Circle any remaining number and cross out all the numbers in the same row and column.
Repeat; circle any remaining number and cross out all the numbers in the same row and column.
Circle the remaining number.
The sum of the circled numbers is 48. Why?

26. a. Fill in the squares in Figure 1.11 by adding the numbers at the head of the rows and columns.

b. Add all the numbers at the head of the columns and rows (6 + 5 + 7 + 9 + 4 + 10 + 2 + 4).

c. Carry out the instructions given in Problem 25 for the completed square.

d. Now explain why the trick works.

Problems for Individual Study

27. *Magic Squares.* Write a short paper about the construction of magic squares. Are there any general methods for finding particular magic squares?

    References: Brandes, Louis, *Yes, Math Can Be Fun!*, pp. 8–11 (Portland, Me.: J. Weston Walch, 1960).

    Fults, John, *Magic Squares* (La Salle, Ill.: Open Court, 1974).

    Gardner, Martin, "Mathematical Games Department," *Scientific American,* January 1976, pp. 118–122.

    Kraitchik, Maurice, *Mathematical Recreations,* pp. 142–192 (New York: Dover Publications, 1953).

    Sawada, Diayo, "Magic Squares: Extensions into Mathematics," *The Arithmetic Teacher,* March 1974, pp. 183–188.

    Williams, Horace, "A Note of Magic Squares," *The Mathematics Teacher,* October 1974, pp. 511–513.

28. *Fibonacci Numbers.* Write a short paper about Fibonacci numbers. You might like to check with *The Fibonacci Quarterly,* particularly "A Primer on the Fibonacci Sequence," Parts I and II, in the February and April 1963 issues. The articles were written by Verner Hoggatt and S. L. Basin. See also a booklet by Verner Hoggatt, Jr., *Fibonacci and Lucas Numbers,* published in the Houghton Mifflin Mathematics Enrichment Series, 1969. An interesting article for teachers, "Fibonacci Numbers and the Slow Learner," by James Curl, appeared in the October 1968 issue (Vol. 6, No. 4) of *The Fibonacci Quarterly.*

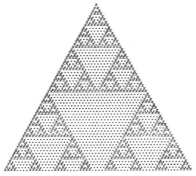

Multiples of 2

29. *Pascal's Triangle.* Do some more research on Pascal's triangle, and see how many properties you can discover. You might begin by answering these questions.

    a. What is a binomial expansion?

    b. How is the binomial expansion related to Pascal's triangle?

    c. How are Fibonacci numbers related to Pascal's triangle?

    d. What relationship do the patterns in Figure 1.12 have to Pascal's triangle?

    Some references you might check are "Mathematical Games Department," *Scientific American,* December 1966 (Vol. 215, No. 6) and January 1967 (Vol. 216, No. 1), and "Probability," by Mark Kac, *Scientific American,* September 1964 (Vol. 211, No. 3). An article with bibliography is "Pascal's Triangle," by Karl Smith, *The Two-Year College Mathematics Journal,* Winter 1973 (Vol. 4).

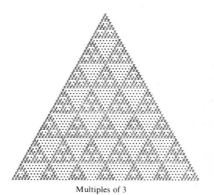

Multiples of 3

FIGURE 1.12

1.3 A GOLDEN PATTERN

In the last section, the Fibonacci sequence

$$1,1,2,3,5,8,13,21,34,55,89,144, \ldots$$

was introduced. Suppose we consider the ratios of the successive terms of this sequence:

$$\frac{1}{1} = 1.000 \qquad \frac{2}{1} = 2.0000$$

$$\frac{3}{2} = 1.500 \qquad \frac{5}{3} \approx 1.667$$

$$\frac{8}{5} = 1.600 \qquad \frac{13}{8} = 1.625$$

$$\frac{21}{13} \approx 1.616 \qquad \frac{34}{21} \approx 1.619$$

$$\frac{55}{34} \approx 1.618 \qquad \frac{89}{55} \approx 1.618$$

If you continue finding these ratios, you will notice that the sequence oscillates about a number approximately equal to 1.618.

Suppose we repeat this same procedure with another sequence, this time choosing *any* two nonzero numbers, say 4 and 7. Construct a sequence by adding terms as was done with the Fibonacci sequence:

$$4,7,11,18,29,47,76,123,199,322, \ldots.$$

Next, form the ratios of the successive terms:

$$\frac{7}{4} = 1.750; \frac{11}{7} \approx 1.571; \frac{18}{11} \approx 1.636; \frac{29}{18} \approx 1.611;$$

$$\frac{47}{29} \approx 1.621; \frac{76}{47} \approx 1.617; \frac{123}{76} \approx 1.618; \frac{199}{123} \approx 1.618.$$

These ratios are oscillating about the same number! There would seem to be something interesting about this number, which is called the *golden ratio* and is sometimes denoted by τ ($\tau \approx 1.6180339885$). This number has been given a special name and symbol because of its frequency of occurrence.

Many everyday rectangular objects have a length-to-width ratio of about 1:1.6, as illustrated in Figure 1.13. Psychologists have tested individuals to determine the rectangles they find most pleasing; the results are

Historical Note

Leonardo Fibonacci (about 1175–1250), also known as Leonardo da Pisa, was called Fibonacci because he was Leonardo, son of Bonaccio. Because his father was a customs manager, Fibonacci visited a number of Eastern and Arabic cities, were he became interested in the Hindu-Arabic numeration system we use today. In Europe at that time, Roman numerals were used for calculation, so Fibonacci wrote Liber Abaci, *in which he strongly advocated the use of the Hindu-Arabic numeration system. In the book* In Mathematical Circles, *Howard Eves tells the following story: "Fibonacci sometimes signed his work with the name Leonardo Bigollo. Now bigollo has more than one meaning; it means both 'traveler' and 'blockhead.' In signing his work as he did, Fibonacci may have meant that he was a great traveler, for so he was. But a story has circulated that he took*

those rectangles whose length-to-width ratios are near the golden ratio. Such rectangles are called *golden rectangles*.

pleasure in using this signature because many of his contemporaries considered him a blockhead (for his interest in the new numbers), and it pleased him to show these critics what a blockhead could accomplish."

From In Mathematical Circles *by Howard Eves (Boston: Prindle, Weber, & Schmidt, 1969, p. 91).*

FIGURE 1.13 A 15-oz box of Kellogg's Sugar Corn Pops is 30 cm × 19 cm, for a ratio of 1.6. A 1-pound box of C & H sugar is 17 cm × 10 cm, for a ratio of 1.7.

A golden rectangle is easy to construct with a straightedge and a compass. Begin with any square (*CDHG* in Figure 1.14). Divide the square into two equal parts (line *AB* in Figure 1.14). Draw an arc, with the center at *A* and a radius of *AC*, so that it intersects the extension of side *AD* at point *E*. Now side *EF* can be drawn. The resulting rectangle *EFGH* is a golden rectangle; *CDEF* is also a golden rectangle.

This golden ratio is important to artists in a technique known as "dynamic symmetry." Albrecht Dürer, Leonardo da Vinci, Georges Seurat, George Bellows, and Pieter Mondriaan all studied the golden rectangle as a means of creating dynamic symmetry in their work (see

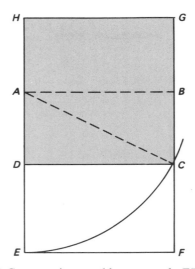

FIGURE 1.14 Constructing a golden rectangle *EFGH*. Another golden rectangle is *CDEF*.

THE GOLDEN RATIO IN NATURE

1. Begin with any square and draw a quarter circle as shown.

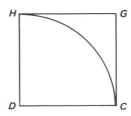

2. Form a golden rectangle (as shown in Figure 1.14).

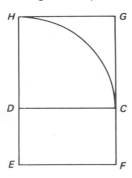

3. Draw a square within the new rectangle CDEF; draw a semicircle as shown.

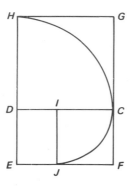

Figure 1.15). In fact, Mondriaan is said to have approached every canvas in terms of the golden ratio. Figure 1.16 shows a study of the human face, by da Vinci, in which the rectangles approximate golden rectangles. Even the proportions of Michelangelo's *David* conform to the golden ratio in, for example, the placement of joints on the fingers and the location of the navel with respect to the height.

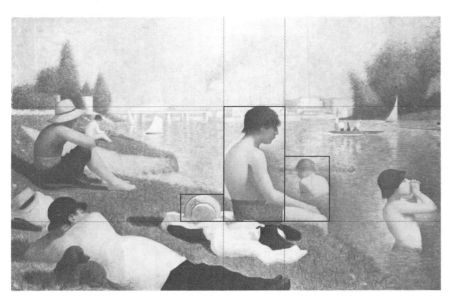

FIGURE 1.15 *Bathers* (1883–1884) by the French impressionist Georges Seurat. Three successive golden rectangles are shown. Can you find others?

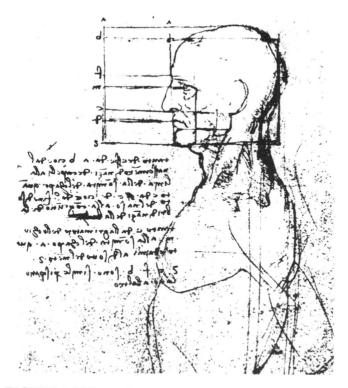

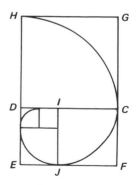

4. Repeat the process.
5. The resulting curve is called a logarithmic spiral and resembles a chambered nautilus.

FIGURE 1.16 Dynamic symmetry of a human face. A study of golden rectangles by Leonardo da Vinci.

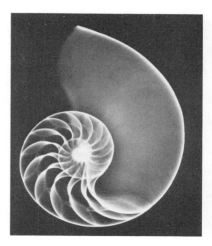

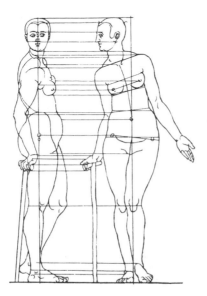

Proportions of the human body by Albrecht Dürer.

The dimensions of the Parthenon at Athens have a length-to-width ratio that almost exactly equals the golden ratio.

In calculating the ratio of the golden rectangle, we used length-to-width ratios. Suppose, instead, we used width-to-length ratios. Surprisingly, we obtain a number approximately equal to 0.618, which differs from τ by 1. That is,

$$\tau = 1 + \frac{1}{\tau}$$

but, since $\tau = 1 + 1/\tau$, we can substitute $1 + 1/\tau$ for τ to obtain

$$\tau = 1 + \cfrac{1}{1 + \cfrac{1}{\tau}}$$

Repeating the process,

$$\tau = 1 + \cfrac{1}{1 + \cfrac{1}{1 + \cfrac{1}{\tau}}}$$

and

$$\tau = 1 + \cfrac{1}{1 + \cfrac{1}{1 + \cfrac{1}{1 + \cfrac{1}{\tau}}}}$$

Thus, the number τ, which is that number which differs from its reciprocal by 1, is approached by the sequence of numbers

$$1 + 1, \quad 1 + \cfrac{1}{1 + 1}, \quad 1 + \cfrac{1}{1 + \cfrac{1}{1 + 1}}, \quad 1 + \cfrac{1}{1 + \cfrac{1}{1 + \cfrac{1}{1 + 1}}}, \ \ldots$$

This sequence is

$$2, \ \frac{3}{2}, \ \frac{5}{3}, \ \frac{8}{5}, \ \ldots,$$

returning us to the successive ratios of the Fibonacci numbers.

This chapter is concerned with the nature of inductive reasoning, and the golden ratio can provide the point of departure for a wide range of problems. The following problem set will give you an opportunity to pursue some of these directions by using inductive reasoning.

The fractions shown here are called continued fractions. *There is a very readable book about them by Charles G. Moore called* An Introduction to Continued Fractions *(Washington, D.C.: National Council of Teachers of Mathematics, 1964). The current price is $2.50.*

PROBLEM SET 1.3

A Problems

In Problems 1–5 use Table 1.1 to fill in the blanks.

1. a. $F_1 = $ _____
   b. $F_1 + F_3 = $ _____
   c. $F_1 + F_3 + F_5 = $ _____
   d. $F_1 + F_3 + F_5 + F_7 = $ _____

2. a. $F_2 = $ _____
   b. $F_2 + F_4 = $ _____
   c. $F_2 + F_4 + F_6 = $ _____
   d. $F_2 + F_4 + F_6 + F_8 = $ _____

3. a. $F_1 + F_2 = $ _____
   b. $F_1 + F_2 + F_3 = $ _____
   c. $F_1 + F_2 + F_3 + F_4 = $ _____
   d. $F_1 + F_2 + F_3 + F_4 + F_5 = $ _____

4. a. $F_1 + F_2 + F_3 = $ _____
   b. $F_2 + F_3 + F_4 = $ _____
   c. $F_3 + F_4 + F_5 = $ _____
   d. $F_4 + F_5 + F_6 = $ _____

5. a. $(F_1)^2 = $ _____
   b. $(F_2)^2 = $ _____
   c. $(F_3)^2 = $ _____
   d. $(F_4)^2 = $ _____

6. If $(F_1)^2$ and $(F_2)^2$ are represented by

   and $(F_3)^2$ by

   how would you represent $(F_4)^2$?

7. Repeat Problem 6 for $(F_5)^2$.

8. If $(F_1)^2 + (F_2)^2$ is written

   $$(F_1)^2 \longrightarrow \longleftarrow (F_2)^2$$

   $(F_1)^2 + (F_2)^2 + (F_3)^2$ is written

   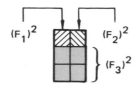

TABLE 1.1 The first 40 Fibonacci Numbers

n	F_n
1	1
2	1
3	2
4	3
5	5
6	8
7	13
8	21
9	34
10	55
11	89
12	144
13	233
14	377
15	610
16	987
17	1597
18	2584
19	4181
20	6765
21	10,946
22	17,711
23	28,657
24	46,368
25	75,025
26	121,393
27	196,418
28	317,811
29	514,229
30	832,040
31	1,346,269
32	2,178,309
33	3,524,578
34	5,702,887
35	9,227,465
36	14,930,352
37	24,157,817
38	39,088,169
39	63,245,986
40	102,334,155

and $(F_1)^2 + (F_2)^2 + (F_3)^2 + (F_4)^2$ is written

then draw a representation for

$$(F_1)^2 + (F_2)^2 + (F_3)^2 + (F_4)^2 + (F_5)^2$$

9. Using Problem 8, represent

$$(F_1)^2 + (F_2)^2 + (F_3)^2 + (F_4)^2 + (F_5)^2 + (F_6)^2$$

10. a. $(F_1)^2 + (F_2)^2 = $ _____
    b. $(F_1)^2 + (F_2)^2 + (F_3)^2 = $ _____
    c. $(F_1)^2 + (F_2)^2 + (F_3)^2 + (F_4)^2 = $ _____
    d. $(F_1)^2 + (F_2)^2 + (F_3)^2 + (F_4)^2 + (F_5)^2 = $ _____

11. If $F_1 \times F_2$ is represented by

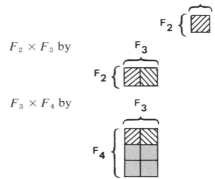

$F_2 \times F_3$ by

$F_3 \times F_4$ by

and $F_4 \times F_5$ by

write out an expression for

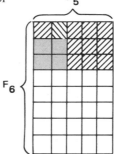

12. a. $F_3 \cdot F_1 - (F_2)^2 = $ _____
    b. $F_4 \cdot F_2 - (F_3)^2 = $ _____
    c. $F_5 \cdot F_3 - (F_4)^2 = $ _____
    d. $F_6 \cdot F_4 - (F_5)^2 = $ _____

13. The Lucas sequence is found in the same way as the Fibonacci sequence, except the first two terms are 1 and 3.

$$1,3,4,7, \ldots .$$

Write out the first ten terms of the Lucas sequence.

B Problems

14. On a planet far, far away, Luke finds himself in a strange building with hexagon-shaped rooms. In his search for the princess, he always moves to an adjacent room and always moves south.
    a. How many paths are there to room 1? room 2? room 3? room 4?
    b. How many paths are there to room 10?
    c. How many paths are there to room n?

15. Make a conjecture about

$$(F_1)^2 + (F_2)^2 + (F_3)^2 + \cdots + (F_n)^2$$

Use Problems 8–11, look for a pattern, and then state this result.

16. In 1963, Leo Moser studied the effect that two face-to-face panes of glass have on light reflected through the panes. If a ray is unreflected, it has just one path through the glass. If it has one reflection, it can be reflected two ways. For two reflections, it can be reflected three ways.
    a. Show the five paths possible for three reflections.
    b. Show the eight paths possible for four reflections.
    c. Make a conjecture about the number of paths for five reflections.
    d. Make a conjecture about the number of paths for n reflections.

17. Find the ratios of the successive terms of the Lucas sequence (see Problem 13), and show that after awhile they oscillate around the golden ratio.

18. a. Pick any two nonzero numbers. Add them to obtain a third number. Construct a sequence of numbers in the Fibonacci fashion by adding the two previous terms to obtain a new term.
    b. Form the ratios of the successive terms, and show that after awhile they oscillate around the golden ratio.

19. Repeat Problem 18 for two other nonzero numbers.

20. Draw a logarithmic spiral by constructing golden rectangles, as shown in Figure 1.14. Use graph paper and begin with a square with sides measuring 34 units.

Mind Bogglers

21. Make a conjecture about $F_2 + F_4 + \cdots + F_{2k}$ using Problem 2.

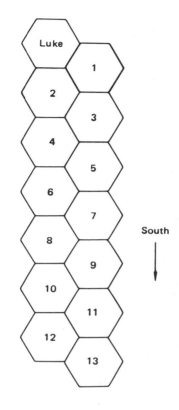

South

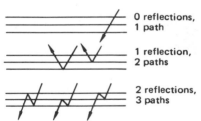

0 reflections, 1 path

1 reflection, 2 paths

2 reflections, 3 paths

22. Make a conjecture about $F_1 + F_3 + F_5 + \cdots + F_{2k-1}$ using Problem 1.

23. Make a conjecture about $F_1 + F_2 + \cdots + F_k$ using Problem 3.

Problems for Individual Study

24. Do some research on the length-to-width ratios of the packaging of common household items. Form some conclusions. Find some examples of the golden ratio in art. Do some research on dynamic symmetry.
   References: Hambridge, Jay, *The Elements of Dynamic Symmetry* (New York: Dover Publications, 1967).
   Huntley, H. E.; *The Divine Proportion: A Study in Mathematical Beauty* (New York: Dover Publications, 1970).
   Razzel, Authur, and K. Watts, *Symmetry* (New York: Doubleday, 1968).

1.4 PATTERNS OF LARGE AND SMALL NUMBERS

Scientists often work with very large numbers. For example, an astronomer might talk of a star cluster that is about 5,865,700,000,-000,000,000 miles away (this is about one million light-years, a light-year being the distance that light, moving at 186,000 miles per second, travels in a year). Such numbers are difficult to comprehend, and it takes a lot of time to write them. So let's turn to patterns to see if there's an easier way to deal with these numbers.

EXPONENTIAL NOTATION

We often encounter numbers that comprise multiplication of the same numbers. For example,

$$10 \cdot 10 \cdot 10 \quad \text{or} \quad 6 \cdot 6 \cdot 6 \cdot 6 \cdot 6 \quad \text{or}$$
$$15 \cdot 15 \cdot 15 \quad 15 \cdot 15 \cdot 15 \cdot 15 \cdot 15 \cdot 15 \cdot 15 \cdot 15 \cdot 15 \cdot 15$$

These numbers can be written more simply by inventing a new notation:

David (1501–1504) by Michelangelo. Relate its proportions to the golden ratio for Problem 24.

$$10^3 = \underbrace{10 \cdot 10 \cdot 10}_{3 \text{ factors}}$$

$$6^5 = \underbrace{6 \cdot 6 \cdot 6 \cdot 6 \cdot 6}_{5 \text{ factors}}$$

$$15^{13} = \underbrace{15 \cdot 15 \ldots 15}_{13 \text{ factors}}$$

We call this system *exponential notation* and define it as follows.

DEFINITION: For any number b and any counting number* n,

$$b^n = \underbrace{b \cdot b \cdot b \ldots b}_{n \text{ factors}}$$

The number b is called the *base*.
The number n is called the *exponent*.
The number b^n is called a *power* or *exponential*.

Definition of exponential notation

EXAMPLES:

1. $10^3 = 10 \cdot 10 \cdot 10$ or 1000
2. $6^2 = 6 \cdot 6$ or 36
3. $7^5 = 7 \cdot 7 \cdot 7 \cdot 7 \cdot 7$ or 16,807
4. $2^{63} = \underbrace{2 \cdot 2 \cdot 2 \ldots 2 \cdot 2}_{63 \text{ factors}}$ or 9,223,372,036,854,775,808

We now use this notation to observe a pattern for powers of 10.

$$10^1 = 10$$
$$10^2 = 10 \cdot 10 = 100$$
$$10^3 = 10 \cdot 10 \cdot 10 = 1000$$
$$10^4 = 10 \cdot 10 \cdot 10 \cdot 10 = 10,000$$

Do you see a relationship between the exponent and the number?

$$10^5 = \underbrace{100,000}_{5 \text{ zeros}}$$

$$\uparrow$$
exponent
is 5

Notice that the exponent and the number of zeros are the same.

Could you write 10^{12} without actually multiplying?

There is a similar pattern for multiplications of any number by a power of 10. Consider the following examples, and notice what happens to the decimal point.

---

*Counting numbers are the numbers we use for counting—namely, 1,2,3,4,.... Sometimes they are also called *natural numbers*.

$$9.42 \times 10^1 = 94.2$$
$$9.42 \times 10^2 = 942.$$
$$9.42 \times 10^3 = 9420.$$
$$9.42 \times 10^4 = 94{,}200.$$

We find these answers by direct multiplication.

Do you see a pattern? If we multiply 9.42×10^5, how many places to the right will the decimal point be moved?

$$9.42 \times 10^5 = 942{,}000.$$

5 places ⟶

This answer is found by observing the pattern, not by directly multiplying.

Using this pattern, can you multiply the following *without direct calculation*?

$$9.42 \times 10^{12}$$

We will investigate one final pattern of 10s, this time looking at smaller values in this pattern.

$$100{,}000 = 10^5$$
$$10{,}000 = 10^4$$
$$1000 = 10^3$$
$$100 = 10^2$$
$$10 = 10^1$$

Continuing with the same pattern:

$$1 = 10^0$$
$$.1 = 10^{-1}$$
$$.01 = 10^{-2}$$
$$.001 = 10^{-3}$$
$$.0001 = 10^{-4}$$
$$.00001 = 10^{-5}$$

By observing this pattern, we're led to the following definition.

Recall that

$$10^n = \underbrace{10 \times 10 \times 10 \times \cdots \times 10}_{n \text{ factors}}$$

provided n is a counting number.

$$10^0 = 1$$
$$10^{-n} = \frac{1}{10^n}$$

where n is a counting number.

We interpret the zero exponent as "decimal point moves 0 places."

When we multiply a power of 10 by some number, a pattern emerges:

$$9.42 \times 10^2 = 942.$$
$$9.42 \times 10^1 = 94.2$$
$$9.42 \times 10^0 = 9.42$$

To continue, direct calculation extends the pattern:

$$9.42 \times 10^{-1} = .942$$
$$9.42 \times 10^{-2} = .0942$$
$$9.42 \times 10^{-3} = .00942$$

These numbers are found by direct multiplication. For example,

$$9.42 \times 10^{-2} = 9.42 \times \frac{1}{100}$$
$$= 9.42 \times .01$$
$$= .0942.$$

Do you see that the same pattern for multiplying by a negative exponent also holds? Can you multiply 9.42×10^{-6} *without direct calculation?* The solution is as follows:

$$9.42 \times 10^{-6} = .00000942$$

moved 6 places to the left

These patterns lead to a useful way for writing large and small numbers, called *scientific notation*.

SCIENTIFIC NOTATION

DEFINITION: The *scientific notation* for a number is that number written as a power of 10 times another number x, such that $1 \le x < 10$.

Definition of scientific notation

EXAMPLES:
1. $123,600 = 1.236 \times 10^5$
2. $.000035 = 3.5 \times 10^{-5}$
3. $1,000,000,000,000 = 10^{12}$
4. $7.35 = 7.35 \times 10^0$ or just 7.35
5. Light travels at about 186,000 miles/second. One year comprises 31,557,600 seconds. Thus light travels about

$$5,869,713,600,000 \text{ miles}$$

in one year. In scientific notation we write

$$5.87 \times 10^{12} \text{ miles/light-year}$$

Notice that a number between 1 and 10 is also in scientific notation, since it can be written as that number times 10^0, as with 7.35×10^0.

6. The estimated age of the earth is about

$$5 \times 10^9 \text{ years}$$

7. The sun loses about

$$4.3 \times 10^9 \text{ kilograms}$$

of solar mass per second.

8. The number of possible combinations of genes that an individual may possess is about

$$10^{9,400,000,000}$$

INVESTIGATING LARGE AND SMALL NUMBERS

We have been looking at some very large numbers. But just how large is large? Most of us are accustomed to hearing about millions and even billions, but do we really understand the magnitude of these numbers? Would you do a better job than Dennis' parents in the cartoon at explaining "How much is a million?"

Historical Note

One of the first times very large numbers were used was by Archimedes in about 250 B.C., when he reportedly computed the number of grains of sand in the universe to be 10^{63}. Archimedes (about 287–212 B.C.) was one of the greatest mathematicians of all times. He once boasted that he could move the earth if he had a place to stand. King Hiero of Syracuse told him to make good his boast by moving a ship (fully loaded) out of dry dock. He did this single-handedly with the use of pulleys. This feat so impressed the king that he involved Archimedes in the defense of Syracuse against the Romans. Archimedes invented catapults, pulleys to raise and smash Roman ships, and devices to set them on fire. The Roman general Marcellus left strict orders to let no harm come to Archimedes during the seige, but nevertheless Archimedes was killed by a Roman soldier.

A million is a fairly modest number, 10^6. Yet if we were to count one number per second, nonstop, it would take us about 278 hours or approximately 11¹/₂ days to count to a million. Not a million days have elapsed since the birth of Christ (a million days is about 2700 years). A million letters in small type would fill a book of about 700 pages. A million bottle caps placed in one line would stretch about 17 miles.

But the age in which we live has been called the age of billions. The U.S. national budget is measured in terms of billions of dollars (soon it will be a trillion-dollar budget). Just how large is a billion? How long would it take you to count to a billion? To get some idea about how large a billion is, let's compare it to some familiar units.

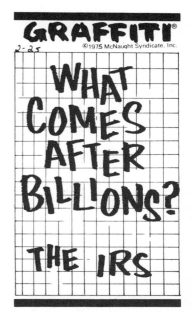

a. Suppose you give away $1000 every day. It would take more than 2700 years to give away a billion dollars.
b. A stack of a billion $1 bills would be more than 59 miles high.
c. At 8% interest, a billion dollars would earn you $219,178.08 interest per day.
d. A billion seconds ago was the bombing of Pearl Harbor.
e. A billion minutes ago Christ was living on Earth.
f. A billion hours ago people had not yet appeared on Earth.

But a billion is only 10^9, a mere nothing compared with the real giants. There is an old story of a king who, being under obligation to one of his subjects, offered to reward him in any way the subject desired. Being of mathematical mind and modest tastes, the subject simply asked for a chessboard with one grain of wheat on the first square, two on the second, four on the third, and so forth. The old king was delighted with this request because he had a beautiful daughter and had feared the subject would ask for her hand in marriage. However, the king was soon sorry he had granted the request. He needed 2^{63} grains of wheat for the last square alone! Now,

$$2^{63} = 9,223,372,036,854,775,808.$$

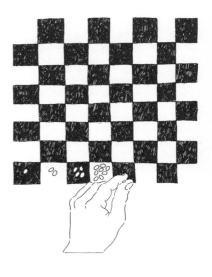

If we purchase some raw wheat and count the number of grains in one cubic inch, we find that there are about 250 grains per cubic inch. We can then compute (using scientific notation) the amount of wheat needed to equal 2^{63} grains. There are 2150 cubic inches per bushel, or 537,500 grains of wheat in a bushel. This means that 2^{63} grains of wheat would equal about 17,383,000,000,000 bushels. For the last several years, the yearly output of wheat in the United States has been about 1.25 billion bushels. At this rate of production, it would take the United States almost 14,000 years to satisfy the requirements for the last square of the chessboard! The story goes no further. The chances are that the king lost his temper and the subject lost his head before the 64th square of the chessboard was reached.

Let's consider a real giant. The answer to Problem 18 of Section 1.2 (about the largest possible number using only three digits) is 9^{99} (certainly larger than 999). Now, 9^{99} means $9^{(99)}$ or 9 raised to the 9^9 power. Now,

$9^9 = 387,420,489$; thus, $9^{99} = 9^{387,420,489}$. To compare this with the "little" number from the chessboard problem, 2^{63}, realize that $9^{387,420,489}$ is greater (a lot greater) than $8^{387,420,489} = (2^3)^{387,420,489} = 2^{1,162,261,467}$, which makes our big number 2^{63} look minuscule.

The number 9^{99} would take 6 million lines to write. It has 369,693,100 digits, and it begins 428,124,773, . . . and ends with 89. That number of bacteria would overflow the Milky Way, and the number of grains of sand of this and every other planet in our solar system would not be so large as this giant.

TABLE 1.2 Some large numbers

million: 1,000,000
billion: 1,000,000,000
trillion: 1,000,000,000,000
quadrillion: 1,000,000,000,000,000
quintillion: 1,000,000,000,000,000,000
sextillion: 1,000,000,000,000,000,000,000
septillion: 1,000,000,000,000,000,000,000,000
octillion: 1,000,000,000,000,000,000,000,000,000
nonillion: 1,000,000,000,000,000,000,000,000,000,000
decillion: 1,000,000,000,000,000,000,000,000,000,000,000
undecillion: 1,000,000,000,000,000,000,000,000,000,000,000,000

$\vdots$

vigintillion: 1,000,000,000,000,000,000,000,000,000,000,000,000,000,000,000,000,000,000,000,000,000

For every large number, there is a corresponding small number. If 9^{99} is very large, then $1/9^{99}$ is very small. Therefore, when we speak of finding large numbers, we are also finding small numbers.

These third-graders at Michigan School for the Blind in Lansing decided to find out what a million something is like. One year later they are seen frolicking on one million bottle caps. The students found the caps. Their teacher, Mrs. Jackie Taylor, said it was difficult to describe what a million bottle caps look like, but "They smell like a very large brewery."

We have been looking at some very large (and very small) numbers, but there are yet larger numbers. Professor Edward Kasner of Columbia University named the "googol" and "googolplex." A googol is 10^{100}, and a googolplex is 10^{googol}. Thus, as Schroeder in the *Peanuts* strip tells us, a googol is 1 followed by 100 zeros. It has been estimated that, if we tried to write out a googolplex on a single line with a typewriter, it would not fit between the earth and the moon.

Have we reached a limit? Are there still larger numbers? What about the following numbers?

$$googol^{googol} \qquad or \qquad googol^{googol^{googol}}$$

These are real giants. Is there a largest number? Clearly not, since, if you think you've found the largest number, you would only have to add 1 to have a larger number. All these numbers are large beyond comprehension, but they are still *finite*. Is it possible to have even larger numbers that are not finite? We'll consider this question later in the text.

PROBLEM SET 1.4

A Problems

1. Consider the number 10^6.
   a. What is the common name for this number?
   b. What is the base?
   c. What is the exponent?
   d. According to the definition of exponential notation, what does the number mean?

2. Consider the number 10^{-1}.
   a. What is the common name for this number?
   b. What is the base?
   c. What is the exponent?
   d. According to the definition of exponential notation, what does the number mean?

Write each of the numbers in Problems 3–19 in scientific notation.

EXAMPLE: 4500. In scientific notation, we write a number between 1 and 10, a multiplication symbol, and exponential notation for a power of 10. Thus, $4500 = 4.5 \times 10^3$.

3. 3200

4. .0004

5. 5629

6. 23.79

7. 35,000,000,000

8. .00000000000000000000035

9. 63,000,000

10. googol

11. .00001

12. $55^{1}/_{2}$

13. Drawn to the scale shown in Figure 1.17, the distance between Earth and Mars (220,000,000 miles) would be .0000025 in.

14. A thermochemical calorie is about 41,840,000 ergs.

15. The velocity of light in a vacuum is about 30,000,000,000 cm/sec.

16. The wavelength of the orange-red line of krypton 86 is about 6100 A.

17. The world's largest library, the Library of Congress, has 59,000,000 items.

18. In 1975, Americans spent $118.5 billion on health care.

19. In 1975, Americans smoked 600 billion cigarettes.

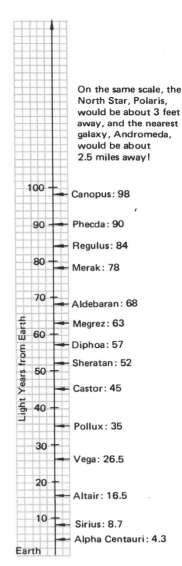

On the same scale, the North Star, Polaris, would be about 3 feet away, and the nearest galaxy, Andromeda, would be about 2.5 miles away!

Light Years from Earth

100 — Canopus: 98

90 — Phecda: 90

— Regulus: 84

80 — Merak: 78

70 — Aldebaran: 68

— Megrez: 63

60 — Diphoa: 57

— Sheratan: 52

50 — Castor: 45

40 —

— Pollux: 35

30 —

— Vega: 26.5

20 —

— Altair: 16.5

10 — Sirius: 8.7

— Alpha Centauri: 4.3

Earth

FIGURE 1.17 Distances of selected stars from Earth in light-years

Write each of the numbers in Problems 20–35 without using exponents.

EXAMPLES:

$$5^3 = 5 \times 5 \times 5 = 125$$
$$4.8 \times 10^5 = 4.8 \times 100,000 = 480,000$$

20. 7^2 21. 2^6 22. 7.2×10^{10} 23. 2.1×10^{-3}

24. 6×10^3 25. 6^3 26. 6.81×10^0 27. 4.1×10^{-7}

28. A kilowatt-hour is about 3.6×10^6 joules.

29. A ton is about 9.07×10^2 kilograms.

30. Saturn is about 8.86×10^8 miles from the sun.

31. The mass of the sun is about 3.33×10^5 times the mass of the earth.

32. The sun develops about 5×10^{23} horsepower per second.

33. The volume of a typical neuron is about 3×10^{-8} cm^3.

34. If the sun were a light bulb, it would be rated at 3.8×10^{25} watts.

35. There are 1.28×10^6 supermarket employees in the United States.

B Problems

36. The newspaper clipping in the margin column describes a trillion. Write the numbers in the article in scientific notation.

37. In 1973 about 1.6 million all-aluminum cans were returned for recycling, which was enough metal to put storm windows in a half-million homes. Installing the storm windows resulted in a savings of over a half-billion kilowatt hours for the year—enough electricity to supply the energy for about 63,000 homes. Write the numbers in the problem in scientific notation.

38. In the 17th century, Christian Huygens estimated the star Sirius to be about 4.2×10^{17} cm away. It is really 20 times farther away. Write both these distances without exponents.

39. The newspaper clipping in the margin column says that our atmosphere weighs 5 quadrillion, 157 trillion tons. Write this number in scientific notation.

40. Approximately how high would a stack of one million $1 bills be? (There are 233 new $1 bills per inch.)

41. Estimate how many pennies it would take to make a stack one inch high. A fistful of pennies and a ruler will help you. Approximately how high would a stack of one million pennies be?

42. There are 2,260,000 grains in a pound of sugar. If the U.S. production of sugar in 1972 was 29,500,000 tons, estimate the number of grains of sugar produced in a year in the United States. Use scientific notation.

WHAT IS A TRILLION?

How much is a trillion? It is 1,000,000,000,000 or a million millions. A trillion inches is more than 15.8 million miles or half the distance to Venus. The moon is about 230,000 miles, so a trillion *inches* is about 68 round trips to the moon. A trillion seconds is about 81,700 years!

SOVIETS WEIGH THE ATMOSPHERE

MOSCOW (UPI)—The atmosphere weighs 5 quadrillion, 157 trillion tons, the Soviets said in an announcement of great import to scientists and connoisseurs of trivia.

The official Tass news agency said an electronic brain known as the Minsk-22 computer had figured out the total weight of Earth's air cover more precisely than had been done before. "The mass of air enveloping our planet is now estimated at 5 quadrillion, 157 trillion tons," Tass said.

"This calculation is essential for research in cosmonautics, space geodesy and gravimetry," Tass said.

There are 2240 lb per ton.

43. Light travels at 186,282 miles per second. A light-year is the distance that light travels in one year. Approximately how many miles are there in one light-year?

44. Use Problem 43 and Figure 1.17 to write the distances in miles of the following stars from Earth.
   a. Alpha Centauri   b. Sirius   c. Phecda

45. a. Use Figure 1.17 to estimate the number of light-years that Polaris is from Earth.
   b. Use Problem 43 to write your answer to part a. in miles.

46. If the entire population of the world moved to California and each person was given an equal amount of area, how much space would be allocated to each person? (Does your answer to this problem correspond to the guess you made to the question in the margin?)

47. The total population of the earth is about 3.5×10^9.
   a. If each person has the room of a prison cell (50 sq ft), and if there are about 2.5×10^7 sq ft in a sq mi, how many people could fit into a square mile?
   b. How many square miles would be required to accommodate the entire population of the earth?
   c. If the total land area of the earth is about 5.2×10^7 sq mi, and if all the land area is divided equally, how many acres of land would each person be allocated? (1 sq mi = 640 acres.)

48. a. Guess what percentage of the world's population could be packed into a cubical box measuring 1/2 mi on each side. (Hint: The volume of a typical person is about 2 cu ft.)
   b. Now calculate the answer to part a. using the earth's population as given in Problem 47.

49. If it took one second to write down each digit, how long would it take to write all the numbers from 1 to 1,000,000?

50. Imagine that you have written down the numbers from 1 to 1,000,000. What is the total number of zeros you have recorded?

Mind Bogglers

51. Arrange the first six counting numbers in the six circles below so that the sum of each side of the triangle adds up to:
   a. 9   b. 10   c. 11   d. 12   e. 13

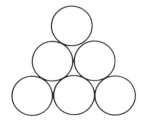

Hint for Problem 43: Approximate the speed of light by $1.86 \cdot 10^5$. In one year there are approximately $(6 \cdot 10) \cdot (6 \cdot 10) \cdot (2.4 \cdot 10) \cdot (3.65 \cdot 10^2)$ seconds. This is about
$$315 \cdot 10^5$$
or
$$3.15 \cdot 10^7$$
seconds per year.

Population Explosion?

What would happen if the *entire world population* moved to California?

California

158,600 square miles

World population about 3,500,000,000!

How much space would each person have (make a guess)?

A.   12 sq. in.
B.   12 sq. ft
C.   125 sq. ft
D.   1250 sq. ft
E.   1 sq. mi

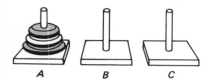

FIGURE 1.18 Tower problem

52. When my daughter was 2, she had a toy that consisted of colored rings of different sizes, as shown in Figure 1.18. Suppose we wish to move the "tower" from stand A to stand C. To make this interesting, let's agree to the following rules: (1) move only one ring at a time; (2) at no time may a larger ring be placed on a smaller ring. According to these rules, we can move the tower from stand A to C in seven moves (try it).

    a. If we add another ring, the tower can be moved in 15 moves. Explain how.

    b. How many moves would be required for five rings? (Hint: Look for a pattern.)

       3 rings: 7 moves and $7 = 2^3 - 1$
       4 rings: 15 moves and $15 = 2^4 - 1$
       5 rings: ? moves

    This is an example of a famous problem called the *Tower of Hanoi*. The ancient Brahman priests were to move a pile of 64 such discs of decreasing size, after which the world would end. This task would require $2^{64} - 1$ moves. Try to estimate how long this would take at the rate of one move per second. (By the way, $2^{64} - 1$ is the number of grains of wheat on the entire chessboard in the story about the king.)

Hint for Problem 53: Go beyond the definition, and fool with fractional exponents.

53. $8^n = 32$. Find n.

Hint for Problem 54: 2^{50} = 1,125,899,906,842,624.

54. A sheet of notebook paper is approximately .003 inch thick. Suppose it were possible to tear *any* sheet of paper in half and place the two halves so that they were on top of each other.

    Tear the sheet in half so that there are 2 sheets. Repeat so that there are 4 sheets. If we repeat again, there will be a pile of 8 sheets. Continue in this fashion until the paper has been halved 50 times. If it were possible to complete this process, how high would you guess the final pile would be? Having guessed, *compute* the height.

Problems for Individual Study

55. Problem 52 on the Tower of Hanoi raises some interesting mathematical questions. Do some research on this problem and answer the following questions.

    a. How soon after you begin to solve the puzzle will you move each of the disks for the first time?

    b. With what frequency will you change a given disk's position after you have moved it the first time?

    c. How many times will you move each of the disks in the course of rebuilding the tower consisting of n disks?

    d. What is the most efficient means of solving this puzzle?

       References: Kritchik, Maurice, *Mathematical Recreations* (New York: Dover Publications, 1953).

       Schuh, Frederick, *The Masterbook of Mathematical Recreations* (New York: Dover Publications, 1968).

Schwager, Michael, "Another Look at the Tower of Hanoi," *The Mathematics Teacher,* September 1977, pp. 528–533.

56. *Population Growth.* The time it takes the population to double is constantly decreasing. Table 1.3 shows the U.S. population for census years 1800–1970.

As we can see from Figure 1.19, the population increases geometrically. Is the same kind of increase true for your own state? Consult an almanac to obtain your information.

TABLE 1.3   U.S. Population during Census Years 1800–1970

Year	Population (in millions)
1800	5.3
1810	7.2
1820	9.6
1830	12.8
1840	17.1
1850	23.2
1860	31.4
1870	39.8
1880	50.2
1890	63.0
1900	76.0
1910	92.0
1920	105.7
1930	122.8
1940	131.7
1950	151.3
1960	179.3
1970	203.2

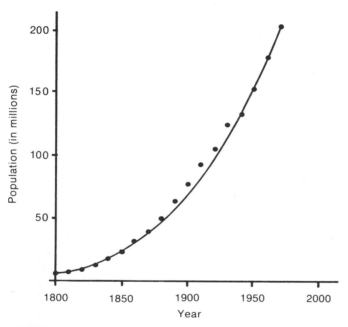

FIGURE 1.19 U.S. population for census years 1800–1970

1.5 METRIC PATTERNS

This chapter is concerned with patterns and inductive reasoning. One of the areas in the near future where Americans will be called upon to use patterns is in the use of the metric system. Officially, the metric system we'll be using is *Le Systeme International d'Unites* (International System of Units), abbreviated SI, and throughout this book the term *metric* will refer to SI.

The most difficult problem in changing from the British system to the metric system in the United States is not mathematical, but psychological. Many people fear that changing to the metric system will require

Historical Note

The system of measurement used in the United States, called the British system, goes back to the Babylonians and Egyptians. Measurements were made in terms of the human body (digit, palm, span, and foot). Eventually, measurements were standardized in terms of the physical measurements of

ONE YARD

certain monarchs. (King Henry I, for example, decreed that one yard was the distance from the tip of his nose to the end of his thumb.)

In 1790, the French Academy of Science was asked by the government to develop a logical system. The original metric system came into being, and by 1900 it was adopted by over 35 major countries. In 1906 there was a major effort to make the conversion to the metric system in the United States mandatory, but it was opposed by big business and the attempt failed. In 1960 the metric system was revised and simplified into what is now known as the SI system. In 1972 and 1973, when the United States was the only major country not using the metric system, further attempts to make it mandatory failed. However, in 1975, Congress declared conversion to the metric system to be "national policy." This time big business supported the drive toward metric conversion, and it appears inevitable that the metric system will come into use in the United States.

complex multiplying and dividing and the use of confusing decimal points. For example, in a recent popular article, James Collier states:

> For instance, if someone tells me it's 250 miles up to Lake George, or 400 out to Cleveland, I can pretty well figure out how long it's going to take and plan accordingly. Translating all of this into kilometers is going to be an awful headache. A kilometer is about 0.62 miles, so to convert miles into kilometers you divide by six and multiply by ten, and even that isn't accurate. Who can do that kind of thing when somebody is asking me are we almost there, the dog is beginning to drool and somebody else is telling you you're driving too fast?
>
> Of course, that won't matter, because you won't know how fast you're going anyway. I remember once driving in a rented car on a superhighway in France, and everytime I looked down at the speedometer we were going 120. That kind of thing can give you the creeps. What's it going to be like when your wife keeps shouting, "Slow down, you're going almost 130"?
>
> But if you think kilometers will be hard to calculate. . . .*

The author of this article has missed the whole point. Why are kilometers hard to calculate? How does he know that it's 400 miles to Cleveland? He knows because the odometer on his car or a road sign told him. Won't it be just as easy to read an odometer calibrated to kilometers or a metric road sign telling him how far it is to Cleveland?

The real advantage of using the metric system will be the ease of conversion from one unit of measurement to another. How many of you remember the difficulty you had learning to change tablespoons to cups? Or pints to gallons? Figure 1.20 shows a page from an 1890 arithmetic book in which pupils were asked to make some English conversions.

*James Lincoln Collier, "Fourth Down, 50 Centimeters to Go," *Kansas City Star Magazine,* January 8, 1978.

140   *FIRST BOOK OF ARITHMETIC.*

13. How many quarts in 10 bushels?

14. How many pints in 3 pecks?

15. How many bushels in 64 quarts?

16. What part of a quart is 1 pint? 3 pints?

17. What part of a peck is 1 quart? 3 quarts?

18. What part of a bushel is 1 peck? 2 pecks?

19. A man sold 1 bu. 3 pk. of clover-seed at 8 cents a quart: how much did he receive?

20. A fruit-dealer paid $7 for 3 bu. 3 pk. of peaches, and sold them at 60 cents a peck: what was his gain?

21. How many pecks of chestnuts can be bought for $15.60, at 40 cents a peck? How many bushels?

LIQUID MEASURE.

Liquid Measure is used in measuring liquids; as, oil, milk, alcohol, etc.

The denominations are *gills, pints, quarts,* and *gallons.*

TABLE.

4 gills (*gi.*) . . . . . are 1 pint . . . . . *pt.*
2 pints . . . . . . are 1 quart . . . . *qt.*
4 quarts . . . . . are 1 gallon . . . . *gal.*
1 gal.   4 qt. = 8 pt. = 32 gi.

FIGURE 1.20 A page from *First Book of Arithmetic for Pupils Uniting Oral and Written Exercises* by Emerson E. White (New York: American Book Co., 1890).

In the metric system there are three basic units of measurement:

length: meter (m)
capacity: liter (ℓ)*
mass: gram (g)

*The liter is a non-SI metric unit, but its use is acceptable with or in additon to SI units.

The BASIC UNITS of metric measurement are meter, liter, and gram; they should be memorized.

Mass and weight are terms that are often used incorrectly. Mass is the quantity of matter in an object, while weight is the force of gravity upon an object. When you step onto a scale you are finding your weight not your mass.

These are used with some common prefixes, which you will need to know, as shown in Table 1.4.

TABLE 1.4   Metric Prefixes

Smaller Unit Prefixes:			Larger Unit Prefixes:		
Prefix	Symbol	Meaning	Prefix	Symbol	Meaning
deci	d	one-tenth	deka	dk	ten
centi	c	one-hundredth	hecto	h	hundred
milli	m	one-thousandth	kilo	k	thousand

There are other prefixes, but these are the most common ones.

MEASURING LENGTH

Even though eventually we will all be thinking in metric units, there will be a period of time during which comparison with the British units is necessary. The *meter* is the basic unit of measurement for length. For comparison, you should remember that a meter is about 3 inches longer than a yard.

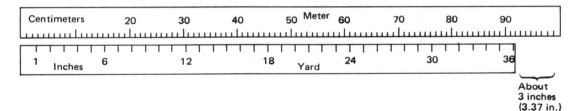

In order to start thinking in terms of meters, you need to be able to estimate distances using meters. Try this little multiple-choice test.

1. The length of a VW bug is about
   A. 3 m   B. 4 m   C. 10 m

2. The length of a football field is
   A. 100 m   B. more than 100 m   C. less than 100 m

3. The height of a kitchen table is usually
   A. 1 m   B. less than 1 m   C. more than 1 m

4. The distance from floor to ceiling in a typical home is about
   A. 2.5 m   B. 3.5 m   C. 4.5 m

5. The height of this textbook is about
   A. .02 m   B. .2 m   C. 2 m

There once was a teacher named Streeter
Who taught stuff like meter and liter
The kids moaned and groaned,
Until they were shown,
That meter and liter were neater!

Many measurements are larger or smaller than a meter. To denote these we use the prefixes given in Table 1.4. Abbreviations are written together; dm stands for decimeter, cm for centimeter, and so on.

One meter equals

1000 mm
100 cm
10 dm

| 1 dm | 1 dm | 1 dm | 1 dm | 1 dm | 1 dm | 1 dm | 1 dm | 1 dm | 1 dm |

1 m

10 dm = 1 meter   dm = decimeter

To change from one unit to another, notice the following pattern.

1 meter equals

$$
\left.\begin{array}{l}
1000 \text{ mm} \\
100 \text{ cm} \\
10 \text{ dm}
\end{array}\right] \quad \text{smaller unit prefixes cause a larger number of units}
$$

$$
1 \text{ m} \quad \big\} \quad \text{basic unit}
$$

$$
\left.\begin{array}{l}
.1 \text{ dkm} \\
.01 \text{ hm} \\
.001 \text{ km}
\end{array}\right] \quad \text{larger unit prefixes cause a smaller number of units}
$$

Therefore, if the height of this book is .225m, then it is also

smaller prefixes
$$
\left[\begin{array}{ll}
225 \text{ mm} & \leftarrow \text{three decimal places} \\
22.5 \text{ cm} & \leftarrow \text{two decimal places} \\
2.25 \text{ dm} & \leftarrow \text{one decimal place}
\end{array}\right.
$$

larger prefixes
$$
\left[\begin{array}{ll}
.0225 \text{ dkm} & \leftarrow \text{one decimal place} \\
.00225 \text{ hm} & \leftarrow \text{two decimal places} \\
.000225 \text{ km} & \leftarrow \text{three decimal places}
\end{array}\right.
$$

Remember, smaller unit prefixes cause a larger number of units, and larger unit prefixes cause a smaller number of units.

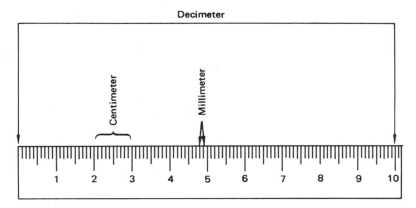

FIGURE 1.21 Actual size of a metric ruler. Metric rulers are usually marked in centimeters and divided into millimeters.

It's customary to use either the meter or the kilometer for measurements larger than a meter.

The number of places the decimal point is moved can be found by counting the number of lines apart the prefixes are found on the following chart:

      milli
      centi
      deci
      BASIC UNIT
      deka
      hecto
      kilo

These prefixes should be memorized.

EXAMPLES:

1. 43 km equals

all prefixes are smaller so units become larger

430 hm	← one decimal place
4300 dkm	← two decimal places
43,000 m	← three decimal places
430,000 dm	← four decimal places
4,300,000 cm	← five decimal places

2. 14.1 cm equals

smaller prefix { 141 mm ← one decimal place

larger prefixes mean smaller units (count the number of decimal places)

.141 m	← two decimal places
.0141 dkm	← three decimal places
.00141 hm	← four decimal places
.000141 km	← five decimal places

HALVE YOUR METER

CBR

© C. B. Rhinehart 1973

♩=120

How much I weigh I'll ask my gram. How far I've gone I'll check my me-ter. But e-ven the big-gest eat-er could hard-ly drink a li-ter. There's

♩=120

de-ci, cen-ti and mil-li. They get so small it is sil-ly; but de-ka, hec-to and ki-lo is large as I now go

MEASURING CAPACITY

Capacity measurement will probably be the easiest for us to use because a liter is almost the same as a quart. Measurements of capacity are defined in terms of the capacity of variously sized cubes, as shown in Table 1.5.

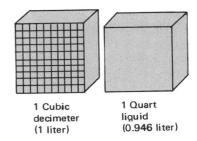

1 Cubic decimeter (1 liter)

1 Quart liquid (0.946 liter)

TABLE 1.5 Measurements of Capacity

Unit	Abbreviation	Definition	Relation to a liter
cubic centimeter	cm³	capacity of a cube having sides of 1 cm	.001ℓ (since it is 1/1000 of a liter, it is also called a *milliliter*)
cubic decimeter	dm³	capacity of a cube having sides of 1 dm	1ℓ
cubic meter	m³	capacity of a cube having sides of 1 meter	1000ℓ (since it is 1000 liter, it is also called a *kiloliter*)

Notice that a volume of liquid of 1 milliliter is exactly the capacity of a cube whose sides are 1 centimeter. Therefore, 1 cubic centimeter is used interchangeably with 1 milliliter, as shown in Figure 1.22.

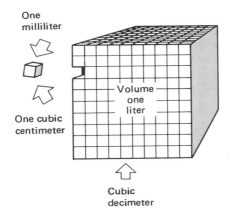

One milliliter

One cubic centimeter

Volume one liter

Cubic decimeter

FIGURE 1.22 Definition of a liter

Even though the liter is not an official SI-unit, it is acceptable and is in common usage. In fact, consumers are already seeing products packaged in liter sizes. The United States Bureau of Alcohol, Tobacco, and Firearms has made metric bottle sizes for liquor mandatory.

Hard-Liquor Bottle Sizes

New—Metric		Current—British	
	oz		oz
50 mℓ	1.7	miniature	1.6
200 mℓ	6.8	half-pint	8.0
500 mℓ	16.99	pint	16.0
750 mℓ	25.4	fifth	25.6
1 ℓ	33.8	quart	32.0
1.75 ℓ	59.2	half-gallon	64.0

In everyday measurements of capacity you will use the milliliter (or cubic centimeter), liter, and cubic meter. A cubic meter is the capacity of a cube measuring 1 m by 1 m by 1 m. It holds 1000 ℓ or 1 kℓ, so for measuring large capacities the usual measurement is cubic meters.

The prefixes for capacity are the same as for length measurements: One liter equals

1000 mℓ	smaller prefixes imply
100 cℓ	larger number of units
10 dℓ	
1 ℓ	
.1 dkℓ	larger prefixes imply
.01 hℓ	smaller number of units
.001 kℓ	

EXAMPLES:

1. 60ℓ equals

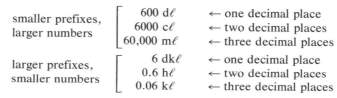

smaller prefixes, larger numbers	600 dℓ	← one decimal place
	6000 cℓ	← two decimal places
	60,000 mℓ	← three decimal places
larger prefixes, smaller numbers	6 dkℓ	← one decimal place
	0.6 hℓ	← two decimal places
	0.06 kℓ	← three decimal places

2. 5.2 cℓ = _____ hℓ (fill in the blank)

     └ larger prefix implies smaller number

   └ decimal point moves four places

  Answer: .00052

3. 1.5 kℓ = _____ ℓ (fill in the blank)

     └ smaller prefix implies larger number

   └ decimal point moves three places

  Answer: 1500

MEASURING MASS

The correct usage is mass, but in everyday usage you will often hear it referred to as weight.

The basic metric unit for measuring the mass of an object is the gram (g). It is also related to the other measurements we've considered. Suppose a cubic centimeter is filled with one milliliter of water. The weight of one milliliter of water is a gram. It is helpful to remember that a gram is about the weight of one paper clip. Since the gram is small, you will probably use the kilogram, which is about 2.2 pounds, for most weight measurements.

Table 1.6 shows some equivalencies for selected masses. Remember, if you say you weigh 155 lb, you probably mean that you stepped onto a scale and it read 155.

TABLE 1.6 Equivalencies (Approximate) between Pounds and Kilograms

Pounds	Kilograms	Pounds	Kilograms
50	23	140	64
55	25	145	66
60	27	150	68
65	29	155	70
70	32	160	73
75	34	165	75
80	36	170	77
85	39	175	80
90	41	180	82
95	43	185	84
100	45	190	86
105	48	195	89
110	50	200	91
115	52	205	93
120	55	210	95
125	57	215	98
130	59	220	100
135	61	225	102

Suppose you wish to know what you weigh in metric units. You could do a mathematical conversion, but, more likely, some day you will buy a metric scale, and when you step onto it you will read 70, which means your mass is 70 kg.

EXAMPLES:

1. 4.8 kg = _____ g (fill in the blank)

   └ smaller unit implies larger number

   └ decimal point moves three places

   Answer: 4800

2. 287 cg = _____ kg (fill in the blank)

   └ larger unit implies smaller number

   └ decimal point moves five places

   Answer: .00287

Even though you can use any metric prefix for measurements of mass, the commonly used ones are gram and kilogram.

Paper clip

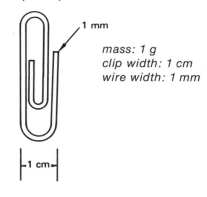

mass: 1 g
clip width: 1 cm
wire width: 1 mm

1 mm

1 cm

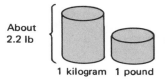

About 2.2 lb

1 kilogram   1 pound

Remember, to find the number of places to move the decimal point, count the lines in
milli
centi
deci
BASIC UNIT
deka
hecto
kilo

PROBLEM SET 1.5

A Problems

When the British changed to metric, their government placed the following newspaper ad: "A meter measures three foot three. It's longer than a yard, you see." A second ad read: "Two and a quarter pounds of jam weigh about a kilogram."

1. Name the most probable unit of measurement you would use to measure each of the following lengths.
   a. the height of a file cabinet
   b. the diameter of a nickel
   c. the distance from New Orleans to Mobile
   d. the length of an automobile
   e. a person's height

2. Name the most probable unit of measurement you would use to measure each of the following capacities.
   a. the volume of a swimming pool
   b. the amount of medication to be injected into a patient
   c. the capacity of a freezer
   d. the amount of salt to be added to the dough for one pie crust
   e. the volume of a kitchen pan

3. Name the most probable unit of measurement you would use to measure each of the following masses.
   a. a letter
   b. a nickel
   c. an automobile
   d. a bag of oranges
   e. a person's mass

Select the most appropriate response for each of Problems 4–9.

4. The length of a new piece of chalk is about:
   A. 1 mm   B. 1 cm   C. 1 dm   D. 1 m

5. The length of the eraser on a new pencil is about:
   A. 1 mm   B. 1 cm   C. 1 dm   D. 1 m

6. The mass of a large egg is about:
   A. 1 g   B. 50 g   C. 500 g   D. 1 kg

7. The mass of an automobile is about:
   A. 2000 g   B. 500 kg   C. 1000 kg   D. 10,000 kg

8. A coffee cup will hold about:
   A. 25 mℓ B. 250 mℓ C. 2.5 ℓ D. 25 ℓ

9. Enough water for a bath is about:
   A. 1 mm³   B. 1 cm³   C. 1 m³   D. 50 ℓ

10. Fill in the blanks.
    a. 1 liter = _____ deciliters
    b. 1 liter = _____ milliliters
    c. 1 liter = _____ kiloliters
    d. 1 kilometer = _____ millimeters
    e. 1 meter = _____ kilometers

THE WIZARD OF ID

11. Fill in the blanks.
 a. 6.23 liters = _____ milliliters
 b. 4.5 meters = _____ centimeters
 c. 6 centimeters = _____ millimeters
 d. 48 millimeters = _____ centimeters
 e. 33 dekaliters = _____ hectoliters

12. Fill in the blanks.
 a. 6.25 hm = _____ cm
 b. 325 dm = _____ cm
 c. 428 hm = _____ dkm
 d. 143 dm = _____ dkm
 e. 25 km = _____ m

13. Fill in the blanks.
 a. 3525 cℓ = _____ mℓ
 b. 23 cm³ = _____ mℓ
 c. 750 mℓ = _____ ℓ
 d. 1 m³ = _____ ℓ
 e. 458 cm³ = _____ ℓ

> *Metrication*
> *Mike Keedy, Purdue University, says*
> *Go Metric—Be a Liter Bug!*

B Problems

14. Use metric measurements to find:
 a. your height
 b. your mass (use Table 1.6)
 c. the width of your thumb
 d. the distance around your waist
 e. the distance from your nose to your thumb

15. Use metric measurements to find:
 a. the height of a standard door
 b. the thickness of a nickel
 c. the length of your house key
 d. the height of a standard beer can
 e. the diameter of an LP record

For Problems 14 and 15, take direct measurements; do not convert from the British system.

   One place where metric measurements will be most obvious is on the highway. Many new cars have speedometers listing both km/h as well as mph. Several states, including Ohio, Florida, and California, are now installing road signs listing kilometers as well as miles. There will be a period of time when it is necessary to convert from one system to another. Remember,

> *from kilometers to miles:*
>   one kilometer is about .6 mile
> *from miles to kilometers:*
>   one mile is about 1.6 kilometers

EXAMPLES:

To change from kilometers to miles (estimation), multiply by .6.

1. .6(600) = 360, so 600 km ≈ 360 mi
2. .6(200) = 120, so 200 km ≈ 120 mi

The exact conversion is more complicated, but for all practical purposes you need only use these estimates. It is interesting to compare this conversion constant for changing kilometers to miles,
      .621 . . . ,
with the golden ratio of Section 1.3:
      .618 . . . .

To change from miles to kilometers (estimation), multiply by 1.6.

3. 1.6(500) = 800.0, so 500 mi ≈ 800 km
4. 1.6(340) = 544.0, so 340 mi ≈ 540 km

16. Estimate the given distances in miles.
    a. 400 km
    b. 100 km
    c. 580 km
    d. 250 km
    e. 2500 km

17. Estimate the given distances in miles.
    a. 300 km
    b. 900 km
    c. 270 km
    d. 640 km
    e. 3500 km

18. Estimate the given distances in kilometers.
    a. 300 mi
    b. 100 mi
    c. 840 mi
    d. 15 mi
    e. 3600 mi

19. Estimate the given distances in kilometers.
    a. 200 mi
    b. 85 mi
    c. 900 mi
    d. 12 mi
    e. 2900 mi

Another place where metric measurements will be obvious is in measuring temperature. The British system uses Fahrenheit and the metric system uses Celsius. Celsius is defined so that 0°C is the freezing point of water and 100°C is the boiling point of water. The three main areas where most of us will need to be concerned with Celsius temperature are atmospheric temperature, body temperature, and oven temperature, which are shown in Figure 1.23.

20. If the temperature outdoors is 32°C, would you most likely find people ice skating or water skiing?

21. What is the normal Celsius body temperature?

22. Would you be more likely to broil steaks at
    100°C, 290°C, or 550°C?

During this transition period, it will be necessary to make some conversions between the British and the metric systems when working in the kitchen. Table 1.7 shows some common conversions for the kitchen.

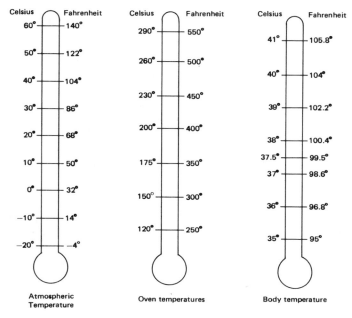

FIGURE 1.23 Temperature conversions between Celsius and Fahrenheit

TABLE 1.7 British-metric kitchen conversions

Volume			Mass		Oven Temperatures	
¹/₄ tsp	1.25	mℓ	1 oz	28 g	225°F	105°C
¹/₂ tsp	2.5	mℓ			250°F	120°C
1 tsp	5	mℓ	¹/₈ lb	57 g	275°F	135°C
1 tbsp	15	mℓ	¹/₄ lb	114 g	300°F	150°C
			¹/₃ lb	151 g	325°F	165°C
¹/₈ cup	30	mℓ	¹/₂ lb	227 g	350°F	180°C
¹/₄ cup	60	mℓ	²/₃ lb	302 g	375°F	190°C
¹/₃ cup	80	mℓ	³/₄ lb	341 g	400°F	205°C
¹/₂ cup	120	mℓ	1 lb	454 g	425°F	220°C
²/₃ cup	160	mℓ			450°F	230°C
³/₄ cup	180	mℓ	2.2 lb	1000 g	475°F	250°C
1 cup	240	mℓ		or	500°F	260°C
				1 kg	525°F	275°C
¹/₄ pt	120	mℓ			550°F	290°C
¹/₂ pt	240	mℓ			575°F	300°C
³/₄ pt	360	mℓ			600°F	315°C
1 pt	475	mℓ				
2 pt	950	mℓ				
1 qt	950	mℓ				
1 gal	3785	mℓ				

This table doesn't always show exact mathematical equivalencies. For example, 250°F is 121°C, but for practical purposes a recipe would call for a 120° oven and not a 121° oven. What this table does show are reasonable substitutions for recipe conversions.

Instead of using exact mathematical equivalents, the quantities in most recipes can be rounded. For example, consider the following two recipes.

APPLE CRISP

6 to 8 apples, sliced (about 1 qt.)
$1/4$ cup water
$3/4$ cup sugar
$1/2$ cup cake flour
1 tsp. cinnamon
6 tblsp. butter
$1/2$ tsp. salt

Peel and slice apples thinly; put them in a baking dish. Add the water. Combine the sugar, flour, cinnamon, and salt; blend in the butter until crumbly in consistency. Pour over the apples and press down. Bake uncovered in 375° oven for about 1 hour.
—Mrs. Rigmor Sovndal

HOT MILK SPONGE CAKE

6 or 7 eggs
5 mℓ vanilla
560 mℓ sifted cake flour
480 mℓ sugar
10 mℓ baking powder
pinch of salt
240 mℓ milk heated to boiling with 114 g butter

Beat eggs with electric mixer until light and fluffy. Gradually add sugar and beat until very light. Add vanilla. Lower mixer to slowest speed and beat in dry ingredients. Don't overbeat. Using same speed, add hot milk and butter. Pour batter into two 23-cm pans lined with wax paper and greased. Bake at 180° for 25 minutes.
—Mrs. Linda Smith

If you were to convert the apple crisp recipe to metric by formula, $1/4$ cup water would equal 59.1 mℓ water. But if you were *given* this recipe in metric, it would no doubt call for 60 mℓ water.

23. Translate the recipe for apple crisp into a metric recipe.

24. Translate the sponge cake recipe into a British measurement recipe.

One teaspoon ≈ 5 cm^3

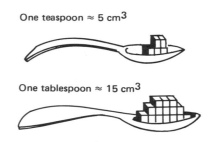

One tablespoon ≈ 15 cm^3

Problems For Individual Study

25. You have, no doubt, seen many examples of metric usage—for example, 100 mm cigarettes, 35 mm cameras, or liter steins of beer. Find some examples of the metric system in advertising or in other aspects of American life today.

26. Figure 1.24 shows a map of the world illustrating the countries that are using the metric system and those that are not. Write a history of the metric system.

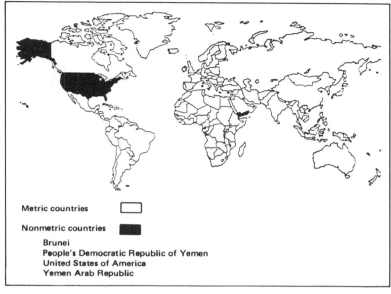

Metric countries ☐

Nonmetric countries ■

    Brunei
    People's Democratic Republic of Yemen
    United States of America
    Yemen Arab Republic

FIGURE 1.24 The metric world

1.6 SUMMARY AND REVIEW

CHAPTER OUTLINE

I. Mathematical Patterns

  A. By looking for patterns, we can often simplify calculations and make conjectures about more complicated results.

  B. Inductive reasoning

    1. Arithmetic patterns
    2. Geometric patterns

  C. Pascal's triangle

  D. Number sequences

    1. Arithmetic progression: common difference
    2. Geometric progression: common ratio
    3. Fibonacci sequence: 1,1,2,3,5,8,13,21,34, . . .

  E. Golden ratio

    1. $\tau \approx 1.6180$
    2. Golden rectangle

II. Large and Small Numbers

  A. Exponential notation

    1. *Definition:* For any number b and any counting number n,

$$b^n = \underbrace{b \cdot b \cdot b \ldots b}_{n \text{ factors}}$$

2. $10^0 = 1$

3. If n is a counting number, then $10^{-n} = 1/10^n$.

B. Scientific notation

1. *Definition:* The **scientific notation** for a number is that number written as a power of 10 times another number x, such that $1 \leqslant x < 10$.

III. Metric System

A. Basic units of measurement

1. Length: meter, m
2. Capacity: liter, ℓ
3. Mass: gram, g

B. Metric prefixes

1. milli (m) ⎫
2. centi (c) ⎬ smaller unit prefixes
3. deci (d) ⎭
4. BASIC UNIT
5. deka (dk) ⎫
6. hecto (h) ⎬ larger unit prefixes
7. kilo (k) ⎭

C. To change prefix

1. Substituting a smaller unit prefix causes a larger number of units
2. Substituting a larger unit prefix causes a smaller number of units
3. The decimal point on the number is moved the number of places the prefixes are separated on the chart listed under III.B. above.

D. Estimate lengths, capacities, and masses in metric

E. Applications

1. Measurements
2. Highway travel
3. Temperature (Celsius)
4. Cooking

You can work these Review Problems as a practice test. All of the answers for Chapter Reviews are found in the answer section of this text.

REVIEW PROBLEMS

1. Compute $(111,111,111) \cdot (111,111,111)$. Do not use direct multiplication. Show all of your work.

2. What do we mean by inductive reasoning?

3. Fill in the blanks.
   a. 10^4, 10^3, 10^2, _____
   b. $42, 32, 22,$ _____
   c. $98, 87, 76,$ _____

4. a. One of the sequences in Problem 3 is a geometric progression. Which one is it?
   b. What is its common ratio?

5. Write in expanded form: a. 5.79×10^{-4} b. 4.01×10^5

6. Write in expanded form: a. 10^{-1} b. 4.321×10^7

7. Write in scientific notation: a. .0034   b. 4,000,300

8. Write in scientific notation: a. 17,400   b. 5

9. What are the three basic metric units?

10. Give the meaning of each of the following prefixes.
    a. centi   b. deka   c. kilo   d. milli   e. deci

11. Tell whether each of the following are measuring length, capacity, or mass.
    a. 15 kg
    b. 38.2 m
    c. 6 m³
    d. 7 mℓ
    e. 68 km

12. Name the metric unit you would most likely use to measure each of the following.
    a. the distance from New York to Chicago
    b. the distance around your waist
    c. the amount of gin to use in a martini
    d. the amount of space in the trunk of your car
    e. the weight of a pencil

13. Fill in the blanks.
    a. 4300 m = _____ km
    b. 1ℓ = _____ m³
    c. 4.8 cm = _____ mm
    d. 2.88 kg = _____ g
    e. 1 mile ≈ _____ km

Choose the most appropriate answer for each of Problems 14–20.

14. The volume of a large upright freezer is about
    A. 8ℓ B. .8 m³   C. 2 m³   D. 8 m³   E. 154 kg

15. On a hot, summer day the temperature is about
    A. 0°C   B. 10°C   C. 40°C   D. 80°C   E. 100°C

16. The height of a tall basketball player might be
    A. 85 cm   B. 1.5 m   C. 72 ℓ D. 4 m   E. 200 cm

17. If you had a fever your temperature might be
    A. 100°C   B. 50°C   C. 40°C   D. 30°C   E. 98.6°C

18. The length of a new pencil is about
    A. 2 km   B. 2 m   C. 2 cm   D. 2 dkm   E. 2 dm

19. The capacity of a 2.5ℓ coffee pot is about
    A. 4 cups   B. 10 cups   C. 15 cups   D. 25 cups   E. 35 cups

20. Water boils at
    A. 0°C   B. 50°C   C. 100°C   D. 200°C   E. 212°C

Try to answer these questions without converting from one system to another—THINK METRIC!

CULTURAL		MATHEMATICAL
-500: Persian Empire flourishes	**-500**	-500: Sulvasūtras—Pythagorean numbers, geometric constructions
		-450: Zeno—paradoxes of motion
-427: Birth of Plato		-425: Theordorus of Cyrene—irrational numbers
-399: Death of Socrates		
-377: Plato's academy founded	**-400**	-380: Plato's Academy—logic
		-340: Aristotle—deductive logic
-323: Alexander the Great conquers known world		-300: Euclid—geometry, perfect numbers
	-300	
-250: Great wall of China		-230: Sieve of Eratosthenes
-218: Hannibal crosses the Alps		-225: Archimedes—circle, pi, curves, series
	-200	-180: Hypsicles—number theory
-146: Destruction of Carthage and Corinth		
	-100	-60: Geminus—parallel postulate
-20: Virgil's *Aeneid*		
0: Birth of Christ	**0**	
64: Nero fiddles as Rome burns		75: Heron—measurement, roots, surveying
100: Plutarch's *Lives*	**100**	100: Nicomachus—number theory
162: Mohammed establishes Islam		150: Ptolemy—trigonometry
	200	250: Diophantus—number theory, algebra
		300: Pappus: *Mathematical Collection*
324: Founding of Constantinople	**300**	
400: Augustine's *Confessions*	**400**	410: Hypatia of Alexandria—first woman mentioned in history of mathematics
476: Fall of Rome		480: Tsu Ch'ung-chi approximated π as 355/113
529: Closing of school in Athens	**500**	
641: Last library at Alexandria burned	**600**	628: Brahmagupta—algebra
732: Moslems defeated at Tours	**700**	710: Bede—calendar, finger arithmetic

2

THE NATURE OF
SETS AND
DEDUCTIVE REASONING

. . . the two great components of the critical movement,
though distinct in origin and following separate paths,
are found to converge at last in the thesis:
Symbolic Logic is Mathematics, Mathematics is Symbolic Logic,
the twain are one.

C. J. Keyser

2.1 DEDUCTIVE REASONING

In the first chapter we studied patterns that allow us to form conclusions using inductive reasoning. Remember, inductive reasoning is the process of forming conclusions that are based on a number of observations of specific instances. The conclusions may be probable, but they are not necessarily true. For example, based on our observations of planetary motion, we may predict that tomorrow the sun will rise at 6:03 A.M. Although it is very probable that this prediction is correct, it is not absolutely certain.

There is, however, a type of reasoning that produces *certain* results. For example, consider the following argument.

Statements 1, 2, and 3 are called an argument.

1. If you take the *Times,* then you are well informed.
2. You take the *Times.*
3. Therefore, you are well informed.

If you accept statements 1 and 2 as true, then you *must* accept 3 as true. Statements 1 and 2 are called the *hypotheses* or *premises* of the argument, and 3 is called the *conclusion.* Such reasoning is called *deductive reasoning,* and, if the conclusion follows from the hypotheses, the reasoning is said to be *valid.*

Hypotheses or premises
Conclusion
Deductive reasoning
Valid reasoning

The purpose of this chapter is to build a logical foundation to aid you not only in your study of mathematics and other subjects but also in your day-to-day contact with others.

Logic is a method of reasoning that accepts no conclusions except those that are inescapable. This is possible because of the strict way in which every concept is defined. That is, everything must be defined in a way that leaves no doubt or vagueness in meaning. Nothing can be taken for granted, and dictionary definitions are not usually sufficient. For example, in English one often defines a sentence as "a word or group of words stating, asking, commanding, requesting, or exclaiming something; conventional unit of connected speech or writing, usually containing a subject and predicate, beginning with a capital letter, and ending with an end mark."* In symbolic logic, a sentence has a much more restricted meaning and is called a statement.

Historical Note

Logic began to flourish during the classical Greek period. Aristotle (384–322 B.C.) was the first person to systematically study the subject, and he and many other Greeks searched for universal truths that were

> DEFINITION: A *statement* is a sentence that is either true or false but not both true and false.

If the sentence is a question or a command, or if it is vague or nonsensical, then it cannot be classified as true or false; thus we would not call it a statement.

*Webster's New World Dictionary, College Edition, World Publishing Company, 1960.

For example:

1. Today is Wednesday.
2. $5 + 6 = 16$.
3. Fish swim.
4. Mickey Mouse is President.

All of these are statements, since they are either true or false. But consider:

5. Go away!
6. What are you doing?
7. This sentence is false.

These are not statements by our definition, since they cannot possibly be either true or false.

Difficulty in simplifying arguments may arise because of their length, the vagueness of the words used, the literary style, or the possible emotional impact of the words used. Consider the following two arguments.

1. If George Washington was assassinated, then he is dead.
   Therefore, if he is dead, he was assassinated.
2. If you use heroin, then you first used marijuana.
   Therefore, if you use marijuana, then you will use heroin.

Logically these two arguments are exactly the same, and both are *invalid* forms of reasoning. Nearly everyone would agree that the first is invalid, but many people see the second as valid. The reason here is the emotional appeal of the words used.

To avoid these difficulties, and to be able to simplify a complicated logical argument, we set up an *artificial symbolic language*. This procedure was first suggested by Leibniz in his search for a *universal characteristic* to unify all of mathematics. What we will do is invent a notational shorthand. We denote simple statements with letters such as $p, q, r, s, \ldots$ and then define certain connectives. The problem, then, is to *translate the English statements into symbolic form, simplify the symbolic form*, and then *translate the simpler form back into English statements*.

The problem of length can be avoided by considering only simple statements connected by certain well-defined operators, such as *not, and, or, neither . . . nor, if . . . then, unless, because*, and so on.

A compound statement is formed by combining simple statements with operators. Because of our basic definition of a statement, we see that the *truth value* of any sentence is either true (T) or false (F).

The truth value of a compound statement will depend only on the truth values of its component parts. It is not sufficient to assume that we know the meanings of the operators *and, or, not*, and so on, even though they

irrefutable. The logic of this period, referred to as Aristotelian logic, is still used today and is based on the syllogism, which we will study later in this chapter.

The second great period for logic came with the use of symbols to simplify complicated logical arguments. This was first done when the great German mathematician Gottfried Leibniz (1646–1716), at the age of 14, attempted to reform Aristotelian logic. He called his logic the universal characteristic *and wrote, in 1666, that he wanted to create a general method in which truths of reason would be reduced to a calculation so that errors of thought would appear as computational errors.*

However, the world took little notice of Leibniz' logic, and it wasn't until George Boole (1815–1864) completed his book An Investigation of the Laws of Thought *that logic entered its third and most important period. Boole considered various mathematical operations by*

separating them from the other commonly used symbols. This idea was popularized by Bertrand Russell (1872–1970) and Alfred North Whitehead (1861–1947) in their monumental Principia Mathematica. *In this work, they began with a few assumptions and three undefined terms and built a system of symbolic logic. From this, they then formally developed the theorems of arithmetic and mathematics.*

TABLE 2.1 Definition of Conjunction

p	q	$p \wedge q$
T	T	T
T	F	F
F	T	F
F	F	F

may seem obvious and simple. The strength of logic is that it does not leave any meanings to chance or to individual interpretation. In defining the truth values of these words, we will, however, try to conform to common usage.

CONJUNCTION

If p and q represent two simple statements, then "p and q" is the compound statement using the operator called *conjunction*. The word *and* is symbolized by $\wedge$. For example,

*I have a penny **and** a quarter in my pocket.*

When will this compound statement be true? There are four possibilities:

1. I have a penny. I have a quarter.
2. I have a penny. I do not have a quarter.
3. I do not have a penny. I have a quarter.
4. I do not have a penny. I do not have a quarter.

If we let p: I have a penny and q: I have a quarter, then the four possibilities can be shown as follows:

	p	q
1.	T	T
2.	T	F
3.	F	T
4.	F	F

If p and q are both true, we would certainly say that the compound statement is true; otherwise, we would say it is false. Thus we define "$p \wedge q$" according to Table 2.1.

It is worth noting that statements p and q need not be related. For example,

Fish swim and Neil Armstrong was on the moon

is a true compound statement.

DISJUNCTION

The operator *or*, denoted by $\vee$, is called *disjunction*. The meaning of this simple word is ambiguous, as we can see by considering the following examples.

1. I have a penny or a quarter in my pocket.
2. Ted is speaking in New York or in California at 7:00 P.M. tonight.*

If

p: I have a penny in my pocket,
q: I have a quarter in my pocket,
n: Ted is speaking in New York at 7:00 P.M. tonight,
c: Ted is speaking in California at 7:00 P.M. tonight.

what do we mean by each of the examples?
Statement 1 may mean:

I have a penny in my pocket.
I have a quarter in my pocket.
I have both a penny and a quarter in my pocket.

Statement 2 may mean:

Ted is speaking in New York at 7:00 P.M. tonight.
Ted is speaking in California at 7:00 P.M. tonight.

It does *not* mean that he will do both.

These statements illustrate different usages of the word *or*. Now we are forced to select a single meaning for the operator, so we will choose a definition that will conform to the first example. That is, "p or q" means "p or q, perhaps both." Thus we define "$p \lor q$" according to Table 2.2.

NEGATION

The operator *not*, denoted by $\sim$, is called *negation*. Table 2.3 serves as a straightforward definition of negation.

EXAMPLE: If t: Otto is telling the truth, then

$\sim t$: Otto is not telling the truth.

We also can translate $\sim t$ as "It is not the case that Otto is telling the truth."

Any statement can be negated, but we need to be careful about the way we form the negation of a compound statement. The negation of the compound statement

---

*We have used a literary shorthand here. These statements are to be interpreted as "I have a penny in my pocket, or I have a quarter in my pocket" and "Ted is speaking in New York at 7:00 P.M. tonight, or Ted is speaking in California at 7:00 P.M. tonight." We will use this type of shorthand throughout the chapter if no ambiguity results.

Part of the difficulty in translating English into logical statements is the inaccuracy of our language. For example, how can "fat chance" and "slim chance" mean the same thing?

TABLE 2.2 Definition of Disjunction

	p	q	$p \lor q$
1.	T	T	T
2.	T	F	T
3.	F	T	T
4.	F	F	F

In logic, the "p or q, perhaps both" that has been defined by Table 2.2 is called the inclusive or. *The second meaning of the word* or *is "p or q, but not both." This is called the* exclusive or, *and in this book will be translated by the words "either . . . or."*

TABLE 2.3 Definition of Negation

p	$\sim p$
T	F
F	T

I have a penny and a quarter in my pocket

is symbolized by $\sim(p \wedge q)$ and translated as

It is not the case that I have a penny and a quarter in my pocket.

The negation is not formed by negating each of the simple statements.

In negating a statement, it is also important that we do not change the statement itself. For example,

Let b: Jack's car is blue,
r: Jack's car is red.

Statement r is not the negation of statement b, even though both statements cannot be true. The negation of b is the statement

$\sim b$: Jack's car is not blue,

and the negation of r is the statement

$\sim r$: Jack's car is not red.

TRANSLATING AND COMBINING OPERATORS

Working with logical arguments requires that we be able to translate from English into symbols and from symbols back into English. For example, the statement

Bonnie and Clyde are cute

can be translated as follows:

Let b: Bonnie is cute,
c: Clyde is cute.

Then we have

$b \wedge c$: Bonnie is cute and Clyde is cute.

Thus, $b \wedge c$ is the symbolic statement.

We have seen that the statement "Ted is speaking in New York or in California at 7:00 P.M. tonight" really means

Either Ted is speaking in New York at 7:00 P.M. tonight,
or Ted is speaking in California at 7:00 P.M. tonight,
but not both.

Let

n: Ted is speaking in New York at 7:00 P.M. tonight.
c: Ted is speaking in California at 7:00 P.M. tonight.

Then the statement would be translated by

$$(n \vee c) \wedge \sim (n \wedge c).$$

In this book, the phrase "it is not the case that" will be used to negate a compound statement.

This is an example of an exclusive or. The words "but not both" may or may not be used, but whenever you see "either ... or" it signals the exclusive or.

This is the symbolic translation for "either n or c."

$(N \lor C) \land \sim (N \land C)$

This would be read *"n or c and not the case that n and c."* This is saying that either *n* or *c* can be true, but not both.

Notice that the use of parentheses here is the same as in algebra: it indicates the order of operations. Thus

$\sim(n \land c)$ means the negation of the statement *"n and c,"*

whereas

$\sim n \land c$ means the negation of *"n"* and the statement *"c."*

Suppose that *n* is T and *c* is F. What is the truth value of the compound statement "Ted is speaking in New York or in California at 7:00 P.M. tonight"? In symbols, we write:

$$(n \lor c) \land \sim(n \land c)$$
$$(T \lor F) \land \sim(T \land F)$$
$$T \quad \land \quad \sim F$$
$$T \quad \land \quad T$$
$$T$$

Thus the statement is true.

EXAMPLE: Test the truth value of $(p \land q) \lor (p \lor r)$ when *p* is T, *q* is F, and *r* is F.

Solution:

$$(p \land q) \lor (p \lor r)$$
$$(T \land F) \lor (T \lor F)$$
$$F \quad \lor \quad T$$
$$T$$

The statement is true. This result does *not* depend on the particular statements *p*, *q*, and *r*. As long as *p* is T, *q* is F, and *r* is F, the result will be the same—namely T.

The procedure requires that we translate not only from English into logical symbols but also from symbols into English. For example, suppose

p: I eat spinach,
q: I am strong.

Then we wish to translate the following statements into words:

1. $p \land q$
2. $\sim p$
3. $\sim(p \land q)$
4. $\sim p \land q$
5. $\sim(\sim q)$ *I AM not (not strong)*

Our translations would be as follows:

$\sim(n \land c)$ is pronounced "it is not the case that n and c."

$\sim n \land c$ is pronounced "not n and c."

Use Tables 2.1, 2.2, and 2.3. For example. $T \lor F$ is found on the second line of Table 2.2 and, as shown on that line, is replaced by T.

1. I eat spinach, and I am strong.
2. I do not eat spinach.
3. It is not the case that I eat spinach and am strong.
4. I do not eat spinach and am strong.
5. I am not not strong, *or* (if we assume that "not strong" is the same as "weak") I am not weak.

EXAMPLES: Suppose that p is T and q is F. Find the truth value for each of the five translations we just made.

1. $p \wedge q$
   T $\wedge$ F
      F   Thus the statement is false.
2. $\sim p$
   $\sim$T
     F   Thus the statement is false.
3. $\sim(p \wedge q)$
   $\sim$(T $\wedge$ F)
     $\sim$ F
      T   Thus the statement is true.
4. $\sim p \wedge q$
   $\sim$T $\wedge$ F
   F $\wedge$ F
     F   The statement is false.
5. $\sim(\sim q)$
   $\sim(\sim$F)
    $\sim$T
    F   The statement is false.

Notice that Examples 3 and 4 prove that $\sim(p \wedge q)$ and $\sim p \wedge q$ do not mean the same thing.

Notice from Example 5 that $\sim(\sim q)$ has the same truth value as q.

PROBLEM SET 2.1

A Problems

1. According to our definition, which of the following are statements?
   a. Hickory, Dickory, Dock, the mouse ran up the clock.
   b. John and Mary are married.
   c. Is John ugly?
   d. John has a wart on the end of his nose.
   e. Today is the first day of the rest of my life.

2. According to our definition, which of the following are statements?
   a. Today is Monday.
   b. Division by zero is impossible.
   c. Logic is not so difficult as I had anticipated.
   d. Eschew obfuscation!
   e. Please do not read this sentence.

3. According to our definition, which of the following are statements?
   a. Tomorrow is the end of the world!
   b. Do you have a cold?

   c. Thomas Jefferson was the 23rd President.
   d. Sit down and be quiet!
   e. If wages continue to rise, then prices will also rise.

4. Let p represent the statement "Prices will rise" and q the statement "Inflation will be controlled." Translate each of the statements into symbols.
   a. Prices will rise, or inflation will not be controlled.
   b. Prices will rise, but inflation will not be controlled.
   c. Prices will rise, and inflation will not be controlled.
   d. Prices will not rise, and inflation will be controlled.
   e. It is not the case both that prices will rise and that inflation will be controlled.

5. Assume that prices rise and inflation is not controlled. Under these assumptions, which of the statements in Problem 4 are true?

6. Let p represent the statement "Paul is peculiar" and q the statement "Paul likes to read mathematics textbooks." Translate each of the following statements into words.
   a. $p \wedge q$
   b. $\sim p \wedge q$
   c. $\sim(p \wedge q)$
   d. $p \vee \sim q$
   e. $\sim p \vee \sim q$

7. Assume that p is T and q is T. Under these assumptions, which of the statements in Problem 6 are true?

8. Let r represent a statement having a truth value F, and let s and t each represent statements having truth values T. Find the truth value of each statement.
   a. $(r \vee s) \vee t$
   b. $(r \wedge s) \wedge \sim t$
   c. $r \wedge (s \vee t)$
   d. $(r \wedge s) \vee (r \wedge t)$
   e. $(\sim r \vee s) \wedge \sim t$

9. Find the truth value when p and q are each T and r is F.
   a. $(p \vee q) \wedge r$
   b. $(p \wedge q) \wedge \sim p$
   c. $p \wedge (q \vee r)$
   d. $(p \wedge q) \vee (p \wedge r)$
   e. $(p \wedge \sim q) \vee (p \wedge q)$

10. Find the truth value when p is F and q and r are each T.
   a. $(p \vee q) \vee (r \wedge \sim q)$
   b. $\sim(\sim p) \vee (p \wedge q)$
   c. $(p \wedge q) \vee (p \wedge \sim r)$
   d. $(p \vee q) \wedge [(r \wedge \sim p) \wedge (q \vee \sim p)]$
   e. $(\sim p \vee q) \wedge \sim p$

11. Define conjunction.

12. Define disjunction.

For the statement in 4b, translate "but" as a conjunction.

Historical Note

Galileo Galilei (1564–1642) is best known for his work in astronomy, but he should be remembered as the father of modern science, in general, and of physics, in particular. Galileo believed mathematics was the basis for science. Using nature as his teacher, he reasoned deductively and described the qualities of matter as quantities of mathematics. He laid the foundation of physics that Newton would build upon almost 50 years later (see the historical note on page 370).

B Problems

13. Does the *B.C.* cartoon in the margin illustrate inductive or deductive reasoning? Explain your answer.

14. Does the following story illustrate inductive or deductive reasoning? Explain your answer.

SIMPLE DEDUCTION

The old fellow in charge of the checkroom in a large hotel was noted for his memory. He never used checks or marks of any sort to help him return hats to their rightful owners.

Thinking to test him, a frequent hotel guest asked him as he received his headgear, "Sam, how do you know this is my hat?"

"I don't, sir," was the calm response.

"Then why did you give it to me?" asked the guest.

"Because," said Sam, "it's the one you gave me, sir."

—*Lucille J. Goodyear*

Problems 15–18 refer to the lyrics of **By the Time I Get to Phoenix.** *Tell whether each answer you give is arrived at inductively or deductively.*

By the Time I Get to Phoenix
Words and Music by Jim Webb

By the time I get to Phoenix she'll be risin'.
She'll find the note I left hangin' on her door.
She'll laugh when she reads the part that says I'm leavin',
'Cause I've left that girl so many times before.

By the time I make Albuquerque she'll be workin'.
She'll probably stop at lunch and give me a call.
But she'll just hear that phone keep on ringin'
Off the wall, that's all.

By the time I make Oklahoma she'll be sleepin'.
She'll turn softly and call my name out low.
And she'll cry just to think I'd really leave her,
'tho' time and time I've tried to tell her so,
She just didn't know,
I would really go.

15. What is the basic direction (north, south, east, or west) the person is traveling?

16. What method of transportation or travel is the person using?

17. What is the probable starting point of this journey?

18. List five facts you know about each person involved.

Problems 19–23 refer to the lyrics of Ode to Billy Joe. *Tell whether each answer you give is arrived at inductively or deductively.*

*Ode to Billy Joe**
Words and Music by Bobbie Gentry

It was the third of June, another sleepy, dusty, delta day.
I was out choppin' cotton and my brother was balin' hay;
And at dinnertime we stopped and walked back to the house to eat,
And mama hollered at the back door, "Y'all remember to wipe your feet."
Then she said, "I got some news this mornin' from Choctaw Ridge,
Today Billy Joe McAllister jumped off the Tallahatchee Bridge."

Papa said to Mama, as he passed around the black-eyed peas,
"Well, Billy Joe never had a lick o' sense, pass the biscuits please,
There's five more acres in the lower forty I've got to plow,"
And Mama said it was a shame about Billy Joe anyhow.
Seems like nothin' ever comes to no good up on Choctaw Ridge,
And now Billy Joe McAllister's jumped off the Tallahatchee Bridge.

Brother said he recollected when he and Tom and Billy Joe,
Put a frog down my back at the Carroll County picture show,
And wasn't I talkin' to him after church last Sunday night,
I'll have another piece of apple pie, you know, it don't seem right
I saw him at the sawmill yesterday on Choctaw Ridge,
And now you tell me Billy Joe's jumped off the Tallahatchee Bridge.

Mama said to me, "Child, what's happened to your appetite?
I been cookin' all mornin' and you haven't touched a single bite,
That nice young preacher Brother Taylor dropped by today,
Said he'd be pleased to have dinner on Sunday, Oh, by the way,
He said he saw a girl that looked a lot like you up on Choctaw Ridge
And she an' Billy Joe was throwin' somethin' off the Tallahatchee Bridge."

A year has come and gone since we heard the news 'bout Billy Joe,
Brother married Becky Thompson, they bought a store in Tupelo,
There was a virus goin' 'round, Papa caught it and he died last spring,
And now Mama doesn't seem to want to do much of anything.
And me I spend a lot of time pickin' flowers up on Choctaw Ridge,
And drop them into the muddy water off the Tallahatchee Bridge.

19. How many people are involved in this story? List them by name and/or description.

20. Who "saw him at the sawmill yesterday"?

21. In which state is the Tallahatchee Bridge located?

22. On what day or days of the week could the death not have taken place? On what day of the week was the death most probable?

23. What was it that she "was throwing off the bridge"?

Translate the statements in Problems 24–28. For each simple statement, be sure to indicate the meanings of the symbols you use.

EXAMPLE: Peace is good and war is bad.

Solution:

Let p: peace is good,
q: war is bad.

The statement is translated: $p \wedge q$; or, if we assume that "bad" is the same as "not good," then we might have

p: peace is good,
w: war is good,

Notice that answers are not unique.

and the translation would be $p \wedge {\sim}w$.

24. W. C. Fields is eating, drinking, and having a good time.

25. Sam will not seek and will not accept the nomination.

26. Jack will not go tonight and Rosamond will not go tomorrow.

27. Fat Albert lives to eat and does not eat to live.

28. The decision will depend on judgment or intuition, and not on who paid the most.

In Problems 29–34, find the truth value when p is T and q and r are each F.

29. $(p \vee q) \wedge {\sim}(p \vee {\sim}q)$

30. $(p \wedge {\sim}q) \vee (r \wedge {\sim}q)$

31. ${\sim}({\sim}p) \wedge (q \vee p)$

32. $(r \wedge p) \vee (q \wedge r)$

33. ${\sim}(r \wedge q) \wedge (q \vee {\sim}q)$

34. $(q \vee {\sim}q) \wedge [(p \wedge {\sim}q) \vee ({\sim}r \vee r)]$

Mind Bogglers

35. What is the largest amount of money you can have in coins and not be able to make change for a dollar? (Silver dollars are not allowed.)

36. Smith received the following note from Melissa. "Dr. Smith, I wish to explain that I was really joking when I told you that I didn't mean what I said about reconsidering my decision not to change my mind." Did Melissa change her mind or didn't she?

37. Their are three errers in this item. See if you can find all three.

2.2 TRUTH TABLES AND EULER CIRCLES

There are two elementary ways of proving simple logical arguments—by using truth tables and by using Euler circles.

A truth table is a device that shows how the truth values of compound statements depend on the operators used and the truth value for the original simple statements. Tables 2.1, 2.2, and 2.3 are fundamental for constructing truth tables. We summarize them in Tables 2.4 and 2.5.

TABLE 2.4 Fundamental Operators

Connective	Symbol	Name	Symbolic Statement	Example
and or not	$\wedge$ $\vee$ $\sim$	conjunction disjunction negation	$p \wedge q$ $p \vee q$ $\sim p$	Fish swim *and* birds fly. Fish swim *or* birds fly. Fish do *not* swim.

TABLE 2.5 Truth Table of the Fundamental Operators

p	q	$p \wedge q$	$p \vee q$	$\sim p$	$\sim q$
T	T	T	T	F	F
T	F	F	T	F	T
F	T	F	T	T	F
F	F	F	F	T	T

For example, let's construct a truth table for

*Alfie did **not** come last night **and** he did **not** pick up his money.*

Let

p: Alfie came last night,
q: Alfie picked up his money.

Then the statement can be written $\sim p \wedge \sim q$. To begin, list all the possible combinations of truth values for the simple statements p and q.

p	q	
T	T	
T	F	
F	T	
F	F	

Historical Note

In regard to the real nature of scientific truth, Einstein said "As far as the laws of mathematics refer to reality, they are not certain, and as far as they are certain, they do not refer to reality." Albert Einstein (1879–1955) was one of the intellectual giants of the 20th century. He was a shy, unassuming man who was told as a child that he would never make a success of anything. However, he was able to use the tools of logic and mathematical reasoning to change our understanding of the universe.

Insert the truth values for $\sim p$ and $\sim q$.

p	q	$\sim p$	$\sim q$
T	T	F	F
T	F	F	T
F	T	T	F
F	F	T	T

Finally, insert the truth values for $\sim p \wedge \sim q$.

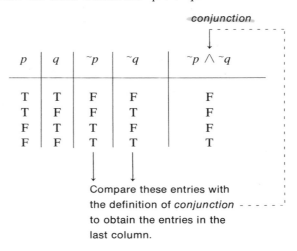

p	q	$\sim p$	$\sim q$	$\sim p \wedge \sim q$
T	T	F	F	F
T	F	F	T	F
F	T	T	F	F
F	F	T	T	T

Compare these entries with
the definition of *conjunction*
to obtain the entries in the
last column.

EXAMPLE: Construct a truth table for $\sim(\sim p)$.

Solution: We begin with p.

p
T
F

Next we fill in the values for $\sim p$ according to the definition of negation.

p	$\sim p$
T	F
F	T

Finally, we fill in the values for $\sim(\sim p)$. The arrows show how we obtain
the entries for this column.

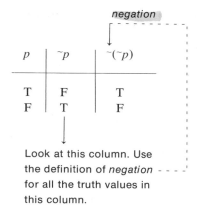

Look at this column. Use
the definition of *negation* - - - -
for all the truth values in
this column.

Notice that ~(~p) and p have the same truth values. **If two propositions have the same truth values, one can replace the other in any logical expression.** This means that the double negative of a proposition is the same as the original proposition.

> ~(~p) **may be replaced by** p **in any logical expression.**

EXAMPLE: Construct the truth table to determine when the following statement is true.

$$~(p \wedge q) \wedge [(p \vee q) \wedge q]$$

Solution: We begin as before and move from left to right, focusing our attention on only two columns at a time (refer to Table 2.5 to find the correct entries).

A	B	C	D	E	F	G
p	q	$p \wedge q$	~$(p \wedge q)$	$p \vee q$	$(p \vee q) \wedge q$	~$(p \wedge q) \wedge [(p \vee q) \wedge q]$
T	T	T	F	T	T	F
T	F	F	T	T	F	F
F	T	F	T	T	T	T
F	F	F	T	F	F	F

Step 1. Attend to the parentheses first. Look at Columns A and B along with the definition of conjunction to fill in the entries in Column C.

Step 2. Use Column C and the definition of negation to fill in the entries of Column D.

Step 3. Use Columns A and B and the definition of disjunction to fill in the entries of Column E.

~(~p) *will help us write some complex statements in simple form.*

There is a story about a logic professor who was telling her class that a double negative is known to mean a negative in some languages and a positive in others (as in English). She continued by saying that there is no spoken language in which a double positive means a negative. Just then she heard from the back of the room a sarcastic "Yeah, yeah."

Step 4. Use Columns E and B and the definition of conjunction to fill in the entries of Column F.

Step 5. Use Columns D and F, which are the left and right sides of the final operation. By using these columns and the definition of conjunction, we obtain the entries of Column G.

A second method for analyzing certain types of logical arguments is to represent the statements of an argument by circles or oval-shaped regions. These figures are called *Euler diagrams* or *Euler circles,* after the famous Swiss mathematician Leonhard Euler. When you're considering certain complicated arguments, Euler diagrams often help you arrive at a valid conclusion.

Let's begin by considering the Euler diagrams for two propositions, p and q. There are several possibilities, as shown in Figure 2.1.

Euler circles

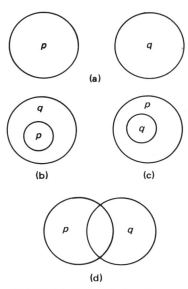

FIGURE 2.1 Euler circles for two propositions

We can use these circles as substitutes for a truth table. The region inside a circle represents the truth of a statement, and the region outside represents the falsity of that statement. Figure 2.1a represents the situation where p and q can't both be true. Figure 2.1b represents the situation where, if p is true, then q must be true. Figure 2.1c is similar: if q is true, then p must be true. Cases b and c can depict what is called *implication,* which we'll discuss later in this chapter. However, we usually work with two arbitrary propositions p and q, so we use an illustration similar to Figure 2.1d, also shown in Figure 2.2.

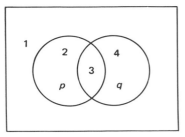

FIGURE 2.2 Euler circles for two
general sets

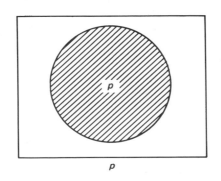

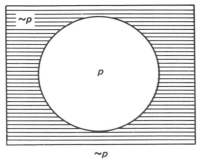

FIGURE 2.3 Euler diagrams for
one proposition, p

In Figure 2.2, region 1 stands for that part of the truth table in which p and q are both false. Region 2 represents that part in which p is T and q is F, while region 4 represents that part where p is F and q is T. Finally, region 3 represents that part of the truth table where p and q are both true. Thus, we see that Euler circles can be used as pictorial representations for truth tables.

The common operators discussed in Section 2.1 can be represented easily by Euler circles. Consider the general diagrams for one proposition in Figure 2.3 and for two propositions in Figure 2.4.

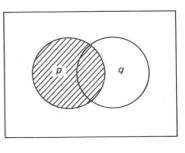

p

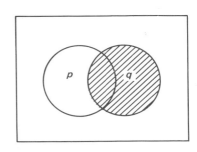

q

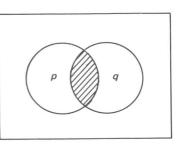

$p \wedge q$

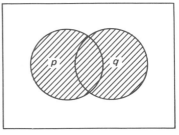

$p \vee q$

FIGURE 2.4 Euler diagrams for two propositions, p and q

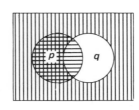

EXAMPLES: Illustrate the following with Euler circles.

1. $p \vee {\sim}q$
   a. Draw two circles in a rectangle.
   b. Shade p using horizontal lines and shade ${\sim}q$ using vertical lines, as shown in the margin column.
   c. The answer is all parts shaded with horizontal **or** ($\vee$) vertical lines, as shown below.

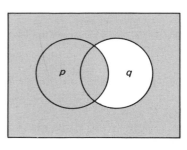

$p \vee {\sim} q$

2. $p \wedge {\sim}q$
   a. Follow steps a, b, and c of Example 1.
   b. Answer is all parts shaded with horizontal **and** ($\wedge$) vertical lines.

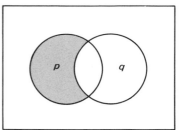

$p \wedge {\sim} q$

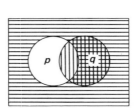

3. $({\sim}p \wedge q) \wedge {\sim}q$
   a. Parentheses first; shade ${\sim}p$ (horizontal lines); shade q (vertical lines). The result is that part shaded with horizontal lines **and** vertical lines. Show this with slant lines, as shown in the margin illustration.
   b. Show q with opposite slant lines, as shown in the margin illustration.
   c. Answer is all parts shaded with left slant lines **and** right slant lines, as shown below.

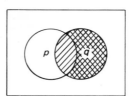

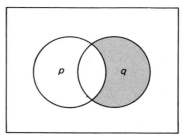

$({\sim}p \wedge q) \wedge {\sim}q$

PROBLEM SET 2.2

A Problems

Construct a truth table for the statements given in Problems 1–14.

1. $\sim p \vee q$

2. $\sim p \wedge \sim q$

3. $\sim(p \wedge q)$

4. $\sim r \vee \sim s$

5. $\sim(\sim r)$

6. $(r \wedge s) \vee \sim s$

7. $p \wedge \sim q$

8. $\sim p \vee \sim q$

9. $(\sim p \wedge q) \vee \sim q$

10. $(p \wedge \sim q) \wedge p$

11. $(\sim p \vee q) \wedge (q \wedge p)$

12. $(p \wedge q) \wedge \sim r$

13. $[(p \vee q) \wedge \sim r] \wedge r$

14. $[p \wedge (q \vee \sim p)] \vee r$

15. Let p: $2 + 3 = 5$,
      q: $12 - 7 = 5$.
      Tell which of the statements in Problems 1–3 are true.

16. Let r: man pollutes the environment,
      s: man will survive.

   Translate each of the statements in Problems 4–6 into words, and tell
   under which conditions each will be true.

> *How are you going to
> teach logic in a world
> where everybody talks
> about the sun setting, when
> it's really the horizon
> rising?—Cal Craig*

B Problems

Illustrate Problems 17–26 by Euler circles.

17. $\sim p \vee q$

18. $\sim p \wedge \sim q$

19. $\sim(p \wedge q)$

20. $\sim r \vee \sim s$

21. $p \wedge \sim q$

22. $\sim p \vee \sim q$

23. $\sim(\sim r)$

24. $(r \wedge s) \vee \sim s$

25. $(\sim p \wedge q) \vee \sim q$

26. $(p \wedge \sim q) \wedge p$

27. a. How many entries are there on a truth table for one proposition, p?
   b. How many entries are there on a truth table for two propositions, p and
       q? (Answer is 4.)
   c. How many entries are there on a truth table for three propositions, p, q,
       and r?
   d. How many entries are there on a truth table for four propositions?
   e. Consider the patterns found in parts a–d. Can you generalize and say
       how many entries there would be on a truth table with n propositions?

Mind Bogglers

28. One often hears the expression "Every coin has two sides." Is this always
    true? Is it possible to have a "coin" with only one side? Before you answer

this question, formulate precisely what you mean by "a coin with one side" and "a coin with two sides."

29. This is an old problem, but it is still fascinating. One day three men went to a hotel and were charged $30 for their room. The desk clerk then realized that he had overcharged them $5, and he sent the refund up with the bellboy. Now, the bellboy, being an amateur mathematician, realized that it would be difficult to split the $5 three ways. Therefore, he kept a $2 "tip" and gave the men only $3. Each man had originally paid $10 and was returned $1. Thus it cost each man $9 for the room. This means that they spent $27 for the room plus the $2 "tip." What happened to the other dollar?

Problem for Individual Study

30. Sometimes statements p and q are described as contradictory, contrary, or consistent. Consult a logic text, and then define these terms using truth tables.

2.3 THE CONDITIONAL AND OTHER OPERATORS

In this section we'll use truth tables to prove certain results, and in the next section we'll illustrate the use of Euler circles in proof.

CONDITIONAL

THERE'S TOO MUCH UNCERTAINTY ABOUT OUR RELATIONSHIP!

if

The statement "if p, then q" is called a *conditional statement*. It is symbolized by $p \rightarrow q$; p is called the *antecedent*, and q is called the *consequent*. There are several ways of using a conditional, as illustrated by the following examples.

1. We can use "if-then" to indicate a *logical* relationship—one in which the consequent follows logically from the antecedent.
   If ~(~p) has the same truth value as p, then p can replace ~(~p).
2. We can use "if-then" to indicate a causal relationship.
   If John drops that rock, then it will land on my foot.
3. We can use "if-then" to report a decision on the part of the speaker.
   If John drops that rock, then I will hit him.
4. We can use "if-then" when the consequent follows from the antecedent by the very definition of the words used.
   If John drives an Oldsmobile, then John drives a car.
5. Finally, we can use "if-then" to make a *material implication*. There is no logical, causal, or definitional relationship between the antecedent and consequent; we use the expression simply to convey humor or emphasis:
   If John got an A on that test, then I'm a monkey's uncle.
   The consequent is obviously false, and the speaker wishes to emphasize that the antecedent is also false.

Our task is to try to devise a definition of the conditional that will apply in all these different types of "if-then" statements. We'll approach the problem by asking under what circumstances a given conditional would be false. Let's consider another example.

Suppose I make you a promise: "If I receive my check tomorrow, then I will pay you the $10 that I owe you." If I keep my promise, we say the statement is true; if I don't, then it is false. Let

  p: I receive my check tomorrow,
  q: I will pay you the $10 that I owe you.

We symbolize the promise by $p \rightarrow q$. There are four possibilities:

   p q

1. T T *I receive my check tomorrow, and I pay you the $10.* In this case, the promise, or conditional, is true.
2. T F *I receive my check tomorrow, and I do not pay you the $10.* In this case, the conditional is false, since I did not fulfill my promise.
3. F T *I do not receive my check tomorrow, but I pay you the $10.* In this case, I certainly didn't break my promise, so the conditional is true.
4. F F *I do not receive my check tomorrow, and I do not pay you the $10.* Here, again, you could not say I broke my promise, so the conditional is true.

No doubt I came into some unexpected good fortune.

Actually the promise was not tested for case 4, since I didn't receive my check. Assume the principle of "innocent until proved guilty."

The only time that I will have broken my promise is in case 2.

The test for a conditional is to determine when it is false. In symbols,

   $p \rightarrow q$ is false whenever $p \wedge \tilde{\ }q$ is true (case 2).

Or,

   $p \rightarrow q$ is true whenever $\tilde{\ }(p \wedge \tilde{\ }q)$ is true.

Let's construct a truth table for $\tilde{\ }(p \wedge \tilde{\ }q)$.

p	q	$\tilde{\ }q$	$p \wedge \tilde{\ }q$	$\tilde{\ }(p \wedge \tilde{\ }q)$
T	T	F	F	T
T	F	T	T	F
F	T	F	F	T
F	F	T	F	T

We use this truth table to define the conditional $p \rightarrow q$ as shown in Table 2.6 (see also Figure 2.5).

EXAMPLES: These examples illustrate the definition of the conditional, as shown in Table 2.6.
1. Case 1: T $\rightarrow$ T
 If 7 < 14, then 7 + 2 < 14 + 2.
 This is a true statement, since both component parts are true.

TABLE 2.6 Definition of conditional

p	q	$p \rightarrow q$
T	T	T
T	F	F
F	T	T
F	F	T

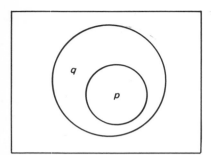

FIGURE 2.5 Euler circles for the conditional $p \rightarrow q$

2. Case 2: T → F
   If 7 + 5 = 12, then 7 + 10 = 15.
   This is a false statement, since the antecedent is true but the consequent is false.
3. Case 3: F → T
   If you have bad breath, then George Washington was President.
   This is true, since the consequent is T (whenever the consequent is true the conditional is true). This example shows that the conditional, in mathematics, does not mean that there is any cause-and-effect relationship. *Any* two statements can be connected with the connective of conditional and the result must be T or F.
4. Case 4: F → F
   If 16 = 8, then 8 = 4.
   This is a true statement, since both component parts are false.

The *if*-part of an implication need not be stated first. All the following statements have the same meaning:

Conditional translation	EXAMPLE
if p, then q	If you are 18, then you can vote.
q, if p	You can vote, if you are 18.
p, only if q	You are 18 only if you can vote.

Some statements, although not originally written as a conditional can be put into *if-then* form. For example, "All ducks are birds" can be rewritten as "If it is a duck, then it is a bird." Thus we add one more form to the list:

all p are q	All 18-year-olds can vote.

This statement appeared on the 1978 Federal income tax form.

EXAMPLE: Translate the following sentence into symbolic form. *If you do not itemize deductions and line 16 is under $15,000, do not complete lines 17 and 18.*

Step 1: Isolate the simple statements and assign them variables.
    Let d: you itemize deductions,
    u: line 16 is under $15,000,
    s: you complete line 17,
    e: you complete line 18.

Your choice of variables is, of course, arbitrary, but you must be careful not to let variables represent compound statements.

Step 2: Rewrite the sentence making substitutions for the variables.
    If $\sim d$ and u, then $\sim(s \wedge e)$.
Step 3: Complete the translation to symbols.
    $(\sim d \wedge u) \rightarrow \sim(s \wedge e)$

CONVERSE, INVERSE, AND CONTRAPOSITIVE

Related to a conditional $p \rightarrow q$ are other statements, which we will now define.

DEFINITION: Given the conditional $p \rightarrow q$, we define:

1. The *converse*: $q \rightarrow p$
2. The *inverse*: $\sim p \rightarrow \sim q$
3. The *contrapositive*: $\sim q \rightarrow \sim p$

Not all these statements are equivalent in meaning.

EXAMPLE: Write the converse, inverse, and contrapositive of the statement: If it is a 280ZX, then it is a car.

$$\text{Let } p: \text{ It is a 280ZX,}$$
$$q: \text{ It is a car.}$$

The statement is written $p \rightarrow q$.
Converse: $q \rightarrow p$: If it is a car, then it is a 280ZX.
Inverse: $\sim p \rightarrow \sim q$: If it is not a 280ZX, then it is not a car.
Contrapositive: $\sim q \rightarrow \sim p$: If it is not a car, then it is not a 280ZX.

TABLE 2.7

p	q	$\sim p$	$\sim q$	Statement $p \rightarrow q$	Converse $q \rightarrow p$	Inverse $\sim p \rightarrow \sim q$	Contrapositive $\sim q \rightarrow \sim p$
T	T	F	F	T	T	T	T
T	F	F	T	F	T	T	F
F	T	T	F	T	F	F	T
F	F	T	T	T	T	T	T

Refer to Table 2.7. Notice that the contrapositive and the original statement always have the same truth values, as do the converse and the inverse. Thus we state the following general principle:

A conditional may always be replaced by its contrapositive without having its truth value affected.

Notice that, if the conditional is true, the converse and inverse are not necessarily true. However, the contrapositive is always true if the original conditional is true.

Law of contraposition

EXAMPLES:
1. Let p: An animal is a bird,
     q: An animal has wings.
   Given $p \rightarrow q$, write its converse, inverse, and contrapositive.

   Solution: Statement: If it is a bird, then it has wings.
   Converse: If it has wings, then it is a bird.
   Inverse: If it is not a bird, then it does not have wings.
   Contrapositive: If it does not have wings, then it is not a bird.

2. Given $p \rightarrow \tilde{\ }q$, write its converse, inverse, and contrapositive.

Solution: Statement: $p \rightarrow \tilde{\ }q$ Converse: $\tilde{\ }q \rightarrow p$
                Inverse: $\tilde{\ }p \rightarrow q$ Contrapositive: $q \rightarrow \tilde{\ }p$

We have replaced the double negative $\tilde{\ }(\tilde{\ }q)$ by q in the inverse and contrapositive. We make this simplification whenever possible.

Assuming that the statement is true, the contrapositive is also true. Consider the following: p: You obey the law.
                                                                  q: You will go to jail.

Statement: $p \rightarrow \tilde{\ }q$	If you obey the law, then you will not go to jail.
Contrapositive: $q \rightarrow \tilde{\ }p$	If you go to jail, then you did not obey the law.
Converse: $\tilde{\ }q \rightarrow p$	If you do not go to jail, then you obey the law.
Inverse: $\tilde{\ }p \rightarrow q$	If you do not obey the law, then you will go to jail.

The converse and inverse are not necessarily true.

3. Given $\tilde{\ }p \rightarrow \tilde{\ }q$, write its converse, inverse, and contrapositive.

Solution: Statement: $\tilde{\ }p \rightarrow \tilde{\ }q$ Converse: $\tilde{\ }q \rightarrow \tilde{\ }p$
                Inverse: $p \rightarrow q$ Contrapositive: $q \rightarrow p$

4. Given $\tilde{\ }p \rightarrow q$, write its converse, inverse, and contrapositive.

Solution: Statement: $\tilde{\ }p \rightarrow q$ Converse: $q \rightarrow \tilde{\ }p$
                Inverse: $p \rightarrow \tilde{\ }q$ Contrapositive: $\tilde{\ }q \rightarrow p$

BICONDITIONAL

As we've just noted, statement $p \rightarrow q$ and its converse $q \rightarrow p$ do not have the same truth values. However, it may be the case that $p \rightarrow q$ *and also* $q \rightarrow p$. In this case we write

$$p \leftrightarrow q$$

and call this operator the *biconditional*. To determine the truth values of the biconditional, we construct a truth table for $(p \rightarrow q) \wedge (q \rightarrow p)$ as shown in Table 2.8. (Why?)

TABLE 2.8

p	q	$p \rightarrow q$	$q \rightarrow p$	$(p \rightarrow q) \wedge (q \rightarrow p)$
T	T	T	T	T
T	F	F	T	F
F	T	T	F	F
F	F	T	T	T

TABLE 2.9 Definition of Biconditional

p	q	$p \leftrightarrow q$
T	T	T
T	F	F
F	T	F
F	F	T

This leads us to define the biconditional so that it is true only when both p and q are true or when both p and q are false (that is, whenever they have the same truth values). (See Table 2.9 and Figure 2.6.)

In mathematics, $p \leftrightarrow q$ is translated in several ways, all of which have the same meaning:

1. p if and only if q.
2. q if and only if p.
3. if p then q, and conversely.
4. if q then p, and conversely.

EXAMPLE: Rewrite the following into one statement:
1. If a polygon has three sides, then it is a triangle.
2. If a polygon is a triangle, then it has three sides.

Solution: A polygon is a triangle if and only if it has three sides.

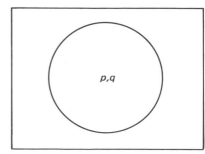

FIGURE 2.6 Euler circle for the biconditional $p \leftrightarrow q$

OTHER OPERATORS

Occasionally, we encounter other operators, and it's necessary to formulate precise definitions of these additional operators. As Table 2.10 shows, they are all defined in terms of our previous operators.

TABLE 2.10 Additional Operators

Operator	Definition	Comments
Either p or q	$(p \vee q) \wedge \sim (p \wedge q)$	This is the "or" we discussed earlier, which says that p occurs, or q occurs, but *not both*.
Neither p nor q	$\sim p \wedge \sim q$ or $\sim (p \vee q)$	Neither p nor q.
p only if q	$p \rightarrow q$	If p occurs, then q occurs.
p unless q	$\sim q \rightarrow p$	This says p unless q. If q does not occur, then p will.
p because q	$(p \wedge q) \wedge (q \rightarrow p)$	This is true only if p and q are true.
No p is q	$p \rightarrow \sim q$	This says that if it is a p, then it cannot be a q.

The truth table for each of these new operators is shown in Table 2.11.

TABLE 2.11 Truth Table for the Additional Operators

p	q	Either p or q $(p \vee q)$ $\wedge \sim(p \wedge q)$	Neither p nor q $\sim p \wedge \sim q$	p only if q $p \rightarrow q$	p unless q $\sim q \rightarrow p$	p because q $(p \wedge q)$ $\wedge (q \rightarrow p)$	No p is q $p \rightarrow \sim q$
T	T	F	F	T	T	T	F
T	F	T	F	F	T	F	T
F	T	T	F	T	T	F	T
F	F	F	T	T	F	F	T

The only if *operator is often used incorrectly in everyday speech. However, it seems reasonable that, if*

$$p \rightarrow q,$$

then it is impossible to have p without having q.

Notice from the truth table that the operator because *is equivalent to conjunction.*

PROBLEM SET 2.3

A Problems

EXAMPLE: You will succeed only if you value others' judgments.

Solution: If you succeed, then you value others' judgments.

1. Write each of the following statements in the "if-then" format.
   a. Everything happens to everybody sooner or later if there is time enough. (G. B. Shaw)
   b. We are not weak if we make a proper use of those means which the God of Nature has placed in our power. (Patrick Henry)
   c. A useless life is an early death. (Goethe)
   d. All work is noble. (Thomas Carlyle)
   e. Everything's got a moral if only you can find it. (Lewis Carroll)

2. Translate the statements of Problem 1 into symbolic form. Indicate the letters used for each simple statement.

3. First decide whether the simple statements are true or false. Then state whether the given compound statement is true or false.
   a. If $5 + 10 = 16$, then $15 - 10 = 3$.
   b. The moon is made of green cheese only if Mickey Mouse is President.
   c. If $1 + 1 = 10$, then the moon is made of green cheese.
   d. If Alfred E. Newman was elected President in 1980, then F. I. Knight was chosen as his chief mathematician.
   e. $3 \cdot 2 = 6$ if and only if water runs uphill.

4. a. Verify the results of Table 2.11 for the operator *unless* by using the definition given in Table 2.10.
   b. Verify the results of Table 2.11 for the operator *neither . . . nor* by using the definition given in Table 2.10.

5. a. Verify the results of Table 2.11 for the operator *either . . . or* by using the definition given in Table 2.10.
   b. Verify the results of Table 2.11 for the operator *because* by using the definition given in Table 2.10.

Construct truth tables for Problems 6–11.

6. a. $[p \wedge (p \vee q)] \to p$ b. $(p \to {}^{\sim}q) \to (q \to {}^{\sim}p)$

7. a. $p \vee (p \to q)$ b. $p \to ({}^{\sim}p \to q)$

8. a. $(p \vee q) \vee (p \wedge {}^{\sim}q)$ b. $(p \wedge q) \wedge (p \to {}^{\sim}q)$

9. a. $(p \wedge q) \to p$ b. $(p \to p) \to (q \to {}^{\sim}q)$

10. a. ${}^{\sim}p \to {}^{\sim}(p \wedge q)$ b. $(p \to q) \to ({}^{\sim}q \to {}^{\sim}p)$

11. a. $(p \wedge q) \leftrightarrow (p \vee q)$ b. $(p \to q) \leftrightarrow ({}^{\sim}p \vee q)$

Write the converse, inverse, and contrapositive of the statements in Problems 12–23.

12. If you break the law, then you will go to jail.

13. If p, then q.

14. $\sim p \rightarrow \sim q$

15. If your car is air-conditioned, then I will go with you.

16. You are happy if the sun shines.

17. I will go on Saturday if I get paid.

18. All cockroaches are ugly.

19. If you brush your teeth with Smiles toothpaste, then you will have fewer cavities.

20. $p \rightarrow q$

21. $\sim r \rightarrow t$

22. $t \rightarrow (a \wedge b)$

23. $\sim t \rightarrow \sim s$

P => Q IFF whenever P is true, then Q must be true; i.e. It is impossible for P to be true and Q false

B Problems

Translate the statements of Problems 24–41 into symbols. For each simple statement, be sure to indicate the meanings of the symbols you use.

24. If I think about logic problems too long, I get a headache.

25. Jacob and Ken are going to be late.

26. If you are reasonable, then Harry is unreasonable.

27. Harry is unreasonable if he has a headache.

28. You are reasonable if you consider the alternatives.

29. All elephants are big.

30. It is not the case that Tom or I was at the scene of the crime.

31. Either you or I was at the scene of the crime.

32. I will deliver the message if it arrives after 9:00 P.M.

33. Mathematicians are ogres.

34. Neither smoking nor drinking is good for your health.

35. I will not buy a new house unless all provisions of the sale are clearly understood.

36. I cannot go with you because I have a previous engagement.

37. No man is an island.

38. Either I will invest my money in stocks or I will put it in a savings account.

39. Be nice to people on your way up 'cause you'll meet 'em on your way down. (Jimmy Durante)

40. No man who has once heartily and wholly laughed can be altogether irreclaimably bad. (Thomas Carlyle)

41. If by the mere force of numbers a majority should deprive a minority of any

P → Q ≡ ∼P ∨ Q

clearly written constitutional right, it might, in a moral point of view, justify revolution. (Abraham Lincoln)

42. "Then you should say what you mean," the March Hare went on.

"I do," Alice hastily replied; "at least—at least I mean what I say—that's the same thing, you know."

"Not the same thing a bit!" said the Hatter. "Why, you might just as well say that 'I see what I eat' is the same thing as 'I eat what I see'!"

Is the Mad Hatter's criticism of Alice justified? Why or why not?

Mind Bogglers

43. One day in a foreign country I met three politicians. Now, all of the politicians of this country belonged to one of two political parties. The first was the Veracious Party, consisting of persons who could tell only the truth. The other party, called the Deceit Party, consisted of persons who were chronic liars. I asked these politicians to which party they belonged. The first said something I did not hear. The second remarked "He said he belonged to the Veracious Party." The third said "You're a liar!" To which party did the third politician belong?

44. Translate the following statements into symbolic form.
    a. Either Alfie is not afraid to go, or Bogie and Clyde will have lied.
    b. Either Donna does not like Elmer because Elmer is bald, or Donna likes Frank and George because they are handsome twins.
    c. Either neither you nor I am honest, or Hank is a liar because Iggy did not have the money.

45. A man is about to be electrocuted but is given a chance to save his life. In the execution chamber are two chairs, labeled 1 and 2, and a jailer. One chair is electrified; the other is not. The prisoner must sit in one of the chairs, but, before doing so, he may ask the jailer one question, to which the jailer must answer yes or no. The jailer is a consistent liar or else a consistent truthteller, but the prisoner does not know which. Knowing that the jailer either deliberately lies or faithfully tells the truth, what question should the prisoner ask?

    Hint 1: Let *p* be the proposition "Number 1 is the hot seat" and let *q* be the proposition "You are telling the truth."

    Hint 2: Since I would like you to try working the problem without using this hint, I'll make you read it backward: "?hturt eht gnillet era uoy fi ylno dna fi taes toh eht si 1 rebmun taht eurt ti sI"

46. If the Roman god Jupiter could do anything, could he make an object that he could not lift?

47. Decide about the truth or falsity of the following: "If wishes were horses, then beggars could ride."

48. There are 16 distinct possible "truth tables" for the propositions *p* and *q*. List all of them, and identify as many of them as possible. Three of the 16 are shown in the margin column, and we have identified two of these.

p	*q*	1.	2.	3.
T	T	T	T	T
T	F	T	T	F
F	T	T	T	F
F	F	T	F	F

This is "$p \wedge q$"

This is "$p \vee q$"

Problems for Individual Study

49. a. Explain the difference between the words *necessary* and *sufficient*.
    b. How do they relate to the conditional?
    c. Put the following statement into "if-then" format: "The presence of oxygen is a necessary condition for combustion."
    d. Put the following statement into "if-then" format: "Combustion is a sufficient condition for the presence of oxygen."
    e. Put the following statements into "if-then" format:
       i. *p* is a necessary condition for *q*.
       ii. *p* is a sufficient condition for *q*.
    f. Now put the statements in parts b–e into the form of an Euler diagram.

50. Between now and the end of the course, look for logical arguments in newspapers, periodicals, and books. Translate these arguments into symbolic form. Turn in as many of them as you can find. Be sure to indicate where you found each argument.

2.4 EULER CIRCLES, VENN DIAGRAMS, AND QUANTIFIERS

EULER CIRCLES

In the last section we noted that one translation for $p \rightarrow q$ is *all p are q*. Euler circles can also be used to represent such propositions. A single diagram, as shown in Figure 2.7, represents several relationships between p and q:*

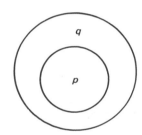

FIGURE 2.7 Euler circles for propositions p and q

1. all *p* are *q*
2. no *p* are *~q*
3. some *p* are *q*
4. some *q* are *~p*
5. some *~q* are *~p*
6. the converses of the last four

In mathematics the word "some" is translated by "at least one and possibly all."

Another Euler relationship is shown in Figure 2.8 and is translated by:

Another possibility is the reversal of p and q. Can you fill in this possibility?

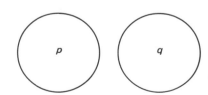

FIGURE 2.8 Euler circles

1. all *p* are *~q*
2. all *q* are *~p*
3. no *p* are *q*
4. some *p* are *~q*
5. some *q* are *~p*
6. the converses of the last four

*Notice that we are not saying these six relationships are the same; we are simply saying that a diagram similar to Figure 2.7 would be used to illustrate these relationships.

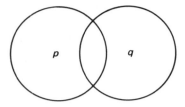

FIGURE 2.9 Euler circles

One difficulty with Euler's original scheme is the fact that all *Euler's diagrams (Figures 2.7–2.9) assert "some ~p are ~q" as one possibility. Apparently it never occurred to Euler that it might sometimes fail to be true.*

The final Euler relationship is shown in Figure 2.9, which has the following translations:

1. some p are q
2. some p are $\sim q$
3. some $\sim p$ are q
4. some $\sim p$ are $\sim q$
5. their four converses

As you can imagine, using a system with so many possibilities is rather tedious and difficult. Many modern-day writers have cleaned up Euler's schemes and have presented modified versions under his name.

VENN DIAGRAMS

The logician John Venn used circle diagrams similar to Euler's, but he was able to improve upon them by using them in a general way. Starting with Figure 2.9, for propositions p and q Venn represented *any* relationship by placing an x in a region known to be occupied and shading in a region known to be empty, as shown in Figure 2.10.

𝕳istorical 𝔑ote

John Venn (1834–1923) used Euler circles in a general way in his work Symbolic Logic *(1881). He was an ordained priest but left the clergy in 1883 to spend all of his time teaching and studying logic. We will use Venn diagrams for relationships of two or three propositions, but in* Symbolic Logic *Venn showed four terms:*

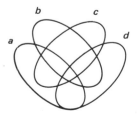

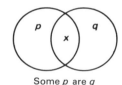 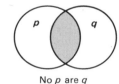 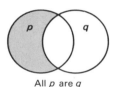

Some p are q No p are q All p are q

FIGURE 2.10 Venn diagrams

The Venn diagram for three propositions is shown in Figure 2.11.

FIGURE 2.11 Venn diagram for three propositions

Modern-day mathematicians have made some improvements on Venn's work. One is the addition of a rectangle around the diagram to allow possibilities such as $\sim p$ are $\sim q$ (the region outside both circles), and another is the use of the symbol $\varnothing$ (for empty) instead of shading.

EXAMPLE: If the regions of a Venn diagram are labeled as shown, describe each region in words.

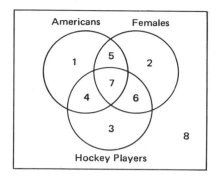

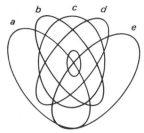

and five terms:

For six terms, Venn suggested two five-term diagrams. He did not go beyond six terms in his book.

Solution:
1. Male Americans who are not hockey players.
2. Non-American females who are not hockey players.
3. Non-American male hockey players.
4. American male hockey players.
5. Female Americans who are not hockey players.
6. Non-American female hockey players.
7. American female hockey players.
8. Non-American males who are not hockey players.

Suppose we wish to test the validity of the following argument:

Premises: 1. All dictionaries are useful.
             2. All useful books are valuable.
Conclusion: All dictionaries are valuable.
           Let D: dictionaries,
               U: useful objects,
               V: valuable objects.

Figure 2.12 shows the Venn diagram for the given premises.

Venn diagram for premise 1:

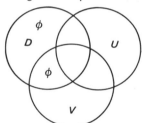

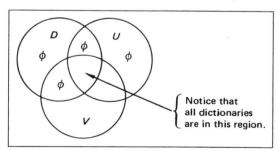

FIGURE 2.12

The conclusion is valid, as shown in Figure 2.12.

Venn diagram for premise 2:

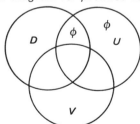

EXAMPLE: Test the validity of the following argument.
Premises: 1. All lions are fierce.
             2. Some lions do not eat hamburger.

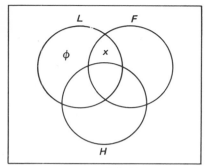

FIGURE 2.13

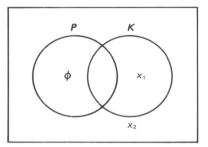

FIGURE 2.14

Conclusion: Some fierce creatures do not eat hamburger.
Solution: Let L: lions,
F: fierce creatures,
H: creatures that eat hamburger.
Draw the Venn diagram as shown in Figure 2.13. Add $\varnothing$ from premise 1 and x from premise 2. The result—some fierce creatures do not eat hamburger—is inescapable, so the argument is valid.

EXAMPLE: Test the validity of the following argument.
Premises: 1. If you like potato chips, then you will like Krinkles.
2. You do not like potato chips.
Conclusion: You do not like Krinkles.
Solution: Let P: people who like potato chips.
K: people who like Krinkles.
Draw the Venn diagram as shown in Figure 2.14. Add $\varnothing$ from premise 1. Premise 2 does not force x into a single region; it could be in two places—labeled x_1 and x_2. Since the conclusion is not forced, the argument is *not valid*.

UNIVERSAL AND EXISTENTIAL QUANTIFIERS

Up to now, we've been discussing statements that are classified as true or false. However, in mathematics we sometimes need to consider three types of sentences: (1) true, (2) false, and (3) open. Examples of each of these types are given:

1. *True sentences*
   a. $2 + 3 = 5$.
   b. $6 < 9$.
   c. George Washington was the first U.S. President.
2. *False sentences*
   a. $2 + 3 = 9$.
   b. $5 < 2$.
   c. Alfred E. Newman was the first U.S. President.
3. *Open sentences*
   a. $2x^2 + 3x = 5$.
   b. $5y < 20$.
   c. He was the first U.S. President.

We see that these open sentences may be true or false depending on the replacements for x, y, and *he*, whereas true sentences must *always* be true and false sentences must *always* be false.

We now wish to apply our study of logic to open sentences. To do this, we must make open sentences into statements. We can do so if we restrict or quantify the variable (x, y, or *he* in our examples) by saying that it is true for all or some of its possible replacements.

For example, $2x + 3x = 5x$ is an open sentence that is true for all possible replacements of x; thus we change this open sentence into a true sentence by saying

For all x, $2x + 3x = 5x$.

EXAMPLES: Change each of the following open sentences into a true sentence.
1. $x + 2 = 2 + x$
2. Two right angles are congruent.
3. $x + y = y + x$.

Solutions:
1. For all x, $x + 2 = 2 + x$.
2. Every pair of right angles is congruent.
3. For all x and for all y, $x + y = y + x$.

Any quantifier of the form *for all, all, for each,* or *for every* is called a *universal quantifier* and is symbolized by an upside-down A:

Universal quantifier

1. $\forall x, x + 2 = 2 + x$
2. $\forall x \forall y, x + y = y + x$

Sometimes we cannot change an open sentence into an equivalent statement by using a universal quantifier. For example, the open sentence $5y < 20$ is not true for all replacements of y. It is true for at least one $y (y = 1$, for example), although it may be true for more than one y. To make the open sentence $5y < 20$ into a true sentence, we would say
   There exists a y such that $5y < 20$.

EXAMPLES: Change each of the following open sentences into a true sentence.
1. $2x + 2 = 5$.
2. $4z < 19$.
3. He was the first U.S. President.
4. Apples are red.

Solutions:
1. There exists an x such that $2x + 2 = 5$.
2. There exists a z such that $4z < 19$.
3. There exists a person who was the first U.S. President.
4. There exists at least one apple that is red. Or, Some apples are red.

Quantifiers of the form *there exists, some,* or *there is at least one* are called *existential quantifiers* and are denoted by a backward E:

Existential quantifier

1. $\exists x$ such that $2x + 2 = 5$.
2. $\exists z$ such that $4z < 19$.

TRUTH VALUES OF QUANTIFIED SENTENCES

We see that in the case of an open sentence, or of an open sentence that has been quantified, some universal set has been either stated or implied. By a *universal set* we mean a set consisting of all the elements in the

Universal set

The elements of a set are the objects contained within a set.

Replacement set

The braces are used to indicate a set; therefore, {1} means the set whose element is 1.

*Definition of the truth value of a statement with a **universal** quantifier*

*Definition of the truth value of a statement with an **existential** quantifier*

universe of discourse. In addition, there is a set of possible replacements for the variable. This set is called the *replacement set*. The truth values of these quantified sentences may vary as the replacements vary.

For example,

$$\forall x, \; x = 1$$

is a statement with a universal quantifier. Suppose we let the universal set be the set of all real numbers. Then we would say that our statement is false. However, if the universal set is the set {1}, the statement is true. What we are saying here is that the truth value of a statement with a universal quantifier depends on the universal set. This leads to the following definition.

> DEFINITION: A statement with a universal quantifier is true if and only if the replacement set of the variable is the same as the universal set for the problem.

On the other hand, when we use the existential quantifier, we mean that there is (at least) *one* replacement for the variable that makes the statement true. For example,

$$\exists x \text{ such that } x^2 - 1 = 0,$$

where the universal set is the set of real numbers, is true, since we can find at least one replacement for x that makes the sentence true. However,

$$\exists x \text{ such that } x^2 + 1 = 0,$$

where the universal set is the set of real numbers, is false, since there are *no* replacements for x that make the statement true.

> DEFINITION: A statement with an existential quantifier is true if and only if there is at least one element in the replacement set of the variable that makes the quantified statements true, and all elements in the replacement set are contained within the universal set.

EXAMPLES: Determine the truth values of the following statements.
1. $\forall x, \; x + 2 = 10$. False, since $x = 3$ makes the sentence false.
2. $\exists x$ such that $x + 2 = 10$. True, since there exists an x (namely $x = 8$) that makes the sentence true.
3. $\forall x, \; 3x + 10x = 13x$. True, since all replacements for x make the sentence true.

For a given statement, the set of logical possibilities for which the given statement is true is called its *truth set*. If its truth set is identical to its universal set, then the statement is a logically true statement and is

called a *tautology*. There are many statements in symbolic logic that fall   *Tautology*
into this category, and we'll consider some of the more important ones in
Section 2.6.

EXAMPLE: Show that $(p \wedge q) \vee (p \to {\sim}q)$ is a tautology.

Solution: Construct the truth table.

p	q	$p \wedge q$	${\sim}q$	$p \to {\sim}q$	$(p \wedge q) \vee (p \to {\sim}q)$
T	T	T	F	F	T
T	F	F	T	T	T
F	T	F	F	T	T
F	F	F	T	T	T

Since the statement is true for every case of its truth table, it is a tautology.

EXAMPLE: Is $(p \vee q) \to ({\sim}q \to p)$ a tautology?

Solution:

p	q	$p \vee q$	${\sim}q$	${\sim}q \to p$	$(p \vee q) \to ({\sim}q \to p)$
T	T	T	F	T	T
T	F	T	T	T	T
F	T	T	F	T	T
F	F	F	T	F	T

Thus, it is a tautology.

If a conditional is a tautology, as in the above example, then it is called
an *implication* and is symbolized by $\Longrightarrow$. That is, the example above   *Implication*
can be written

$$(p \vee q) \Longrightarrow ({\sim}q \to p).$$

The implication symbol $p \Longrightarrow q$ is pronounced "p implies q."
A biconditional statement $p \leftrightarrow q$ that is also a tautology is a *logical*   *Logical equivalence*
equivalence, written $p \Longleftrightarrow q$ and read "p is logically equivalent to q."

EXAMPLE: Show $(p \to q) \Longleftrightarrow {\sim}(p \wedge {\sim}q)$.

p	q	$p \to q$	${\sim}q$	$p \wedge {\sim}q$	${\sim}(p \wedge {\sim}q)$	$(p \to q) \leftrightarrow {\sim}(p \wedge {\sim}q)$
T	T	T	F	F	T	T
T	F	F	T	T	F	T
F	T	T	F	F	T	T
F	F	T	T	F	T	T

Since all possibilities are true, it is a logical equivalence and we write

$$(p \rightarrow q) \Longleftrightarrow \,\tilde{}(p \wedge \,\tilde{}q)$$

PROBLEM SET 2.4

A Problems

1. Determine the truth values of each statement.
   a. $\forall x, x + 3x = 4x$.
   b. $\forall x, x - 7 = 10$.
   c. $\exists x$ such that $3x - 4 = 11$.
   d. $\exists x$ such that $2x < 13$.
   e. Every number is divisible by itself and 1.

2. Write each of the following open sentences as a true statement by using the most general quantifier.
   a. $5x + 7 = 13$
   b. $3y < 10$
   c. $5x \cdot 1 = 5x$
   d. $xy = yx$
   e. $a(b + c) = ab + ac$

3. Write an appropriate symbol for the universal or existential quantifier used in each statement.
   a. Some apples are rotten.
   b. All women are beautiful.
   c. There exist counting numbers.
   d. Some rectangles are squares.
   e. All squares are rectangles.

Show each of the statements in Problems 4–13 using Euler circles.

4. $\tilde{}p \wedge q$

5. $\tilde{}p \wedge \tilde{}q$

6. $p \rightarrow \tilde{}q$

7. $p \vee \tilde{}q$

8. $\tilde{}p \vee \tilde{}q$

9. $q \rightarrow \tilde{}p$

10. $p \vee (q \vee r)$

11. $(p \vee q) \vee r$

12. $p \vee (q \wedge r)$

13. $(p \vee q) \wedge (p \vee r)$

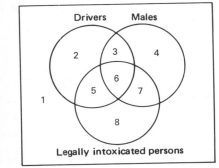

B Problems

14. Consider the Venn diagram in the margin. Describe the eight regions in words.

Problem 15 proves what is called the law of double negation. Problem 17 proves the law of contraposition from Section 2.3.

15. Show $p \Longleftrightarrow \tilde{}(\tilde{}p)$

16. Show $[(p \rightarrow q) \wedge \tilde{}q] \Longrightarrow \tilde{}p$

17. Show $(p \rightarrow q) \Longleftrightarrow (\tilde{}q \rightarrow \tilde{}p)$

Use Figure 2.15 to solve Problems 18–22.

18. If p is T and q is F, which regions are represented?

19. If p and q are both T but r is F, which regions are represented?

20. If q is T and r is F, which regions are represented?

21. If $(p \wedge q) \wedge r$ is true, which regions are represented?

22. If $p \wedge (q \vee r)$ is true, which regions are represented?

23. Assume that the following four statements are all true.
    All savings and loan companies are safe.
    Some savings and loan companies are large companies.
    All large companies pay taxes.
    Some large companies are not safe.
Which of the following statements *must* be true and which *must* be false?
    a. All savings and loan companies that are large are safe.
    b. Some large companies that are safe are not savings and loan companies.
    c. All savings and loan companies pay taxes.
    d. Some savings and loan companies that pay taxes are not large companies.
    e. Some savings and loan companies that pay taxes are not large.

24. In a certain geographical region, Alfie sold more policies than Ernie and Calvin together; Bogie sold more policies than Darin and Ernie; Darin sold more policies than Alfie and Ernie; Ernie sold more policies than Calvin. Rank the salesmen according to their ability to sell policies.

Use Venn diagrams to check the validity of the arguments in Problems 25–34.

25. All mathematicians are eccentrics.
    All eccentrics are rich.
    Therefore, all mathematicians are rich.

26. Some beautiful women are blond.
    Blonds have more fun.
    Therefore, some beautiful women have more fun.

27. All bachelors are handsome.
    Some bachelors do not drink lemonade.
    Therefore, some handsome men do not drink lemonade.

28. No students are enthusiastic.
    You are enthusiastic.
    Therefore, you are not a student.

29. Some women are tall.
    No men are strong.
    Therefore, some tall people are not strong.

30. No politicians are honest.
    Some dishonest people are found out.
    Therefore, some politicians are found out.

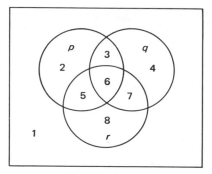

FIGURE 2.15

Historical Note

Problems 25–34 are patterned after the English mathematician Charles Dodgson's (1832–1898) famous logic problems. Dodgson, probably better known by his pseudonym, Lewis Carroll, was the author of Alice in Wonderland *and* Through the Looking Glass *as well as numerous mathematics textbooks. Queen Victoria was said to have been so taken by his children's books that she requested copies of every book he had written. Imagine her surprise when she received a pile of mathematics books! Dodgson stammered and felt most at ease with children, which accounts for his interest in children's books. The fascinating thing about his children's books is the fact that they can also be appreciated from a mathematical standpoint. For example,* Through the Looking Glass *is based on a game of chess, and, in* Alice in Wonderland, *Alice's size and proportions form a closed set of projective transformations.*

31. All candy is fattening.
    All candy is delicious.
    Therefore, all fattening food is delicious.

32. No professors are ignorant.
    All ignorant people are vain.
    Therefore, no professors are vain.

33. No monkeys are soldiers.
    All monkeys are mischievous.
    Therefore, some mischievous creatures are not soldiers.

34. All lions are fierce.
    Some lions do not drink coffee.
    Therefore, some creatures that drink coffee are not fierce.

Mind Bogglers

35. Every Gleb is a Zonk.
    Half of all Zonks are Glebs.
    Half of all Frebs are Zonks.
    No Freb is a Gleb.
    There are 60 Frebs and 40 Glebs.
    How many Zonks are neither Glebs nor Frebs?

36. As we saw in the text, a general Venn diagram with two circles divides the universe into four regions. Into how many regions will the universe be divided with
    a. four circles?
    b. five circles?
    c. *n* circles?

Problem For Individual Study

37. Research and study Lewis Carroll's *Game of Logic*. After learning the game, make a report to the class.
    References: Carroll, Lewis, *Game of Logic* (New York: Dover Publications, 1958).

2.5 SETS

The procedure for handling propositions is closely related to one of the most underlying concepts of mathematics—the idea of sets. You are no doubt familiar with the idea of a set from your previous work, but in mathematics the notion of a set is taken as an undefined term.

Two special sets, introduced in the last section, will now be formally defined.

The fact that we do not define "set" doesn't keep us from having an intuitive grasp of the word. However, we don't define "set"

> **DEFINITION:** The *empty set* contains no elements and is denoted by { } or ∅. The *universal set* contains all the elements in the universe of discourse.

because any attempt to define it would be circular or would require that we accept other undefined terms. For example,
set: a collection of objects
collection: an accumulation
accumulation: collection or pile or heap
pile: a heap
heap: a pile

SUBSETS

> **DEFINITION:** A set A is a *subset* of a set B, denoted by $A \subseteq B$, if every element of A is an element of B.

Consider the following sets:

$$U = \{1,2,3,4,5,6,7,8,9\}$$
$$A = \{2,4,6,8\}$$
$$B = \{1,3,5,7\}$$
$$C = \{5,7\}$$

Now, A, B, and C are subsets of the universal set U (all sets we consider are subsets of the universe, by definition of the universal set). $C \subseteq B$, since every element in C is also an element of B.

However, C is not a subset of A. To substantiate this claim, we must show that there is some member of C that is not a member of A. In this case, we merely note that $5 \in C$ and $5 \notin A$.

Do not confuse the notions of "element" and "subset." That is, 5 is an *element* of C, since it is listed in C; $\{5\}$ is a *subset* of C but is not an element, since we do not find $\{5\}$ contained in C (if we did, C might look like $\{5,\{5\},7\}$).

Consider all possible subsets of the set C:

The notation "$5 \in C$" means that 5 is an element of C, and "$5 \notin A$" means that 5 is not an element of A.

$\{5\}$ This is a subset, since $5 \in C$.
$\{7\}$ This is a subset, since $7 \in C$.
$\{5,7\}$ This is a subset, since both 5 and 7 are elements of C.
$\{ \}$ This is a subset, since all of its elements belong to C. Stated a different way, if it were not a subset of C, then we would have to find an element of $\{ \}$ that is not in C. Since we could not find such an element, we say that the empty set is a subset of C (and, in fact, of every set). Sometimes we name $\{ \}$ by using the symbol ∅. That is, $\emptyset = \{ \}$.

We see that there are four subsets of C but that C has only two elements. Certainly, then, we must be careful to distinguish between a subset and an element (remember that 5 and $\{5\}$ mean different things).

The subsets of C can be classified into two categories, called *proper* and *improper*, as shown in Table 2.12. Every set has but one improper

TABLE 2.12

Proper Subsets of C	Improper Subset of C
{5} {7} { }	{5,7}

subset, and that is the set itself. Verify that set A has 15 proper subsets and 1 improper subset. The improper subset of A is {2,4,6,8}. Some of the proper subsets are the following:

$$\{ \}, \{2\}, \{4\}, \{6\}, \{8\}, \{2,4\}, \{2,6\}, \{2,8\}, \ldots$$

The notion of subset can be used to define the equality between two sets.

This definition says that two sets are equal if they are exactly the same.

> DEFINITION: Sets A and B are equal, denoted by $A = B$, if $(A \subseteq B) \wedge (B \subseteq A)$.

OPERATIONS ON SETS

There are three common operations that are performed on sets: union, intersection, and complementation.

Union

Union is an operation for sets A and B in which a set is formed that consists of all the elements that are in A or B or both. The symbol for the operation of union is $\cup$, and we write $A \cup B$.

Definition of union

The union of sets can be defined in terms of disjunction. You may even have noticed the similarity of their notations:
$\cup$ *and* $\vee$.

> DEFINITION: The *union* of sets A and B, denoted by $A \cup B$, is the set consisting of all elements of A or B or both.

EXAMPLES: Let $U = \{1,2,3,4,5,6,7,8,9\}$
$A = \{2,4,6,8\}$
$B = \{1,3,5,7\}$
$C = \{5,7\}$

1. $A \cup C = \{2,4,5,6,7,8\}$. That is, the union of A and C is that set consisting of all elements in A or in C or in both.
2. $B \cup C = \{1,3,5,7\}$. Notice that, even though the elements 5 and 7 appear in both sets, they are listed only once. That is, the sets $\{1,3,5,7\}$ and $\{1,3,5,5,7,7\}$ are equal (exactly the same).
3. $A \cup B = \{1,2,3,4,5,6,7,8\}$.

4. $(A \cup B) \cup \{9\} = \{1,2,3,4,5,6,7,8,9\} = U$. Here we are considering the union of three sets. However, the parentheses indicated the operation that should be performed first. Notice also that, because the solution

We can use Venn diagrams to illustrate union. In Figure 2.16, we first shade A and then shade B. *The union is all parts that have been shaded at least once.*

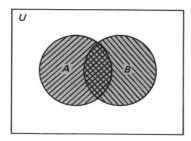

FIGURE 2.16 Venn diagram showing the union of two sets. Solutions are indicated by darker screens throughout.

Intersection

A second operation for sets is called intersection.

DEFINITION: The *intersection* of sets A and B, denoted by $A \cap B$, is the set consisting of all elements common to A and B.

Definition of intersection

The intersection ($\cap$) can be defined in terms of conjunction ($\wedge$).

EXAMPLES: Let $U = \{a,b,c,d,e\}$
$A = \{a,c,e\}$
$B = \{c,d,e\}$
$C = \{a\}$
$D = \{e\}$

1. $A \cap B = \{c,e\}$. That is, the intersection of A and B is that set consisting of elements in both A and B.
2. $A \cap C = C$, since $A \cap C = \{a\}$ and $\{a\} = C$; we write down the name for the set that is the intersection.
3. $B \cap C = \varnothing$, since B and C have no elements in common (they are disjoint).
4. $(A \cap B) \cap D = \{c,e\} \cap \{e\}$
$\quad\quad\quad\quad\quad\quad = \{e\}$. The parentheses told us which operation to do first; $\{e\}$ is the set of elements in common to the sets $\{c,e\}$ and $D = \{e\}$.

Intersection of sets can also be easily shown using a **Venn** diagram as shown in Figure 2.17. To find the intersection of two sets A and B, first shade A and then shade B; *the intersection is all parts shaded twice.*

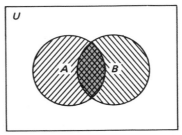

FIGURE 2.17 Venn diagram showing the intersection of two sets

For more than two sets, the procedure is the same. For example,

$$(A \cap B) \cap C$$

is found as follows.

1. Shade $A \cap B$ (we do the operation in parentheses first); the result is shown in Figure 2.18a.

2. Shade C; the result after this step is shown in Figure 2.18b.

In your work, you would not show three separate steps. Your work would look like Figure 2.18 c only.

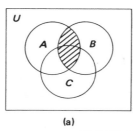
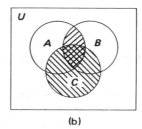
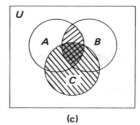

(a)   (b)   (c)

FIGURE 2.18 Venn diagram showing $(A \cap B) \cap C$

3. The result is the intersection of the two sets; that is, the part that has been shaded twice is the answer, which is shown in Figure 2.18c.

Complementation

Complementation is an operation on a set that must be performed in reference to the universal set.

Definition of complementation

DEFINITION: The *complement* of a set A, denoted by $\bar{A}$, is the set of all elements in U that are not in the set A.

EXAMPLES: Let U = {people in California}
A = {people who are over 30}
B = {people who are 30 or under}
C = {people who own a car}

The complementation can, of course, be related to negation.

1. $\overline{C}$ = {Californians who do not own a car}.
2. $\overline{A}$ = B and $\overline{B}$ = A.
3. $\overline{U}$ = $\varnothing$ and $\overline{\varnothing}$ = U.

Complementation can be shown using a Venn diagram. The unshaded part of Figure 2.19 shows the complement of A. In a Venn diagram, *the complement is everything that is not in the set under consideration* (in this case, everything not in A).

Combining Set Operations

The above operations can be combined. Remember that parentheses indicate the operations to be performed first.

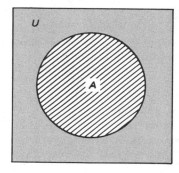

FIGURE 2.19 Venn diagram showing the complement of a set A

EXAMPLES: Let U = {m,a,t,h,i,s,f,u,n}.
A = {m,a}
B = {t,h,i,s}
C = {i,s}
D = {n,u,t,s}

1. $\overline{A \cup B}$. This is a combination of the operations of union and complementation. It is asking for the complement of A union B. First we must find $A \cup B$: {m,a,t,h,i,s}. Now

$$\overline{A \cup B} = \overline{\{m,a,t,h,i,s\}} = \{f,u,n\}.$$

2. $\overline{A} \cup \overline{B}$. This is asking for the union of the complements. First we find the complements:

$$\overline{A} = \{t,h,i,s,f,u,n\}$$
$$\overline{B} = \{m,a,f,u,n\}$$

Then we take their union:

$$\overline{A} \cup \overline{B} = \{t,h,i,s,f,u,n\} \cup \{m,a,f,u,n\}$$
$$= \{m,a,t,h,i,s,f,u,n\} = U$$

Therefore, $\overline{A} \cup \overline{B} = U$.

3. $\overline{A} \cap \overline{B}$. Find the complements (see Example 2), and then find the intersection:

$$\overline{A} \cap \overline{B} = \{t,h,i,s,f,u,n\} \cap \{m,a,f,u,n\} = \{f,u,n\}.$$

Compare Examples 1 and 2, and notice that $\overline{A \cup B} \neq \overline{A} \cup \overline{B}$.

By comparing Examples 1 and 3, we see that $\overline{A \cup B} = \overline{A} \cap \overline{B}$. We can try to *prove* this result for any sets X and Y:

$$\overline{X \cup Y} = \overline{X} \cap \overline{Y}$$

This statement is called De Morgan's law, named in honor of Augustus De Morgan (1806–1872).

Historical Note

Not only was Augustus De Morgan a distinguished mathematician and logician, but in 1845 he suggested the slanted line we sometimes use when writing fractions such as 1/2 or 2/3. There is a story told about the time when De Morgan was explaining to an actuary what the probability was that a certain group would be alive at a given time, and he used a formula involving π. Recall that π is defined as the ratio of the circumference of a circle to its diameter. The actuary was astonished and responded: "That must be a delusion. What can a circle have to do with the number of people alive at a given time?" Of course, actuaries today are more sophisticated. The number π is used in a wide variety of applications.

We proceed using Venn diagrams. We draw two diagrams, one for each side of the equals sign. If the final results (shown for both sides in Figure 2.20) are identical when we are finished, then we have proved the above result. A similar result can be proved for logical propositions by using truth tables. (See Problem Set 2.5.)

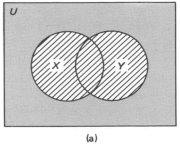

(a)

$\overline{X \cup Y}$ (parts not shaded)
This represents the left side of the equivalence.

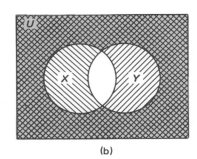

(b)

$\overline{X} \cap \overline{Y}$ (parts shaded twice)
This represents the right side of the equivalence.

FIGURE 2.20 Venn diagram proof of $\overline{X \cup Y} = \overline{X} \cap \overline{Y}$

DE MORGAN'S LAWS:	
Sets	*Logic*
$\overline{X \cup Y} = \overline{X} \cap \overline{Y}$	$\sim(p \vee q) \Longleftrightarrow (\sim p \wedge \sim q)$
$\overline{X \cap Y} = \overline{X} \cup \overline{Y}$	$\sim(p \wedge q) \Longleftrightarrow (\sim p \vee \sim q)$

4. $\overline{D} \cap B$. Here we wish to find the intersection of the complement of D and the set B. First find $\overline{D} = \{m,a,h,i,f\}$. Then

$$\overline{D} \cap B = \{m,a,h,i,f\} \cap \{t,h,i,s\} = \{h,i\}.$$

5. Find $(A \cup C) \cap \overline{C}$, and draw the corresponding Venn diagram.

Solution: Parentheses first: $A \cup C = \{s,i,a,m\}$.
Then find $\overline{C}$:

$$\overline{C} = \{f,u,n,m,a,t,h\}.$$

Finally, consider their intersection:

$$\begin{aligned}(A \cup C) \cap \overline{C} &= \{s,i,a,m\} \cap \{f,u,n,m,a,t,h\} \\ &= \{m,a\} \\ &= A\end{aligned}$$

The Venn diagram is shown in Figure 2.21. The result is the part that has been shaded twice. This Venn diagram is a picture of $(A \cup C) \cap \overline{C}$ for *any* sets A and C.

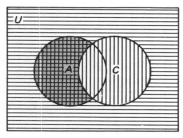

$(A \cup C) \cap \overline{C}$

First draw $A \cup C$ (vertical lines).
Then draw $\overline{C}$ (horizontal lines).
The result is the part that is shaded twice.

FIGURE 2.21 Venn diagram for Example 5

6. Find the solution set and Venn diagram for $\overline{A \cup B} \cap D$. This problem involves three sets, A, B, and D.

Solution: $A \cup B = \{m,a,t,h,i,s\}$
$\overline{A \cup B} = \{f,u,n\}$
$\overline{A \cup B} \cap D = \{f,u,n\} \cap \{n,u,t,s\}$
$= \{n,u\}$

The Venn diagram is shown in Figure 2.22.

Venn diagrams can sometimes be used for survey problems. Suppose a survey indicates that 45 students are taking mathematics and 41 are taking English. How many students are taking math or English? At first, it might seem that all we need to do is add 41 and 45, but such is not the case, as you can see by looking at Figure 2.23a.

Suppose, further, that there are 12 students taking both math and English. Then we can fill in the number in the intersection first and finish up by using subtraction for the other sections, as shown in Figure 2.23b. There are $33 + 12 + 29 = 74$ students enrolled in mathematics or English.

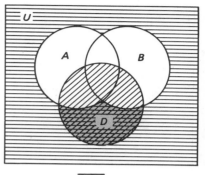

$\overline{A \cup B} \cap D$

First draw $\overline{A \cup B}$ (parts that are not in $A \cup B$).

Then draw D.

The result is the part that is shaded twice.

FIGURE 2.22 Venn diagram for Example 6

To find out how many students are taking math and English, we need to know the number in the intersection.

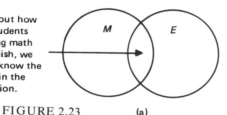

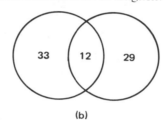

FIGURE 2.23   (a)   (b)

For three sets, the situation is a little more involved, as illustrated by the next example. Remember, the overall procedure is to fill in the number in the innermost set first and work your way out through the Venn diagram by using subtraction.

EXAMPLE: A survey of 100 randomly selected students gave the following information:

    45 students are taking mathematics.
    41 students are taking English.
    40 students are taking history.
    15 students are taking math and English.
    18 students are taking math and history.
    17 students are taking English and history.
     7 students are taking all three.

a. How many are taking only mathematics?
b. How many are taking only English?
c. How many are taking only history?
d. How many are not taking any of these courses?

Solution: Draw a Venn diagram and fill in the various regions. Let
  $M = \{\text{persons taking mathematics}\}$
  $E = \{\text{persons taking English}\}$
  $H = \{\text{persons taking history}\}$

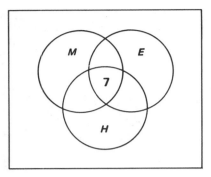

Step 1

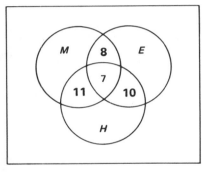

Step 2

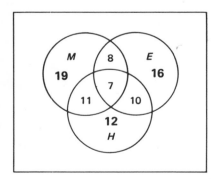

Step 3

FIGURE 2.24 Survey problem

Step 1: Fill in the innermost section first—namely $M \cap E \cap H$; there are 7 in this subset.

Step 2: Fill in the other inner portions by subtraction.

$E \cap H$ is given as 17, but 7 have previously been accounted for, so an additional 10 are added to the Venn diagram (Figure 2.24).

$M \cap H$ is given as 18; fill in an additional 11 members.

$M \cap E$ is given as 15; fill in 8.

Step 3: Fill in the other regions of the Venn diagram.

H is given as 40, but 28 have previously been accounted for, so we need an additional 12 members.

E is given as 41; we need to fill in 16 members.

M is given as 45; we need 19 members.

Step 4: We add all of the numbers in the diagram and see that 83 members have been accounted for. Since 100 students were surveyed, we see that 17 are not taking any of the three courses. We now have all the answers to the questions directly from the Venn diagram: a. 19, b. 16, c. 12, d. 17.

PROBLEM SET 2.5

A Problems

1. Explain the difference between the terms *element of a set* and *subset of a set*. Give examples.

2. List all the subsets of $\{o,w,n\}$.

3. Explain the difference between the terms *subset* and *proper subset*.

4. Explain the difference between universal set and empty set.

5. In your own words, define union, intersection, and complement. Illustrate with examples.

In Problems 6–13:

Let $U = \{1,2,3,4,5,6,7\}$
$A = \{1,2,3,4\}$
$B = \{1,2,5,6\}$
$C = \{3,5,7\}$

a. *List all the members of each of the sets.*
b. *Draw a Venn diagram.*

EXAMPLE: $A \cap \overline{B \cup C}$

Solution: a. $B \cup C$
$= \{1,2,3,5,6,7\}$
$\overline{B \cup C} = \{4\}$
$A \cap \{4\} = \{4\}$
$A \cap \overline{B \cup C}$
$= \{4\}$

b.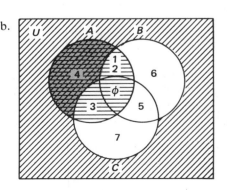

6. $A \cup B$

7. $B \cap C$

8. $\overline{A} \cap C$

9. $\overline{A \cap C}$

10. $(A \cap B) \cap C$

11. $A \cap (B \cup C)$

12. $(A \cap B) \cup (A \cap C)$

13. $\overline{A \cap C} \cup B$

In Problems 14–23:

Let $U = \{1,2,3,4,5,6,7,8,9,10\}$
   $A = \{2,4,6,8\}$
   $B = \{5,9\}$
   $C = \{2,5,8,9,10\}$

a. *List the members of each of the following.*
b. *Draw a Venn diagram.*

14. $(A \cup B) \cup C$

15. $(A \cap B) \cap C$

16. $A \cup (B \cap C)$

17. $(A \cup B) \cap (A \cup C)$

18. $\overline{A} \cap (B \cup C)$

19. $A \cap \overline{B} \cup C$

20. $\overline{A \cap (B \cup C)}$

21. $\overline{A} \cup (B \cap C)$

22. $\overline{A \cup (B \cap C)}$

23. $A \cap (\overline{B} \cup C)$

Draw Venn diagrams for the sets in Problems 24–37.

24. $\overline{A} \cap C$

25. $\overline{A \cap C}$

26. $A \cap (B \cup C)$

27. $(A \cap B) \cup (A \cap C)$

28. $A \cap B$

29. $A \cup B$

30. $A \cap (B \cap C)$

31. $\overline{A}$

32. $\overline{A \cup B}$

33. $\overline{A} \cap B$

34. $\overline{(A \cup B) \cup C}$

35. $\overline{A} \cap \overline{B}$

36. $A \cap \overline{B} \cup C$

37. $\overline{A \cup B} \cup C$

B Problems

In Problems 38–41:

Let $U = \{s,e,c,o,n,d,a,r,y\}$
   $A = \{e,a,s,y\}$
   $B = \{d,a,y\}$
   $C = \{c,a,n,d,o,r\}$

38. Find $\overline{A} \cup \overline{B}$.

39. Find $\overline{A \cap B}$.

40. Does $\overline{A} \cup \overline{B} = \overline{A \cup B}$?

41. Does $\overline{A} \cup \overline{B} = \overline{A \cap B}$?

42. In a recent survey of 100 women, the following information was gathered:

   59 use shampoo A
   41 use shampoo B
   35 use shampoo C
   24 use shampoos A and B

EXAMPLE: $A \cap \overline{B} \cup C$

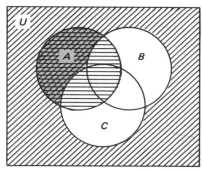

The answer is the part that is shaded twice.

EXAMPLE: Find $\overline{A \cup B} \cap C$.
$A \cup B$
$= \{e,a,s,y,d\}$
$\overline{A \cup B} = \{c,o,r,n\}$
$\overline{A \cup B} \cap C$
$= \{c,o,r,n\}$

19 use shampoos A and C
13 use shampoos B and C
11 use all three

Let A = {women who use shampoo A }; B = {women who use shampoo B }; and C = {women who use shampoo C }. Use a Venn diagram to show how many women are in each of the eight possible categories.

43. Matt E. Matic was applying for a job. To see if he could handle the job, the personnel manager sent him out to poll 100 people about their favorite types of TV shows. His data were:

59 preferred comedies
38 preferred variety shows
42 preferred serious drama
18 preferred comedies or variety programs
12 preferred variety or serious drama
16 preferred comedies or serious drama
 7 preferred all types
 2 didn't like any TV shows

If you were the personnel manager, would you hire Matt on the basis of this survey?

44. In an interview of 50 students,

12 liked Proposition 8 and Proposition 13;
18 liked Proposition 8 but not Proposition 2;
 4 liked Proposition 8, Proposition 13, and Proposition 2;
25 liked Proposition 8;
15 liked Proposition 13;
10 liked Proposition 2 but not Proposition 8 or Proposition 13;
 1 liked Proposition 13 and Proposition 2 but not Proposition 8.

a. Of those surveyed, how many did not like any of the three propositions?
b. How many liked Proposition 8 and Proposition 2?
c. Show the completed Venn diagram.

In Problems 45-49, use Venn diagrams to prove or disprove the given expression.

For Problems 45–49, draw a Venn diagram for the left side and another for the right side; if the final shaded portions are the same, then you have proved the result. If the final shaded portions are not identical, then you have disproved the result.

45. $\overline{A \cup B} = \overline{A} \cup \overline{B}$

46. $\overline{A} \cap \overline{B} = \overline{A \cup B}$

47. $(A \cup B) \cup C = A \cup (B \cup C)$

48. $A \cup (B \cup C) = (A \cup B) \cup (A \cup C)$

49. $A \cap (B \cup C) = (A \cap B) \cap (A \cap C)$

50. Prove De Morgan's law: $\overline{X \cap Y} = \overline{X} \cup \overline{Y}$. (Hint: Use Venn diagrams.)

51. Prove De Morgan's law for logic by using truth tables.
a. $\sim(p \vee q) \Longleftrightarrow (\sim p \wedge \sim q)$
b. $\sim(p \wedge q) \Longleftrightarrow (\sim p \vee \sim q)$

52. The letter illustrated says "Cross out in order the letters that are common to both [sets]. . . . This leaves TBMY EPEA EVIL. . . ." Describe this operation using set-operation symbols.

$$Let\ U = \{A,B,E,H,I,K,L,M,N,O,R,S,T,V,Y\}$$
$$\mathcal{A} = \{T,H,E,B,R,A,S,S,M,O,N,K,E,Y\}$$
$$\mathcal{B} = \{S,E,E,H,E,A,R,S,P,E,A,K,N,O,E,V,I,L\}$$

JAY E. COSSEY
2021 LARCHWOOD AVENUE
WILMETTE, ILLINOIS 60091

28 January 1974

Advertising Manager
Heublein, Inc.
Hartford, Conn. 06101

Dear Sir:

Re your recent advertisements concerning the Brass Monkey --- your conjectures are interesting, but misguided. The agent's name was Pete Yale and you could have found him sitting at table number 6 at the Brass Monkey any night.

You were right about the coaster being the tip-off, except that you went at it the wrong way. Take the name THE BRASS MONKEY and the motto SEE HEAR SPEAK NO EVIL, and cross out in order the letters that are common to both:

        T H̶E̶ B̶R̶A̶S̶S̶ M̶O̶N̶K̶E̶Y

        S̶E̶E̶ H̶E̶A̶R̶ S̶P̶E̶A̶K̶ N̶O̶ EVIL

This leaves TBMY EPEA EVIL, which you arrange in three rows:

T B M Y

E P E A   and then draw an X through the rows:

E V I L

Using the letters in each section, you get PETE YALE - BM (Brass Monkey) VI (Roman numeral 6).

Pete had a sense of humor, as you might guess. The letters BM are also the term commonly used on maps for "bench mark".

        Sincerely yours,

53. If $A \subseteq B$, describe:
    a. $A \cup B$ b. $A \cap B$ c. $\overline{A} \cup B$ d. $A \cap \overline{B}$

54. Another operation on sets can be defined as follows:

Difference

> **DEFINITION:**   The *difference* of two sets A and B, written $A - B$, is $A \cap \overline{B}$.

Show that this definition is equivalent to the following definition:

$$A - B = \overline{\overline{A} \cup B}.$$

55. For the sets U, A, B, and C of Problem 6, find
  a. $U - A$
  b. $A - B$
  c. $B - A$

56. For the sets U, A, B, and C of Problem 14, find
  a. $U - A$
  b. $A - B$
  c. $B - A$

57. Describe $A - B$ in another way.

Mind Boggler

58. Make a conjecture about the number of subsets that can be found in a set consisting of n elements.

Problem for Individual Study

59. A set is either a member of itself or not a member of itself. For example, suppose in the small California town of Ferndale it is the practice of many of the men to be shaved by the barber. Now, the barber has a rule, which has come to be known as The Barber's Rule: *he shaves those men and only those men who do not shave themselves.* The question is: "Does the barber shave himself?" If he does shave himself, then, according to The Barber's Rule, he does not shave himself. On the other hand, if he does not shave himself, then, according to The Barber's Rule, he shaves himself.

This problem of paradoxes in set theory has been studied by many famous mathematicians, including Zermelo, Frankel, von Neumann, Bernays, and Poincaré. These studies have given rise to three main schools of thought concerning the foundations of mathematics. Do some investigation of these paradoxes and these schools of thought.

EXAMPLE:
Let $A = \{1,4,9,16,25\}$
  $B = \{1,9,10,11,12\}$
  Then,
  $A - B = \{4,16,25\}$
  and
  $B - A = \{10,11,12\}$.

𝕳istorical 𝕹ote

About the time that Cantor's work began to gain acceptance, certain inconsistencies began to appear. One of these inconsistencies, called Russell's paradox, *is discussed in Problem 59.*

2.6 THE NATURE OF PROOF

One of mathematics' greatest strengths is its concern with the logical proof of its propositions. Any logical system must start with some undefined terms, definitions, and postulates or axioms. We have seen several examples of each of these. From here, other assertions can be made. These assertions are called theorems, which must be proved using the rules of logic. In this course, we are concerned not so much with "proving mathematics" as with investigating the nature of mathematics. It seems appropriate, then, to speak of the nature of proof, since the idea of proof does occupy a great portion of a mathematician's time.

Consider the following exercise:

a. Draw a large triangle.
b. Measure the three angles and add their measures together.
c. Repeat for an obtuse triangle.
d. Repeat for a right triangle.
e. Make a conjecture (that is, make a general statement about all triangles).

This exercise illustrates the creative stage of mathematics—the experimentation, guessing, and looking for patterns. Finally, a generalization is made; in this case, it is that the sum of the measures of any triangle in a plane is 180°. The next step is the formal stage. Our statement should be proved using the rules of logic. Once a conjecture has been proved, it is called a theorem. Often, several years will pass from the conjectural stage to the final proved form. Certain definitions must be made, and certain axioms or postulates must be accepted. Perhaps the proof of the conjecture will require the results of some previous theorems.

Two of the major types of reasoning used by mathematicians are direct and indirect reasoning.

"I think you should be more explicit here in step two."

DIRECT REASONING

Direct reasoning is the drawing of a conclusion from two premises. Let's consider the following pattern, called a *syllogism*:

Syllogism

$p \rightarrow q$	If you receive an A on the final, then you will pass the course.
p	You receive an A on the final.
$\therefore q$	You pass the course.

This argument consists of two *premises*, or *hypotheses*, and a *conclusion*; the argument is valid if

Premise, Conclusion

$$[(p \rightarrow q) \land p] \Longrightarrow q$$

is always true. By means of Table 2.13 we can see that the argument is valid.

TABLE 2.13 Truth Table for Direct Reasoning

p	q	$p \rightarrow q$	$(p \rightarrow q) \land p$	$[(p \rightarrow q) \land p] \Longrightarrow q$
T	T	T	T	T
T	F	F	F	T
F	T	T	F	T
F	F	T	F	T

Venn diagrams could also be used to prove direct reasoning:

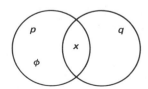

The conclusion, q, is forced.

The pattern of argument is illustrated as follows:

Direct reasoning

This reasoning is also sometimes called modus ponens, law of detachment, *or* assuming the antecedent. *Recall from geometry that* ∴ *means "therefore."*

Major premise:	$p \to q$
Minor premise:	p
Conclusion:	$\therefore q$

EXAMPLES:

1. If you play chess, then you are intelligent.
    You play chess.

    $\therefore$ You are intelligent.
2. If you are a logical person, then you will understand this example.
    You are a logical person.

    $\therefore$ You understand this example.

We say that these arguments are *valid*, since we recognize them as direct reasoning.

INDIRECT REASONING

This method is also known as reasoning by denying the consequent or as modus tollens.

The following syllogism illustrates what we call *indirect reasoning*.

$p \to q$	If you receive an A on the final, then you will pass the course.
$\tilde{~}q$	You did not pass the course.
$\therefore \tilde{~}p$	$\therefore$ You did not receive an A on the final.

We can prove this by direct reasoning as follows:

$$[(\tilde{~}q \to \tilde{~}p) \wedge \tilde{~}q] \implies \tilde{~}p.$$

Also, since $\tilde{~}q \to \tilde{~}p$ is logically equivalent to the contrapositive $p \to q$, we have

$$[(p \to q) \wedge \tilde{~}q] \implies \tilde{~}p.$$

We can also prove indirect reasoning by using a truth table, as shown in Table 2.14.

The Venn diagram proof for indirect reasoning is much simpler than the truth table. Can you supply it?

TABLE 2.14 Truth Table for Indirect Reasoning

p	q	$\tilde{~}p$	$\tilde{~}q$	$p \to q$	$(p \to q) \wedge \tilde{~}q$	$[(p \to q) \wedge \tilde{~}q] \implies \tilde{~}p$
T	T	F	F	T	F	T
T	F	F	T	F	F	T
F	T	T	F	T	F	T
F	F	T	T	T	T	T

The pattern of argument is illustrated as follows:

Argument form for indirect reasoning

Major premise:	$p \rightarrow q$
Minor premise:	$\tilde{\;}q$
Conclusion:	$\therefore \tilde{\;}p$

EXAMPLES:

1.   If the cat takes the rat, then the rat will take the cheese.
     The rat does not take the cheese.

     ∴ The cat does not take the rat.
2.   If you received an A on the test, then I am Napoleon.
     I am not Napoleon.

     ∴ You did not receive an A on the test.

TRANSITIVITY

Sometimes we must consider some extended arguments. Transitivity allows us to reason through several premises to some conclusion.
The argument form is given as follows:

Argument form for transitivity

Premise:	$p \rightarrow q$
Premise:	$q \rightarrow r$
Conclusion:	$\therefore p \rightarrow r$

Table 2.15 shows the proof for transitivity.

TABLE 2.15 Truth Table for Transitivity

You are asked to prove transitivity with Venn diagrams in Problem 13.

p	q	r	$p \rightarrow q$	$q \rightarrow r$	$p \rightarrow r$	$(p \rightarrow q)$ $\wedge (q \rightarrow r)$	$[(p \rightarrow q) \wedge (q \rightarrow r)] \Rightarrow (p \rightarrow r)$
T	T	T	T	T	T	T	T
T	T	F	T	F	F	F	T
T	F	T	F	T	T	F	T
T	F	F	F	T	F	F	T
F	T	T	T	T	T	T	T
F	T	F	T	F	T	F	T
F	F	T	T	T	T	T	T
F	F	F	T	T	T	T	T

EXAMPLE:

If you attend class, then you will pass the course.	$p \rightarrow q$
If you pass the course, then you will graduate.	$q \rightarrow r$
∴ If you attend class, then you will graduate.	$\therefore p \rightarrow r$

Transitivity can be extended so that a chain of several "if-then" sentences is connected together. For example, we could continue:

If you graduate, then you will get a good job.
If you get a good job, then you will meet the right people.
If you meet the right people, then you will become well known.

∴ If you attend class, then you will become well known.

Recall that the poem "For Want of a Nail a Kingdom Was Lost" used this type of reasoning to conclude: "If a nail is lost, then a kingdom is lost." (See Problem 7, Problem Set 2.6.)

EXAMPLE: Using transitivity, we can show that, if pesticide is sent to Borneo, then there is a real danger of plague and the roofs will cave in.

If the World Health Organization sends pesticide to Borneo, then it will be used to kill mosquitoes.
If pesticide is used to kill mosquitoes, then their bodies will become contaminated.
If the mosquitoes' bodies become contaminated, then the roaches that eat them will have an accumulation of pesticide in their bodies.
If the roaches have an accumulation of pesticide in their bodies, then the lizards that eat them will also become contaminated.
If the lizards become contaminated, then the cats that eat the lizards will die from a buildup of the pesticide in their bodies.
If the cats die, then the rats will go unchecked in Borneo.
If the rats go unchecked in Borneo, then there is a real danger of plague.

∴ If the World Health Organization sends pesticide to Borneo, then there will be a real danger of plague.

Also:

If the World Health Organization sends pesticide to Borneo, then it will kill a type of parasite that feeds on caterpillars.
If the parasite that feeds on caterpillars is killed, then the caterpillars will multiply in the huts where they live.
If the caterpillars multiply, then they will eat away too much of the roof thatching.
If the caterpillars eat away too much of the roof thatching, then the roofs will cave in.

∴ If the World Health Organization sends pesticide to Borneo, then the roofs will cave in.*

* Dr. Lamont C. Cole of Cornell University uses this story to show the interrelationships of nature.

Several of these argument forms may be combined into one argument. Remember:

---

1. Translate into symbols.
2. Simplify the symbolic argument. You might replace a statement by its contrapositive, use direct or indirect reasoning, or use transitivity.
3. Translate the conclusion back into words.

EXAMPLE: Form a valid conclusion using all these statements:

1. If I receive a check for $500, then we will go on vacation.
2. If the car breaks down, then we will not go on vacation.
3. The car breaks down.

Solution: First, change into symbolic form:

1. $c \rightarrow v$ where c = I receive a $500 check,
2. $b \rightarrow {\sim}v$ v = we will go on vacation.
3. b b = car breaks down.

Next, simplify the symbolic argument. In this example, we must rearrange the premises.

2. $b \rightarrow {\sim}v$ second premise
1. $c \rightarrow v$ first premise
3. b

Now $c \rightarrow v$ is the same as ${\sim}v \rightarrow {\sim}c$ (replace the statement by its contrapositive). That is, if we replace (1) by (1') ${\sim}v \rightarrow {\sim}c$, we obtain the following argument.

2. $b \rightarrow {\sim}v$ second premise
1'. $\underline{{\sim}v \rightarrow {\sim}c}$ contrapositive of first premise
     $\therefore b \rightarrow {\sim}c$ transitive
3. $\underline{b}$ third premise
     $\therefore {\sim}c$ direct reasoning

Finally, we translate the conclusion back into words: "I did not receive a check for $500."

Notice that, when presenting the argument, we state a reason for each step. With a little practice we'll be able to give reasons for all our arguments.

PROBLEM SET 2.6

A Problems

1. Name the type of reasoning illustrated by each of the following.
   a. If I inherit $1000, I will buy you a cookie.
      I inherit $1000.
      Therefore, I will buy you a cookie.

Hint: Assume that a is a counting number. If a counting number is odd, then it is not even.

Hint: Rewrite the premises in "if-then" form.

   b. If a^2 is even, then a must be even.
      a is odd.
      Therefore, a^2 is odd.
   c. All snarks are fribbles.
      All fribbles are ugly.
      Therefore, all snarks are ugly.

2. Using symbolic form, state the following.
   a. direct reasoning   b. indirect reasoning   c. transitivity

In Problems 3–11, form a valid conclusion using all the statements for each argument.

3. If you can learn mathematics, then you are intelligent.
   If you are intelligent, then you understand human nature.

4. If I am idle, then I become lazy.
   I am idle.

5. All trebbles are frebbles.
   All frebbles are expensive.

6. If we interfere with the publication of false information, we are guilty of suppressing the freedom of others. We are not guilty of suppressing the freedom of others.

7. If a nail is lost, then a shoe is lost.
   If a shoe is lost, then a horse is lost.
   If a horse is lost, then a rider is lost.
   If a rider is lost, then a battle is lost.
   If a battle is lost, then a kingdom is lost.

8. If you climb the highest mountain, you will feel great.
   If you feel great, then you are happy.

9. If $b = 0$, then $a = 0$.
   $a \neq 0$.

10. If $a \cdot b = 0$, then $a = 0$ or $b = 0$.
    $a \cdot b = 0$.

11. If I eat that piece of pie, I will get fat.
    I will not get fat.

12. Prove indirect reasoning using Venn diagrams.

13. Prove transitivity using Venn diagrams.

In Problems 14–23, some of the arguments illustrate a valid form of reasoning, but others do not. If the problem illustrates a valid argument, name the type of reasoning used.

14. If you understand logic, then you will enjoy this sort of problem.
    You do not understand logic.
    Therefore, you will not enjoy this sort of problem.

15. If you understand a problem, it is easy.
    The problem is not easy.
    Therefore, you do not understand the problem.

16. If I don't get a raise in pay, I will quit.
    I don't get a raise.
    Therefore, I quit.

17. If Fermat's Last Theorem is ever proved, then my life is complete.
    My life is not complete.
    Therefore, Fermat's Last Theorem is not proved.

18. All that glitters is not gold.
    Therefore, if it is gold, then it does not glitter.

19. If you eat Krinkles cereal, you will have "extra energy."
    Therefore, if you want "extra energy," eat Krinkles.

20. All mathematicians are eccentrics.
    All eccentrics are rich.
    Therefore, all mathematicians are rich.

21. All candy is fattening.
    All candy is delicious.
    Therefore, all fattening food is delicious.

22. No students are enthusiastic.
    You are enthusiastic.
    Therefore, you are not a student.

23. No professors are ignorant.
    All ignorant people are vain.
    Therefore, no professors are vain.

Problems 20–23 also appeared in Section 2.4, where you were asked to work them using Venn diagrams. In this problem set you are asked to analyze these problems in a different way.

Remember, no p are q is defined as $p \rightarrow \sim q$.

B Problems

In Problems 24–31, form a valid conclusion using all of the statements for each argument.

24. If we win first prize, we will go to Europe;
    If we are ingenious, we will win first prize.
    We are ingenious.

25. If I can earn enough money this summer, I will attend college in the fall.
    If I do not participate in student demonstrations, then I will not attend college.
    I earned enough money this summer.

26. If I am tired, then I cannot finish my homework.
    If I understand the material, then I finish my homework.

27. If you go to college, then you will get a good job.
    If you get a good job, then you will make a lot of money.
    If you do not obey the law, then you will not make a lot of money.
    You go to college.

28. All puppies are nice.
    This animal is a puppy.
    No nice creatures are dangerous.

29. Babies are illogical.
    Nobody is despised who can manage a crocodile.
    Illogical persons are despised.

30. Everyone who is sane can do logic.
    No lunatics are fit to serve on a jury.
    None of your sons can do logic.

31. No ducks waltz.
    No officers ever decline to waltz.
    All my poultry are ducks.

Hint for Problem 32: You do not need to use all the premises.

32. If the butler was present, he would have been seen; if he was seen, he would have been questioned. If he had been questioned, he would have replied; if he had replied, he would have been heard. But the butler was not heard. If the butler was neither seen nor heard, then he must have been on duty; if he was on duty, then he must have been present. Therefore, the butler was questioned. Is the conclusion valid?

33. There is a logic game available called WFF 'N PROOF, which I highly recommend. It is suitable for all levels of maturity. In its advertising, the following statements are made:

> 1. There are three numbered statements in this box.
> 2. Two of these numbered statements are not true.
> 3. The average increase in I.Q. scores of those people who learn to play WFF 'N PROOF is more than 20 points.

Is statement 3 true?

Mind Bogglers

34. In the story in the *Buz Sawyer* cartoon, Lucille is able to figure out the boys' fish story. The solution has been blacked out. See if you can determine how many fish were caught by Rosco, Jake, and Elmo.

35. If Alice shouldn't, then Carol would. It is impossible that the following statements can both be true at the same time:

   a. Alice should.   b. Betty couldn't.

   If Carol would, then Alice should and Betty could.
   Therefore, Betty could. Is the conclusion valid?

36. Form a valid conclusion using all of the following statements.
   No kitten that loves fish is unteachable.
   No kitten without a tail will play with a gorilla.
   Kittens with whiskers always love fish.
   No teachable kitten has green eyes.
   No kittens have tails unless they have whiskers.

37. Form a valid conclusion using all of the following statements.
   When I work a logic problem without grumbling, you may be sure it is one
     that I can understand.
   These problems are not arranged in regular order, like the problems I am
     used to.

No easy problem ever makes my head ache.

I can't understand problems that are not arranged in regular order, like those I am used to.

I never grumble at a problem unless it gives me a headache.

38. Form a valid conclusion using all of the following statements.

Every idea of mine that cannot be expressed as a syllogism is really ridiculous.

None of my ideas about rock stars is worth writing down.

No idea of mine that fails to come true can be expressed as a syllogism.

I never have any really ridiculous idea that I do not at once refer to my lawyer.

All my dreams are about rock stars.

I never refer any idea of mine to my lawyer unless it is worth writing down.

39. *Liar Problem*. On the South Side, one of the mob had just knocked off a store. Since the Boss had told them all to lay low, he was a bit mad. He decided to have a talk with the boys.

From the boys' comments, can you help the Boss figure out who committed the crime? Assume that the Boss is telling the truth.

40. Daniel Kilraine was killed on a lonely road, two miles from Pontiac, Michigan, at 3:30 A.M., March 3, 1967. Otto, Curly, Slim, Mickey, and the Kid were arrested a week later in Detroit and questioned. Each of the five made four statements, three of which were true and one of which was false. One of these men killed Kilraine. Whodunit? Their statements were:

Otto said: "I was in Chicago when Kilraine was murdered. I never killed anyone. The Kid is the guilty man. Mickey and I are pals."

Curly said: "I did not kill Kilraine. I never owned a revolver in my life. The Kid knows me. I was in Detroit the night of March 3rd."

Slim said: "Curly lied when he said he never owned a revolver. The murder was committed on March 3rd. One of us is guilty. Otto was in Chicago at the time."

Mickey said: "I did not kill Kilraine. The Kid has never been in Pontiac. I never saw Otto before. Curly was in Detroit with me on the night of March 3rd."

The Kid said: "I did not kill Kilraine. I have never been in Pontiac. I never saw Curly before. Otto lied when he said I am guilty."

41. I found the following article in the *Reader's Digest*. Answer the questions asked in the article.

WHODUNIT?

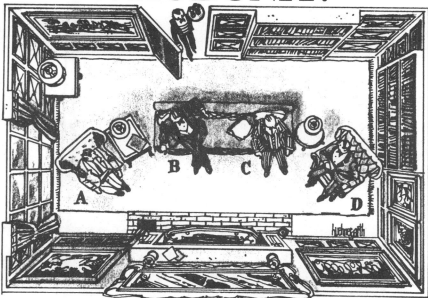

A brain-busting puzzle for mystery fans

Condensed from "MURDER INK"   PERPETRATED BY DILYS WINN

BOGGS HAS BEEN FOUND DEAD in the club lounge, his wine poisoned. Four men, seated on a sofa and two chairs in front of the fireplace as shown above, are discussing the foul deed. Their names are Howell, Scott, Jennings and Wilton. They are, not necessarily respectively, a general, schoolmaster, admiral and doctor.

(a) The waiter pours a glass of whiskey for Jennings and a beer for Scott.

(b) In the mirror over the fireplace, the general sees the door close behind the waiter. He turns to speak to Wilton, next to him.

(c) Neither Howell nor Scott has any sisters.

(d) The schoolmaster is a teetotaler.

(e) Howell, who is sitting in one of the chairs, is the admiral's brother-in-law. The schoolmaster is next to Howell on his left.

Suddenly a hand moves stealthily to put something in Jennings' whiskey. It is the murderer. No one has left his seat and no one else is in the room.

What is the profession of each man, where is he sitting, and who is the murderer? You'll find the solution on page 153. But don't weaken till you've worked out the answers on your own!

2.7 SUMMARY AND REVIEW

CHAPTER OUTLINE

  I. Simple and Compound Statements

    A. *Definition:* A **statement** is a sentence that is either true or false but not both true and false.

      1. *Simple statement*—no connective

      2. *Compound statement*—one or more connectives

B. Common operators
1. **Conjunction:** *and*; $\wedge$
2. **Disjunction:** *or*; $\vee$
3. **Negation:** *not*; $\sim$
4. **Conditional:** *if . . . then*; $\rightarrow$
   a. Statement: $p \rightarrow q$
   b. Converse: $q \rightarrow p$
   c. Inverse: $\sim p \rightarrow \sim q$
   d. Contrapositive: $\sim q \rightarrow \sim p$
   e. Equivalent statements
      i. The conditional and contrapositive are equivalent.
      ii. The converse and inverse are equivalent.
5. **Biconditional:** *if and only if*; $\leftrightarrow$

C.

p	q	*not* $\sim p$	*and* $p \wedge q$	*or* $p \vee q$	*conditional* $p \rightarrow q$	*biconditional* $p \leftrightarrow q$
T	T	F	T	T	T	T
T	F	F	F	T	F	F
F	T	T	F	T	T	F
F	F	T	F	F	T	T

D. Additional operators
1. Either p or q; $(p \vee q) \wedge \sim(p \wedge q)$
2. Neither p nor q; $\sim p \wedge \sim q$
3. p unless q; $\sim q \rightarrow p$
4. p because q; $(p \wedge q) \wedge (q \rightarrow p)$
5. No p is q; $p \rightarrow \sim q$
E. Quantifiers
1. Ways of changing open mathematical sentences into statements.
2. **Universal quantifiers:** *for all;* $\forall$
3. **Existential quantifiers:** *there exists;* $\exists$
F. Tautologies
1. **Implication:** $p \implies q$
2. **Equivalence:** $p \iff q$
II. Sets
A. Notation and definitions
B. Special sets
1. Universal set
2. Empty set
C. Relationships between sets
1. **Subset:** A set A is a *subset* of B if every element of A is an element of B (denoted by $A \subseteq B$).
2. **Equality:** Two sets are *equal* if they are the same set (denoted by $A = B$).
3. **Proper subset:** A set A is a *proper subset* of B if every element of A is an element of B and $A \neq B$.

4. **Union:** The *union* of two sets A and B is the set consisting of elements of A or B or both (denoted by $A \cup B$).

5. **Intersection:** The *intersection* of two sets A and B is the set consisting of all elements in common to A and B (denoted by $A \cap B$).

6. **Complementation:** The *complement* of a set A is the set consisting of all elements in the universe not in A (denoted by $\overline{A}$).

III. Methods of Proof
   A. Truth tables
   B. Euler circles
   C. Venn diagrams
   D. Logical form
     1. Translate the English into symbols.
     2. Simplify the symbolic argument.

       a.   Direct reasoning:

$$\begin{array}{c} p \to q \\ p \\ \hline \therefore q \end{array}$$

       b.   Indirect reasoning:

$$\begin{array}{c} p \to q \\ \sim q \\ \hline \therefore \sim p \end{array}$$

       c.   Transitivity:

$$\begin{array}{c} p \to q \\ q \to r \\ \hline \therefore p \to r \end{array}$$

       d.   An implication may be replaced by its contrapositive.

     3. Translate the conclusion back into English.

REVIEW PROBLEMS

1. Does the story in the margin illustrate inductive or deductive reasoning? Explain your answer.

2. Complete the following truth table.

p	q	$p \wedge q$	$p \vee q$	$\sim p$	$\sim q$	$p \to q$	$p \leftrightarrow q$
T	T						
T	F						
F	T						
F	F						

3. a. Construct a truth table for $\sim(p \wedge q)$.
   b. Show $\sim(p \wedge q) \Longleftrightarrow (\sim p) \vee (\sim q)$.

4. Let p: C. D. Money is trustworthy,
     q: C. D. Money inherits a fortune.
   Translate the following symbols into written statements.
   a. $p \to q$ b. $\sim q \to p$

Q: What has 18 legs and catches flies?

A: I don't know, what?

Q: A baseball team. What has 36 legs and catches flies?

A: I don't know that either.

Q: Two baseball teams. If the United States has 100 senators, and each state has 2 senators, what does . . .

A: I know this one!

Q: Good. What does each state have?

A: Three baseball teams!

5. Consider the statement "All computers are incapable of self-direction."
   a. Translate this statement into symbolic form.
   b. Write the contrapositive of the statement.

6. Consider the statement "If I don't go to college, then I will not make a lot of money." Write the converse, inverse, and contrapositive.

7. State and prove the principle of direct reasoning.

8. Give an example of indirect reasoning.

9. Is the following valid?

$$p \rightarrow q$$
$$\frac{\sim q}{\therefore \sim p}$$

Can you support your answer?

10. Write an appropriate symbol for the universal or existential quantifier.
    a. Some of these logic problems stink.
    b. There exists a non-negative integer.
    c. All of these problems are rotten.
    d. $x + 2 = 15$.
    e. $ab = ba$.

11. Let $U = \{1,2,3,4,5,6,7,8,9,10\}$
    $A = \{1,3,5,7,9\}$
    $B = \{2,4,6,9,10\}$

    a. Find $A \cup B$.   b. Find $A \cap B$.   c. Find $\overline{B}$.

12. a. For the sets given in Problem 11, find $A \cup \overline{B}$.
    b. Find $\overline{A} \cap (B \cup A)$.

13. For the sets given in Problem 11, find $\overline{A \cup B} \cap B$.

14. Indicate whether the following are true or false:
    a. $\varnothing \in \varnothing$ b. $\varnothing \subseteq \varnothing$ c. $\varnothing \subseteq X$, for all sets X

15. Indicate whether the following are true or false:
    a. $\varnothing = \varnothing$ b. $\varnothing = \{\varnothing\}$ c. $\varnothing = 0$

16. Prove or disprove:
    a. $A \cup (B \cap C) = (A \cup B) \cap C$
    b. $(A \cup B) \cap C = (A \cap C) \cup (B \cap C)$

In Problems 17–20, form a valid conclusion using all the statements for each argument.

17. No person going to Hawaii fails to get a tan.
    Mr. Gent went to Hawaii.

18. All organic food is healthy.
    All artificial sweeteners are unhealthy.
    No prune is nonorganic.

19. If I attend to my duties, I will be rewarded.
    If I am lazy, I will not be rewarded.
    I am lazy.

20. If we go on a picnic, then we must pack a lunch.
    If we pack a lunch, then we must buy bread.
    If we buy bread, then we must go to the store.

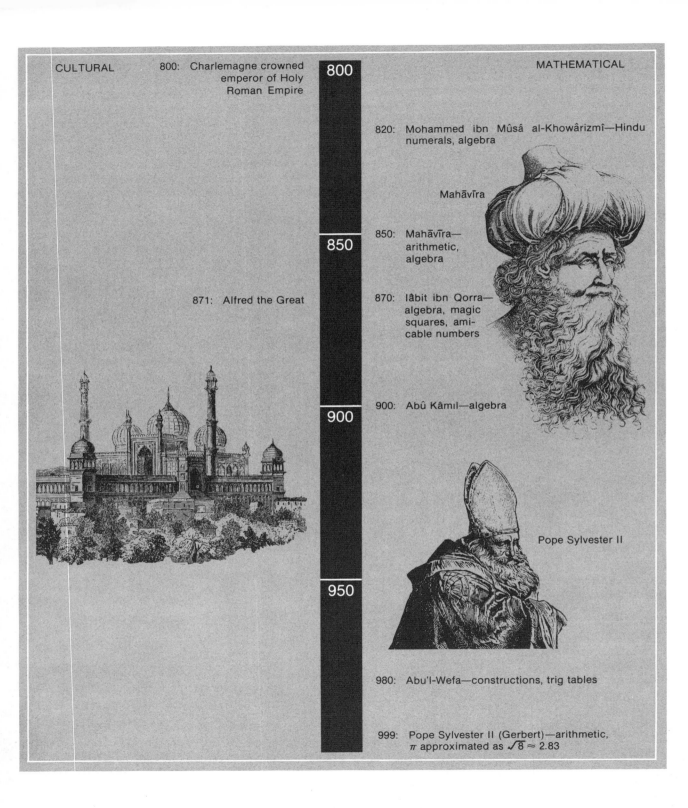

CULTURAL

MATHEMATICAL

800: Charlemagne crowned emperor of Holy Roman Empire

800

820: Mohammed ibn Mûsâ al-Khowârizmî—Hindu numerals, algebra

Mahāvīra

850

850: Mahāvīra—arithmetic, algebra

871: Alfred the Great

870: Iâbit ibn Qorra—algebra, magic squares, amicable numbers

900: Abû Kâmıl—algebra

900

Pope Sylvester II

950

980: Abu'l-Wefa—constructions, trig tables

999: Pope Sylvester II (Gerbert)—arithmetic, π approximated as $\sqrt{8} \approx 2.83$

3

THE NATURE OF
CALCULATORS
AND COMPUTERS

To err is human, but to really foul things up
requires a computer.

Anonymous

3.1 HISTORY OF COMPUTING DEVICES

In his book *Future Shock,* Alvin Toffler divided humanity's time on earth into 800 lifetimes. The 800th lifetime, in which we now live, has produced more knowledge than the previous 799 combined, and this has been made possible because of computers.

Most people still think of computers in the way they were portrayed in the 1950s—as giant brains full of wires, chips, and integrated circuits with the potential for turning us into second-class citizens. It is astounding that computers, which influence our lives to a considerable extent every day, can remain misunderstood by the majority of our population.

To function intelligently in our society, we need to understand how and why computers influence our lives. Every educated person should know something about computers and, if possible, have some experience with one. As computers diligently carry out assigned and generally tedious tasks, people are freed to do creative thinking. It is mathematics, not computation, that is fascinating. Prolonged computations lead to weariness, errors, and a distaste for arithmetic. It is not surprising that we have developed mechanical devices to perform our calculations.

The first "aid" to arithmetic computations is finger counting (see Figure 3.1). It has advantages of low cost and instant availability. You are familiar with addition and subtraction on your fingers, but here is a method of multiplication by nine. Place both hands as shown.

"It says, 'don't fold, spindle, or mutilate.'"

To multiply 4 × 9, simply bend the fourth finger from the left.

The answer is read as 36, the bent finger serving to distinguish between the first and second digits in the answer.

What about a two-digit number times 9? There is a procedure, provided the tens digit is smaller than the ones digit. For example, to multiply 36 × 9, separate the third and fourth fingers from the left (as shown), since 36 has 3 tens.

Next bend the sixth finger from the left, since 36 has 6 units. Now the answer can be read directly from the fingers. The answer is 324.

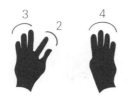

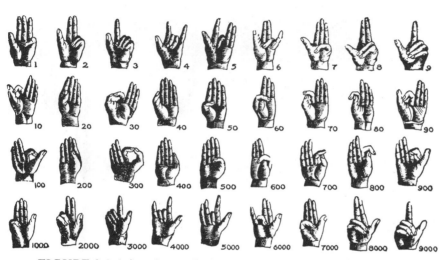

FIGURE 3.1 Aristophanes devised a complicated finger-calculating system in about 500 B.C., but it was very difficult to learn. The finger symbols illustrated here come from a manual published about two thousand years later, in 1520.

Numbers can also be represented by stones, slip knots, or beads on a string. These devices eventually evolved into the abacus (Figure 3.2). They were used thousands of years ago and are still used. In the hands of an expert, they even rival mechanical (but not electronic) calculators in speed. An abacus consists of rods containing sliding beads, four or five in the lower section and one or two in the upper. One in the upper equals five in the lower, and two in the upper equals one in the lower section of the next higher denomination. The abacus is useful today both for figuring business transactions and for teaching mathematics to youngsters. One can actually "see" the "carry" in addition and the "borrowing" in subtraction.

The 17th century marked the beginnings of modern calculating machines. John Napier invented a device in 1614 that was similar to a

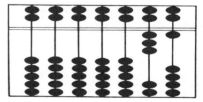

FIGURE 3.2 Abacus. The number shown is 31.

multiplication table with movable parts (see Figure 3.3), but the first real step was taken in 1642 by Blaise Pascal.

FIGURE 3.3 Napier's rods (1624)

Historical Note

John Napier (1550–1617) invented these rods, or, as they were sometimes called, Napier's bones. *Napier also invented logarithms, which led to the development of the slide rule (shown in Figure 3.8). Napier considered himself a theologian and not a mathematician. In 1593 he published a commentary on the Book of Revelation and argued that the Pope at Rome was the anti-Christ. He also predicted that the end of the world would come near the end of the 17th century. Napier believed his reputation would rest solely on this work, but instead we remember him because of what he considered his minor mathematical diversions.*

Pascal was mentioned earlier in reference to Pascal's triangle. Born in 1623, he was a brilliant child and by the age of 13 was introduced to many of the intellectuals of France. He studied physics, mathematics, and religion. However:

> . . . the young Pascal was almost lost to history. At the age of one, he became seriously ill with either tuberculosis or rickets. The doctors found that the sight of water made him hysterical, and he would often throw tantrums when he saw his father and mother together.
>
> Medicine was still quite primitive, and witchcraft was looked upon as the doer of physical harm. Blaise's illness was diagnosed as a sorcerer's spell. To remove the curse, it was necessary to find the witch, and the town turned against a poor old woman living on the charity of Mrs. Pascal and a frequent visitor to the Pascal house. This pathetic woman was under constant persecu-

tion. Although she believed herself innocent, to relieve this endless pressure, she threw herself before Étienne Pascal and promised to reveal all if he would not have her hanged.

Her story was full of mysticism. She said that she had earlier asked Blaise's father, a lawyer, to defend her against an unjust suit, and, in retaliation for his denying her request, she had put a death spell on the child. Although the charm could not be withdrawn, she said it could be transferred to bring about some other animal's death. A horse was first offered, but it was determined that a cat would be less costly. Descending the stairs with a cat in hand, she met two friars, who scared the old lady, and in her fright she threw the animal out of a window. Unlike the story of nine lives, this cat hit the pavement and died. Time was to be the healer, and Blaise did eventually recover.

Blaise Pascal (1623–1662)

When Pascal was 19, he began to develop a machine that was supposed to add long columns of figures (see Figure 3.4). The machine was essentially like those of today. He built several versions, and, since all were unreliable, he considered his project a failure; but with it he introduced basic principles that are used in modern mechanical calculators.

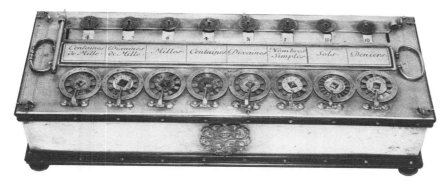

FIGURE 3.4 Pascal's calculator (1642)

The next advance in calculating devices came from Germany in about 1672, when the great mathematician Gottfried Leibniz studied Pascal's calculators, made improvements on them, and drew up plans for a mechanical calculator (see Figure 3.5). In 1695 a machine was actually built, but the resulting calculator was also unreliable. Yet Leibniz:

> . . . almost captured one of the most important aspects of modern-day computing but failed to see its application to mathematics. The binary system, so important in present computers, was envisioned by Leibniz, but he saw it in terms of religious significance. He saw God represented by 1 and nothing represented by 0. Leibniz used the binary scale as proof that God had created the world (1) out of nothing (0). In fact, he used the binary system as proof to convert the emperor of China to accept a God who could create a universe out of nothing.

Gottfried Leibniz (1646–1716)

Detail of one of the punched cards

FIGURE 3.5 Leibniz' reckoning machine (1695)

The next advance came, oddly enough, from the weaving industry. In about 1725 a man named Basile Bouchon invented an endless loop of *punched paper tape* to aid in the weaving of designs with the hand loom, and in 1728 a Frenchman, M. Falcon, used *punched paper cards* for the same task. The device was employed on a Jacquard loom (see Figure 3.6). Although it is generally called the Jacquard card, it was invented by Falcon and was the forerunner of modern punched cards. Falcon took a card and punched holes in it where he wanted the needles to be directed; to change the pattern, he needed only to change the card.

FIGURE 3.6 Jacquard loom with cards (1801)

Punched cards led to the idea that the arithmetic functions of a calculating machine could be "programmed" in advance. This concept was proposed by the eccentric Englishman Charles Babbage (1792–1871), who was years ahead of his time. His calculating machine was grandiose, with thousands of gears, shafts, ratchets, and counters. He called it his "difference engine" (see Figure 3.7).

FIGURE 3.7 Babbage's difference engine (1812)

Four years later Babbage had not yet built his difference engine but had designed a much more complicated machine—one capable of accuracy to 20 decimal places—but it could not be completed, since the technical knowledge to build it was not far enough advanced. This machine, the "Analytic Engine," was the forerunner of modern computers.

Babbage continued for 40 years trying to build his Analytic Engine, always seeking financial help to subsidize his project. However, his eccentricities often got him into trouble. For example, he hated organ-grinders and other street musicians, and for this reason people ridiculed him. He was a no-nonsense man and would not tolerate inaccuracy. Were he willing to compromise his ideas of perfection with the technical skill then available, he might have developed a workable machine. Yet his desire for precision was not limited to his work. After Lord Tennyson wrote "The Vision of Sin," Babbage sent this note to the poet:

Charles Babbage (1792–1871)

Sir,

In your otherwise beautiful poem there is a verse which reads
"Every minute dies a man,
Every minute one is born."
It must be manifest that if this were true, the population of the world would be at a standstill. In truth the rate of birth is slightly in excess of that of death. I would suggest that in the next edition of your poem you have it read—
"Every moment dies a man,
Every moment $1^1/_{16}$ is born."
Strictly speaking, this is not correct; the actual figure is so long that I cannot fit it into one line, but I believe that the figure $1^1/_{16}$ will be sufficiently accurate for poetry.

I am, Sir, yours, etc.

Babbage's Analytical Engine was all mechanical. Modern calculating machines are powered electrically and can perform additions, subtractions, multiplications, and divisions. For raising to a power or taking roots, the device usually used is a slide rule (see Figure 3.8), which was also invented by Napier.

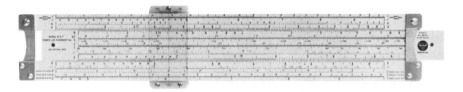

FIGURE 3.8 A slide rule

The answers given by a slide rule are only visual approximations and do not have the precision that is often required. Electronic pocket calculators, which perform all these operations quickly and accurately, represent the latest advance in calculators. Since they are so useful, as well as inexpensive and readily available, we will discuss their use in the next section.

The devices discussed thus far in this section would all be classified as calculators. With some of the new programmable calculators, the distinction between a calculator and a computer is less well defined than it was in the past. The first generally recognized automatic-sequence electromechanical computer was the Mark I, developed by Howard Aiken at Harvard in 1944. The first fully electronic digital computer was built in 1946 at the University of Pennsylvania and was called ENIAC (Electronic Numerical Integrator and Calculator). This and other "first-generation" computers used vacuum tubes, which took up a large amount of space, consumed enormous amounts of electricity, and were not very efficient. The ENIAC, for example, filled a space 30 × 50 feet, weighed 30 tons, had 18,000 tubes, and used enough electricity to operate three 150-kilowatt radio stations.

In the second generation of computers, tubes were replaced by transistors, which reduced space and power requirements (see Figure 3.9). The advantage of the transistor was that it did away with the separate materials (carbon, ceramics, tungsten, and so on) used in fabricating components and physically joined the materials into one structure (usually silicon). Nevertheless, the problem remained of soldering individual transistors and diodes into a circuit.

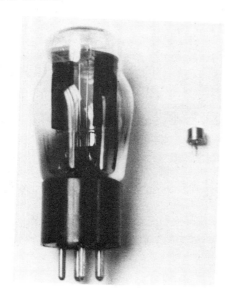

FIGURE 3.9 Size comparison between a vacuum tube and a transistor

The problem of individually connecting the transistors was solved by assembling several transistors, diodes, resistors, and capacitors onto a single circuit board that could be plugged into a computer (TV manufacturers now use this type of circuitry).

The next thing designers began to worry about was speed. For an electrical pulse to travel through 8 inches of wire, it takes one-billionth of a second, or a *nanosecond*. (Light travels about 1 foot in 1 nanosecond.) To us this seems like a small amount of time, but for the computer designer, it is long enough to cause real problems.

More compact circuits were designed, and transistors, diodes, and resistors were all formed on a single chip of silicon called an *integrated circuit*. This microminiaturization of components led to what we call the third generation of computers (see Figures 3.10 and 3.11).

In the computer, the basic operations can be done within the order of a

NANOSECOND

One thousandth of a millionth of a second.

Within the half second it takes this spilled coffee to reach the floor, a fairly large computer could—

(given the information in magnetic form)

Debit 2000 checks to 300 different bank accounts,

and *examine the electrocardiograms of 100 patients and alert a physician to possible trouble,*

and *score 150,000 answers on 3000 examinations and evaluate the effectiveness of the questions,*

and *figure the payroll for a company with a thousand employees,*

and a few other chores.

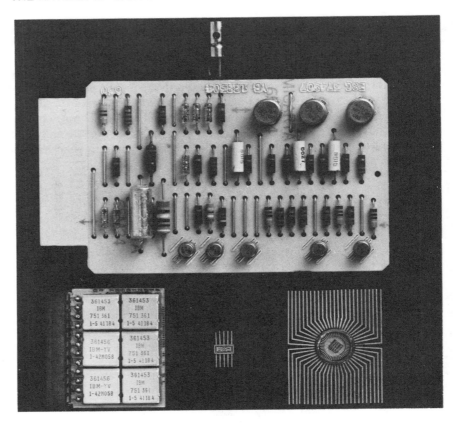

FIGURE 3.10 The miniaturization of computer components

First Generation—Vacuum tube, 1946 (not shown)

Second Generation—Transistor, 1952 (top)

Circuit board, 1960 (directly below). The one shown here incorporates 45 electronic devices

Solid state logic (bottom left); it provides for 40 devices in six tiny cans

Third Generation—Integrated circuit. 1965 (bottom center)

The device at the bottom center contains 91 transistors and resistors; at bottom right an integrated circuit containing 864 devices is shown

Fourth Generation—Microcomputers, 1975 (not shown)

These computers are not essentially different from the earlier computers except that their size and cost have made mass production and mass consumption of these computers a reality.

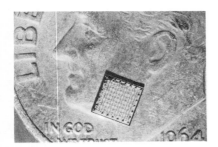

FIGURE 3.11 International Business Machines integrated circuit. A typical integrated circuit like this contains many individual circuits each containing transistors, diodes, and resistors. The density of an integrated circuit like this one might exceed 90,000 devices per square inch.

Today's computers are built to perform millions of operations per second. The state of the industry is advancing so rapidly that textbooks are outdated before they can be published. According to the 1976 edition of the *Guinness Book of World Records,* the fastest computer is the Star-100, which can operate at a rate of 97.9 *million* results per *second.* With fourth-generation computers, we saw a rapid decrease in the size and cost of computers. *Minicomputers* are generally a cubic meter or less in size and cost under $10,000. Home computers are now being advertized and marketed by several companies (see Figure 3.12).

FIGURE 3.12 Radio Shack TRS-80 Microcomputer; beginning level computer from $599

This ruler is 8 cm long. Imagine a computer in a box 8 cm X 8 cm X 10 cm.

What does the future hold? A design goal set by International Business Machines Corporation is to fit an entire computer, including all its memory, into a box 8 centimeters by 8 centimeters by 10 centimeters. That is less than half the volume of this book! Computers will no longer consist of transistors or other semiconductor components but will instead consist of superconductors called Josephson-junction devices. This new generation of computer will shrink the time necessary to execute a single command from 50 nanoseconds (for present-day computers) to about 4 nanoseconds.

PROBLEM SET 3.1

A Problems

1. Use finger multiplication to do the following calculations:

   a. 3×9 b. 7×9 c. 27×9 d. 48×9 e. 56×9

2. Describe some of the computing devices that were used before the invention of the electronic computer.

3. In what way was Jacquard's automated loom related to computers?

4. What do we mean when we refer to first-, second-, third-, and fourth-generation computers?

5. What is a nanosecond?

6. Distinguish between an electronic calculator and a computer.

7. It has been said that "computers influence our lives increasingly every year, and the trend will continue." Do you see this as a benefit or a detriment to mankind? Explain your reasons.

8. In the next section we'll discuss the use of electronic calculators. Much has been written about the merits of using these calculators in the elementary grades. Do you see calculators as a benefit or a detriment to mathematical education? Explain your answer.

B Problems

9. What do you mean by "thinking" and "reasoning"? Try to formulate these ideas as clearly as possible, and then discuss the following question: "Can computers think or reason?"

10. The following are some basic definitions for "thinking." *According to each of the given definitions,* answer the question "Can computers think?"

    a. to remember
    b. to subject to the process of logical thought
    c. to form a mental picture of
    d. to perceive or recognize
    e. to have feeling or consideration for
    f. to create or devise

11. In his article "Toward an Intelligence beyond Man's" (*Time,* February 20, 1978), Robert Jastrow claims that by the 1990s computer intelligence will match that of the human brain. He quotes Dartmouth president John Kemeny, a leading mathematician and computer scientist, as saying that he "sees the ultimate relation between man and computer as a symbiotic union of two living species, each completely dependent on the other for survival." He calls the computer a new form of life. Do you agree or disagree with the hypothesis that a computer could be "a new form of life"?

MONOPOLY MAY NEVER RECOVER

WASHINGTON (UPI)—When 18-year-old Lloyd Treinish says don't buy Boardwalk he knows what he is talking about. His computer told him.

Treinish, a high school senior, has developed several computer programs for the game of Monopoly showing which are the best properties and how they should be developed. It can tell you the probability of success of any move in the game.

"Boardwalk and Park Place are just about the worst properties you can own," he said. "It takes a large amount of capital to make any money from them. They do not have a high efficiency rating."

"The best properties are the orange ones, followed by the red if the player has a fair amount of money or, if not, the light blue ones."

"If you have both the red and the orange you are practically unbeatable," he said.

Railroads are fine early in the game but are not of great value later, Treinish said.

His program showed that Reading Railroad is the most frequently landed on property but that Shortline is the least visited. Chance is landed on more often than any other spot on the board.

12. Several classics from literature, as well as contemporary movies and books, deal with attitudes toward computers and automated machinery. For example, HAL in *2001: A Space Odyssey* tries to take command of the spacecraft, and PROTEUS IV in *Demon Seed* even attempts to procreate a living child with a human "bride." Books such as *Frankenstein, Brave New World,* and *1984* precede computers but offer warnings about future supremacy of technology over humans. In light of these words and the recent developments in computer science, what moral responsibility do you think scientists must take for their creations?

References: Ascher, Marcia, "Computers in Science Fiction" *Harvard Business Review* (November–December 1963).

Clarke, Arthur C., *2001: A Space Odyssey* (New York: New American Library, 1968).

Heinlein, Robert A., *The Moon is a Harsh Mistress* (New York: G. P. Putnam's Sons, 1966).

Huxley, Aldous, *Brave New World* (New York: Harper & Row, 1932).

Orwell, George, *1984* (New York: Harcourt Brace Jovanovich, 1949).

Shelley, Mary, *Frankenstein* (New York: Macmillan, 1970).

Vonnegut, Kurt, Jr., *Player Piano* (New York: Holt, Rinehart & Winston, 1952).

He has programmed the computer to play the game, not only with people but with other computers. It will buy and sell, handle money, roll the dice and catch people cheating. And it really knows how to trade.

"The computer would immediately reject a deal if it was going to be cheated. If it accepts your deal then you've probably been cheated," Treinish said Thursday.

Trenish started working on his program about a year ago because he was a Monopoly fan and wanted to know what the probabilities of success were with different property purchases. He had already taught himself some computer language and was using computers to do his homework in physics, math and chemistry.

"I was looking for a project to enter in the Westinghouse Talent Search," he said, and his program has made him one of 300 semi-finalists.

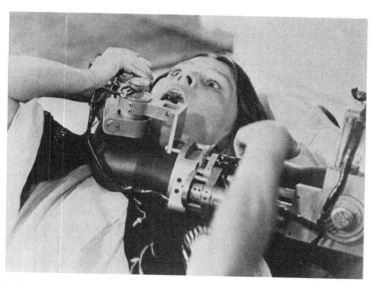

Julie Christie is paralyzed with terror as a humanoid mechanical arm of the computer PROTEUS IV forces her to do the computer's bidding in MGM's *Demon Seed.*

13. A method of finger calculation called Chisanbop has created a great deal of interest since it was demonstrated on the Johnny Carson show. It is described by one of its developers, Edwin Lieberthal, and the Director of the

Psychological Research Laboratory, William Lamon, in ''Chisanbop Finger Calculation'' *California MathematiCs Journal,* Vol. 3, No. 2, October 1978, pp. 2–10):

Chisanbop Finger Calculation is a human extension of the abacus and is based soundly on the decimal system. At times remarkable in performance by a child, the Chisanbop method uses human hands to create a calculator capable of performing with speed and accuracy. Observations conducted in some of the elementary schools in Mount Vernon, New York, reveal that this psychomotor experience tends to establish in a child not only a sense of security with numbers at the earliest grade levels, but also a significant improvement in the mathematical performance of the child. Furthermore, this improvement is seen in below average and handicapped students, as well as in average students. Progress varies from individual to individual, but the variation does not appear to depend on age. For, as with other learning challenges such as language, generally the earlier one becomes exposed to Chisanbop, the more rapidly the knowledge is gained. It is for this reason that the advocates of the system urge its introduction to children of pre-school age for optimum results later. Teaching Chisanbop to children at a point in their development when they just begin to grasp rudimentary number concepts makes it possible for them to escape the limited boundaries that heretofore confined them to using their fingers simply to count to 10.

Chisanbop assigns a value to each finger and thumb. The fingers on the right hand each have a value of one; the right thumb has a value of five. The fingers on the left hand each have a value of ten, and the left thumb has a value of 50.

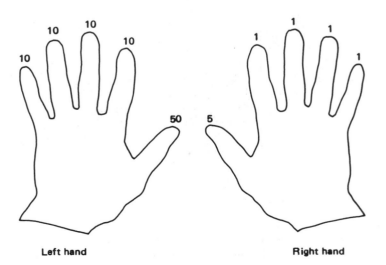

Left hand   Right hand

An example of addition is taken from the article cited above:

Example:	Manipulation	Left	Right	Accumulated Total
3	Press 3		🖐	3
4	Press 5, clear 1		🖐	7
2	Press 2		🖐	9
1	Press 10, clear 9	🖐	🖐	10
3	Press 3	🖐	🖐	13
+ 2	Press 5, clear 3	🖐	🖐	15

However, in a more advanced stage, sets of two numbers are combined and their combined value is pressed; in the above example, only three manipulations would occur:

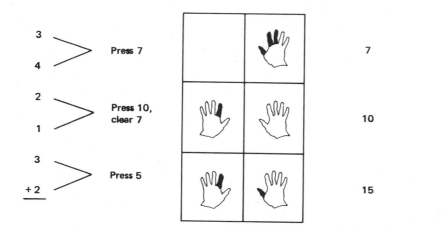

Perform the following using Chisanbop calculation.

a.	4	b.	12	c.	43	d.	4	e.	7
	3		+23		+21		5		9
	+6						7		8
							3		7
							+2		+3

14. Chisanbop multiplication (see Problem 13) is done by repeated addition. Find a source explaining Chisanbop, and prepare a demonstration showing Chisanbop addition and multiplication. (For references see the article mentioned in Problem 13. You might also write to Chisanbop Enterprises, Mount Vernon, N.Y. 10550.)

Mind Bogglers

15. Consider the operation

$$25 \cdot 625$$

Notice that we can write this as

$$5^2 \cdot 5^4$$
$$= 5^{2+4}$$
$$= 5^6$$

This *multiplication problem* ($25 \cdot 625$) was solved by doing an *addition problem* ($2 + 4$). All of the parts of this problem were worked by *adding the exponents* instead of multiplying the numbers. John Napier was the first to use this procedure when he worked with logarithms. We use the word *logarithm* to mean exponent. When we write

$$b^x$$

the number x is called the *exponent* and b is called the *base*. Consider the following patterns:

base 5:	Number	Power Notation	Logarithm
	5	5^1	1
	25	5^2	2
	125	5^3	3
	625	5^4	4
	3125	5^5	5
		⋮	

base 2:	Number	Power Notation	Logarithm
	2	2^1	1
	4	2^2	2
	8	2^3	3
	16	2^4	4
	32	2^5	5
		⋮	

Historical Note

The following is an interesting anecdote about Napier from Howard Eves' In Mathematical Circles:

"It is no wonder that Napier's remarkable ingenuity and imagination led some to believe he was mentally unbalanced and others to regard him as a dealer in the black art. Many stories, probably unfounded, are told in support of these views. Thus there was the time he announced that his coal black rooster would identify to him which one of his servants was stealing from him. He put his rooster in a box in a darkened room and instructed the servants to enter one by one and to place a hand on the rooster's back. Napier assured his servants that his rooster would expose the culprit to him at the completion of these performances. Now, unknown to the servants, Napier had coated the bird's back with lampblack. The innocent servants, having nothing to fear, did as they were bidden, but the guilty one decided to protect himself by not touching the bird. In this way he was exposed, for he was the only servant to return from the darkened room with clean hands.

base 10: Number	Power Notation	Logarithm
10	10^1	1
100	10^2	2
1000	10^3	3
10,000	10^4	4
100,000	10^5	5
	$\vdots$	

As you can see, it is easier to work with base 10 than with base 2 or base 5. For this reason, we call logarithms with base 10 *common* logarithms.

DEFINITION: We write

$$\log N = x$$

when $10^x = N$.

Thus, log 10   = 1, since $10^1 =$ 10;
   log 100   = 2, since $10^2 =$ 100;
   log 1000 = 3, since $10^3 =$ 1000.

For other bases, we use some notation that includes a *subscript* to remind us of the base.

DEFINITION: We write

$$\log_b N = x$$

when $b^x = N$.

This means $\log_5 25$ = 2, since $5^2 =$ 25;
   $\log_5 625$ = 4, since $5^4 =$ 625;
   $\log_2 64$ = 6, since $2^6 =$ 64.
As we have stated, multiplication of numbers can be solved by adding the exponents. This means

$$\log MN = \log M + \log N.$$

Use the definition of logarithms to show this property.

16. Use the definition of logarithms to show the following properties, and explain in your own words what each property is saying.

   a. $\log M/N = \log M - \log N$ b. $\log M^p = p \log M$

17. Use the definition of logarithms to write each equation in exponential form.

   a. $\log_2 8 = 3$
   b. $\log_3 81 = 4$
   c. $\log 100 = 2$
   d. $\log_4 2 = 0.5$
   e. $\log_k h = c$
   f. $\log_x y = z$

18. Use the definition of logarithms to write each equation in logarithmic form.

a. $125 = 5^3$

b. $243 = 3^5$

c. $9 = 3^2$

d. $1/4 = (1/2)^2$

e. $a = b^c$

f. $u = v^w$

19. Determine the values of the variables.

a. $\log_3 x = 2$

b. $\log_5 x = 3$

c. $\log_b 1 = 0$

d. $\log 100 = n$

e. $\log 1000 = m$

f. $\log 10,000 = y$

Problems for Individual Study

20. *Abacus*. Write a short paper or prepare a classroom demonstration on the use of an abacus. Build your own device as a project.
    References: Gardner, Martin, "Mathematical Games," *Scientific American,* January 1970.

21. *Charles Babbage*. Write a short sketch of Charles Babbage after reading "The Strange Life of Charles Babbage," by Philip and Emily Morrison, in *Mathematics in the Modern World* (San Francisco: W. H. Freeman, 1968).

22. *Logarithms*. What are logarithms? How were they discovered? How are the tables of logarithms computed? What is most useful about logarithms? How are logarithms used to build a slide rule?
    Exhibit suggestions: computing with logarithms, graphs of exponential equations on logarithmic graph paper, slide rule made from logarithmic graph paper, problems solved with logarithms, the relationship of logarithms to the large numbers discussed in Section 1.4.
    References: Bakst, Aaron, *Mathematics, Its Magic and Mastery* (2nd ed.), Chap. 18 (Princeton, N.J.: Van Nostrand, 1952).
        Glenn, William, and Donovan Johnson, *Computing Devices (Exploring Mathematics on Your Own Series)* (Sacramento: California State Dept. of Education, 1965).

23. Build a working model of Napier's rods.

24. In 1956 there were fewer than 1000 computers in the United States. By 1967 there were over 30,000. Illustrate graphically the growth of computers in the last 30 years, and make some projections into the future.

25. Do some research and then present some figures on the change in cost of computational time in the last 20 to 25 years.

3.2 POCKET CALCULATORS

A pocket calculator is necessary for working the problems in this section.

In the last few years pocket calculators have been one of the fastest growing items in the United States. There are probably two reasons for this increase in popularity. Most people (including mathematicians!) don't like to do arithmetic, and a good four-function calculator can be purchased for under $20. (You might even find one on sale for around $10.)

Calculators are classified according to their ability to do different types of problems, as well as by the type of logic they use to do the calculations. The problem of selecting a calculator is further compounded by the multiplicity of brands from which to choose. Therefore, to choose a calculator and to learn to use it require some sort of instruction.

The different levels of calculators are distinguished primarily by their price.

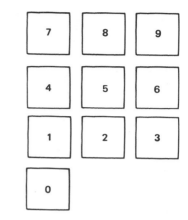

FIGURE 3.13 Standard calculator keyboard. Additional keys depend on brand and model.

1. *Four-function calculators (Under $10).*
   These calculators have a keyboard consisting of the numerals and the four arithmetic operations, or functions: addition $\boxed{+}$, subtraction $\boxed{-}$, multiplication $\boxed{\times}$, and division $\boxed{\div}$.
2. *Four-function calculators with memory ($10–$20).*
   Usually no more expensive than four-function calculators, these offer a memory register, $\boxed{M}$, $\boxed{STO}$, or $\boxed{M^+}$. The more expensive models may have more than one memory register. Memory registers allow you to store partial calculations for later recall. Some checkbook models will even remember the total when they are turned off.
3. *Scientific calculators ($20–$50).*
   These calculators add additional mathematical functions, such as square root $\boxed{\sqrt{}}$, trigonometric $\boxed{SIN}$, $\boxed{COS}$, and $\boxed{TAN}$, and logarithmic $\boxed{LOG}$, and $\boxed{EXP}$. Depending on the particular brand, a scientific model may have other keys as well.
4. *Special-purpose calculators ($40–$400).*
   Special-use calculators for business, statistics surveying, medicine, or even gambling and chess are available.
5. *Programmable calculators ($50–$600).*
   These calculators allow the insertion of different cards that "remember" the sequence of steps for complex calculations.

For most nonscientific purposes, a four-function calculator with memory will be sufficient for everyday usage. If you anticipate taking several mathematics and/or science courses, then you'll find a scientific calculator to be a worthwhile investment.

There are essentially three types of logic used by these calculators: arithmetic, algebraic, and RPN. In algebra you learn to perform multiplication before addition, so that the correct value for

$$2 + 3 \times 4$$

is 14 (multiply first). An algebraic calculator will "know" this fact and will give the correct answer, whereas an arithmetic calculator will simply work from left to right to obtain the incorrect answer 20. Therefore, if you have an arithmetic-logic calculator, you will need to be careful about the order of operations. Some arithmetic-logic calculators provide parentheses $\boxed{(}$ $\boxed{)}$ so that operations can be grouped as in

$$\boxed{2} \; \boxed{+} \; \boxed{(} \; \boxed{3} \; \boxed{\times} \; \boxed{4} \; \boxed{)} \; \boxed{=}$$

but then you must remember to insert the parentheses.

When you're purchasing a calculator, try this test problem:

$$2$$
$$+$$
$$3$$
$$\times$$
$$4$$
$$=$$

If the answer is 20, it is an arithmetic-logic calculator. If the answer is 14 (the correct answer), then it is an algebraic-logic calculator.

The last type of logic is RPN. A calculator using this logic is charac-terized by ENTER or SAVE keys and does not have an equal key =. With an RPN calculator the operation symbol is entered after the num-bers have been entered. These three types of logic can be illustrated by the problem 2 + 3 × 4.

Arithmetic *logic*	*Algebraic* *logic*	*RPN* *logic*
3	2	2
×	+	ENTER
4	3	3
=	×	ENTER
+	4	4
2	=	×
=		+

The Sharp calculator is an example of a calculator with arithmetic logic.

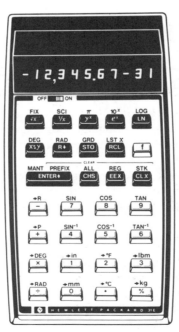

The Hewlett-Packard calculator uses RPN logic.

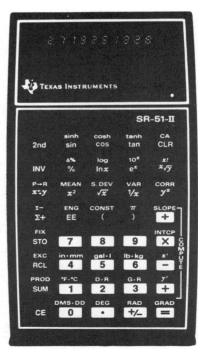

This Texas Instruments SR-51-II is an example of a calculator with algebraic logic.

In this book, we will illustrate the examples using algebraic logic, and, if you have a calculator with RPN logic, you can use your owner's manual to change the examples to RPN. We will also indicate the keys to be pushed by drawing boxes around the numbers and operational signs as shown above, and we'll use only the keys on a four-function calculator with memory. Numerals for calculator display will be designated as

0123456789

Regardless of the type of logic your calculator uses, it is a good idea to check your owner's manual for each type of problem since there are many different brands of calculators on the market.

ELEMENTARY OPERATIONS ON A CALCULATOR

You should work each of these problems on your own calculator. If you don't obtain exactly the same display as shown in the examples, consult your owner's manual to find out how your calculator differs from the one used in these examples.

EXAMPLE 1: *addition:* $14 + 38$

Be sure to turn your calculator on, or clear the machine if it is already on. A clear button is designated by $\boxed{C}$, and the display will show $\boxed{}$ after the clear button is pushed.

Action	Display
$\boxed{\text{on}}$	0.
$\boxed{1}$	1.
$\boxed{4}$	14.
$\boxed{+}$	14.
$\boxed{3}$	3.
$\boxed{8}$	38.
$\boxed{=}$	52.

You will need to check these steps every time you use your calculator, but after a while it becomes automatic. We will not remind you of this on each example.

Your calculator should also have a button labeled $\boxed{\text{CE}}$. This means *clear entry* and is used if you make a mistake keying in a number and don't want to start over with the problem. For example, if you want $2 + 3$ and push

$\boxed{2}$ $\boxed{+}$ $\boxed{4}$

you can then push

$\boxed{\text{CE}}$ $\boxed{3}$ $\boxed{=}$

to obtain the correct answer. This is especially helpful if you are in the middle of a long problem.

Some calculators have a button $\boxed{CE/C}$, which combines the clear and clear entry keys.

EXAMPLE 2:   *addition with decimals:*

$$14.6 + 38.9 + 6 + 12.817$$

Action	Display
1	1.
4	14.
.	14.
6	14.6
+	14.6
38.9	38.9
+	53.5
6	6.
+	59.5
12.817	12.817
=	72.317

For purposes of notation, we'll show each number in a single box, which will mean you key in one numeral at a time, as shown here. From now on, this will be indicated by

14.6   14.6

Notice that every time you push + , a subtotal is shown in the display. Also notice that you do not need to key in the decimal point for a whole number (such as 6 in the example above).

EXAMPLE 3:   *mixed operations:* $4 + 3 \times 5 - 7$

Action	Display
4	4.
+	4.
3	3.
×	3.
5	5.
−	19.
7	7.
=	12.

If you have an algebraic-logic calculator, your machine will perform the correct order of operations. If it is an arithmetic-logic calculator, it will give the incorrect answer of 28.

EXAMPLE 4:   *multiplication with decimals:* $4.95 × 6

Action	Display
4.95	4.95
×	4.95
6	6.
=	29.7

Notice that you do not need to push 6.00 *or even* 6

The result is \$29.70.

Notice from Example 4 that your calculator may not display zeros to the right of the last nonzero digit in the answer.

    There is also a limit to the accuracy of your calculator. You may have a calculator with a 6-, 8-, or 10-digit display. You can test the accuracy with the following problem.

EXAMPLE 5:   *division:* 2 ÷ 3

Action	Display
2	2.
÷	2.
3	3.
=	.6666667

There may be some discrepancy between this answer and the one you obtain on your calculator. Some machines will not round the answer as shown here but will show the display

$$.6666666$$

Others will show a display such as

$$6.6666-01$$

This is a number in scientific notation and should be interpreted as

$$6.6666 \times 10^{-1} \quad \text{or} \quad 0.66666$$

    Most calculators will also use scientific notation when the numbers become larger than that allowed by their display register.

Some calculators will simply show an overflow and won't accept larger numbers. Every calculator has some upper limit beyond which it will not function. You'll need to find out what this limit is for your machine.

EXAMPLE 6:   *repeated multiplication:* 50^6

Check your owner's manual to see if your calculator has an exponent key or a constant key. You can use these keys to work this example more easily than is shown below; however, the calculator we are using is only a four-function machine.

Action	Display
50	50.
×	50.
50	50.
×	2500.
50	50.
×	125000.
50	50.
×	6250000.
50	50.
×	3. 125 08
50	50.
=	1.5625 10

The maximum size of the display for this calculator has been reached, so it automatically switches to scientific notation. The point at which a calculator will do this varies from one type or brand to another.

The answer is

$$1.5625 \times 10^{10} = 15,625,000,000.$$

EXAMPLE 7:   Sometimes problems can be inverted so that the answer can be easily checked. For example, to determine an ancient Arab proverb, calculate

$$5 \times .9547 + 2 \times 353.$$

If you have correctly handled the order of operations your display should be

$$710.7735$$

The answer is 710.7735, but the check is found by turning the calculator over to see if it forms a word.

Now turn the display upside-down to read the answer:

SELL 'OIL

PROBLEM SET 3.2

A Problems

1. Briefly describe the three different types of calculator logic.

Work Problems 2–11 on a calculator. After completing the calculation, use the question in the margin to check your answer by turning your calculator over.

2. 14×351 3. 218×263

4. $3478.06 + 2256.028 + 1979.919 + 0.00091$

5. $57,300 + 0.094 + 32.27 + 2.09 + 0.0074$

6. $(1979 \times 1356) \div 452$ 7. $(515 \times 20,600) \div 200$

8. $2^4 \times 1245 \times 277$ 9. $3^3 \times 182$

10. $(0.14)(19) + (197)(25.08)(19)$

11. $1 \div (0.005 + 0.02)$

2. *What is the opposite of low?*
3. *You step on them.*
4. *Where do you live on the pipeline?*
5. *How do you look taller?*
6. *What is above your feet?*
7. *What is below your feet?*
8. *How do you like your calculator?*
9. *How are you feeling?*
10. *How will you raise the money?*
11. *This is fun!*

12. In 1975 *Godfather II* won the Academy Award for the best picture. A little calculator exercise will tell you why. Suppose the studio paid $5210 for each of 42 advertisements (multiply 5210 by 42), gave 212 parties costing an average of $2824 (add the product of 212 and 2824), and spent $8784.70 for postage (add that to previous total). The reason for the picture's success becomes apparent when you divide this result by the number of voting members of the academy (150) and turn the calculator over. What you see is why the godfather always wins.

13. Superman is faster than a speeding bullet, so put 4000 feet per second (the rate of a speeding bullet) into your calculator. He is more powerful than a locomotive, so add the average weight of a locomotive (438,000 pounds). He is able to leap tall buildings with a single bound, so add 1472 (the height of the Empire State Building). Now send mild-mannered Clark Kent into a phone booth (add 67,242 for the number of phone booths in Metropolis), and push the equal-sign button. If you turn your calculator over, you will see what a red-faced Superman says when Lois Lane catches him in the phone booth without his cape.

14. What do Congress and belly dancers have in common? Multiply the prime number 2417 by the number of months in the year, divide by the number of letters in the word "congress," and then multiply by the number of letters in "George Washington." Turn the machine around to read the answer. For greater precision, add .7956 and subtract .1776.

B Problems

15. Zerah Colburn (1804–1840) toured America when he was 6 years old in order to display his calculating ability. He could instantaneously give the square and cube roots of large numbers. It is reported that it took him only a few seconds to find 8^{16}. Use your calculator to estimate this answer (do it exactly, if possible).

Historical Note

From time to time there have been people who had the ability to perform in their heads calculations that would rival today's electronic calculators. Some of them are described in Problems 15–18. Most calculating prodigies toured the country to display their abilities. Some were idiot savants, who were brilliant in handling immense numbers and stupid in everything else (for example, see Jedidiah Buxton in Problem 16). Others had powers that fell off as they grew older (see Zerah Colburn in Problem 15), while others were bright and had many mental skills (for example, see George Bidder in Problem 17). Another brilliant calculating prodigy was Truman Safford (1836–1901), who was an astronomer but never exhibited his talents publicly. When he was 10 years old he was asked to find the surface area of a regular pyramid with a slant height of 17 ft and a pentagon-shaped base with sides of 33.5 ft. He gave the answer, 3354.5558 sq ft, in two minutes.

EXAMPLE FOR PROBLEM 19:

(1)   41
(2)   16
        ———   Add these
(3)   57
(4)   73
(5)   :
        :

For Problem 20, you can use Table 1.1 on page 29.

EXAMPLE FOR PROBLEM 21:

123

123,123

        17,589
7 / 123,123
        :
        :

16. Jedidiah Buxton (1707–1772) never learned to write and was described as slow witted, but given any distance he could tell you the number of inches, and given any length of time he could tell you the number of seconds. If he listened to a speech or a sermon, he could tell the number of words or syllables in it. It reportedly took him only a few moments to mentally calculate the number of cubic inches in a right-angle block of stone 23,145,789 yards long, 5,642,732 yards wide, and 54,465 yards thick. Estimate this answer on your calculator. You will use scientific notation because of the limitations of your calculator, but remember Jedidiah gave the *exact* answer by working the problem in his head.

17. George Bidder (1806–1878) not only possessed exceptional power at calculations but also went on to obtain a good education. He could give immediate answers to problems of compound interest and annuities. One question he was asked was "If the moon is 123,256 miles from the earth and sound travels at the rate of 4 miles per minute, how long would it be before the inhabitants of the moon could hear the battle of Waterloo?" By calculating *mentally,* he gave the answer in less than one minute! Use your calculator to give the answer in days, hours, and minutes, to the nearest minute.

18. *Favorite Number Trick.* Enter your favorite counting number from 1 to 9. Multiply by 259; then multiply this result by 429. What is your answer? Try it for three different choices.

19. *Magic 0.618.* Pick any two numbers (1) and (2), and add them to obtain a third number (3). Add the sum (3) to the second number (2) to obtain a fourth (4). Continue to add in this fashion until you have obtained 20 numbers. Divide the last two numbers (in either order), and write down the first three digits after the decimal point. Repeat for another two starting numbers, and compare the final results.

20. Problem 19 is related to the golden ratio introduced in Section 1.3. Use a calculator to find F_n/F_{n-1} for several n's. Make a conjecture (guess) about this ratio for very large values of n.

21. Pick any three-digit number, and repeat the digits to make a six-digit number. Divide by 7, then 11, and then 13. What is the result? Repeat for three different numbers.

22. *Factorial* is an operation of repeated multiplication: $n!$ (pronounced "n factorial") means

$$n(n - 1)(n - 2) \cdot \cdot \cdot 3 \cdot 2 \cdot 1$$

where $1! = 1$ and $0! = 1$. For example,

$$5! = 5 \cdot 4 \cdot 3 \cdot 2 \cdot 1$$
$$= 120.$$

Find

a. 4!   b. 6!   c. 8!   d. 10!   e. $\dfrac{10!}{8!}$

If the exact decimal representation for $^1/_3$ *is* .333 . . . *and not* .33333333, *find the exact decimal representation for each of the numbers given in Problems 23–28.*

23. $^1/_6$

24. $^1/_{11}$

25. $^1/_7$

26. $^1/_{41}$

27. $^1/_{271}$

28. $^1/_{37}$

29. Use your calculator to write each fraction as a decimal.

    a. $^1/_7$ b. $^2/_7$
    c. $^3/_7$ d. $^4/_7$
    e. $^5/_7$ f. $^6/_7$

30. Use Problem 29 to describe a pattern.

Mind Bogglers

31. Use your calculator to find

    a. $^1/_{17}$ b. $^2/_{17}$ c. $^3/_{17}$ d. $^4/_{17}$

    e. If the pattern of parts a–d is the same as that described in Problem 30, can you guess the entire decimal representation for $^1/_{17}$ even though it exceeds the accuracy of the calculator?

Cryptic Arithmetic. Cryptic arithmetic is a mathematical puzzle in which letters have been replaced by the digits of numbers. Replace each letter by a digit (the same digit for the same letter throughout; different digits for different letters), and the arithmetic will be performed correctly. See if you can do these puzzles:

32.   SEND
    +MORE
    ―――――
    MONEY

33.   THIS
    IS
    +VERY
    ―――――
    EASY

3.3 THE ELECTRONIC "BRAIN"

Computers are used for a variety of purposes. *Data processing* involves large numbers of data on which relatively few calculations need to be made. For example, a credit-card company could use the system to keep track of the balances of its customer's accounts. *Informational retrieval* involves locating and displaying material from a description of its content. For example, a police department may want information about a suspect or a realtor may want listings that meet the criteria of a client. *Pattern recognition* is the identification and classification of shapes, forms, or relationships. For example, a chess player will examine the chessboard carefully to determine a good next move. Finally, a computer can be used for *simulation* of a real situation. For example, a prototype of a new spacecraft can be tested under various circumstances to determine its limitations.

Remember, your calculator does not give exact answers. For example,

$$\frac{1}{3} = .333\ldots.$$

If you find

    [1] [÷] [3]

on any calculator, it will give you an approximate answer correct to a certain number of significant digits. That is,

$$\frac{1}{3} \neq .3333333333.$$

These problems are called alphametrics. The rules for constructing these problems state:

1. *Each letter must stand for a unique digit, and no digit is represented by more than one letter.*
2. *The leftmost letter of any word may not equal zero.*
3. *Each numerical combination is a word in the English language.*

Many cryptic-arithmetic problems can be found in The Journal of Recreational Mathematics.

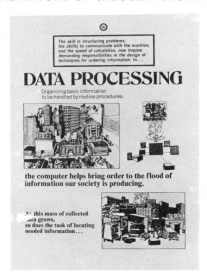

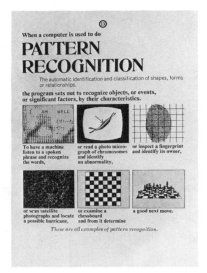

The computer carries out these tasks by accepting information, performing mathematical and logical operations with the information, and then supplying the results of the operations as new information. Since there are two ways of dealing with numbers and quantities (counting and measuring), there are generally two types of computers presently in use: (1) digital computers (counting) and (2) analog computers (measuring). The fundamental difference between the two types of computers is that digital computers work with discrete numbers, whereas analog computers work with continuous information. The turnstile counter at a fair represents a digital device, since the indicator shows only discrete numbers from 00000 to 99999. On the other hand, the odometer (mileage indicator) on many cars serves as an example of an analog device. It shows continuous change and may stop at any place. For example, it might show:

TABLE 3.1   Analog and Digital Devices

Analog	Digital
Odometer	Turnstile
Thermometer	Telephone dial
Burning candle	Calendar
Spring balance	Adding machine

Table 3.1 lists some familiar analog and digital devices.

In this text we will concentrate on digital computers. Every computing system is made up of five main components: input, storage or memory, accumulator or arithmetic unit, control, and output. (See Figure 3.14.)

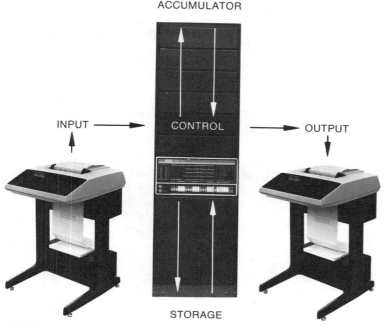

FIGURE 3.14 Relationships among the five main components of a digital computer

INPUT

The input mechanism is used to enter data and instructions into the machine. This is most commonly done with the familiar punched card consisting of 80 columns and 12 rows (only 10 of the rows have numbers printed on the cards). A punched card is shown in Figure 3.15a. A small computer may read about 200 of these cards a minute, but many computers are capable of reading over 1000 such cards per minute.

Punched cards are prepared on machines called *keypunches,* which have typewriter-like keyboards and can produce new cards or duplicate old ones. An organization with a large input of data, such as a bank or insurance company, may have 50 or 60 keypunches and keypunch operators all preparing cards for one computer.

Paper tape is also used to input information to the computer, but it is generally being replaced by magnetic or casette tape. The paper tape can be read at a high or a low speed. A low-speed reader will read about 600 characters per minute and a high-speed reader about 18,000 per minute. Some computers can read as many as 100,000 characters per minute. An example of paper tape is shown in Figure 3.15b.

Much faster is magnetic tape, which can be used to enter hundreds of thousands of characters per second (see Figure 3.15c.)

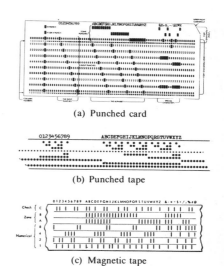

(a) Punched card

(b) Punched tape

(c) Magnetic tape

FIGURE 3.15 Methods for inputting information

The method of inputting information that you will probably use is a *teletype* or *CRT keyboard* (see Figure 3.16). You type in your instructions or messages manually, just as on a typewriter, and the computer responds on the teletype or video screen.

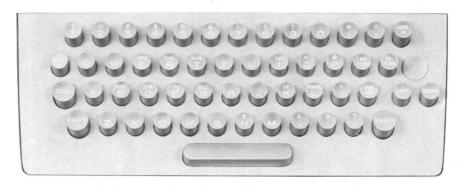

FIGURE 3.16 Teletype or CRT keyboard. Note that: (1) All the letters are capitals only. (2) Unlike on the typewriter, we do not use the letter "el" to mean the number "one." Also, we do not use the letter "oh" to mean the number "zero." (3) There are some additional keys, such as Return, Control, Shift, *, ↑, and /.

Notice from Figure 3.16 that teletype or CRT keyboards are similar to standard typewriter keyboards, but with certain differences. The most noticeable difference is that all the letters are uppercase. The shift key is activated to type certain other symbols, which you'll learn about later. Also notice some other special keys:

RETURN: This key is the same as a carriage return key, and it is also used to designate the end of a message to the computer.

LINE FEED: This key advances the roller to a new line without a carriage return.

RUB OUT: This key is used to correct a mistake in typing. Pressing the key will "erase" the last character that was typed. Depressing the rub-out key twice will erase the last two symbols typed. Remember that spaces also need to be erased on the computer.

CONTROL: This key is pressed simultaneously with another key, the same way you use a shift key. For example, on some computers, depressing the control and C keys at the same time will interrupt a program (symbolized by CTL/C).

Some computers have unconventional input devices, as shown in Figure 3.17. The General Motors advertisement illustrates a "seeing" computer, the UPI newspaper article talks of a "speaking" computer, and the third article mentions a "hearing" computer.

SCIENTIST SAYS COMPUTERS THAT TALK A REALITY

PASADENA (UPI)—Talking computers are a reality, according to a scientist at the California Institute of Technology.

Computers can now be taught to speak, including such niceties as stressing important words and pausing for punctuation, said Dr. John R. Pierce, a communications expert and former executive director of Bell Laboratories communications sciences division.

"Computers that imitate human speech with a remarkable degree of accuracy already are a reality." Pierce said Monday in a lecture at Cal Tech.

Noriko Umeda, a Japanese linguist, has succeeded in synthesizing human speech from a computer equipped with a pronouncing dictionary in its memory system, Pierce said.

Some computers are now equipped to give voice responses in some circumstances, by triggering a tape recorded message. But teaching the computer itself to talk leads to greater accuracy—because the computer will no longer repeat a mistake made by a human announcer—and "economy, rapidity and flexibility," Pierce said.

"Tape recorded answers have a very limited range and only give a few responses. A computer would have a large body of information stored in it alphabetically. The computer would be able to put these words together in an intelligible form which could answer questions." Pierce said.

Although talking computers are still imperfect, the idea is now workable, he said.

Talking computers can give instructions to workmen, he said, and could be linked to a telephone to provide advice services, such as counseling housewives on what dishes could be prepared from leftovers.

COMPUTERS TO 'HEAR' COMMANDS

By Edward O'Brien
Newhouse News Service

WASHINGTON—Pentagon research sometimes wanders into fields far removed from the battlefield. Now Defense officials predict confidently that in two or three years they will have developed computers that take instruction from a human voice.

"The use of natural spoken English as an input language to computers will revolutionize the effectiveness and utility of computer systems," Stephen Lukasik, director of defense Advanced Research Projects, has reported to Congress.

The study is at the midpoint and has produced computer systems that understand continuous speech with a vocabulary of about 100 words. This is much more sophisticated than anything done by government or private industry.

The next 900 words will come more easily because of progress already made in teaching the computers to break speech into separate words, find individual sounds within words, and use clues provided by the vocabulary, grammar, and subject matter.

"Speech is the most natural means for a person to express himself, and we will be able to integrate the computer much more closely to our daily workings if we can develop means for it to understand natural speech," Lukasik said.

"Fast typists (feeding information into computers) can normally type 60 to 70 words per minute, while nontypists may be able to achieve only six to 10 words.

"Almost everybody, however, talks in excess of 100 words per minute, and no special manual skill or equipment is required . . . with voice access, we can expect computers to be as easy to use as telephones are today."

Voice control of computers, however, may not be the ultimate.

A computer that can see gets its first real job.

A "seeing" computer developed by the General Motors Research Laboratories has recently become the first of its kind to go to work on a U.S. automotive production line. The employer: GM's Delco Electronics Division.

FIGURE 3.17 This page contains examples of computer input/output devices that simulate human senses.

ADDRESS 1	INFORMATION
ADDRESS 2	INFORMATION
ADDRESS 3	IS
ADDRESS 4	STORED
ADDRESS 5	IN
ADDRESS 6	EACH
ADDRESS 7	NUMBERED
ADDRESS 8	LOCATION

Historical Note

You are no doubt aware of the special numerals on the bottom of your bank checks.

0 1 2 3 4 5 6 7 8 9
⑊: ⁗ ⑊ ▮

These are called MICR (Magnetic Ink Character Recording) numerals and are printed in magnetic ink so they can be read mechanically. Although they may look futuristic, modern-day computers do not use MICR, which was designed in the late '50s, but instead use an alphabet called OCR (Optical Character Recognition).

ABCDEFGHIJKLM
NOPQRSTUVWXYZ
0123456789
.,:;=+/?'"
Ɏ♪⌐⫝̸%|&*{}▮

STORAGE OR MEMORY

The storage unit stores information until it is needed for some purpose. Storage might be visualized as a long corridor in a hotel. Each door is numbered, and behind each door is a limited number of "slots" to hold information. Each compartment (door) is capable of holding one piece of information or one of the instructions in a program.

The number associated with each of these slots is called the location or *address*. The information inside the box is called a *word*. During the process of solving a problem, the storage unit contains:

1. a set of instructions that specify the sequence of operations required to complete the problem;
2. the input data
3. the intermediate results for as long as they are required in the development of the problem; and
4. the final result before it is sent to the output unit.

In practice, each number, letter of the alphabet, punctuation mark or other symbol, and special computer instruction is "coded" using only two symbols (represented as 0 and 1). This is called a *binary coding* and was chosen because it assumes only two states (on–off, hole–no hole, right–left, and so on). (See Figure 3.18.) That is, a possible coding might be:

$$0 = 000\ 000 \qquad\qquad A = 100\ 000$$
$$1 = 000\ 001 \qquad\qquad B = 100\ 001$$
$$2 = 000\ 010 \qquad\qquad C = 100\ 010$$
$$3 = 000\ 011$$
$$\vdots \qquad\qquad\qquad\qquad \vdots$$
$$\vdots \qquad\qquad\qquad\qquad S = 110\ 011$$
$$\qquad\qquad\qquad\qquad\quad T = 110\ 100$$
$$\qquad\qquad\qquad\qquad\quad \vdots$$

$$\text{SPACE} = 111\ 111$$

FIGURE 3.18 Two-state device. Light bulbs serve as a good example of two-state devices. The 1 is symbolized by "on," and the 0 is symbolized by "off." The number coded into this figure is 11001 in base two, or 25. (See Problems 21–26 in Problem Set 3.3.)

To store the information "3 cats" in a computer, we would need 6 locations or addresses:

Address numbers	Information
1000	000011
1001	111111
1010	100010
1011	100000
1100	110100
1101	110011

Notice that an address number has nothing to do with the contents of that address. The early computers stored these individual digits (0 or 1), called *bits*, with switches, vacuum tubes, or cathode ray tubes. In the "3 cats" example, each word contains 6 bits. In practice, however, a word may contain 36 or more bits, which allows the use of much larger numbers. That is, a location might look like the following:

Address 00101   000111110000000011011011000000001101

Modern memory devices fall into one of two categories—moving-surface devices or electronic devices. The moving-surface devices are adapted from audio-cassette technology and are called magnetic disk or drum memories. They provide information storage in localized areas of a thin magnetic film that is coated onto a nonmagnetic supporting surface. Information is stored in the form of tiny magnetized spots in the magnetic film. The magnetic film and the read/write head move in relation to each other to bring a storage site into position for writing or reading of information.

The newest type of memory is an electronic device, made possible by modern semiconductor technology, called a random-access memory (RAM). Random-access memories are characterized by the fact that they have the same access time to any storage location. In the 1950s and 1960s electronic memories were arrays of cores or rings of a ferrite material strung by the thousands on a grid of wires. These have been replaced by magnetic bubbles, superconducting tunnel-junction devices (called Josephson-junction devices), as well as devices accessed by laser beams. The RAM shown in Figure 3.19 provides storage for 16,384 bits. The time required to read or write one bit in any arbitrary location is about 150 nanoseconds. This type of memory has considerably reduced the cost of computer memory. The one shown, which is manufactured by the Mostek Corporation, sells for $16 each at quantities of 100 pieces.

Another modern optical scanner is coming to your supermarket, using what is called a bar code. All packages will have a bar code:

All the grocery clerk will need to do is wave a wand over the bar code, and the correct amount will be charged and the tax recorded. The computer will keep track of the store inventory by noting that the item has been purchased.

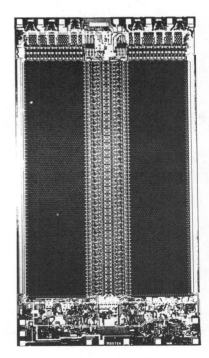

FIGURE 3.19 RAM MK 4116 measures 2.8 mm × 5.1 mm

Magnetic drum

Magnetic disk

ACCUMULATOR

The accumulator is the part of the computer in which addition, subtraction, multiplication, or other operations are carried on. It is also called the *processor* or *arithmetic unit*. The accumulator keeps a running tally of whatever is put into it. This is accomplished with logic circuits, which enable it to do these operations. We need not be concerned with how these operations occur; we will merely assume that the computer "knows how" to carry them out.

One problem that frequently arises in computer applications is that of number scale. Assume that a computer is capable of working with eight-place accuracy, and you wish to write a very large number—say a googol. We have discussed the use of scientific notation to represent such large numbers. Computers use a form of scientific notation called *floating-point notation*. In floating-point notation, the power of 10 is expressed as E + exponent. Thus,

$$650 = 6.5 \times 10^2 = 6.5E + 2$$
$$650,000 = 6.5 \times 10^5 = 6.5E + 5$$
$$.000065 = 6.5 \times 10^{-5} = 6.5E - 5$$

Therefore, $4.7E + 17$ means $470,000,000,000,000,000$, and a googol can be written as

$$1 \times 10^{100} = 1.0E + 100.$$

Numbers such as 650, ⁻489.5678, and 470 are said to be in *fixed-point form*.

EXAMPLES: Express the given numbers in floating-point notation.
1. 745

Solution: Write $745 = 7.45 \times 10^2 = 7.45E + 2$

2. .00573

Solution: Write $.00573 = 5.73 \times 10^{-3} = 5.73E - 3$

EXAMPLES: Express the given numbers in fixed-point notation.
3. $1.23E + 6$

Solution: "E + 6" indicates that we must multiply the given number, 1.23, by 10^6. Thus, $1.23E + 6 = 1.23 \times 10^6 = 1,230,000$.

4. $6.239E - 7$

Solution: $6.239E - 7 = 6.239 \times 10^{-7} = .0000006239$.

CONTROL

The control unit of the computer directs the flow of data to and from the various components of the system. It obtains instructions, one at a time, from the storage unit and causes the arithmetic unit to carry out the operations.

When data and instructions are read into the computer, control will usually direct them to storage. Both the data and the instructions can be placed anywhere in storage, although they are usually put apart from each other and kept in some predetermined sequence.

In solving a problem, the computer is directed to start at the first instruction of the program stored at a given location. Once the first instruction has been executed, the control circuits usually get the next instruction from the next higher storage position.

OUTPUT

The output unit is the means by which the operator obtains the results of the computations or data processing. The output unit usually consists of one or more of the devices discussed for input: teletypewriter, line printer, tape, and so on.

COMPUTER VENDING MACHINES?
You are all familiar with vending machines dispensing everything from soda pop to pantyhose. But did you know that you can also buy computer time on a vending machine? In 1972 a coin-operated mini-computer was installed in a library in Monterey, California. In Menlo Park, California, at People's Computing Center you can rent a computer terminal for $3 an hour. You can use it to write programs, play games, do personal calculations, or help with your math homework. At the Lawrence Hall of Science in Berkeley, California, you can use a computer for $1 an hour. Perhaps one day in the not-too-distant future a computer vending machine may be as accessible as a cigarette vending machine is today.

PROBLEM SET 3.3

A Problems

1. Explain the difference between an analog and a digital computer.

2. Name the five main components of a computer, and briefly explain the function of each component.

3. Name three methods for inputting information into a computer.

4. Explain what we mean by "address," "word," and "bit" when we speak of computers.

5. What is floating-point notation?

Express the numbers in Problems 6–10 in floating-point notation.

6. a. 546   b. 1700.82

7. a. .0000537   b. 3.14159

8. a. 23,000   b. 45.2

9. a. 378,596   b. 500.2

10. a. .000035   b. 45,000,000,000

Express the numbers in Problems 11–16 in fixed-point notation.

11. a. 4.56E + 3   b. 1.79×10^{-4}

12. a. 5.800E − 7   b. 3.98×10^{4}

13. a. 3.05E + 4   b. 7.056E + 2

14. a. 7.9E + 14   b. 6.02E − 5

15. a. 5.9326E + 6   b. 3.217E − 6

16. a. 5.72E + 7   b. 5.87E − 11

B Problems

17. a. Think of five additional examples of analog devices to add to Table 3.1.
     b. Think of five additional examples of digital devices.

18. *GIGO*. There is a slogan among computer operators: *garbage in, garbage out*. This phrase reflects the fact that a computer will do exactly what it is told to do. Consider the following humorous story.

> A very large computer system had been built for the military. It was built and staffed by the best computer people in the country.
> "The system is now ready to answer questions," said the spokesperson for the project.
> A four-star general bit off the end of a cigar, looked whimsically at his comrades and said, "Ask the machine if there will be war or peace."
> The machine replied: YES.
> "Yes *what*?" bellowed the general.
> The operator typed in this question, and the machine answered: YES SIR.

In terms of what you know about logic from the last chapter, why did the computer answer the first question YES correctly?

19. The locations given in the marginal note about computer vending machines are in California, near the home of the author. Locate the facility nearest your home where anyone can gain access to a computer. Visit the facility and prepare a report.

20. Give examples of:

     a. data processing
     b. information retrieval
     c. pattern recognition
     d. simulation

Mind Bogglers

Binary Numeration System. A two-state device can be used to represent numbers quite easily if the numbers are represented in what is called a *binary numeration system*. It is similar to our decimal system, except that instead of

---

A PROGRAMMER'S LAMENT

I really hate this damned machine;
I wish that they would sell it.
It never does quite what I want,
But only what I tell it.

10 symbols there are only two (0 and 1). The following code is established (look for a pattern).

Decimal Numeral	Binary Numeral
0	0
1	1
2	10
3	11
4	100
5	101
6	110
7	111
8	1000
9	1001
10	1010
11	1011
12	1100
13	1101
14	1110
15	1111
16	10000
⋮	⋮

So that there is no confusion between decimal and binary numerals, when the context does not make it clear which is being represented, a subscript "two" is used on the binary numeral. For example, if you see 101 it could mean one hundred one *or* one zero one base two *(the decimal number five). In such a case we write 101 for* one hundred one *and* 101_{two} *for the binary numeral.*

Study the code until you understand the pattern. Write each of the numbers given in Problems 21–23 as a decimal numeral.

21. a. 1101_{two} b. 1001_{two}

22. a. 11101_{two} b. 10111_{two}

23. a. 1100011_{two} b. 10111000_{two}

Notice that, in a binary numeral, a 1 in a certain column has a different meaning from a 1 in another column.

```
           thirty-two
           | sixteen
           | | eight
           | | | four
           | | | | two
           | | | | | units
   . . .   | | | | | |
           1 1 1 1 1 1
```

To change a number from decimal to binary, it is necessary to find how many units (1 if odd, 0 if even), how many twos, how many fours, and so on are contained in that given number. This can be accomplished by repeated division by 2.

EXAMPLE: Change 47 to a binary numeral.

```
                 0   r. 1      there is one thirty-two
              2 / 1  r. 0      there is zero sixteen
              2 / 2  r. 1      there is one eight
              2 / 5  r. 1      there is one four
              2 / 11 r. 1      there is one two
              2 / 23 r. 1      there is one unit
Start Here → 2 / 47
```

(Work Up / Read Down)

If you read down the remainders, you obtain the binary numeral 101111, which represents

one thirty-two + zero sixteen + one eight + one four + one two + one unit

=   32   +   0   +   8   +   4   +   2   +   1

= 47.

Write each of the numbers given in Problems 24–26 as a binary numeral.

24. a. 35   b. 46

25. a. 615   b. 256

26. a. 795   b. 512

Problems for Individual Study

27. Find out what local, state, and federal governments have stored in their computers about you and your family. Find out what you can see and what others can see. This will provide you with an interesting intellectual journey, if you wish to take it.

28. Read and report on several recent articles dealing with computers. Do these articles reveal anything about the public's feelings toward computers? I have listed some articles below, but by the time you read this book they will be out of date. Be sure to seek out current articles.
     References: "Behold the Computer Revolution," *National Geographic,* November 1970.
     "Business Takes a Second Look at Computers," *Business Week,* June 5, 1971.
     "Man and Machine," *Psychology Today,* April 1969.
     "Spacewar: Fanatic Life and Symbolic Death among the Computer Bums," *Rolling Stone,* December 2, 1972.

29. Build and operate a simple analog or digital computer. It is not as difficult as you might imagine. See, for example, *Popular Electronics,* January 1975. Some additional sources are listed below.
     References: Bergamini, David, *Mathematics,* pp. 26–38 (New York: Time, Inc., 1963).
     Berkeley, Edmund C., *Giant Brains or Machines That Think* (New York: Wiley, 1949).
     Campbell, Robert, "How the Computer Gets the Answer," *Life,* November 27, 1967.

This lion was made entirely of computer components.

Computer-Oriented Mathematics (Washington, D.C.: National Council of Teachers of Mathematics, 1963).

McCarthy, John, "Information," *Scientific American,* September 1966 (Vol. 215, No. 3), pp. 64–73.

3.4 IT'S BASIC

WHAT IS PROGRAMMING?

To get a computer to operate, a complete set of step-by-step instructions must be provided. These sets of instructions are known as *programs*, and the process of setting them up is known as *programming*.

Each step or instruction for the solution of a problem must be provided for the computer in detail. For example, if a teacher asks a pupil to add 7 and 4, the student will generally say 11; the programmer, however, not only must ask the computer to add 7 and 4 but also must ask it to indicate the answer.

To write a program, we should carry out the following four steps:

1. Analyze the problem and break it down into its basic parts.
2. Prepare a flow chart.
3. Put the flow chart into a language that the machine can "understand."
4. "Debug" the program. That is, conduct a trial run of your program to see if it does what it is supposed to do.

FLOW CHARTS

A flow chart is an outline that lists the steps to be performed in sequence. The directions are usually listed within squares, circles, or other geometric figures (as shown in Figure 3.20). The direction, or flow, of the process is shown by arrows.

Edwin S. Cohen, Assistant Secretary of the Treasury for tax policy, marvels over the prowess of the Internal Revenue Service in using computers. Still, he was disturbed to find that the computers sent him two sets of 1040 forms, one to his current address and the other to his Charlottesville, Va., home. Cohen wasn't eager to pay his taxes twice, so he asked the IRS computer people to fix things up. Sorry, Cohen says he was told, there was no way to correct the error because it was impossible for it to have happened.

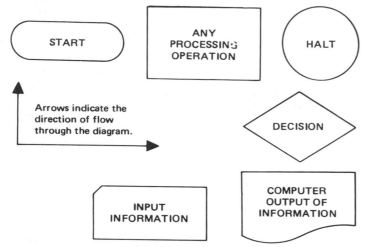

FIGURE 3.20 Common flow chart symbols

The flow chart provides a way of checking the overall analysis of a process or problem, and it also provides a way for describing the process to someone else once the flow chart is completed. Using the flow chart as a guide, the details of the program are easier to write.

Flow charts can direct us in other types of activities besides writing computer programs. For example, the General Information Leaflet of The General Library of The University of California, Davis, uses the flow chart shown in Figure 3.21 to explain how to find library materials. Note that no detail should be overlooked in a flow chart.

The basic steps for finding library materials

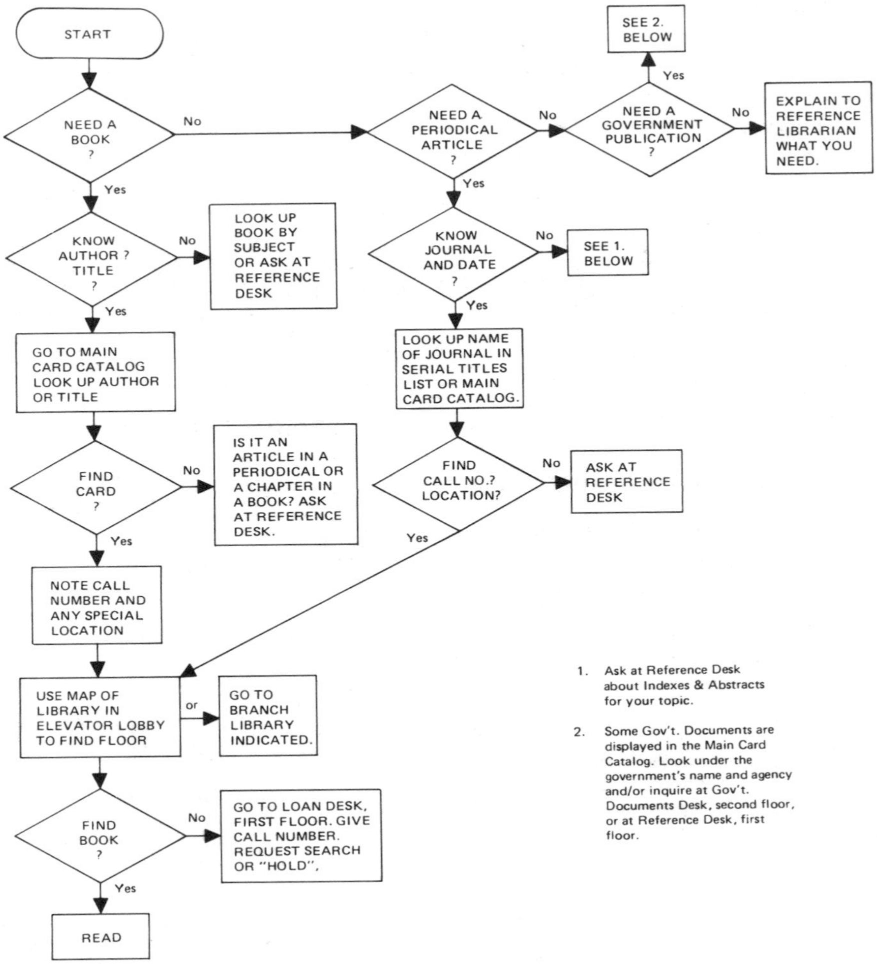

FIGURE 3.21 The basic steps for finding library materials

EXAMPLE: Follow the directions given by the following flow chart.

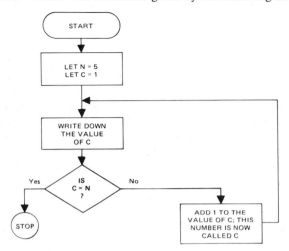

Solution: Write down the following: 1
2
3
4
5

EXAMPLE: Draw a flow chart that will direct someone to write down the square numbers up to 50.

Solution:

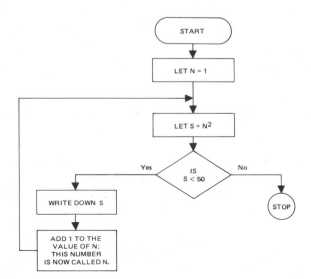

INTRODUCTION TO BASIC

After preparing a flow chart, the next step for writing a program is to put the flow chart into a language that the machine can "understand." There are three types of programming languages: (1) *machine language,* (2) *symbolic language,* and (3) *algorithmic language.* Machine language is the "native language" of the computer, and any other language that the computer "understands" must first be translated into machine language before the computer can carry out the given commands. Machine language consists of coded sequences of binary digits for the letters of the alphabet, numerals, symbols, and so on. Symbolic languages use mnemonic codes rather than numerical ones. Since machine and symbolic languages are used primarily by professional programmers, we will focus our attention on an algorithmic language.

These are the types of computer languages; do not confuse them with the names of the languages themselves.

Mnemonic refers to an easily remembered code word that stands for one or more computer instructions.

Algorithmic languages, or procedure-oriented languages, are much more conversational than machine or symbolic languages. In symbolic language a numeral is replaced by a mnemonic code, but in algorithmic language a single word can replace an entire list of numerals. For the computer to accept an algorithmic command, a special program called the *compiler* must be used. The compiler translates the conversational algorithmic command into machine language.

Some widely used algorithmic languages are FORTRAN (FOR-mula TRANslator), ALGOL (ALGOrithmic Language), BASIC (Beginner's All-purpose Symbolic Instruction Code), FOCAL (FOrmula CALculator), and COBOL (COmmon Business Oriented Language).

In this course, you may want to use a computer to help in solving problems. So that you will be able to communicate with a computer, we will now learn a very conversational language called BASIC.

BASIC consists of short, easy-to-learn English commands. We will talk to the computer on a CRT or teletype keyboard, and the computer will respond on that same device.

It is important that we talk to the computer in the proper format:

LINE NUMBER   COMMAND   VARIABLE or NUMERICAL EXPRESSION

The *line number* is some counting number between 0 and 1000 that causes the computer to do the commands in a specified order. It will do the lowest line number first and then move to the next higher number. When writing a program, we usually use the line numbers 10, 20, 30, and so on, instead of 1,2,3, ... We do this so that we can insert new statements between old statements. The *command* tells the computer what to do. The *variables* or *numerals* are those quantities acted upon by the particular command.

To the performer of STALKS AND TREES AND DROPS AND CLOUDS:

You need two sets of instruments.

Set A, for STALKS AND TREES: Seven non-reverberating instruments,

dry, dead, explosive, whipping, rattling, wooden, etc. Preferably,

but not necessarily, just <u>one</u> of each of seven <u>very</u> different kinds.

The seven symbols for Set A, and the number of times each appears in

the score:

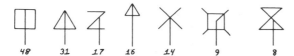

48   31   17   16   14   9   8

Set B, for DROPS AND CLOUDS: Six reverberating instruments, resounding,

ringing, vibrating, sizzling, noisy, whispering, etc. Preferably, but

not necessarily, several of each of six <u>very</u> different kinds. The six

symbols for Set B, and the number of times each appears in the score:

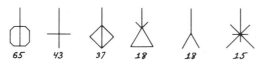

65   43   37   18   18   15

During the first preparatory stage (see General Preface) you have to

determine the particular kind of instrument, which each symbol will

represent throughout the piece.

COMPUTERS AND MUSIC

This example of music composition by computer depicts instructions to the performer and a part of the score from Herbert Brun's *Stalks and Trees and Drops and Clouds* (1967). The complete score consists of 31 pages, and it would last about 7$\frac{1}{2}$ minutes in performance. It was composed on an IBM 7094 computer.

In another effort, two Southwestern College professors used an IBM 360/22 computer to compose music. John Bibbo, mathematics and computer programming instructor at the college, assisted Dr. Victor Saucedo, music instructor, in a computerized approach to musical composing. Dr. Saucedo was asked to present his composition "RAN. I.X. For Solo Clarinet and Tape" at the Western Regional Conference of the American Society of University Composers held at Fullerton State University in November, 1976.

SOME FIRST PROGRAMS

You are now ready for a simple program. Every computer has some sort of a log-in procedure. These procedures vary from machine to machine and facility to facility, and they must be kept relatively secret so that not just anyone can use the computer. You will need to find out the log-in procedure for the computer or terminal you will be using. All the examples in this book will assume that you have correctly followed the log-in procedure for your particular computer.

Suppose you wish to have the computer type the phrase "I love you." In order to do this, you will use a command called PRINT.

Any counting number under 1000; this is called the *line number* of the command; this is your (the programmer's) choice.

10   PRINT "I LOVE YOU."

Whatever is enclosed in quotation marks will be printed by the computer.

Command

20   END

The end of the program is designated by END (yes, even one this short). Any line number larger than 10 can be used.

This program, along with another example, is shown below as it would look on the CRT.

To prevent confusion, computers use Ø for "zero" and O for "oh."

Program A	Program B
10 PRINT"I LOVE YOU."	10 PRINT "I AM ROBBIE THE ROBOT."
20 END	20 PRINT "I LOVE YOU."
	30 PRINT "GOODBYE"
	40 END

Now that we've written a program, the next step is to run the program. The RUN command causes the computer to begin the program. When you tell the computer to RUN, it will begin at the lowest line number and execute in consecutive order all commands in its core memory. In the course of the program, the computer will probably output data or information; it may ask you questions or do some calculations. When it is finished, it will type

READY

to signify that it has finished the program.

EXAMPLES: Run Programs A and B.
1. Type RUN and return (which we'll denote by RUN↵), and the computer will respond with

I LOVE YOU.

READY

Notice that the quotation marks are not printed and that, when the computer reaches the END command, it types READY to signify that it has finished the program.

2. RUN↵

I AM ROBBIE THE ROBOT.
I LOVE YOU.
GOODBYE

READY

Notice that each new line number generates a new line of output.

If you wish to use a computer for arithmetic problems (which is usually foolish when a calculator is available) you will use * for multiplication, / for division, and ↑ for exponentiation, as shown in Table 3.2.

EXAMPLE: Write a program to calculate $5 \times 3 - 6$.

Solution: 10   PRINT 5*3 − 6
              20   END

EXAMPLE: Write a program to calculate the area of a circle with radius 6.3.

Solution: 10   PRINT 3.1416 * 6.3↑2
              20   END

Notice that no quotation marks are used when using the PRINT command to do calculations.

Recall the formula for the area of a circle is $A = \pi r^2$, where π is approximately 3.1416.

TABLE 3.2 Mathematical Symbols in BASIC

Symbol		Math Notation	BASIC
↑	Exponentiation	3^4	3↑4
*	Multiplication	$3 \cdot 4$	3*4
/	Division	$3 \div 4$ or $^3/_4$	3/4
+	Addition	$3+4$	3+4
−	Subtraction	$3-4$	3−4

DEBUGGING A PROGRAM

Debugging means checking and revising your program to make sure it does what it is supposed to do. The programs in this section will be easy to debug, but if your program is more complex and involves several decisions and options, it will be much more difficult to check.

There are two types of mistakes. The first is a logic error, which is an error in your thinking. A flow chart may help you overcome this type of error if the problem is complicated. The second type of error is a coding error, which is a clerical error. See the example about an $18 million error in the historical note on page 172.

When you are debugging your program, the computer may type out an error message such as ILLEGAL CHARACTER IN LINE *n*. Most of the error messages are self-explanatory.

The computer may also type out an error message when you are typing in a program. Most of these are also self-explanatory. For example, if the computer cannot understand the command just given, it will type WHAT? Or, if the line number is outside the range of 1 to 999, the computer will type ILLEGAL LINE NUMBER.

In the next section, we will introduce the BASIC commands you will use when writing a program. We will also write some simple programs.

A hairdresser created a pouf
So high that it looked like a goof!
He said, while a-tugging,
"Madame, I am debugging
This artless and curlycued roof!"

Historical Note

In 1962 an Atlas-Agena rocket blasted off from Cape Kennedy; it was to have been the first U.S. spacecraft to fly by Venus. When it was about 90 miles above the earth, it had to be destroyed. The reason for the failure was later determined to have been due to a mathematician who left out a hyphen when writing the flight plan. That missing hyphen cost the United States about $18,500,000.

PROBLEM SET 3.4

A Problems

1. What is a flow chart?

2. What is the difference between machine language and symbolic language?

3. Discuss the steps to follow when writing a program.

4. In your own words, describe what is meant by debugging a program.

5. Draw the symbol used in a flow chart to denote

   a. start
   b. stop
   c. a decision
   d. input data
   e. output data

6. Consider the following program.

   ```
   10   PRINT "GOOD MORNING."
   20   PRINT "CAN YOU DO THIS PROBLEM CORRECTLY?"
   30   END
   ```

   Without using a computer, explain what the computer will type after you give a RUN command.

7. Suppose you have just written the program shown in Problem 6. Then you type

   $$25 \quad \text{PRINT } 2 + 3$$

   Without using a computer, explain what the computer will type after you give a RUN command.

In Problems 8–14, write each BASIC expression in ordinary algebraic notation.

EXAMPLE: $2 * X \uparrow 2 - 14/5$ means $2x^2 - \dfrac{14}{5}$

8. a. $4*X + 3$ b. $(5/4)*Y + 14\uparrow2$

9. a. $35*X\uparrow2 - 13*X + 2$ b. $6*X - 7$

10. a. $5*X\uparrow2 - 3*X + 4$ b. $17*X\uparrow3 - 13*X\uparrow2 + (15/2)$

11. a. $6.29\uparrow14 - 7$ b. $(X - 5)*(2*X + 4)\uparrow2$

12. $4*(X\uparrow2 + 5)*(3*X\uparrow3 - 3)\uparrow2$

13. $(16.34/12.5)*(42.1\uparrow2 - 64)$

14. $47*X\uparrow3 + (13/2)*X\uparrow2 + (15/4)*X - 17$

In Problems 15–20, write each given sentence in BASIC language notation.

EXAMPLE: $^2/_3 x^2 + 17$ would be written in BASIC as

$$(2/3)*X\uparrow2 + 17.$$

15. a. $\dfrac{2}{3}x^2$

    b. $3x^2 - 17$

16. a. $5x^3 - 6x^2 + 11$

    b. $14x^3 + 12x^2 + 3$

17. a. $12(x^2 + 4)$

    b. $\dfrac{15x + 7}{2}$

18. a. $(5 - x)(x + 3)^2$

    b. $6(x + 3)(2x - 7)^2$

19. $(x + 1)(2x - 3)(x^2 + 4)$

20. $\dfrac{1}{4}x^2 - \dfrac{1}{2}x + 12$

21. Follow the directions shown by the flow chart in Figure 3.22. Is the answer predicted in the flow chart correct?

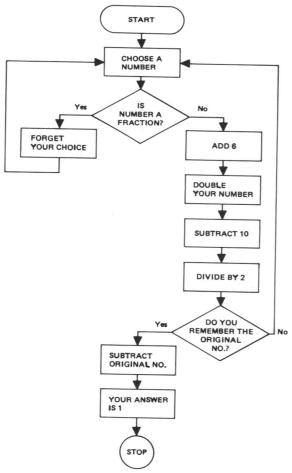

FIGURE 3.22   Flowchart for Problem 21

MACHINE TRANSLATION

In the 1960s there was a big push to use computers to do foreign-language translation. Computers were supplied with a small bilingual dictionary with the corresponding words in the two languages. It soon became apparent that word-for-word translation was virtually useless. The addition of a dictionary of phrases brought only marginal improvement.

These translators can be tested by translating English to Russian and then Russian back to English. Hopefully, one should end up with about the same as one started. Using this method, the maxim, "Out of sight, out of mind" ended up as "The person is blind, and is insane."

Another example was, "The spirit is willing but the flesh is weak." It was translated as "The wine is good but the meat is raw."

Needless to say, computer translation is presently used very little. And it is doubtful that it will be useful in the near future, even though there are now calculators that will translate individual words from one language to another.

22. Follow the directions given by the following flow chart.

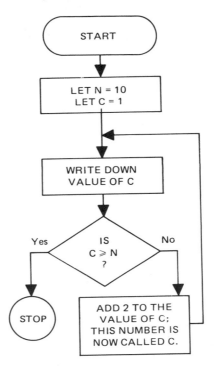

B Problems

Write a simple BASIC program to carry out the tasks given in each of Problems 23–27.

23. Say "I'm a happy computer."

24. Say "I will demonstrate my computational skill."
    Then calculate $\dfrac{53 + 87}{2}$.

25. Calculate $(1 + .08)^{12}$.

26. Calculate $5{,}000(1 + .06)^9$.

27. Calculate $\dfrac{23.5^2 - 5(61.1)}{2}$.

28. Write a flow chart to describe how to multiply two numbers on an algebraic-logic calculator.

29. Write a flow chart that will tell someone how to multiply two numbers.

30. Prepare a flow chart to describe the process of taking attendance by calling roll. You must take into account the possibilities of both tardy and absent persons.

31. Write a flow chart that will direct someone to write down the Fibonacci numbers less than 100.

32. Write a flow chart that will direct someone to write down the first 50 Fibonacci numbers.

33. It has been said that anyone who does not know how to program a computer is functionally illiterate. Do you agree or disagree? Give some arguments to support your answer. Project this question into the future; do you think it will ever be true?

Fibonacci numbers were discussed in Sections 1.2 and 1.3. A Fibonacci number is a number in the sequence 1,1,2,3,5,8,13,21,34,55, ...

Computer Career Opportunities
Honeywell Corporation

Future career opportunities in the rapidly growing world of computers seem to be practically limitless. These opportunities may be direct, as in the case of those who manufacture and operate computers, or indirect, as in the case of businessmen, scientists, and others who use computer systems.

Increasingly great numbers of skilled personnel will be needed by the computer industry itself:

Designers and manufacturers of systems
Engineers and scientists for research and development
Sales personnel skilled in marketing methods
Systems analysts to analyze and meet special requirements of customers
Programmers who prepare programs to meet customers' needs
Computer operators to run systems
Personnel for clerical and data preparation jobs
Managers of computer operations
Management interpreters of computer systems, needs, opportunities
Specialists in areas such as business, science, education, and government
Interdisciplinarians—those who can understand and meet the needs of
    persons from varied professions united on mutual projects.

More than a thousand colleges and universities in the U.S.A. and Canada, according to a recent survey, now offer courses in the computer sciences and data processing. Computer usage is being taught in many high schools and even in some grammar schools. Many independent training schools exist for high school and college graduates.

The use of remote terminals, that connect to a central computer system sometimes thousands of miles away, is becoming commonplace. Industry experts say it's only a matter of time and cost reduction before the use of household terminals, for a variety of purposes ranging from information services to entertainment, becomes as ordinary as the use of the telephone.

Economists predict that by the end of the century, or earlier, the computer industry and directly associated industries will be the largest American business.

Mind Bogglers

34. Arrange four checkers as shown. Try to interchange the checkers so the black ones are at the left and the red ones at the right.

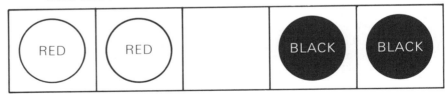

Rules:
  i. You may move only one checker at a time.
 ii. You may jump over only one checker.
iii. Black checkers may be moved only to the left; red checkers may be moved only to the right.
 iv. No two checkers may occupy the same space at the same time.
What is the minimum number of moves required to complete the game?

35. Repeat Problem 34 with three black checkers, three red checkers, and seven squares. What is the minimum number of moves required to complete the game?

36. Generalize the results of Problems 34 and 35. That is, look for a pattern, and try to predict the minimum number of moves required to complete the game with *n* black checkers and *n* red checkers, using $2n + 1$ squares.

37. Write a flow chart for Problem 34 (You might try 35 instead. Or, if you really feel like a Whiz Kid, write a program for Problem 36 instead of 34 or 35.)

38. Write a flow chart for playing a game of tic-tac-toe when you are making the first play. You must take into account all possible moves by your opponent. *Hint:* Number the spaces as shown.

1	2	3
8	0	4
7	6	5

Problems for Individual Study

39. The shortage of computer programmers has never been greater. In 1977 there were 100,000 persons employed as programmers, and there were openings for at least 50,000 more. Check out the job opportunities for programmers in your area.

40. Do you think that a flow chart for checkers or chess would be possible? Read some of the current literature to help you answer this question.

3.5 COMMUNICATING WITH A COMPUTER

In this section, we'll introduce some of the more common BASIC commands and write some programs. If you have access to a computer, it will be helpful to try running some of these programs.

BASIC treats all numbers as decimal numbers; if the input number does not have a decimal, the computer assumes a decimal point after the last numeral. Also, if the answer to an arithmetic operation is $1/3$, the computer would write this as .333333. We also use variables in BASIC, which are formed by a single letter or by a letter followed by a digit.

SCR Command*

To begin with a clean slate, we must first erase ("scratch") from its memory any program the computer is storing. Therefore, before we write a program, we would probably give the computer an SCR command.

LIST Command

To find out what is in the computer's core memory at any stage, we use a LIST command, which tells the computer to list everything stored in its core. The following are variations of the list command:

LIST n,   This command causes the computer to list line n only.
LIST n,m This command causes the computer to list lines n through m, inclusive.

END Command

The command with the largest line number in every program must be an END command. This signifies that the program is complete. If a program has a command with a higher line number, or if there are two or more END commands, an error code will result.

PRINT Command

As you saw in the last section, the PRINT command carries out more than one function. The first is to perform arithmetical calculations on the computer. For example, suppose we have erased any previous programs with a SCR command. Next we type the following:

```
10   PRINT 7 + 4
50   END
```

---

*BASIC has a number of variations, most of which are minor and depend on the computer being used. You should check with your instructor for possible variations in the system you are using.

As you learn about programming, keep in mind some of the deception that is carried out in the name of computers. People seem to think that "If a computer did it, then it must be correct." This is false, and the fact that a computer is involved has no effect on the validity of the results. Have you even been told that

"The computer requires that . . ."

"There is nothing we can do; it is done by computer."

"The computer won't permit it." What they really mean is that the program doesn't permit it or that they don't want to write the program to permit it.

Notice that the END command tells the computer that the program is complete at this point. Now, if we give the computer a RUN command, the computer will output the answer.

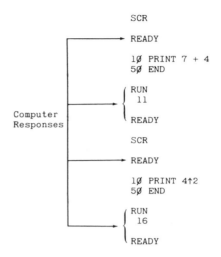

```
                                    SCR

                                 →  READY

                                    1Ø  PRINT 7 + 4
                                    5Ø  END

                               ⎧    RUN
                               ⎪      11
              Computer         ⎨
              Responses        ⎪    READY
                               ⎩
                                    SCR

                                 →  READY

                                    1Ø  PRINT 4↑2
                                    5Ø  END

                               ⎧    RUN
                               ⎪      16
                               ⎨
                               ⎩    READY
```

Several computations can be done with one program. Each new line in the program will require that the answer be on a new line. For example,

```
10    PRINT 55 + 11
20    PRINT 55*11
30    PRINT 55 − 11, 55/11
50    END
```

will cause the computer to output the answers for 55 + 11 and 55*11 on successive lines, but the answers for 55 − 11 and 55/11 will be recorded on the same line.

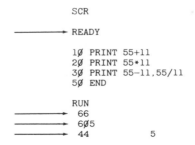

```
                    SCR

                →   READY

                    1Ø  PRINT 55+11
                    2Ø  PRINT 55*11
                    3Ø  PRINT 55−11,55/11
                    5Ø  END

                    RUN
                →     66
                →     6Ø5
                →     44              5
```

Remember, a single PRINT statement can contain more than one expression, but the results will be printed on the same line (up to five;

more than five are automatically printed on the next line). If a PRINT statement contains more than one expression, those expressions must be separated by commas.

The PRINT command can also be used for spaces in the output. For example, the command PRINT followed by no variables or numerals will simply generate a line feed.

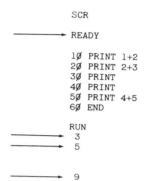

```
            SCR

            READY

            10  PRINT 1+2
            20  PRINT 2+3
            30  PRINT
            40  PRINT
            50  PRINT 4+5
            60  END

            RUN
             3

             5


             9
```

If the answer that we are calling for is too large, the computer will automatically print it out in *floating-point form* (see Section 3.3).

```
            SCR

            READY

            10  PRINT 1*2
            20  PRINT 2*3
            30  PRINT 6*4
            40  PRINT 24*5
            50  PRINT 120*6
            60  PRINT 720*7
            70  PRINT 5040*8
            80  PRINT 40320*9
            90  PRINT 362880*10
            100  END

            RUN
             2
             6
             24
             120
             720
             5040
             40320
             362880
             3.628800E+6

            READY
```

Notice that when the answers get too large, the computer will automatically switch to floating-point form.

In the last section we also used the PRINT command to type messages. For example, suppose we would, like the computer to tell us "Hello." This can be accomplished by giving a PRINT command followed by the word enclosed in quotation marks.

```
10   PRINT "HELLO"
20   END
```

A run of this program would simply cause the computer to say "HELLO."

```
RUN
HELLO
```

Let's look at this use of the PRINT command more carefully by studying the following program:

```
10   PRINT "4 + 7 =", 4 + 7
20   END
```

The difference is the quotation marks.

What is the difference between "4 + 7 =" and 4 + 7 in the program? The "4 + 7 =" cause the computer to type out

$$4 + 7 =$$

and does no arithmetic. The second part, 4 + 7, tells the computer to do the arithmetic (which causes the computer to type out 11).

```
RUN
4 + 7 =        11
```

LET Command

We assign values to locations by using a LET command. For example, we might say

We usually write flow charts only for the more involved programs, but the flow chart for this program is shown in Figure 3.23. Notice that each direction on the flow chart is translated into a separate line in the program.

```
10   LET A = 2
20   LET B = 5
30   LET P = 3.1416
40   PRINT A,B
50   PRINT P
60   PRINT A + B
70   PRINT
80   PRINT P*B↑A
100  END
```

INPUT Command

Suppose we wish to compute the amount of money present after one year for various interest rates. The simple interest formula is

$$A = P(1 + rt)$$

where P = principle (amount invested)
r = interest rate (written as a decimal)
t = time (in years)
A = amount present (after t years)

For purposes of this example, we will let $P = \$10,000$ and $t = 1$. We must have some way of telling the computer the value of r without rewriting the program each time. A BASIC command called INPUT will be used. It is used to supply data while the program is running. When the computer receives an INPUT command, it will type out a question mark, ?, and wait for the operator to input the data. When you are finished inputting data, depress the return key (⤸). Consider the following BASIC program (the flow chart is shown in Figure 3.24 on page 182).

```
10   INPUT R
20   LET A = 10000*(1 + R)
30   PRINT A
40   END
```

When a RUN command is given, the computer will type

?

The operator must then type the value for r (three runs of the program are shown for interest rates 5%, 5³/₄%, and 7%):

```
RUN
?  .05              Notice that interest rates are input as decimals.
 10500
```

```
READY              Can you tell which of these are operator typed and which are
                   computer typed?
RUN
?  .0575
 10575
```

```
READY

RUN
?  .07
 10700
```

READY

The programmer may find it helpful to remind those people using the program what is being asked for with a question mark; so the program can be modified as follows (notice that even the spaces must be included inside the quotation marks):

```
5   PRINT "I WILL CALCULATE THE AMOUNT PRESENT FROM A $10,000 ";
6   PRINT "INVESTMENT AFTER"
7   PRINT "1 YEAR AT SIMPLE INTEREST. WHAT IS THE INTEREST RATE";
24  PRINT
25  PRINT "THE AMOUNT PRESENT IS $";
```

This second version of the program has several nonessentials from a programming standpoint, but it makes the computer much more conversational.

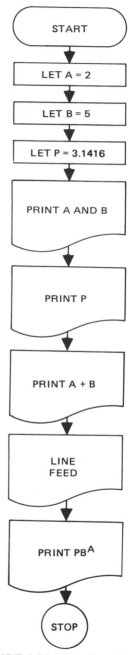

FIGURE 3.23 Flow chart for the example program on page 180

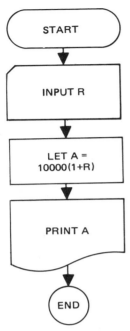

FIGURE 3.24 Flow chart for *A* = *P*(1 + *rt*) where *P* = $10,000 and *t* = 1.

Can you add PRINT commands to this program to make it more conversational?

This first run of the program is calculating the simple interest formula for one year at 7% with $5000 invested.

Can you interpret what is being calculated in the second two runs of this program?

```
 3    PRINT
 5    PRINT "I WILL CALCULATE THE AMOUNT PRESENT FROM A $10,000 ";
 6    PRINT "INVESTMENT AFTER"
 7    PRINT "1 YEAR AT SIMPLE INTEREST. WHAT IS THE INTEREST RATE";
10    INPUT R
20    LET A = 10000*(1 + R)
24    PRINT
25    PRINT "THE AMOUNT PRESENT IS $";
30    PRINT A
40    END
```

RUN
I WILL CALCULATE THE AMOUNT PRESENT FROM A $10,000 INVESTMENT AFTER 1 YEAR AT SIMPLE INTEREST. WHAT IS THE INTEREST RATE? .05

THE AMOUNT PRESENT IS $ 10500

READY

Notice the significance of the semicolon at the end of lines 5, 7, and 25. It means that there should be *no line feed* when going to the next line of the program.

It is possible to input several variables at once. Suppose we wish to input not only R but also P, as in the command

INPUT P, R

We have the following variation:

```
10    INPUT P, R
20    LET A = P*(1 + R)
30    PRINT A
40    END
```

When we run this program, two variables must be input after the question mark.

```
10    INPUT P, R
20    LET A = P*(1 + R)
30    PRINT A
40    END
```

RUN
? 5000,.07
  5350

READY

RUN
? 30000,.075
  32250

READY

RUN
? 30000,.09
  32700

READY

EXAMPLE: Predict the run for the following program.

```
10   INPUT A,B,C
20   PRINT A + B + C
30   END
```

Solution: Three variables must be input after the question mark.

```
RUN
? 2,3,4
 9

READY
```

This run shows what happens if we push the carriage return after each response instead of inputting all three values on the same line.

```
RUN
? 5,10,8
 23

READY

RUN
? 2
? 3
? 5
 10

READY
```

COMPUTER GRAPHICS

Computers can be programmed to create a wide variety of designs. They can draw charts and maps in both two and three dimensions. They can also be used to simulate art.

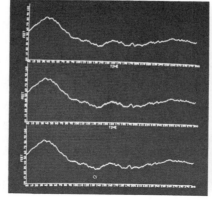

INTERACTIVE EDITING

Several other commands are used in writing programs, and some of them will help you when you make a mistake.

When more than one possibility is given, each command occurs on a different computer system. The commands will probably not all work on your system. You will need to check these out on the computer system that you'll be using.

RUB-OUT or CTL/H	This will erase a typing error. It deletes the preceding space. If RUB-OUT is depressed twice, it will delete the preceding two characters.
STOP or CTL/C or CTL/Y or ABORT	This will stop the running of a program. The computer will type STOP or C and then reply READY.
EDIT 50 [a]	This will allow the user to search line 50 for the character a.
DELETE 50	This will delete the line number indicated (50 in this example).

PROBLEM SET 3.5

A Problems

1. Give an example of each of the three uses for the PRINT command.

2. What is the significance of the semicolon at the end of a line beginning with a PRINT command?

3. What is the command that allows us to put information into the computer?

4. You have fed the following program into a computer:

```
10    PRINT 6 + 5
20    PRINT 6 − 5
30    PRINT 6*5, 6/3
40    END
```

What will the computer type if you give a RUN command?

5. You have fed the following program into a computer:

```
10    PRINT 4 + 5
20    PRINT "HELLO, ";
30    PRINT "I LIKE YOU."
40    PRINT "4 + 5 ="; 4 + 5
50    END
```

What will the computer type if you give a RUN command?

6. You have fed the following program into a computer:

```
10    PRINT 5 + 10
20    PRINT 5*10
30    PRINT 10/5, 5/10
40    END
```

What will the computer type if you give a RUN command?

7. You have fed the following program into a computer:

```
10   PRINT 7 + 12
20   PRINT 7 − 12, 7*12
30   END
```

What will the computer type if you give a RUN command?

8. Find the error in the following program:

```
10   PRINT "I WILL COMPUTE (2*X ↑ 2)*(X + 1)"
20   INPUT X
30   PRINT (2*X↑2)*(X + 1)
```

9. Find the error in the following program:

```
10   INPUT X
20   PRINT (X↑2 − 3)(2*X + 1)
30   END
```

B Problems

10. You have fed the following program into a computer:

```
10   LET A = 2
20   PRINT "WORKING"
30   LET B = 3
40   PRINT A + B
50   END
```

What will the computer type if you give a RUN command?

11. You have fed the following program into a computer:

```
10   "I WILL CALCULATE IQ."
20   "WHAT IS MENTAL AGE";
30   INPUT M
40   PRINT "WHAT IS CHRONOLOGICAL AGE";
50   INPUT C
60   PRINT
70   PRINT "IQ IS";
80   PRINT (M/C)*100
```

What will the computer type if the program is run for a mental age of 12 and a chronological age of 10?

12. What will the computer type if you RUN the following program for a 4 by 5 rectangle?

```
10    PRINT "I WILL CALCULATE THE PERIMETER AND ";
20    PRINT "AREA OF A RECTANGLE."
30    PRINT "WHAT IS THE LENGTH";
40    INPUT L
50    PRINT "WHAT IS THE WIDTH";
60    INPUT W
70    PRINT "THE PERIMETER IS ";
80    PRINT 2*(L + W);
90    PRINT "AREA IS ";
100   PRINT L*W
110   END
```

COMPUTERS IN MEDICINE

Computers are becoming an important tool in medicine and medical research. For example, at the Texas Institute for Rehabilitation in Houston, scientists are using an IBM computer to produce highly accurate, three-dimensional measurements of the human body for studies of problems ranging from spinal deformities in children to weight loss in astronauts. Overlapping photographs of the body are taken simultaneously, and then a plotting device identifies as many as 40,000 reference points, which are then fed into the computer. With a computer-driven plotter, a contour map of the body (like the one shown here) is then drawn to assist doctors in their diagnosis and treatment.

13. Write a BASIC program to ask for a value for x, and then use this value to evaluate $5x^3 + 17x - 128$. Also, show the flow chart.

14. Write a BASIC program to ask for a value of x, and then use this value to evaluate $12.8x^2 + 14.76x + 6^1/_3$. Also, show the flow chart.

15. *Computer Problem.* Write a BASIC program to ask for a value of x, and then use this value to evaluate $24x^3 + 3x^2 - x + 6$ for $x = 1$, $x = 10$, and $x = 4.63$.

Problems that require a computer are designated by the words "Computer Problem."

16. Suppose you are considering a job with a company as a sales representative. You are offered $100 per week plus 5% commission on sales. Write a program that will compute your weekly salary if you input the total amount of sales. Be sure to include a flow chart.

17. A manufacturer of sticky widgets wishes to determine a monthly cost of manufacturing on an item that costs $2.68 per item for materials and $14 per item for labor. The company also has fixed expenses of $6000 per month (this includes advertising, taxes, plant facilities, and so on). Write a program that will compute monthly cost if you input the number of items manufactured. Be sure to include a flow chart.

18. *Computer Problem.* The Indians sold Manhattan Island to the Dutch in 1626 for goods worth about $24. (It has often been said that the Dutch took advantage of the Indians.) Suppose the Indians had put the $24 in a savings account at 7% interest. The formula for calculating

$$A = P(1 + i)^n$$

where A is the present amount, P is the principal ($24 for this problem), i is the interest written as a decimal (.07 for this problem), and n is the number of years. For example, after one year:

$$A = 24(1 + .07)$$
$$= 25.62$$

After 2 years:

$$A = 24(1 + .07)^2$$
$$= 27.48$$

Manhattan Island was once bought for $24 worth of beads from the Indians, but the city's real estate is now valued at $82.3 billion.

Find the value of $24 in 1976 if it had been deposited in 1626 at 7% interest. Write your answer in scientific notation and compare it with the amount given in the newspaper article.

Mind Bogglers

19. *Computer Problem.* Write a BASIC program to compute the sum of the first N even counting numbers.

20. *Computer Problem.* Think of any counting number. Add 9. Square. Subtract the square of the original number. Subtract 61. Multiply by 2. Add 24. Subtract 36 times the original number. Take the square root. What is the result?

  a. Show algebraically why everyone who works the problem correctly gets the same answer.
  b. Write a BASIC program to solve the puzzle.

21. *Computer Problem*. Write a BASIC program for obtaining the pattern shown.

```
XXXXX   XXXXX   XXXXX   XXXXX   XXXXX
XX  X   X       X   X   X       X
XXXXX   XXXXX   XXXXX   X       XXXXX
XX      X       X   X   X       X
XX      X       X   X   X       X
XX      XXXXX   X   X   XXXXX   XXXXX
```

Problem for Individual Study

22. *Computer Problem*. Write a BASIC program for playing a game of tic-tac-toe.

3.6 LOOPS

Repetition of arithmetic operations is a tedious task that can be assigned to computers. Suppose we wish to add the first 100 counting numbers:

$$1 + 2 + 3 + 4 + \cdots + 97 + 98 + 99 + 100.$$

We could write a program similar to the ones of the previous section, but these methods would still be tedious. Instead, we will use a *loop*, which repeats a sequence of operations. The loop is eventually completed (we hope!), and the sequence of instructions proceeds to the stopping point.

GOTO Command

This command transfers the control to a line number other than the next higher one. The flow chart in Figure 3.25 illustrates the GOTO command. This command can cause the computer to go into a loop, as shown by the program below.

```
10   LET N = 1
20   PRINT N
30   LET N = N + 1
40   GOTO 20
50   END
```

If we type RUN, this program will continue forever typing out the successive counting numbers. Can you follow the steps of this program? The loop is set up using the GOTO statement. The problem is, however, how to "get out" of the loop. To do this, we introduce two additional commands.

FOR-NEXT Commands

The FOR and NEXT commands are used in conjunction as follows:

	Line Number	Command
FOR command starts the loop	15	FOR I = 1 TO 10
Other commands in the loop are repeated	⋮	⋮
NEXT command ends the loop and increments the variable I	40	NEXT I

GOTO is pronounced as two words: "go to."

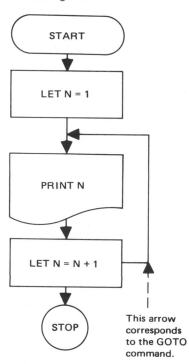

This arrow corresponds to the GOTO command.

FIGURE 3.25 Flow chart illustrating a GOTO command

These commands provide a means of repeating a loop a given number of times. For example, suppose we wish to write out the numbers from 1 to 10. (See the flow chart in Figure 3.26). We would use the following program:

```
10   LET N = 1
15   FOR I = 1 TO 10
20   PRINT N
30   LET N = N + 1
40   NEXT I
50   PRINT "THIS PROGRAM IS COMPLETE."
60   END
```

When the computer is given a RUN command, the program would print out the numbers 1,2, . . . , 10 and the statement

```
                THIS PROGRAM IS COMPLETE.
```

Let's analyze this program more carefully.

	Program	Description	Location N	Location I	Output
10	Let $N = 1$	Sets $N = 1$	1	unknown	
15	FOR $I = 1$ TO 10	Sets $I = 1$	1	1	
20	PRINT N	Prints "1"	1	1	1
30	LET $N = N + 1$	Sets $N = 2$	2	1	
40	NEXT I	Increments I and then begins back at the line following the FOR command	2	2	
20	PRINT N	Prints "2"	2	2	2
30	LET $N = N + 1$	Sets $N = 3$	3	2	
40	NEXT I	Increment I. It will continue to repeat until I has reached 10. Then it continues with the rest of the program.	3	3	

Considering our original problem of adding the first 100 numbers, we see that we will need to use the FOR and NEXT commands

```
10   LET A = 0
20   FOR I = 1 TO 100
30   LET A = A + 1
40   NEXT I
```

FIGURE 3.26 Flow chart illustrating the FOR-NEXT commands

This arrow shows the loop. When the NEXT command is reached, two things happen:
1. The computer transfers back to the FOR command.
2. The value of I is changed so that it is the *next* value. When I reaches the last value named (10 in this example), the NEXT command is skipped and the computer completes the rest of the program.

```
50   PRINT A
60   END

RUN
5050

READY
```

We can easily modify this program so that it will find the sum of the first *N* counting numbers.

```
 5   PRINT
10   LET A = 0
20   PRINT "I WILL FIND THE SUM OF THE FIRST N ";
25   PRINT "COUNTING NUMBERS."
28   PRINT
30   PRINT "WHAT IS THE VALUE OF N";
35   INPUT N
40   FOR I = 1 TO N
50   LET A = A + I
60   NEXT I
70   PRINT "THE SUM IS:"; A
80   END
```

Can you predict what the computer will do if you type RUN?

IF-THEN or IF-GOTO Commands

Sometimes we will want to transfer from one point of the program to another only if certain conditions are met. The IF-THEN or IF-GOTO command gives the computer its basic decision-making ability. Consider the following example:

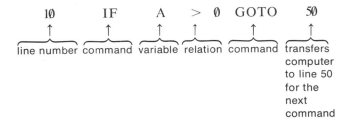

```
10        IF     A    >   0   GOTO    50
```
line number   command   variable   relation   command   transfers computer to line 50 for the next command

TABLE 3.3 Relation Symbols That Can Be Used with IF-THEN Commands

Math Symbol	BASIC Symbol
=	=
>	>
≥	> =
<	<
≤	< =
≠	<>

Some different relations that can be used are shown in Table 3.3.

See if you can understand the following program:

```
10   PRINT "INPUT TWO NUMBERS, AND I WILL TELL YOU WHETHER THEIR ";
20   PRINT "PRODUCT IS "
30   PRINT "POSITIVE, NEGATIVE, OR ZERO."
40   PRINT "WHAT ARE THE NUMBERS";
50   INPUT A, B
60   PRINT
70   PRINT "THE PRODUCT IS ";
80   IF A*B = 0 GOTO 150
90   IF A*B > 0 GOTO 200
```

On a flow chart conditional transfers are symbolized by a diamond-shaped box.

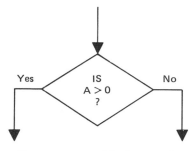

One arrow into the diamond-shaped box

Two arrows out of the box; one is labeled "Yes" and the other is labeled "No." The condition is written inside the box.

See if you can follow the flow chart in Figure 3.27.

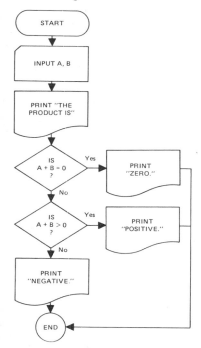

FIGURE 3.27 Flow chart for the example illustrating the IF-THEN or IF-GOTO command

```
100   IF A*B < 0 GOTO 250
150   PRINT "ZERO."
160   GOTO 300
200   PRINT "POSITIVE."
210   GOTO 300
250   PRINT "NEGATIVE."
300   END
```

Here are some sample outputs for this program:

```
RUN
INPUT TWO NUMBERS, AND I WILL TELL YOU WHETHER THEIR PRODUCT IS
POSITIVE, NEGATIVE, OR ZERO.
WHAT ARE THE NUMBERS? 4,-5

THE PRODUCT IS NEGATIVE.

READY

RUN
INPUT TWO NUMBERS, AND I WILL TELL YOU WHETHER THEIR PRODUCT IS
POSITIVE, NEGATIVE, OR ZERO.
WHAT ARE THE NUMBERS? -4,-5

THE PRODUCT IS POSITIVE.

READY

RUN
INPUT TWO NUMBERS, AND I WILL TELL YOU WHETHER THEIR PRODUCT IS
POSITIVE, NEGATIVE, OR ZERO.
WHAT ARE THE NUMBERS? 0,123

THE PRODUCT IS ZERO.

READY
```

GOSUB Command

Sometimes it is convenient or necessary to transfer to a different group of line numbers. To do this a GOSUB (pronounced "go sub") command can be used. When the computer receives a RETURN command, it will return to the place where the GOSUB command was given and then continue in sequence. For example,

```
10    PRINT "THIS ";
20    PRINT "EXAMPLE ";
30    GOSUB 100
40    PRINT "DID YOU DO IT ";
50    PRINT "CORRECTLY?"
60    GOTO 150
100   PRINT "ILLUSTRATES";
110   PRINT " A NEW ";
120   PRINT "COMMAND."
130   RETURN
150   END
```

On receiving a RUN command, the computer will execute the following sequence of line numbers:

10,20,30   100,110,120,130   40,50,60   150

transfers to subroutine   return to original point in program   unconditional transfer to new location   end of program

After being given a RUN command, the computer will type:

```
THIS EXAMPLE ILLUSTRATES A NEW COMMAND.
DID YOU DO IT CORRECTLY?
```

"Only once in every generation is there a computer that can write poetry like this."

® DATAMATION ®

COMPUTER POETRY

This poem is a translation of a computer-written German poem. It was done at Stuttgart's Technical College from a program written by Professor Max Bense. Professor Bense has been trying to discover if it is possible to mathematically formulate problems in esthetics.

THE POEM

The joyful dreams rain down
The heart kisses the blade of grass
The green diverts the tender lover
Far away is a melancholy vastness
The foxes are sleeping peacefully
The dream caresses the lights
Dreamy sleep wins an earth
Grace freezes where this glow dallies
Magically the languishing shepherd dances.

These four computer-written poems were produced by Margaret Masterman and Robin McKinnon Wood using human-machine interaction at the Cambridge Language Research Unit. The program is written in the TRAC language, and the schematic for the third poem is shown.

1 POEM eons deep in the
ice
I paint all time in
a whorl
bang the sludge
has cracked

2 POEM eons deep in the
ice
I see gelled time
in a whorl
pffftt the sludge
has cracked

3 POEM all green in the
leaves
I smell dark
pools in the trees
crash the moon
has fled

4 POEM all white in the
buds
I flash snow
peaks in the
spring
bang the sun has
fogged

Table 3.4 provides us with a summary of the BASIC commands we've considered in this chapter.

TABLE 3.4 Summary of BASIC Commands

	Command	Example	Explanation
This is a direct command. →	SCR	SCR	Erases or scratches the program in the core memory.
This is a direct command. →	LIST	LIST	Lists the entire program.
		LIST 30	Lists line 30.
This is a direct command. Recall that "compile" means that it will first change the program to machine language.		LIST 30,50	Lists lines 30 to 50, inclusive.
	RUN	RUN	Compiles and runs the program.
	STOP	S	Causes the program to stop.
	CTL/C	Keys C and CONTROL pushed simultaneously	Causes the program to stop.
These are direct commands. →	PRINT	PRINT 4 + 4	Prints out the value of the specified arguments, which may be variables, text, or blank (to generate line feed).
The rest are not direct commands but program commands.	END	END	Last statement in *every* program; signals the completing of the program.
	LET	LET A = 5	Assigns the variable "A" the value "5".
	INPUT	INPUT A,B	Causes a typeout of a "?" to the operator and waits for him or her to supply the variables —in this case, A and B.
	GOTO	GOTO 10	Transfers control to line 10, and the program will continue from there.
	FOR	FOR I = 1 TO 10	Used to implement loops. The variable I is set to equal 1. From this point the loop cycle is completed, after which 1 is incremented to 2 and the cycle is completed again. This continues until it reaches 10.
	NEXT	NEXT I	Used to tell the computer to return to the FOR command and execute the loop again until it reaches the last number. Then the program continues.
	IF-THEN	IF A = 2 THEN 50	These commands are the same. If $A = 2$, then the program will be transferred to line number 50; if not, it will continue in the regular sequence.
	IF-GOTO	IF A = 2 GOTO 50	

PROBLEM SET 3.6

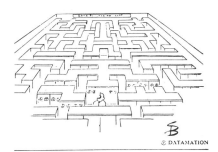

"LET ME OUT!"

A Problems

1. What is a loop?

2. What procedure can be used to terminate a loop?

3. Consider the following program:

```
10   PRINT "H";
20   PRINT "I";
30   PRINT " ";
40   GOTO 70
50   PRINT "END";
60   GOTO 110
70   PRINT "F";
80   PRINT "R";
90   PRINT "I";
100  GOTO 50
110  PRINT ".";
120  END
```

What will the computer type if you give a RUN command?

4. Consider the following program:

```
10   PRINT "T";
20   PRINT "H";
30   FOR I = 1 TO 2
35   PRINT "IS ";
40   NEXT I
50   PRINT "C";
60   PRINT "O";
70   FOR I = 1 TO 2
75   PRINT "R";
80   NEXT I
90   PRINT "ECT";
100  PRINT ".";
110  END
```

Look at line 30 carefully to make sure you can distinguish between the I and the 1: this line reads "eye" equals one to two.

What will the computer type if you give a RUN command?

5. The program in Problem 3 is inefficient from a programmer's standpoint, since the same output can be generated by using only two commands. Refine the program in Problem 3 by writing a new program that is more efficient.

By "refine" we mean write a better program that will generate the same output.

6. The program in Problem 4 is inefficient from a programmer's standpoint, since the same output can be generated by using only two commands. Refine the program in Problem 4 by writing a new program that is more efficient.

7. Suppose you are given the following program.

```
10   LET A = 7
20   IF A > 100 GOTO 80
30   IF A < 1 GOTO 60
40   PRINT "A IS BETWEEN 1 AND 100 INCLUSIVE."
50   GOTO 90
60   PRINT "A IS LESS THAN 1."
70   GOTO 90
80   PRINT "A IS GREATER THAN 100."
90   END
```

What will the computer type if you give a RUN command?

8. Repeat Problem 7 for $A = 107$.

9. Repeat Problem 7 for $A = -10$.

Don't forget the case where $A = B$.

10. Write a program that will compare two numbers, A and B, and then print the larger number.

B Problems

11. Suppose you are given the following program:

```
10   LET A = 0
20   FOR I = 1 to 5
30   LET A = A + I
40   NEXT I
50   PRINT A
60   END
```

What will the computer type if you give a RUN command?

12. What would be the difference in output (if any) if lines 40 and 50 of Problem 11 were interchanged?

13. What would be the difference in output (if any) if lines 10 and 20 of Problem 11 were interchanged?

14. What would be the difference in output (if any) if lines 50 and 60 of Problem 11 were interchanged?

15. You have fed the following program into a computer:

```
10   LET A = 1
20   FOR I = 1 TO 6
30   LET A = A*I
40   NEXT I
50   PRINT A
60   END
```

What will the computer type if you give a RUN command?

16. What would be the difference in output (if any) if lines 40 and 50 of Problem 15 were interchanged?

17. What would be the difference in output (if any) if lines 10 and 20 of Problem 15 were interchanged?

18. What would be the difference in output (if any) if lines 50 and 60 of Problem 15 were interchanged?

19. Suppose you are given the following program.

```
10   LET A = 3
20   IF A ≥ 10 GOTO 70
30   FOR I = 1 to A
40   PRINT I
50   NEXT I
60   GOTO 80
70   PRINT "THIS PROBLEM IS TOO HARD."
80   END
```

What will the computer type if you give a RUN command?

20. Repeat Problem 19 where $A = 13$.

21. The old song "100 Bottles of Beer on the Wall" is quite tedious because of the lyrics. Write a computer program that will generate an output:

```
100   BOTTLES OF BEER ON THE WALL.
 99   BOTTLES OF BEER ON THE WALL.
 98   BOTTLES OF BEER ON THE WALL.
                    ⋮
  2   BOTTLES OF BEER ON THE WALL.
  1   BOTTLE OF BEER ON THE WALL.
      THAT'S ALL FOLKS!
```

22. Write a BASIC program that will print out the first one hundred numbers backwards:

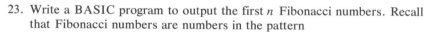

$$100, 99, 98, 97, \ldots 3, 2, 1.$$

23. Write a BASIC program to output the first n Fibonacci numbers. Recall that Fibonacci numbers are numbers in the pattern

$$1, 1, 2, 3, 5, 8, 13, 21, \ldots .$$

In Problem 31 on page 175 you were asked for a flow chart for this problem.

24. What will the computer type if you RUN the following program?

```
 10   GOSUB 200
 20   PRINT "TO ";
 30   GOTO 60
 40   PRINT "TOO ";
 50   GOTO 140
 60   PRINT "THE SEVENTH ";
 70   PRINT "POWER IS ";
 80   FOR I = 1 TO 6
 90   GOSUB 200
100   PRINT "TIMES";
110   NEXT I
115   PRINT "TWO"
120   PRINT "AND THIS IS ";
130   GOTO 40
140   PRINT "MUCH.";
150   GOTO 300
200   PRINT "TWO ";
210   RETURN
300   END
```

This program assumes that the input value for N is an odd number.

Recall the quadratic formula:
If $ax^2 + bx + c = 0$, $a \neq 0$, then

$$x = \frac{-b \pm \sqrt{b^2 - 4ac}}{2a}$$

A flow chart is shown in Figure 3.28.

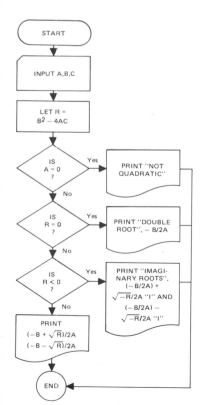

FIGURE 3.28 Flow chart for Problem 29

25. *Computer Problem.* We wish to find a pattern for the sum of the odd numbers $1 + 3 + 5 + \cdots + N = ?$ where N is an odd number. Use the following program to help you find a replacement for the question mark.

```
10   PRINT "WHAT IS N";
20   INPUT N
30   IF N < = 5 GOTO 100
40   LET S = 0
50   FOR I = 1 TO (N + 1)/2
60   LET S = S + (2*I − 1)
70   NEXT I
80   PRINT "THE SUM OF THE FIRST N ODD NUMBERS IS",S
90   END
```

26. *Computer Problem.* We wish to find a pattern for the sum of the even numbers $2 + 4 + 6 + \cdots + N = ?$ where N is an even number. Write a BASIC program similar to the one given in Problem 25 to help you find a replacement for the question mark.

Mind Bogglers

27. *Computer Problem.* Write a program similar to the one in Problem 25 for finding the following sum:

$$1^2 + 3^2 + 5^2 + 7^2 + \cdots + (2n − 1)^2$$

Using this program, make a conjecture about this sum for any value of n.

28. *Computer Problem.* Write a program similar to the one in Problem 25 for finding the following sum:

$$2^2 + 4^2 + 6^2 + 8^2 + \cdots + (2n)^2$$

Using this program, make a conjecture about this sum for any value of n.

29. *Computer Problem.* Write a BASIC program to find the roots of the quadratic equation $ax^2 + bx + c = 0$. (In BASIC the command PRINT SQR(2) is used to find the square root of the number 2.)

Problem for Individual Study

30. *Computer Problem.* Write a BASIC program for obtaining the pattern

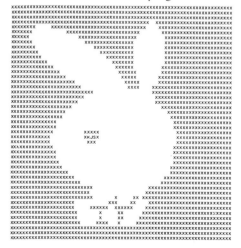

COMPUTERS AND ARCHITECTURE

When architectural drawings require a large amount of repetition, computers can be used quite effectively. This computer drawing shows the furniture, telephone, and electrical layouts for a floor of the Sears Tower in Chicago. Drawings were made for four separate plans for each of the building's 110 stories.

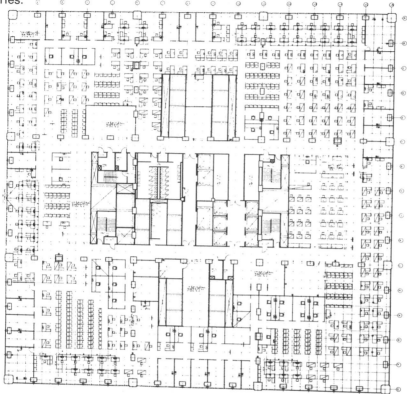

3.7 SUMMARY AND REVIEW

CHAPTER OUTLINE

   I. History of Computing Devices
      A. Ancient methods of calculating
         1. Finger multiplication
         2. Abacus
      B. The dawning of computers
         1. Leibniz
         2. Babbage
         3. Jacquard
         4. Mark I, ENIAC
         5. Pocket calculators

   C. Modern computers
       1. First-generation computers
       2. Second-generation computers
       3. Third-generation computers
       4. Fourth-generation computers

 II. The Pocket Calculator
   A. Types of calculators
       1. Four-function
       2. Four-function with memory
       3. Scientific
       4. Special purpose
       5. Programmable
   B. Calculator logic
       1. Arithmetic
       2. Algebraic
       3. RPN
   C. Elementary operations

III. The Electronic Brain
   A. Types of electronic computers
       1. Analog
       2. Digital
   B. The main components of a computer
       1. Input
           a. punched card
           b. paper tape
           c. magnetic tape
       2. Storage or memory
           a. word
           b. address
           c. bit
           d. types of storage
       3. Accumulator
           a. arithmetic performed here
           b. floating-point and fixed-point representation of a number
       4. Control
       5. Output

IV. BASIC Programming
   A. What is programming?
   B. Flow charts
   C. Types of languages
       1. Machine language—the computer's native language, which is in binary
           code
       2. Symbolic language—uses mnemonic codes rather than numerical
           ones
       3. Algorithmic language—BASIC is an example of an algorithmic
           language

D. Computer Commands
   1. Writing a program (see Table 3.4 on page 192 for a summary)
   2. Using established format:

   line number   command   variable or numerical expression

E. Loops
   1. GOTO command
   2. FOR-NEXT commands
   3. IF-THEN commands
   4. GOSUB command

REVIEW PROBLEMS

1. Briefly discuss some of the events leading up to the invention of the computer.

2. Use a calculator to find

   a. $63 + 28.7 - 6.8 \times 7$

   b. $\dfrac{41.3 + 6.85}{16}$

3. *Calculator Problem.* To find out what to do when the market is dropping, calculate

$$7700 + 7.01 \times 5$$

   and turn your calculator over to read the answer.

4. *Calculator Problem.* The child prodigy Jacques Inaudi (1867–1923) instantly gave the correct answer to the following problem by doing all the arithmetic mentally:

$$\frac{4811^2 - 1}{6}$$

   Use a calculator to find the answer.

5. What do we mean when we speak of a third-generation computer?

6. Name the five main components of a computer, and briefly explain the function of each.

7. What is a floating-point number? Why are floating-point numbers sometimes needed in working with computers?

8. Express in floating-point notation:

   a. 576,000,000,000   b. .000002

9. Express in fixed-point notation:

   a. $3.8E + 8$ b. $5.74E - 8$

10. Briefly explain what we mean when we speak of computer ''programming.''

11. What are the three *types* of programmed languages? Briefly describe each.

12. Write a BASIC language program to ask the value of x, and then compute $18x^2 + 10$. Include a flow chart.

13. Write 5*(X + 2)↑2 in ordinary mathematical notation.

14. Write $3(x + 4)(2x - 7)^2$ in BASIC notation.

15. Draw a flow chart to compute $15x^3 - 3x^2 + 7$ in BASIC.

16. Write a BASIC program to compute $15x^3 - 3x^2 + 7$.

17. Suppose you are given the following BASIC program:

```
 10   PRINT "TH";
 20   PRINT "E";
 30   PRINT " ";
 40   PRINT "M";
 50   FOR I = 1 TO 2
 60   PRINT "ISS";
 70   NEXT I
 80   GOTO 100
 90   PRINT "H";
100   PRINT "IPP";
110   GOTO 130
120   PRINT "Y ";
130   PRINT "I ";
140   PRINT "I";
150   PRINT "S ";
160   PRINT "WET";
170   GOTO 190
180   PRINT "NOODLE";
190   PRINT ".";
200   END
```

What will the computer type if you give a RUN command?

18. The program given in Problem 17 can be shortened considerably. Refine it.

19. What will the computer type when given a RUN command for the following program?

```
 10   PRINT "NOT ";
 20   GOSUB 130
 30   PRINT ", ";
 40   PRINT "BUT THE ";
 50   GOSUB 130
 60   PRINT " OF ";
 70   GOSUB 130
 80   PRINT ", "
 90   PRINT " IS THE ";
```

```
100   GOTO 150
110   PRINT "KNOWLEDGE";
120   GOTO 170
130   PRINT "IGNORANCE ";
140   RETURN
150   PRINT "DEATH OF ";
160   GOTO 110
170   PRINT "."
180   END
```

20. Write a BASIC program to calculate the areas of various circles; the radii are the input values. (Use 3.1416 for the computer approximation of π.)

The formula for the area of a circle is $A = \pi r^2$.

1000

1003: Leif Ericson crosses the Atlantic to "Vinland"

1020: Al-Karkhî—algebra but everything is written out in words, even the names of numbers. Later critics will consider this refusal to abbreviate to be a step backward on the road toward symbolical algebra.

1025

1028: School of Chartres

1042: Edward the Confessor

1050

This is an age of translation, preservation, and compilation. It will be many centuries before mathematics again attains the level it once had at the hands of such men as

Eudoxus Euclid Archimedes Apollonius Diophantos

But, at its modest level, mathematics begins to show an astonishing variety of subject matter, drawn from Greek, Hindu, and Islamic sources, widely separated in place and time. This variety opens the door to future progress.

1066: Norman Conquest

1075

1096: First Crusade

1100: Paper is used for the first time in Europe; it came to Sicily and Spain from the Near East

1100

1110: Omar Khayyam—cubic equations, Pascal's triangle

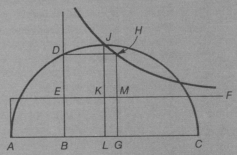

1125

1146: Second Crusade

1150

Construction used by Omar Khayyam to solve the cubic equation $x^3 + b^2 x + a^3 = cx^2$

1175

1186: The Domesday Book is published; it is a census of all the manors of England and their tax value—ordered by William the Conqueror

4

THE NATURE OF NUMBER

In most sciences one generation tears down what another has built
and what one has established another undoes.
In mathematics alone each generation builds a new story
to the old structure.

Hermann Hankel

4.1 EARLY NUMERATION SYSTEMS

A numeration system is a set of basic symbols and some rules for making other symbols from them, the purpose of the whole game being the identification of numbers. The invention of a precise and "workable" system is one of the greatest inventions of humanity. It is certainly equal to the invention of the alphabet, which enabled humans to carry the knowledge of one generation to the next. It is simple for us to use the symbol 17 to represent this many objects:

XXXXX XXXXX

XXXXX XX

However, it took us centuries to arrive at this stage of symbolic representation. It wasn't the first and probably won't be the last numeration system to be developed. Here are some of the ways that 17 has been written:

Tally	⋈ |||| |||| ||
Egyptian	∩ |||||||
Roman	XVII
Mayan	≟
Linguistic	seventeen
	siebzehn
	dix-sept

The concept represented by each of these symbols is the same, but the symbols differ. The concept or idea of "seventeenness" is called a *number*; the symbol used to represent the concept is called a *numeral*. The difference between *number* and *numeral* is analogous to the difference between a person and his or her name, as is illustrated by the *Peanuts* cartoon.

Two of the earliest civilizations known to use numerals were the Egyptian and Babylonian. We shall examine these systems for two reasons. First, it will help us more fully to understand our own numeration (decimal) system; second, it will help us to see how the ideas of these other systems have been incorporated into our system.

THE EGYPTIAN NUMERATION SYSTEM

Perhaps the earliest type of written numeration system developed was that of a *simple grouping system*. The Egyptians used such a system by the time of the first dynasty, around 2850 B.C. The symbols of the Egyptian system were part of their hieroglyphics and are shown in Table 4.1.

Any number is expressed by using these symbols additively; each symbol is repeated the required number of times, but with no more than nine repetitions.

TABLE 4.1   Egyptian Hieroglyphic
Numerals

Our Numeral (decimal)	Egyptian Numeral	Descriptive Name
1	|	Stroke
10	∩	Heel bone
100	ͻ	Scroll
1000	ℒ	Lotus flower
10,000	⌐	Pointing finger
100,000	⌒	Polliwog
1,000,000	𝕏	Astonished man

EXAMPLE: $12{,}345$ = ⌐ ℒ ℒ ͻͻ∩∩ / ͻ ∩∩ / \|\|\|\|\|

The position of the individual symbols is not important. That is,

ⅅℒℒ ͻͻ∩∩ ͻ ∩∩ \|\|\|\|\| = ℒℒ ͻͻ ͻ ⅅ ∩∩ ∩∩ \|\|\|\|\|

= \|\|\|\|\| ⅅ ℒ ͻͻ ∩∩∩∩ ℒ ͻ

The Egyptians had a simple repetitive-type arithmetic. Addition and subtraction were performed by repeating the symbols and by regrouping:

EXAMPLES:

1.   245   ͻͻ ∩∩ / ∩∩ \|\|\|\|\|

     $+\ 457$ $+$ ͻͻ∩∩∩∩\|\|\|\|\| / ͻͻ ∩∩ \|\|

                ͻͻͻ∩∩∩∩∩ / ͻͻͻ∩∩∩∩∩ \|\|\|\|\|\|\|\|\|\|\|\|

     Regroup: ͻͻͻ∩∩∩∩∩∩ / ͻͻͻ ∩∩∩∩∩ \|\|

     Regroup again: ͻͻͻͻ / ͻͻͻ \|\|

2.   142   ͻ ∩∩ / ∩∩ \|\|

     $-\ 67$ $-$ ∩∩∩ / ∩∩∩ \|\|\|\|\|\|\|

     Regroup: ∩̶∩̶∩̶∩∩∩ / ∩̶∩̶∩̶∩∩ ǂǂǂǂǂ\|\|\|\|\|

                $-$ ∩̶∩̶∩̶ / ∩̶∩̶∩̶ ǂǂǂǂǂ

                ∩∩∩∩ / ∩∩∩ \|\|\|\|\|

Multiplication and division were performed by successions of additions that did not require the memorization of a multiplication table and could easily be done on an abacus-type device (see Problem 33 of Problem Set 4.1).

𝕳istorical 𝕹ote

There are two important sources for our information about the Egyptian numeration system. The first is the Moscow papyrus, which was written in about 1850 B.C. and contains 25 mathematical problems. The second is the Rhind papyrus, which has been dated to approximately 1650 B.C. It contains 85 problems and was a type of mathematical handbook for the Egyptians. From these two sources, we conclude that most of the problems were of a practical nature (although some were of a theoretical nature).

Rhind papyrus. This is a papyrus roll about 30 cm high and 5.5 m long. It was bought in 1858 in a Nile resort town by an antiquary, Henry Rhind. It is also sometimes called the Ahmes papyrus because it was written by Ahmes in about 1650 B.C.

The Egyptians also used unit fractions ($1/n$) in their computations. These were indicated by placing the symbol $\bigcirc$ over the numeral for the denominator. Thus

$$\overset{\frown}{\text{III}} = \frac{1}{3}$$

$$\overline{||||} = \frac{1}{4}$$

$$\overline{\cap} = \frac{1}{10}$$

$$\overline{\cap\cap\cap}_{|||} = \frac{1}{33}$$

The fraction $\frac{1}{2}$ was an exception:

$$\bigcirc_{||} \quad \text{or} \quad \boxed{} = \frac{1}{2}$$

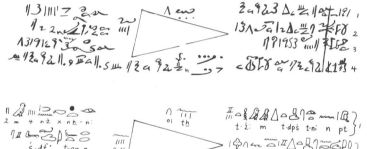

This is a portion of the Rhind papyrus showing the area of a triangle.

Historical Note

From the development of the arithmetical calculations of finding two-thirds of any number, the Egyptians were able to determine π with surprising accuracy, to solve first- and second-degree equations, and to sum arithmetic and geometric progressions. They also found the formula for the volume of a frustum of a square pyramid, as well as the formula for the surface area of a hemisphere. They were able to do all this with very few mathematical tools.

Fractions that were not unit fractions were represented as sums of unit fractions. For example,

$$\frac{2}{7} \text{ would be expressed as } \frac{1}{4} + \frac{1}{28}$$

(Repetitions were not allowed; that is, $^2/_7$ would not be written as $^1/_7 + ^1/_7$ but could be written as $^1/_{28} + ^1/_4$.)

$\dfrac{3}{5}$ would be expressed as $\dfrac{1}{2} + \dfrac{1}{10}$

$\dfrac{2}{99}$ would be expressed as $\dfrac{1}{66} + \dfrac{1}{198}$

The fraction $^2/_3$ was the only nonunit fraction not written as a sum:

 or ⌇ = $\dfrac{2}{3}$

The Rhind papyrus includes a table for all such decompositions for odd denominators from 5 to 101. This papyrus also uses a symbolism for addition and subtraction. Addition was indicated by a pair of legs walking to the left, and subtraction was a pair of legs walking to the right.

THE BABYLONIAN NUMERATION SYSTEM

The Babylonian numeration system differed from the Egyptian in several respects. Whereas the Egyptian system was a simple grouping system, the Babylonians employed a much more useful *positional system*. Since they lacked papyrus, they used mostly clay as a writing medium, and thus the Babylonian cuneiform was much less pictorial than the Egyptian system. They employed only two wedge-shaped characters, which date from 2000 B.C. and are shown in Table 4.2.

Notice from the table that, for numbers 1 through 59, the system is repetitive. However, unlike in the Egyptian system, the position of the symbols was important. The ⟨ *must* appear to the left of any ▼s to represent numbers smaller than 60. For numbers larger than 60, the symbols ⟨ and ▼ are to the left of ⟨, and any symbols to the left of the ⟨ have a value 60 times their original value. That is,

$$\text{▼⟨⟨▼▼▼}$$

means $(1 \times 60) + 35$.

EXAMPLES:

1. ▼▼▼⟨ ▼▼▼ $= (3 \times 60) + 59 = 239$

2. ▼▼▼ ⟨▼ $= (5 \times 60) + 11 = 311$

3. ⟨⟨▼▼▼⟨▼▼▼ $= (23 \times 60) + 16 = 1396$

TABLE 4.2 Babylonian Cuneiform Numerals

Our Numeral (Decimal)	Babylonian Numeral
1	▼
2	▼▼
9	▼▼▼▼▼▼▼▼▼
10	⟨
59	⟨⟨⟨⟨▼▼▼▼▼ ⟨⟨ ▼▼▼▼

Sumerian Clay Tablet

This system is called a *sexagesimal system* and uses the principle of position. However, the system is not fully positional, since only numbers larger than 60 use the position principle; numbers within each basic 60-group are written by a simple grouping system. A true positional sexagesimal system would require 60 different symbols.

The Babylonians carried their positional system a step further. If any numerals were to the left of the second 60-group, they had the value of 60×60 or 60^2. Thus

$$= (2 \times 60^2) + (45 \times 60) + 24$$
$$= 7200 + 2700 + 24$$
$$= 9924$$

The Babylonians also made use of a subtractive symbol, $\ulcorner$. That is, 38 could be written

or

EXAMPLES:

1. Change ▼▼▼◁◁⌐ to a decimal numeral. This number means

$$(3 \times 60) + 20 - 1 = 199.$$

2. Change

to a decimal numeral. This number means $(2 \times 60^2) + (23 \times 60) + 16$

$$= 7200 + 1380 + 16$$
$$= 8596$$

3. Change 1234 to Babylonian numerals. Now, $60 \times 20 = 1200$, so we write $(20 \times 60) + 34 = 1234$. That is,

(Note an ambiguity: $(20 \times 60) + 34$ or   54?)

4. Change 4571 to Babylonian numerals.

$$\begin{aligned}
1 \times 60^2 &= 3600 \\
16 \times 60 &= 960 \\
11 \times 1 &= 11 \\
\hline
&4571
\end{aligned}$$

Thus,

is the way 4571 was written.

Although this positional numeration system was in many ways superior to the Egyptian system, it suffered from the lack of a zero or placeholder symbol. For example, how is the number 60 represented? Does ▼▼ mean 2 or 61? In Example 3 the value of the number had to be found from the context. (Scholars tell us that such ambiguity can be resolved only by a careful study of the context.) However, in later Babylonia, around 300 B.C., records show that there is a zero symbol, ⊻, and this idea was later used by the Hindus.

Arithmetic with the Babylonian numerals is quite simple, since there are only two symbols. A study of the Babylonian arithmetic will be left for the student (see Problems 21–25 of Problem Set 4.1).

PROBLEM SET 4.1

A Problems

1. Explain the difference between *number* and *numeral*. Give examples of each.

2. Discuss the similarities and differences of:
   a. a simple grouping system.
   b. a positional system.
   Give examples of each.

3. a. Does ⌡ ∩∩ represent the same number as ∩∩⌡ ?
   b. Do ▼◁◁, ◁▼◁, and ◁◁▼ represent the same number?
   Explain your answers.

4. Write 258 and 852:
   a. in Egyptian hieroglyphics.
   b. in Babylonian cuneiform symbols.

5. Write the following numbers in Egyptian hieroglyphics.
   a. 47   b. 75   c. 521   d. 1976   e. 5492

6. Write the numbers given in Problem 5 using the Babylonian numeration system.

7. What do you regard as the shortcomings and contributions of the Egyptian numeration system?

8. What do you regard as the shortcomings and contributions of the Babylonian numeration system?

Write the numerals in Problems 9–15 in decimal numerals. (Decimal numerals refer to the usual system. Thus, 5, 248, and ⅓ are considered decimal numerals.)

9. a. ⌡ 99 ∩∩∩ ∩∩ IIIIII   b. ◯ 99

10. a. ⟐ b.  ⌇∩

11. a.  ⊲⊲⊲▼▼▼▼ b.  ▼▼▼▼⊲⊲⊲▼

12. a.  ⊲⊲⊲ / ▼▼▼ ⊲⊲ b.  ⊲⊲▼▼▼▼

13. a.  ϙ∩∩∩𝗅𝗅𝗅 ⌐ b.  ⬭ / ϙ

14. a.  𝗅𝗅∩ b.  ∩∩𝗅𝗅𝗅𝗅𝗅 ⌐

15. a.  ▼▼⊲⊲⊲⊲▼▼ b.  ▼⊲⊲⊲▼⊲⊲▼

16. Place a check mark in the appropriate columns to indicate the presence of the designated property.

	Egyptian	Babylonian
Grouping system		
Positional		
Repetitive		
Additive		
Subtractive		
Zero symbol		

> *In the same vein as Eden's fall is the story of a woman who decided she wanted to write to a prisoner in a nearby state penitentiary. However, she was puzzled over how to address him, since she knew him only by a string of numerals. She solved her dilemma by beginning her letter: "Dear 5944930, May I call you 594?"*

B Problems

Perform the indicated operations in Problems 17–24.

17.  𝒳 ∩∩ ∩∩ 𝗅𝗅
 + ϙϙ ∩∩∩∩ ∩∩∩ 𝗅𝗅𝗅𝗅𝗅𝗅

18. 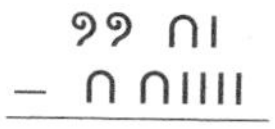 ϙϙ ∩𝗅
 − ∩ ∩𝗅𝗅𝗅𝗅𝗅

19. 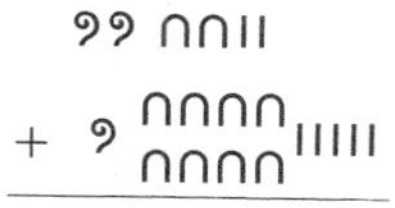 ϙϙ ∩∩𝗅𝗅
 + ϙ ∩∩∩∩ ∩∩∩∩ 𝗅𝗅𝗅𝗅𝗅

20. 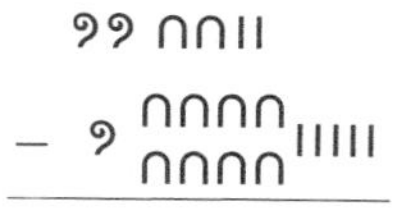 ϙϙ ∩∩𝗅𝗅
 − ϙ ∩∩∩∩ ∩∩∩∩ 𝗅𝗅𝗅𝗅𝗅

21. 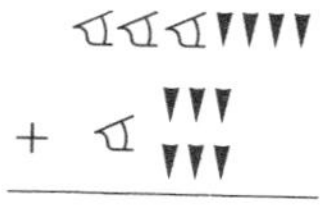 ⊲⊲⊲▼▼▼▼
 + ⊲ ▼▼▼ ▼▼▼

22. 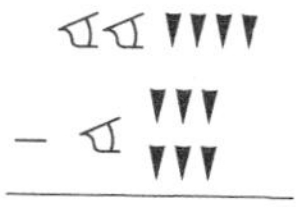 ⊲⊲ ▼▼▼▼
 − ⊲ ▼▼▼ ▼▼▼

23.   ʈ◁ ◁ʈʈʈʈ   24.   ʈ◁ʈʈ
    + ◁◁◁ʈʈʈʈ   − ◁◁◁ʈʈʈʈ
       ◁ ʈʈʈʈ   ◁ ʈʈʈ

In 1977 an Ohio man tried to change his name legally to

1069.

He said he picked 10 as his first name because of its wholeness and 69 as his last name because of its ever-present nature. What special qualities can you find about this number?

25. Discuss addition and subtraction for Babylonian numerals. Show examples.

26. One of the problems found on the Rhind papyrus is sometimes translated as follows:
    In each of 7 houses are 7 cats;
    Each cat kills 7 mice;
    Each mouse would have eaten 7 ears of spelt (wheat);
    Each ear of spelt would have produced 7 kehat (half a peck) of grain.
    Query: How much grain is saved by the 7 houses' cats?
    Can you provide an answer to the question?

27. Problem 26 taken from the Rhind papyrus, reminds us of an 18th-century Mother Goose rhyme:
    As I was going to St. Ives
    I met a man with seven wives.
    Every wife had seven sacks,
    Every sack had seven cats,
    Every cat had seven kits.
    Kits, cats, sacks, and wives.
    How many were there going to St. Ives?
    Read this rhyme very carefully, and then answer the question.

Mind Bogglers

28. *Calculator Problem.* Another problem on the Rhind papyrus asks: "A quantity and its two thirds and its half and its one seventh together make 33. Find the quantity." The answer given on the papyrus is

$$14 + \frac{1}{4} + \frac{1}{56} + \frac{1}{97} + \frac{1}{194} + \frac{1}{388} + \frac{1}{679} + \frac{1}{776}$$

   a. Use a calculator to approximate the answer given on the papyrus in decimal form.
   b. Is the answer given on the papyrus correct? Find the exact answer in fractional form.

A unit fraction (a fraction with a numerator 1) is sometimes called an Egyptian fraction. On page 207 we said that the Egyptians expressed their fractions as a sum of distinct (different) unit fractions. How might the Egyptians have written the following fractions?

29. $3/4$ 30. $47/60$ 31. $67/120$ 32. $7/17$

Problems for Individual Study

33. Write a paper discussing the Egyptian method of multiplication.
    References: Newman, James, *The World of Mathematics,* Vol. I, pp. 170–178 (New York: Simon and Schuster, 1956).
        Eves, Howard, *Introduction to the History of Mathematics* (3rd ed.) (New York: Holt, Rinehart, and Winston, 1969).

34. Write a short paper discussing the development of our numeration system. Refer to places and important dates. You might also like to do research on the Roman, Chinese-Japanese, Greek, and Mayan numeration systems.

35. *Egyptian Fractions.* We mentioned that the Egyptians wrote their fractions as sums of unit fractions. Show that every positive fraction less than 1 can be written as a sum of unit fractions.
   Reference: "Egyptian Fractions," by Bernhardt Wohlgemuth, *Journal of Recreational Math,* Vol. 5, No. 1, 1972, pp. 55–58.

36. Read Chapters 2 and 3 (pp. 9–47) of *A History of Mathematics* by Carl B. Boyer (New York: Wiley, 1968) and report on the Egyptian and Babylonian numeration systems. Answer Problems 7 and 8 in this problem set as part of your report.

37. *Ancient Computing Methods.* How do you multiply with Roman numerals? What is the scratch system? The lattice method of computation? What changes in our methods of long multiplication and long division are being suggested in some new arithmetic textbooks? How is the abacus used for computation? How are Napier's bones used for multiplication? How did the old computing machines work? How were logarithms invented? Who invented the slide rule?
   Exhibit suggestions: charts of sample computations by ancient methods, ancient number representations such as pebbles, tally sticks, tally marks in sand, Roman number computations, abaci, Napier's bones, old computing devices.
   References: Bergamini, David, *Mathematics,* Chaps, 1–3 (New York: Time, Inc., Life Science Library, 1963).
   Swain, Robert, *Understanding Arithmetic,* Chap. 5 (New York: Holt, Rinehart, and Winston, 1965).

This example of Mayan numerals comes from the Dresden codex. See if you can "decode" any of the numerals.

4.2 HINDU-ARABIC NUMERATION SYSTEM

The numeration system in common use today has ten symbols, the digits 0,1,2,3,4,5,6,7,8, and 9. The selection of ten digits was no doubt a result of our having ten fingers (digits).

The symbols originated in India in about 300 B.C. However, because the early specimens do not contain a zero or use a positional system, this numeration system offered no advantage over other systems then in use in India.

The date of the invention of the zero symbol is not known. The symbol did not originate in India but probably came from the late Babylonian period via the Greek world.

By the year 750 A.D. the zero symbol and the idea of a positional system had been brought to Baghdad and translated into Arabic. We are not certain how these numerals were introduced into Europe, but they likely came via Spain in the 8th century. Gerbert, who later became Pope Sylvester II, studied in Spain and was the first European scholar known to have taught these numerals. Because of their origins, these numerals

FIGURE 4.1 This picture depicts a contest between one man computing with a form of abacus and another man computing with Hindu-Arabic numerals.

are called *Hindu-Arabic numerals*. Since ten basic symbols are used, the Hindu-Arabic numeration system is also called the *decimal numeration system,* from the Latin word *decem*, meaning "ten."

Although we now know that the decimal system is very efficient, its introduction met with considerable controversy. Two opposing factions, the "algorists" and the "abacists," arose. Those favoring the Hindu-Arabic system were called algorists, since the symbols were introduced into Europe in a book called (in Latin) *Liber Algorismi de Numero Indorum,* by the Arab mathematician Al-Khwarizmi. The word *algorismi* is the origin of our word *algorism*. The abacists favored the status quo—using Roman numerals and doing arithmetic on an abacus. The battle between the abacists and the algorists lasted for 400 years (see Figure 4.1). The Roman church exerted great influence in commerce, science, and theology. Roman numerals were easy to write and learn, and addition and subtraction with them were easier than with the "new" Hindu-Arabic numerals. Those not using Roman numerals were criticized for using "heathen" numerals. It seems incredible that our decimal system has been in general use only since about the year 1500.

Let's examine the Hindu-Arabic or decimal numeration system a little more closely.

1. It uses 10 symbols, called digits: 0,1,2,3,4,5,6,7,8, and 9.
2. Larger numbers are expressed in terms of powers of 10.
3. It is positional.
4. It is additive.

Now let's review how we count objects.

/	1
//	2
///	3
////	4
/////	5
//////	6
///////	7
////////	8
udienter///	9
//////////	

At this point we could invent another symbol, as the Egyptians did, or we could reuse the digit symbols by repeating them or by altering their position. That is, 10 will mean 1 group of /////////// . The symbol 0 was invented as a placeholder to show that the 1 here is in a different position from the 1 representing the / .

| ///////// | / | 11 | This means 1 group of ///////// and 1 extra. |

12   This means 1 group and 2 extra.

34   This means 3 groups (let's call each group a "ten") and 4 extra.

In the last instance we have a group of groups, or ten groups (or ten-tens, if you like). Let's call this group of groups a "10 · 10" or "10^2" or a "hundred." We again use position and repeat the symbol 1 with a still different meaning: 100.

Thus, what does 134 represent? It means we have the following:

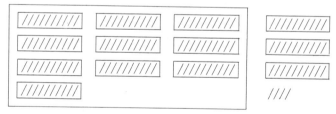

We could denote this more simply by writing;

$$(1 \times 10^2) + (3 \times 10) + 4$$

— These represent the name of the group.

— These represent the number of groups.

This leads us to the name

one hundred, three tens, four ones,

which is read "one hundred thirty-four." A number written in this fashion is in *expanded notation*. As we pointed out earlier, mathematicians often look for patterns; so we notice that the expanded notation for 31,452 suggests a pattern:

$$(3 \times 10^4) + (1 \times 10^3) + (4 \times 10^2) + (5 \times 10) + 2$$

We write $10 = 10^1$ and $1 = 10^0$. Recall that we defined $10^0 = 1$. Then the pattern is complete:

$$(3 \times 10^4) + (1 \times 10^3) + (4 \times 10^2) + (5 \times 10^1) + (2 \times 10^0)$$

A period, called a *decimal point* in the decimal system, is used to separate the fractional parts from the whole parts. We repeat—and generalize—the definition first given on page 33.

Definition of exponent

> DEFINITION: For any nonzero number b and any counting number n,
>
> $$b^n = \underbrace{b \times b \times b \times \cdots \times b}_{n \text{ factors}}$$
>
> $$b^0 = 1$$
>
> $$b^{-n} = \frac{1}{b^n}$$

For example,

$$\frac{1}{10} = 10^{-1}$$

$$\frac{1}{100} = 10^{-2}$$

$$\frac{1}{1000} = 10^{-3}, \text{ since } 10^{-3} = \frac{1}{10^3} = \frac{1}{10 \cdot 10 \cdot 10} = \frac{1}{1000}$$

EXAMPLE: Write 479.352 using expanded notation.

Solution:

$$(4 \times 10^2) + (7 \times 10^1) + (9 \times 10^0) + (3 \times 10^{-1}) + (5 \times 10^{-2}) + (2 \times 10^{-3})$$

Recall that 10 is called the *base*. Notice the pattern of the exponents on the base.

We often fail to see all the great ideas that are incorporated into this system, and it is difficult to understand the problems inherent in developing a numeration system. In the next section, we'll change the focus from the *numerals* used to the kind of *numbers* existing independent of the particular system used. We will, of course, develop the various *number systems* using Hindu-Arabic *numerals*.

To help you realize some of these problems, you are asked in Problem 17 to invent your own numeration system.

PROBLEM SET 4.2

A Problems

1. Let b be any nonzero number and n any counting number. Define:
   a. b^n b. b^0 c. b^{-n}

Write the numbers in Problems 2–7 in decimal notation.

2. a. 10^{-4} b. 5×10^3

3. a. $.01 \times 10^5$ b. 8×10^{-4}

4. a. $(1 \times 10^4) + (0 \times 10^3) + (2 \times 10^2) + (3 \times 10^1) + (4 \times 10^0)$
   b. $(6 \times 10^1) + (5 \times 10^0) + (0 \times 10^{-1}) + (8 \times 10^{-2}) + (9 \times 10^{-3})$

5. a. $(5 \times 10^5) + (2 \times 10^4) + (1 \times 10^3) + (6 \times 10^2) + (5 \times 10^1) + (8 \times 10^0)$
   b. $(6 \times 10^7) + (4 \times 10^3) + (1 \times 10^0)$

6. a. $(5 \times 10^5) + (4 \times 10^2) + (5 \times 10^1) + (7 \times 10^0) + (3 \times 10^{-1}) + (4 \times 10^{-2})$
   b. $(7 \times 10^6) + (3 \times 10^{-2})$

7. $(3 \times 10^3) + (2 \times 10^1) + (8 \times 10^0) + (5 \times 10^{-1}) + (4 \times 10^{-2}) + (6 \times 10^{-3})$
   $+ (2 \times 10^{-4})$

Write the numbers in Problems 8–14 in expanded notation.

8. a. 741   b. 728,407

9. a. 0.06421   b. 27.572

10. a. 521   b. 6245

11. a. 428.31   b. 2,345,681

12. a. 47.03215   b. 100,000.001

13. a. .00000527   b. 5245.5

14. a. 678,000.01   b. 57,285.9361

15. What is the meaning of the 5 in each of these numerals?
    a. 805   b. 508   c. 0.00567

16. Illustrate the meaning of 123 by showing the appropriate groupings.

B Problems

17. Invent your own numeration system. You may use the *ideas* of the Egyptian, Babylonian, and Hindu-Arabic numeration system (such as repetitive, positional, and so on), but don't use the specifics, such as the symbols used or the names for those symbols, of these (or any other) systems.

18. *Galley Multiplication.* A clever device for doing multiplication of large numbers that requires only that we remember basic multiplication facts is

Historical Note

The Hindus often stated their problems poetically, since the problems were frequently solved just for the fun of it. As an example, consider the following problem adapted from an early Hindu writer and quoted in Howard Eves' In Mathematical Circles:

The square root of half the number of bees in a swarm has flown out upon a jessamine bush; $^8/_9$ of the swarm has remained behind; one female bee flies about a male that is buzzing within a lotus flower into which he was allured in the night by its sweet odor, but is now imprisoned in it. Tell me, most enchanting lady, the number of bees.

In connection with his problems, the mathematician Brahmagupta said: "These problems are proposed simply for pleasure; the wise man can invent a thousand others, or he can solve the problems of others by the rules given here. As the sun eclipses the stars by his brilliancy, so the man of knowledge will eclipse the fame of others in assemblies of the people if he proposes algebraic problems, and still more if he solves them."

Historical Note

The Hindus carried out addition and multiplication much as we do today, except that they wrote their numbers with the smaller units on the left and then worked from left to right, as we do when reading a word or a sentence. However, when it came to multiplication, they used a variety of devices to help them (compare with the pencil-box multipliers used in elementary school). One such device was the use of galley multiplication (described in Problem 18).

called "galley multiplication." In this system we write multiplication statements like $8 \times 7 = 56$ in the form

For larger numbers, such as 638×7, we simply write each product separately in three boxes, as shown.

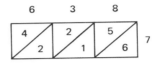

Next we add along the diagonal lines for the answer.

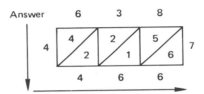

The answer is 4466. The method can be further expanded as shown for 638×729.

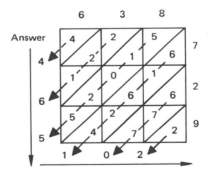

The answer is 465,102. Perform the following operations using galley multiplication.

a. 428×78 b. 624×287 c. 4823×6157

19. Notice the following pattern for multiplication by 11:

$$14 \times 11 = 1 \qquad 4 \text{ original two digits}$$
$$= 1\ 5\ 4$$
$$\uparrow\!\!\text{— insert the sum of the two digits}$$
$$52 \times 11 = 5 \qquad 2$$
$$= 5\ 7\ 2$$

Use expanded notation to show why this pattern "works."

Mind Bogglers

20. Can you find the pattern?
    0,1,2,10,11,12,20,21,22,100, . . .

21. Insert appropriate operation signs (+ , − , ×, or ÷) between each digit so that the following becomes a true sentence:

$$1 \quad 2 \quad 3 \quad 4 \quad 5 \quad 6 \quad 7 \quad 8 \quad 9 = 100$$

22. A farmer has to get a fox, a goose, and a bag of corn across a river in a boat that is large enough only for him and one of these three items. If he leaves the fox alone with the goose, the fox will eat the goose. If he leaves the goose alone with the corn, the goose will eat the corn. How does he get all the items across the river?

Problems for Individual Study

23. *Galley Division*. Related to the galley multiplication of Problem 18 is a process called galley division. Do some research on this topic, and present a report on galley division to the class.
    Reference: Boyer, Carl, *The History of Mathematics* (New York: Wiley, 1968).

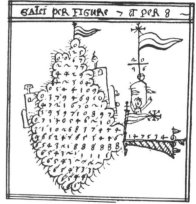

16th-century galley division

24. *The Role of Mathematics in Western Civilization*. What are some of the significant events in the development of mathematics? Who are some of the famous people who have contributed to mathematical knowledge? What are some of the amusing anecdotes from the lives of famous mathematicians? How has mathematics been involved in science, politics, military campaigns, architecture, transportation, philosophy, art, music, and literature?
    Exhibit suggestions: charts and models depicting the inventions and contributions of such men as Archimedes, Newton, and Gauss; a time line showing the significant developments in mathematics.
    References: Bell, E. T., *Men of Mathematics* (New York: Simon and Schuster, 1937).
       Eves, Howard, *In Mathematical Circles,* Vols. 1 and 2 (Boston: Prindle, Weber, & Schmidt, 1969).
       Eves, Howard, *Mathematical Circles Revisited* (Boston: Prindle, Weber, & Schmidt, 1971).
       Kline, Morris, *Mathematics in Western Culture* (New York: Oxford University Press, 1953).

4.3 NATURAL NUMBERS

The most basic set of numbers used by any society is the set of numbers used for counting:

$$\{1,2,3,4, . . . \}$$

This set is called the set of counting numbers or *natural numbers*. Let's assume that you understand what the numbers in this set represent and that you understand the operation of addition, + .

Natural numbers

The word "commute" can mean to travel back and forth from home to work; this back-and-forth idea can help you remember that the commutative property applies if you read from left to right or right to left.

The word "associate" can mean to connect, join, or unite; with this property you associate two of the added numbers.

9
8
7
6
5
4
3
2
1
___

There are a few self-evident properties of addition for this set of natural numbers. They are called self-evident because they almost seem too obvious to be stated explicitly. For example, if we jump into the air, we expect to come back down. That assumption is well founded in experience and is also based on an assumption that jumping has certain undeniable properties. But astronauts have found that some very basic assumptions are valid on earth and false in space. Recognizing these assumptions (properties, axioms, laws, or postulates) is important.

When we add any two natural numbers, we know that we will obtain a counting number as an answer. This "knowing" is an assumption based on experience, but we have actually experienced only a small number of cases for all the possible sums of numbers. The scientist—and the mathematician in particular—is very skeptical about making assumptions too quickly. There is a story about a mathematician riding in a car with a friend. The friend remarked "Oh look, the Joneses have painted their house." The mathematician responded "Yes, at least this side."

The first self-evident property of the natural numbers concerns the order in which they are added. It is called the *commutative property for addition* and states that the order in which two numbers are added makes no difference; that is (if we read from left to right),

$$a + b = b + a$$

for any two natural numbers a and b. The commutative property allows us to rearrange numbers; it is called the property of order. Together with the second property, called the associative property, it is used in calculation and simplification.

The *associative property* allows us to group numbers for addition. Suppose you wish to add three numbers, say 2, 3, and 5:

$$2 + 3 + 5$$

In order to add these numbers we must first add two of them and then add this sum to the third. The associative property tells us that, no matter which two are added first, the final result is the same. If parentheses are used to indicate the numbers to be added first, then this property can be symbolized by

$$(2 + 3) + 5 = 2 + (3 + 5)$$

This associative property for addition holds for *any* three or more natural numbers.

Add the column of numbers in the margin. How long did it take? Five seconds is long enough if you use the associative and commutative properties for addition:

$$(9 + 1) + (8 + 2) + (7 + 3) + (6 + 4) + 5$$
$$=\quad 10\quad +\quad 10\quad +\quad 10\quad +\quad 10\quad + 5$$
$$=\quad 45.$$

However, it takes much longer if you don't rearrange and regroup the numbers:

$$(9 + 8) + (7 + 6 + 5 + 4 + 3 + 2 + 1)$$
$$= (17 + 7) + (6 + 5 + 4 + 3 + 2 + 1)$$
$$= (24 + 6) + (5 + 4 + 3 + 2 + 1)$$
$$= (30 + 5) + (4 + 3 + 2 + 1)$$
$$= (35 + 4) + (3 + 2 + 1)$$
$$= (39 + 3) + (2 + 1)$$
$$= (42 + 2) + 1$$
$$= 44 + 1$$
$$= 45 \quad \text{(whew!)}$$

The properties of associativity and commutativity are not restricted to the operation of addition. They are general properties that can be applied to any operation for any given set.

Multiplication is defined as repeated addition.

DEFINITION: $a \times b$, $(b \neq 0)$, means

$$\underbrace{a + a + a + \cdots + a}_{b \text{ addends}}$$

If $b = 0$, then $a \cdot 0 = 0$.

Definition of multiplication

Multiplication is denoted by a cross ($\times$), as in $3 \times a$; a dot, as in $3 \cdot a$; parentheses, as in $3(a)$; or juxtaposition, as in $3a$.

We can now check commutativity and associativity for multiplication in the set of natural numbers.

COMMUTATIVITY: $2 \cdot 3 \stackrel{?}{=} 3 \cdot 2$

ASSOCIATIVITY: $(2 \cdot 3) \cdot 4 \stackrel{?}{=} 2 \cdot (3 \cdot 4)$

The question mark above the equal sign signifies that we should not assume the conclusion (namely that they are equal) until we check the arithmetic. Even though we can check these properties for particular natural numbers, it is impossible to check them for *all* natural numbers, so we will accept the following axioms.

For any natural numbers a, b, and c:

Commutative Properties

    Addition: $a + b = b + a$
    Multiplication: $ab = ba$

Associative Properties

    Addition: $(a + b) + c = a + (b + c)$
    Multiplication: $(ab)c = a(bc)$

Historical Note

It was difficult to decide where to put the historical note about Leonhard Euler (1707–1783), since Euler's name is attached to every branch of mathematics. He was the most prolific writer on the subject of mathematics, and his mathematical textbooks were masterfully written. His writing was not at all slowed down by his total blindness for the last 17 years of his life. He possessed a phenomenal memory, had almost total recall, and could mentally calculate long and complicated problems. The story of Euler's death is told by Howard Eves in his book In Mathematical Circles *(Boston: Prindle, Weber & Schmidt, 1969, Vol. 2, p. 54).*

"Euler retained vigor and power of mind up to the moment of his death, which occurred in his seventy-seventh year, on September 18, 1783. He had amused himself in the afternoon calculating the laws of ascent of balloons. He then dined with Lexell and his family, and outlined the calculation of the orbit of the recently discovered planet Uranus. A short time later he begged that his grandson be brought in. While playing with the youngster and sipping some tea, he suffered a stroke. His pipe fell from his hand and he uttered, 'I die.' At that instant, in the words of Condorcet, 'Euler ceased to live and calculate.'"

It is possible to have a set with more than one operation. For example, we can add and multiply in the set of counting numbers. Are there properties in this set that involve both operations? Consider an example.

EXAMPLE: Suppose you are selling tickets for a raffle, and the tickets cost $2 each. You sell 3 tickets on Monday and 4 tickets on Tuesday. How much money did you collect?

Solution 1.   You sold a total of $3 + 4 = 7$ tickets, which cost $2 each, so you collected $2 \cdot 7 = 14$ dollars. That is,

$$2 \times (3 + 4) = 14$$

Solution 2.   You collected $2 \cdot 3 = 6$ dollars the first day and $2 \cdot 4 = 8$ dollars the second day for a total of $6 + 8 = 14$ dollars. That is,

$$(2 \times 3) + (2 \times 4) = 14$$

Since these solutions are equal, we say:

$$2 \times (3 + 4) = (2 \times 3) + (2 \times 4)$$

Do you suppose this would be true if the tickets cost a dollars and you sold b tickets on Monday and c tickets on Tuesday? Then the above equation would look like:

$$a \times (b + c) = (a \times b) + (a \times c)$$

or simply

$$a(b + c) = ab + ac$$

This example illustrates the *distributive property*.

Distributive Property for Multiplication over Addition

$a(b + c) = ab + ac$

In the set of counting numbers, is addition distributive over multiplication? We wish to check:

$$3 + (4 \times 5) \overset{?}{=} (3 + 4) \times (3 + 5)$$

Checking:

$$3 + (4 \times 5) = 3 + 20 \qquad \text{and} \qquad (3 + 4) \times (3 + 5) = 7 \times 8$$
$$= 23 \qquad\qquad\qquad\qquad\qquad\qquad = 56$$

Thus, addition is not distributive over multiplication in the set of counting numbers.

The distributive property can also help to simplify arithmetic. Suppose we wish to multiply 9 by 71. We can use the distributive property to think:

$$9 \times 71 = 9 \times (70 + 1) = (9 \times 70) + (9 \times 1)$$
$$= 630 + 9$$
$$= 639$$

This allows us to do the problem quickly and simply in our heads.

Since these properties hold for the operations of addition and multiplication, we might reasonably ask if they hold for other operations. Subtraction is defined as the opposite of addition.

DEFINITION: $a - b = x$ means $a = b + x$.

Definition of subtraction

To verify the commutative property for subtraction, we check a particular example:

$$3 - 2 \overset{?}{=} 2 - 3$$

Now, $3 - 2 = 1$, but $2 - 3$ doesn't even exist in the set of natural numbers. What does it mean to say that a result doesn't exist in the set of natural numbers? It means that, to provide the result of the operation of subtraction for $2 - 3$, we must find a number added to 3 that gives the result 2. But there is *no such natural number*. In mathematics we give a name to such an occurence. We say that a set is *closed* for an operation if every possible result of performing that operation is a number contained in the given set. Thus, the set of natural numbers is *closed* for addition and multiplication but *not closed* for subtraction.

From the definition of subtraction,

means

$$2 - 3 = \Box$$
$$2 = \Box + 3$$

so we need to find a number that, when added to 3, gives the result 2.

Notice that we use the words not closed *instead of* open.

CLOSURE PROPERTY: The set of natural numbers is closed for

1. *addition,* since $a + b$ is a natural number for any natural numbers a and b, and
2. *multiplication,* since ab is a natural number for any natural numbers a and b.

Closure property for addition in the set of natural numbers
Closure property for multiplication in the set of natural numbers

EXAMPLES:

1. The set $A = \{1,2,3,4,5,6,7,8,9,10\}$ is *not closed* for the operation of addition, since $7 + 8 = 15$. Since $15 \notin A$, we see that the set is not closed.
2. The set $B = \{0,1\}$ is closed for the operation of multiplication, since all possible products are in B.

You need find only one *counterexample to show that a property does not hold.*

$$0 \times 0 = 0$$
$$0 \times 1 = 0$$
$$1 \times 0 = 0$$
$$1 \times 1 = 1$$

Although $2 - 3$ doesn't exist in the set of natural numbers, it is possible to invent new numbers and add them to the set, so that it becomes a closed set for subtraction. We will perform every possible subtraction. If the result is a natural number (as in $3 - 2 = 1$), we will do nothing. If the result is not a natural number (as in $2 - 3$), we will invent a new number (call it x for the moment) so that $2 - 3 = x$, and we will then add x to the set of natural numbers to create a new set. After we have done this for all possible subtractions, we will have a new set that is closed for subtraction.

Historically, an agricultural-type society would need only natural numbers, but what about a subtraction like

$$5 - 5 = ?$$

Did the Egyptians or Babylonians have such a number?

Certainly society would have a need for a number representing $5 - 5$, so a new number, called zero, was invented, so that $5 = 5 + 0$ (remember the definition of subtraction). If this new number is annexed to the set of natural numbers, the set is called the set of *whole numbers:*

$$W = \{0,1,2,3,4, \ldots\}$$

This one annexation to the existing numbers satisfied society's needs for several thousands of years.

However, as the society evolved, the need for bookkeeping advanced, and eventually the need for answers to problems like $5 - 6 = ?$ arose. The question that we need to answer is this: "Can we annex new numbers to the set W so that it is possible to carry out *all* subtractions?" That is, can we annex elements to W so that it becomes a closed set for subtraction?

Let's add some new numbers to the set in the following fashion:
1. Find some number so that $0 - 1$ makes sense. That is,

$$1 + \square = 0$$

There should be some number to put into this box so that the equation becomes a true sentence. Is there any such number in the set $\{0,1,2,3, \ldots\}$? Clearly not. So let's invent a number, call it Φ, and place it in the set along with the elements of W to give us

$$\{\Phi,0,1,2,3,4, \ldots\}$$

We shall give this number, Φ, the one property for which it was invented, which is that $1 + \Phi = 0$.
2. What about $0 - 2$?

$$2 + \square = 0$$

Again, we include a new element ς in the set to obtain

$$\{ \varsigma,\Phi,0,1,2,3, \ldots\}$$

so that $2 + \varsigma = 0$.

3. We would also like to find numbers so that

$$0 - 3$$
$$0 - 4$$
$$0 - 5$$
$$0 - 6$$
$$\vdots$$

make sense. If we continue as before, we would soon be at a loss for new symbols. So let's rename Φ and $\S$ as follows.

Φ: $0 - 1$ is designated by $^-1$
$\S$: $0 - 2$ is designated by $^-2$
            $0 - 3$ is designated by $^-3$
            $0 - 4$ is designated by $^-4$

$$\vdots$$

$0 - n$ is designated by ^-n

The number represented by $^-2$ will be called the *opposite* or the *additive inverse* of 2. Now we have a set including W with all these new numbers:

$$\{. . . , ^-5,^-4,^-3,^-2,^-1,0,1,2,3,4,5, . . .\}$$

This is called the set of *integers* and will be discussed in the next section.

This serves as a definition of these new numbers, which are called opposites.

PROBLEM SET 4.3

In order to make sure you understand the properties discussed in this section, this set of problems will focus on the properties of closure, commutativity, associativity, and distributivity rather than the set of natural numbers and operations of addition, multiplication, and subtraction. Since you are so familiar with the set of natural numbers and with these operations, you could probably answer questions about them without much reflection on the concepts involved. Therefore, in this problem set, certain new operations (other than addition, multiplication, and subtraction) will be defined by table. In using a table, rows are horizontal and columns are vertical. Thus, 3×4 on a multiplication table would be found in the third row and fourth column.

A Problems

In Problems 1–10, classify each as an example of the commutative property, the associative property, or both.

1. $3 + 5 = 5 + 3$

2. $6 + (2 + 3) = (6 + 2) + 3$

3. $6 + (2 + 3) = 6 + (3 + 2)$

4. $6 + (2 + 3) = (2 + 3) + 6$

5. $6 + (2 + 3) = (6 + 3) + 2$

To distinguish between the commutative and associative properties, remember the following.
1. When the commutative property is used, the order in which the elements appear from left to right is changed, but the grouping is not changed.
2. When the associative property is used, the elements are grouped differently, but the order in which they appear is not changed.

6. $(4 + 5)(6 + 9) = (4 + 5)(9 + 6)$

7. $(4 + 5)(6 + 9) = (6 + 9)(4 + 5)$

8. $(3 + 5) + (2 + 4) = (3 + 5 + 2) + 4$

9. $(3 + 5) + (2 + 4) = (3 + 5) + (4 + 2)$

10. $(3 + 5) + (2 + 4) = (3 + 4) + (5 + 2)$

11. Is the operation of putting on your shoes and socks commutative?

12. "Isn't this one just too sweet, dear?" asked the wife as she tried on a beautiful diamond ring.
"No," her husband replied. "It's just too dear, sweet."
Does this story remind you of the associative or the commutative property?

13.

Which properties are illustrated by the cartoon above?

14. In the English langauge some groupings of words are associative and others are not. For example the words

BIG RED APPLE

would be associative, since a

(BIG RED) APPLE

is the same as a

BIG (RED APPLE)

On the other hand, the words

HIGH SCHOOL STUDENT

would not be associative, since a

(HIGH SCHOOL) STUDENT

is not the same as a

HIGH (SCHOOL STUDENT)

Decide whether each of the following groups of words is associative.
a. MULTIPLE CHOICE TEST
b. RED FIRE ENGINE
c. BARE FACTS PERSON
d. BROWN SMOKING JACKET
e. SHAGGY DOG STORY

15. Think of three nonassociative word triples as shown in Problem 14.

16. Consider the set $A = \{1,4,7,9\}$ with an operation $*$ defined by Table 4.3. Find each of the following.
a. $7*4$ b. $9*1$ c. $1*7$ d. $9*9$

17. Consider the set $F = \{1,^-1,i,^-i\}$ with an operation $\times$ defined by Table 4.4. Find each of the following.
a. $(^-1) \times i$ b. $i \times i$ c. $(^-i) \times (^-1)$ d. $(^-i) \times (^-i)$

18. Consider the set A and the operation $*$ from Problem 16.
a. Is the set A closed for $*$? Give reasons for your answer.
b. Is $*$ associative for the set A? Give reasons.
c. Is $*$ commutative for the set A? Give reasons.

19. Consider the set F and the operation $\times$ from Problem 17.
a. Is the set F closed for $\times$? Give reasons for your answer.
b. Is $\times$ associative for set F? Give reasons.
c. Is $\times$ commutative for set F? Give reasons.

20. Let A be the process of putting on a shirt; let
     B be the process of putting on a pair of socks; and let
     C be the process of putting on a pair of shoes.
Let $\varnothing$ be the operation of "followed by."

a. Is $\varnothing$ commutative for $\{A,B,C\}$?
b. Is $\varnothing$ associative for $\{A,B,C\}$?

TABLE 4.3

*	1	4	7	9
1	9	7	1	4
4	7	9	4	1
7	1	4	7	9
9	4	1	9	7

TABLE 4.4

×	1	$^-1$	i	^-i
1	1	$^-1$	i	^-i
$^-1$	$^-1$	1	^-i	i
i	i	^-i	$^-1$	1
^-i	^-i	i	1	$^-1$

21. Consider the set of counting numbers and an operation Σ, which means "select the first of the two." That is,

$$4 \Sigma 3 = 4$$
$$3 \Sigma 4 = 3$$
$$5 \Sigma 7 = 5$$
$$6 \Sigma 6 = 6$$

a. Is the set of counting numbers closed for Σ? Give reasons.
b. Is Σ associative for the set of counting numbers? Give reasons.
c. Is Σ commutative for the set of counting numbers? Give reasons.

22. Consider the operation $\blacktriangledown$ defined by Table 4.5.
a. Find $\square \blacktriangledown \triangle$.   b. Find $\triangle \blacktriangledown \bigcirc$.
c. Does $\bigcirc \blacktriangledown \square = \square \blacktriangledown \bigcirc$?
d. Does $(\bigcirc \blacktriangledown \triangle) \blacktriangledown \triangle = \bigcirc \blacktriangledown (\triangle \blacktriangledown \triangle)$?
e. Is $\blacktriangledown$ a commutative operation? Give reasons.

TABLE 4.5

$\blacktriangledown$	$\square$	$\triangle$	$\bigcirc$
$\square$	$\bigcirc$	$\square$	$\triangle$
$\triangle$	$\square$	$\triangle$	$\bigcirc$
$\bigcirc$	$\triangle$	$\square$	$\bigcirc$

23. Let @ mean select the smaller number and Σ mean select the first of the two. Is Σ distributive over @ in the set of counting numbers?

24. Do the following problems mentally using the distributive property.
a. 6×82 b. 8×41 c. 7×49 d. 5×99 e. 4×88

B Problems

Let $U = \{1,2,3,4,5,6,7,8,9,10\}$
$P = \{1,4,7\}$
$Q = \{2,4,9,10\}$
$R = \{6,7,8,9\}$
Does $(P \cup Q) \cup R = P \cup (Q \cup R)$?
Check: $(P \cup Q) \cup R$
$= \{1,2,4,7,9,10\} \cup \{6,7,8,9\}$
$= \{1,2,4,6,7,8,9,10\}$
$P \cup (Q \cup R)$
$= \{1,4,7\} \cup \{2,4,6,7,8,9,10\}$
$= \{1,2,4,6,7,8,9,10\}$
Thus, for these sets,
$(P \cup Q) \cup R = P \cup (Q \cup R)$.

25. Is the operation of union for sets a commutative operation? HINT: The example shown in the margin illustrates the question for *one* example, but it doesn't *prove* that union is associative. To prove this result, use Venn diagrams for *any* sets X, Y, and Z. That is, prove

$$(X \cup Y) \cup Z = X \cup (Y \cup Z)$$

26. Is the operation of intersection for sets a commutative operation?

27. Is the operation of intersection for sets an associative operation?

28. Is the set of even numbers closed for the operation of addition? Of multiplication?

29. Is the set of odd numbers closed for the operation of addition? Of multiplication?

F means "don't move from your present position." It does not mean "return to your original position."

30. Consider the soldier facing in a given direction (say north). Let us denote "left face" by L, "right face" by R, "about face" by A, and "stand fast" by F. Then we define $H = \{L,R,A,F\}$ and an operation $\varnothing$ meaning "followed by." Thus, L $\varnothing$ L = A means "left face" followed by "left face" is the same as the command "about face."
Also, L $\varnothing$ A = R
A $\varnothing$ R = L
L $\varnothing$ F = L
Make a table summarizing all possible movements. Use the headings L, R, A, and F.

31. Using Problem 30, find:
    a. $R \oslash R$ b. $A \oslash A$ c. $L \oslash L$ d. $R \oslash L$ e. $L \oslash A$

32. Using Problem 30:
    a. Is H closed for the operation $\oslash$?
    b. Does $\oslash$ satisfy the associative property for the set H?
    c. Does $\oslash$ satisfy the commutative property for the set H?

33. Let S be a set of stairs in a given stairway in some building. Number the stairs in ascending order: $s_1, s_2, s_3, \ldots, s_n$. Then $S = \{s_1, s_2, \ldots, s_n\}$, where the elements of S are individual steps in the stairway. Consider the operation of walking up the staircase one step at a time. If we denote this operation by $\uparrow$, we have

$$s_1 \uparrow s_2 = s_3 \quad \text{and} \quad s_2 \uparrow s_3 = s_4,$$

meaning that, given two consecutive steps, the result is the next step in ascending order. Similarly, we define a descending operation $\downarrow$:

$$s_5 \downarrow s_4 = s_3; \qquad s_{17} \downarrow s_{16} = s_{15}.$$

The steps must be consecutive, so $s_3 \uparrow s_{19}$ or $s_7 \downarrow s_5$ is not defined. If S is a set of stairs in a typical building, the operations of $\uparrow$ and $\downarrow$ are not closed (why?). However, if we consider the set of stairs

$$S = \{s_1, s_2, \ldots, s_{45}\}$$

depicted by the Escher lithograph, is the set S closed for $\uparrow$ and $\downarrow$?

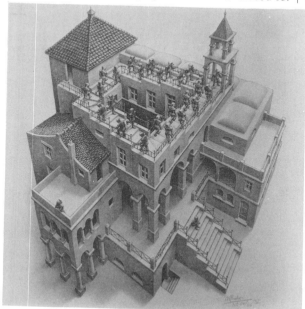

M. C. Escher's *Ascending, Descending*

34. Let $S = \{1, 2, 3, \ldots, 99, 100\}$. Define an operation $\boxtimes$: $a \boxtimes b = 2a + b$, where $+$ refers to ordinary addition. Check the commutative and associative properties.

35. *Computer Problem.* In the text, we indicated that the commutative property could be verified by checking all possibilities. If the set is not too big, we can do this on a computer as follows: Let $S = \{1,2,3,4,5,6,7,8,9,10\}$ for the operation of multiplication. We wish to see whether $a \cdot b = b \cdot a$.

```
10    FOR A = 1 TO 10
20    FOR B = 1 TO 10
30    IF A*B = B*A GOTO 80
40    PRINT "NOT COMMUTATIVE, SINCE I HAVE FOUND A";
45    PRINT " COUNTEREXAMPLE WHEN"
50    PRINT "A = ";A;
60    PRINT ", AND B = "; B
70    STOP
80    NEXT B
90    NEXT A
100   PRINT "THE SET IS COMMUTATIVE FOR MULTIPLICATION."
110   END
```

Write a program similar to this one to see if S is commutative for the operation of subtraction.

36. *Computer Problem.* As noted in the text, testing the associative property can be a tedious task, since all possibilities must be checked. Let $S = \{1,2,3,4,5,6,7,8,9,10\}$ for the operation of multiplication. We wish to see whether $(a \cdot b) \cdot c = a \cdot (b \cdot c)$. We can write a BASIC program as follows:

```
10    FOR A = 1 TO 10
20    FOR B = 1 TO 10
30    FOR C = 1 TO 10
40    IF (A*B)*C = A *(B*C) GOTO 100
50    PRINT "NOT ASSOCIATIVE, SINCE I HAVE FOUND A";
55    PRINT " COUNTEREXAMPLE WHEN"
60    PRINT "A = ";A;
70    PRINT ", B = ";B;
80    PRINT ", AND C = ";C
90    STOP
100   NEXT C
110   NEXT B
120   NEXT A
130   PRINT "SET IS ASSOCIATIVE FOR MULTIPLICATION."
140   END
```

Write a program to see if S is associative for the operation of addition.

37. *Computer Problem.* Check associativity for multiplication in the set $C = \{0,1\}$. Write a program that will print out all the possibilities.

Mind Boggler

38. *The Vanishing Leprechaun.* The following puzzle, available from W. A. Elliott Company, 212 Adelaide St. W., Toronto, Canada, M5H 1W7, consists of three pieces, as shown in Figure 4.2.

FIGURE 4.2 The Vanishing Leprechaun. © 1968.

If we place the pieces together as shown in the top part of the figure, we count 15 leprechauns. However, if we *commute* pieces 1 and 2 as shown in the bottom part of the figure, we count 14 leprechauns. Thus, if we let A and B represent the locations of pieces 1 and 2, respectively, then

$$AB \neq BA.$$

Can you explain where the Vanishing Leprechaun went?

4.4 INTEGERS

In the last section, we introduced the set of integers

$$Z = \{\ldots, {}^-3, {}^-2, {}^-1, 0, {}^+1, {}^+2, {}^+3, \ldots\}.$$

It is customary to refer to certain subsets of Z as follows:

1. Positive integers: $\{{}^+1, {}^+2, {}^+3, \ldots\}$
2. Zero: $\{0\}$
3. Negative integers: $\{{}^-1, {}^-2, {}^-3, \ldots\}$

Historical Note

Historically, the negative integers were developed quite late. There are indications that the Chinese had some knowledge of negative numbers as early as 200 B.C., and in the seventh century A.D. the Hindu Brahmagupta stated the rules for operations with positive and negative numbers. The Chinese represented negatives by putting them in red (compare with the present-day accountant), and the Hindus represented them by putting a circle or a dot over the number.

However, as late as the 16th century, some European scholars were calling numbers such as 0 − 1 absurd. In 1545 Cardano (1501–1576), an Italian scholar who presented the elementary properties of negative numbers, called the positive numbers "true" numbers and the negative numbers "fictitious" numbers. However, they did become universally accepted; as a matter of fact, the word integer *that we use to describe the set is derived from "numbers with integrity."*

Now that we have this new enlarged set of numbers, will we be able to carry out all possible additions, subtractions, and multiplications? Before we answer this question, let's review the processes by which we operate within the set of integers.

ADDITION OF INTEGERS

To see how to add integers, we use the number line.

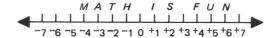

We shall make moves along the number line. Suppose you have the following sums:

$$^+3 + {}^+4 \quad \text{and} \quad {}^+3 + {}^-4$$

Start at $^+3$; add $^+4$ by moving four units to the right, and add $^-4$ by moving four units to the left.

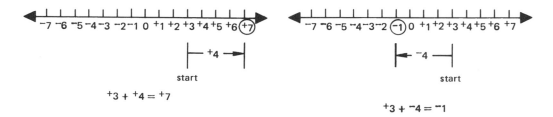

$$^+3 + {}^+4 = {}^+7$$

$$^+3 + {}^-4 = {}^-1$$

Examples 1–3 illustrate what is called directed distance. A three-unit move to the right is different from a three-unit move to the left. However, if you simply want to indicate the distance between two points, you speak of the undirected distance.

EXAMPLES:

1. $^-3 + {}^-4 = ?$

Therefore, $^-3 + {}^-4 = {}^-7$.

2. $^+5 + {}^-6 = ?$

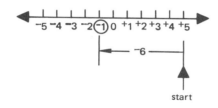

Therefore, $^+5 + {}^-6 = {}^-1$.

3. $^-3 + {}^+4 = ?$

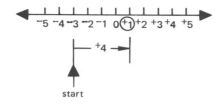

Therefore, $^-3 + {}^+4 = {}^+1$.

Because it is not always convenient or practical to add on a number line, we make some generalizations. We are interested in looking at the numerical value of a quantity, independent of direction or sign. To do this, we define the notion of *absolute value* of a number, which represents the distance of that number from the origin when plotted on a number line.

> DEFINITION: The *absolute value* of x is its undirected distance from zero. We use the symbol $|x|$ for absolute value.

Absolute value

EXAMPLES:

1. $|^+5|$ is 5   2. $|^-5|$ is 5   3. $|^-(^-3)| = 3$

Now notice that, if we're adding numbers with the same sign, the result is the same as the sum of the absolute values except for a plus or a minus sign (since their directions on the number line are the same). If we're adding numbers with different signs, their directions are opposite, so the net result is the difference of the absolute values. This can be summarized as follows (see also Figure 4.3):

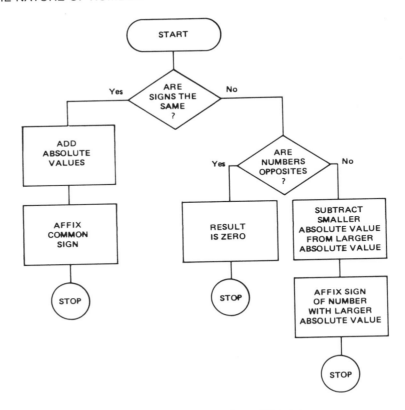

FIGURE 4.3 Flow chart for adding integers

Procedure for adding integers

ADDITION OF INTEGERS $x + y$

1. If signs are the same:
   positive + positive = positive
   negative + negative = $\underbrace{negative}_{\substack{sign \\ part}}$ $\underbrace{|x| + |y|}_{\substack{whole \\ number\ part}}$

2. If signs are different:
   positive + negative =
   negative + positive = $\underbrace{\left\{\begin{array}{l} \text{sign is the} \\ \text{same as one} \\ \text{with larger} \\ \text{absolute value} \end{array}\right\}}_{sign\ part}$ $\underbrace{\left\{\begin{array}{l} |x| - |y| \text{ or } |y| - |x| \\ \text{so answer is a} \\ whole \text{ number} \end{array}\right\}}_{whole\ number\ part}$

Notice that all integers consist of two parts: a sign part and a whole number part.

EXAMPLES:

1. $^+41 + {}^+13 = {}^+54$
2. $^-41 + {}^-13 = {}^-54$
3. $^+41 + {}^-13 = {}^+28$
4. $^-41 + {}^+13 = {}^-28$

It is now easy to verify that the set of integers is closed for addition. We know that the whole numbers are closed for addition and that the sum of two integers will be a whole number or the opposite of a whole number and thus will be an integer (this is not proof and must be accepted without proof).

Since the set of positive integers is the set of natural numbers, let's agree that whenever we write a natural number 1,2,3,4, . . ., we could also write $^+1,^+2,^+3,^+4,$ . . . . That is, $1 = {}^+1$, $2 = {}^+2$, $3 = {}^+3$, ...

Sometimes we will still use the positive-integer form for clarity or emphasis.

MULTIPLICATION OF INTEGERS

For the whole numbers, multiplication can be defined as repeated addition, since we say that $4 \cdot 5$ means

$$\underbrace{4 + 4 + 4 + 4 + 4}_{5 \text{ addends}}$$

However, we cannot do this for the integers, since $4 \cdot (^-5)$, or

$$\underbrace{4 + 4 + \cdots + 4}_{^-5 \text{ addends}}$$

doesn't "make sense." We cannot repeat the addition process a negative number of times. Thus, the best we can do is seek out patterns to help us choose a "good" definition for the multiplication of integers.

There are four cases to consider:

1. positive $\cdot$ positive
2. positive $\cdot$ negative
3. negative $\cdot$ positive
4. negative $\cdot$ negative

1. Suppose we wish to multiply two positive integers:

$$^+3 \cdot {}^+4$$

We agreed that $^+3 = 3$ and $^+4 = 4$; therefore,

$$^+3 \cdot {}^+4 = 3 \cdot 4$$

Now 3 and 4 are natural numbers, and we know how to multiply natural numbers. Thus, since $3 \cdot 4 = 12 = {}^+12$, we say that $^+3 \cdot {}^+4 = {}^+12$. We can then conclude:

The product of two positive numbers is a positive number.

2. Next consider

$$^+3 \cdot ^-4$$

To see how to find this product, we make use of a very important characteristic of mathematics discussed in Section 1.1—the discovery of a new concept by looking at some patterns.

$$^+3 \cdot ^+4 = ^+12$$
$$^+3 \cdot ^+3 = ^+9$$
$$^+3 \cdot ^+2 = ^+6$$

Would you know what to write next? Here it is:

$$^+3 \cdot ^+1 = ^+3$$
$$^+3 \cdot 0 = 0$$

What comes next? *Answer this question before reading further.*

$$^+3 \cdot ^-1 = ^-3$$
$$^+3 \cdot ^-2 = ^-6$$
$$^+3 \cdot ^-3 = ^-9$$
$$^+3 \cdot ^-4 = ^-12$$
$$\vdots$$

Do you know how to continue? Try building a few more such patterns using different numbers. What did you discover about the product of a positive number and a negative number?

The product of a positive number and a negative number is a negative number.

3. Consider the problem of a negative times a positive:

$$^-3 \cdot ^+4$$

Since you know that $^+4 \cdot ^-3 = ^-12$, you should also know that $^-3 \cdot ^+4 = ^-12$ if we are to retain all the basic properties for numbers that we have previously discussed. That is, we want the commutative property to hold for the multiplication of integers. Therefore:

The product of a negative number and a positive number is a negative number.

4. Consider the final example, the product of two negative integers:

$$^-3 \cdot ^-4$$

Let's build another pattern.

$$^-3 \cdot ^+4 = ^-12$$
$$^-3 \cdot ^+3 = ^-9$$
$$^-3 \cdot ^+2 = ^-6$$

What would you write next? Here it is:

$$^-3 \cdot \,^+1 = \,^-3$$
$$^-3 \cdot \;\; 0 = \;\;\; 0$$

What comes next? *Answer this question before reading further.*

$$^-3 \cdot \,^-1 = \;\,^+3$$
$$^-3 \cdot \,^-2 = \;\,^+6$$
$$^-3 \cdot \,^-3 = \;\,^+9$$
$$^-3 \cdot \,^-4 = \,^+12$$
$$\vdots$$

Thus, as the pattern indicates:

> *The product of two negative numbers is a positive number.*

We can summarize our discussion as follows (see also Figure 4.4):

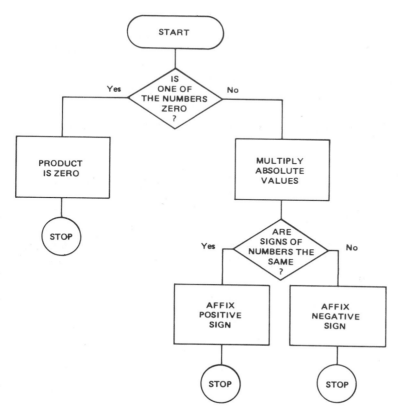

FIGURE 4.4 Flow chart for multiplication of integers

Procedure for multiplying integers

MULTIPLICATION OF INTEGERS $x \cdot y$

1. positive · positive = positive
2. positive · negative = negative
3. negative · positive = negative
4. negative · negative = positive

$$|x| \cdot |y|$$

sign part   *whole number part*

An ususual way of remembering the rules for the signs of products is the following, where positive is good *and negative is* bad.

If you do something good *to someone* good, *that's* good.

If you do something bad *to someone* good, *that's* bad.

If you do something good *to someone* bad, *that's* bad.

If you do something bad *to someone* bad, *that's* good!

EXAMPLES:

1. $(^+41)(^+13) = ^+533$
2. $(^-41)(^-13) = ^+533$
3. $(^+41)(^-13) = ^-533$
4. $(^-41)(^+13) = ^-533$

SUBTRACTION OF INTEGERS

On a number line, subtraction is "going back" in the opposite direction instead of "going ahead" as in addition. For two positive numbers, the process seems quite natural. For example:

$$^+4 - {}^+3 = {}^+1$$

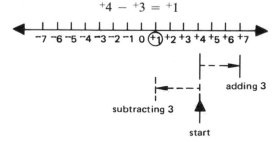

But what about subtracting negative numbers? Negative already indicated "going back," or to the left. Does subtraction of a negative mean "going ahead," or to the right? Consider the following pattern:

$$^+4 - {}^+4 = 0$$
$$^+4 - {}^+3 = {}^+1$$
$$^+4 - {}^+2 = {}^+2$$
$$^+4 - {}^+1 = {}^+3$$
$$^+4 - 0 = {}^+4$$
$$^+4 - {}^-1 = ?$$
$$^+4 - {}^-2 = ?$$
$$^+4 - {}^-3 = ?$$

It does seem as if smaller numbers in each case should give larger differ-ences. On a number line, $^+4 - {}^-3$ means "going in the opposite direction from $^-3$."

$$^+4 - {}^-3 = 7$$

Guided by these results, we make the following definition for subtraction.

$$x - y = x + (^-y)$$

To subtract, add the opposite (of the number being subtracted).

Definition of subtraction of integers. More simply, we say that, to subtract an integer, you add its opposite.

EXAMPLES:

1. $^+41 - {}^+13 = {}^+41 + {}^-13 = {}^+28$

add the opposite

complete the addition of $^+41 + {}^-13$

2. $^+41 - {}^-13 = {}^+41 + {}^+13 = {}^+54$
3. $^-41 - {}^+13 = {}^-41 + {}^-13 = {}^-54$
4. $^-41 - {}^-13 = {}^-41 + {}^+13 = {}^-28$

Let's take an overview of what has been done in the last few sections. We began with the natural numbers, which are closed for addition and multiplication. Next, we defined subtraction and created a situation where it was impossible to subtract some numbers from others. We then "created" a new set (called the integers) by adding to the set of whole numbers each of their opposites.

Now, since the subtraction of integers is defined in terms of addition, you can easily show that the integers are closed for subtraction. You are asked to do this in Problem 24 of this section.

DIVISION OF INTEGERS

The last of the basic operations is division, which is defined as the opposite operation of multiplication.

a divided by b is written

$a \div b$ or $\dfrac{a}{b}$.

DEFINITION OF DIVISION: If a, b, and z are integers, where $b \neq 0$,

$$\frac{a}{b} = z \text{ means } a = b \cdot z$$

See also Figure 4.5. You can use this definition in order to divide integers.

$$12 \div 6 = 2 \qquad \text{and} \qquad 2 \cdot 6 = 12$$

$$\frac{10}{5} = 2 \qquad \text{and} \qquad 10 = 2 \cdot 5$$

$$\frac{65}{13} = 5 \qquad \text{and} \qquad 65 = 5 \cdot 13$$

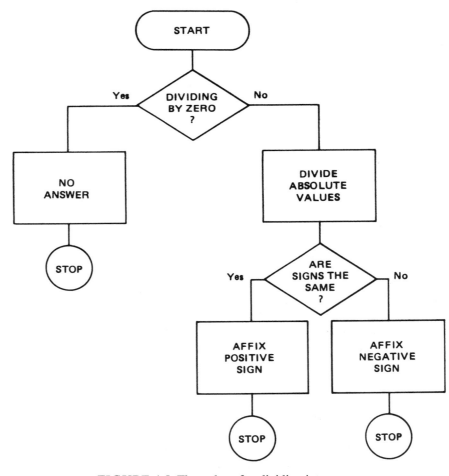

FIGURE 4.5 Flow chart for dividing integers

Notice that we required that $b \neq 0$. Why do we not allow division by zero? We will consider two cases.

1. Division of a nonzero number by zero.

$$a \div 0 \text{ or } \frac{a}{0} = x$$

Is there such a number so that $a \div 0$ makes sense? We see that any number x would have to be such that $a = 0 \cdot x$. Now, $0 \cdot x = 0$; so if $a \neq 0$, we see that such a case is impossible. That is, $a/0$ is not a number.

2. Division of zero by zero.

$$\frac{0}{0} = x$$

Is there such a number x? We see that *any* x makes this true, since $0 \cdot x = 0$ for all x. But this leads to certain absurdities. If $0/0 = 2$ and $0/0 = 5$, then $2 = 5$, since numbers equal to the same number are equal to each other.

EXAMPLES:

1. $^+16/^+4 = {}^+4$, since $(^+4) \cdot (^+4) = {}^+16$.
2. $^+16/^-4 = {}^-4$.
3. $^-16/^+4 = {}^-4$.
4. $^-16/^-4 = {}^+4$.

DIVISION OF INTEGERS

(positive) ÷ (positive) = positive
(negative) ÷ (positive) = negative
(positive) ÷ (negative) = negative
(negative) ÷ (negative) = positive
If you do something bad to someone. . . .

Is the set of integers closed for division? Certainly one can find many examples in which an integer divided by an integer is an integer. Does this mean that the set of integers is closed? What about $1 \div 2$ or $4 \div 5$? These numbers do not exist in the set of integers; thus, the set is *not* closed for division. Now, as long as a society has no need for such division problems, the question of inventing new numbers will not arise. However, as the need to divide 1 into 2 or more parts arises, some new numbers will have to be invented so that the set will be closed for division. The problem in inventing such new numbers is that it must be done in such a way that the properties of the existing numbers are left unchanged. That is, closure for addition, subtraction, and multiplication should be retained.

PROBLEM SET 4.4

A Problems

1. Suppose we are given the following number line:

a. Explain why a move from H to N is described by $^+7$.
b. Describe a move from T to F.
c. Describe a move from U to A.

2. Add the following integers on a number line.

   a. $^+5 + {}^+3$
   b. $^-5 + {}^+3$
   c. $^+4 + {}^-7$
   d. $^+3 + {}^+5$
   e. $^-3 + {}^+5$
   f. $^-2 + {}^-4$

3. Find the sum.

   a. $7 + {}^-3$
   b. $^-9 + 5$
   c. $^-10 + {}^-4$
   d. $162 + {}^-27$
   e. $^-12 + (^-4 + {}^-3)$
   f. $^-5 + (^-6 + 10)$

4. Find the sum.

   a. $^-15 + 8$
   b. $(5 + {}^-7) + (5 + {}^-4)$
   c. $^-14 + 27$
   d. $42 + {}^-121$
   e. $62 + {}^-62$
   f. $(6 + {}^-8) + (6 + 8)$

5. Find the difference.

   a. $10 - 7$
   b. $7 - 10$
   c. $6 - {}^-4$
   d. $^-5 - {}^-10$
   e. $5 - {}^-5$
   f. $7 - {}^-3$

6. Find the difference.

   a. $0 - {}^-15$
   b. $^-46 - {}^-46$
   c. $^-7 - {}^-18$
   d. $^-4 - 8$
   e. $62 - {}^-112$
   f. $5 - {}^-416$

7. Simplify.

   a. $(^-31) + (^-16)$
   b. $^-14 + {}^-21$
   c. $^-9 + 16 + {}^-11$
   d. $^-3 - (^-6 - {}^+4)$
   e. $5 + {}^-19 + 15$
   f. $|^-8| + {}^-8$

8. Simplify.

   a. $^-9 - (4 - 5)$
   b. $^-(^-23 + 14)$
   c. $^-7 - (6 - 4)$
   d. $^-(^-18) + |^-18|$
   e. $|^-3| - (^-(^-2))$
   f. $18 - 4 - 12 - {}^-3$

9. Simplify.

   a. $(^+3)(^-6)$
   b. $(^-5)(^+4)$
   c. $\dfrac{^-4}{^-2}$
   d. $\dfrac{^-6}{^-3}$
   e. $(^-5)(^+7) - (^-9)$
   f. $(^-2)(^+3) - (^-8)$

10. Simplify.

    a. $14 \cdot {}^-5$
    b. $^-14 \cdot {}^-5$
    c. $^-5 \cdot (8 - 12)$
    d. $(^-5 \cdot 8) - 12$
    e. $\dfrac{^-12}{4}$
    f. $\dfrac{^-63}{^-9}$

11. Simplify.

    a. $\dfrac{12}{^-4}$
    b. $^-6 \cdot \dfrac{14}{^-2}$
    c. $\dfrac{^-528}{^-4}$
    d. $45 - (25 \cdot {}^-5)$
    e. $(^-1)^3$
    f. $10 \cdot \left(\dfrac{^-8}{^-2}\right)$

12. Perform the following operations.

    a. $15 - {}^-7$
    b. $^-6 - {}^-6$
    c. $^-18 - 5$
    d. $^-8 + 7 + 16$
    e. $14 + {}^-10 - 8 - 11$
    f. $6 + {}^-8 - 5$

13. Perform the following operations.
  a. $(^-2)(^-3) + (^-1)(^+6)$
  b. $(^-5)(^-1) + (^-9)(^+2)$
  c. $(^-32) \div (^-8) - 5 - (^-7)$
  d. $(^-15) \div (^-5) - 4 - (^-8)$
  e. $(^-5)(^+2) + (^-3)(^-4) - (^+6)(^-7)$
  f. $(^-3)(^-7 + 2)$

14. Perform the following operations.
  a. $6 - ^-2$
  b. $^-5 - ^-3$
  c. $^-15 - ^-6$
  d. $^-4 + 6 - 8$
  e. $15 - ^-3 - 4 - 11$
  f. $^-12 + ^-7 - 10 - 14$

15. Perform the following operations.
  a. $(7)(^-8)$
  b. $(^-5)(15)$
  c. $^-5 \cdot ^-6$
  d. $\dfrac{^-42}{3}$
  e. $(^-54 \div ^-9) \div 3$
  f. $^-54 \div (^-9 \div 3)$

Parts e and f of Problem 15 provide a counterexample for some property. Can you name this property?

16. Perform the following operations.
  a. $(5)(2)$
  b. $(^-6)(14)$
  c. $^-2 \cdot ^-3$
  d. $\dfrac{34}{^-2}$
  e. $(48 \div ^-6) \div ^-2$
  f. $48 \div (^-6 \div ^-2)$

17. Draw a Venn diagram showing the following sets.
$$N = \{\text{natural numbers}\}$$
$$W = \{\text{whole numbers}\}$$
$$Z = \{\text{integers}\}$$

18. What is the intersection of the set of positive integers and the set of negative integers? Is the union of these two sets the set of integers?

B Problems

19. a. State the commutative property.
  b. Is addition or subtraction commutative in the set of integers?
  c. Is multiplication or division commutative in the set of integers? Support your answers.

20. Explain, in your own words, the difference between $0 \div 5$ and $5 \div 0$.

21. a. State the associative property.
  b. Is addition or subtraction associative in the set of integers?
  c. Is multiplication or division associative in the set of integers? Be sure to give reasons for your answers.

22. Perform the indicated operations.
  a. $(42 \div ^-7) \div ^-2$
  b. $\dfrac{\frac{42}{^-7}}{^-2}$
  c. $^-56 \div (14 \div 2)$
  d. $\dfrac{^-56}{\frac{14}{2}}$

23. Perform the indicated operations.

    a. $(^-2)^4$

    b. $(^-1)^{67}$

    c. $(^-1)^{1972}$

    d. $(^-1)^{2k}$, where k is any natural number

24. Show that the set of integers is closed for subtraction.

25. Do you think any finite subset of the set of integers will be closed under multiplication?

Mind Bogglers

26. *Four Fours.* B.C. has a mental block against fours, as we can see in the cartoon. See if you can handle fours by writing the numbers from 1 to 10 using four 4s for each. Here are the first three completed for you:

$$\frac{4}{4} + 4 - 4 = 1 \qquad \frac{4}{4} + \frac{4}{4} = 2 \qquad \frac{4 + 4 + 4}{4} = 3$$

27. *Calculator Problem.* Multiply 1234567×9999999:
    a. by using a calculator.   b. by looking for a pattern.

28. What is wrong with the following "proof"?

i.	12 eggs	=	1 dozen
ii.	24 eggs	=	2 dozen
iii.	6 eggs	=	$^1/_2$ dozen
iv.	144 eggs	=	1 dozen

    v. 12 dozen = 1 dozen

Multiply both sides of step i by 2.
Divide both sides of step i by 2
Multiply step iii times step ii
  (equals times equals are equal).
Substitute, since 144 eggs = 12 dozen.

4.5 RATIONAL NUMBERS

Historically, the need for a closed set for division came before the need for closure for subtraction. We need to find some number k so that

$$1 \div 2 = k.$$

As we saw in Section 4.1, the ancient Egyptians limited their fractions by requiring the numerator to be 1. The Romans avoided fractions by the use

𝔥istorical 𝔑ote

The sign ÷ was first used by Johann Rahn in 1659.

of subunits: feet were divided into inches and pounds into ounces, and a twelfth part of the Roman unit was called an *uncia*.

However, people soon felt the practical need to obtain greater accuracy in measurement and the theoretical need to close the number system with respect to the operation of division. In the set of integers, some divisions are possible:

$$\frac{10}{5}, \frac{^-4}{2}, \frac{^-16}{^-8}, \ldots$$

However, certain others are not:

$$\frac{1}{2}, \frac{^-16}{5}, \frac{5}{12}, \ldots$$

Just as we extended the set of natural numbers by creating the concept of opposites, we can extend the set of integers. That is, the number $^5/_{12}$ is defined to be that number obtained when 5 is divided by 12. This new set, consisting of the integers as well as the quotients of integers, is called the set of rational numbers.

DEFINITION: The *rational numbers*, denoted by Q, are the set of all numbers of the form p/q, where p and q are integers and $q \neq 0$.

Definition of rational numbers

p is called the numerator *and q is called the* denominator.

The rational number $^4/_5$ is interpreted as the quotient obtained when 4 is divided by 5.

Thus,

$$\frac{5}{9}, \frac{^-11}{13}, \frac{5}{1}, \frac{0}{17}, \frac{^-10}{5}, \quad \text{and} \quad \frac{25}{5}$$

are all elements of Q. With this definition, notice that 5 has many representations:

$$\frac{5}{1}, \frac{25}{5}, \frac{^-10}{^-2}, \ldots$$

To ascertain which numeral expressions represent the same numbers, we wish to know when $a/b = c/d, b, d \neq 0$. If $a/b = c/d$, then $(a/b)(bd) = (c/d)(bd)$, since we can multiply both sides by bd. By the definition of division and the inverse and identity properties, we see that $ad = cb$. Thus,

$$\frac{a}{b} = \frac{c}{d} \quad \text{means} \quad ad = bc$$

Equality of rational numbers

Therefore, we see that $25/5 = {}^-10/{}^-2$, since

$$\frac{25}{5}\,(5)({}^-2) \;=\; \frac{{}^-10}{{}^-2}\,(5)({}^-2)$$

$$(25)({}^-2) \;=\; 5({}^-10)$$

$${}^-50 \;=\; {}^-50$$

Also, we see that $125/70 \neq 73/41$, since

$$\frac{125}{70}\,(70)(41) \;\overset{?}{=}\; \frac{73}{41}\,(70)(41)$$

$$125(41) \;\overset{?}{=}\; 70(73)$$

$$5125 \neq 5110$$

Now the next problem to solve is the question of closure for the operations. That is, is the set of rational numbers closed for addition, subtraction, multiplication, and division? Let's first review the definitions of these operations.

ADDITION OF RATIONALS

We are familiar with the idea of adding fractions with common denominators, but what if the given fractions don't have common denominators? For example, consider:

$$\frac{a}{b} + \frac{c}{d}$$

We can multiply each fraction by the identity element, 1:

$$\frac{a}{b}\left(\frac{d}{d}\right) + \frac{c}{d}\left(\frac{b}{b}\right)$$

$$\frac{ad}{bd} + \frac{bc}{bd}$$

This procedure motivates the following definition for the sum of any two rational numbers.

Addition of rational numbers

> *If a/b and c/d are rational numbers, then*
>
> $$\frac{a}{b} + \frac{c}{d} = \frac{ad + bc}{bd}$$

EXAMPLES:

1. $\dfrac{1}{3} + \dfrac{1}{2} = \dfrac{1 \cdot 2 + 3 \cdot 1}{3 \cdot 2} = \dfrac{5}{6}$

2. $\dfrac{4}{5} + \dfrac{7}{9} = \dfrac{4 \cdot 9 + 5 \cdot 7}{5 \cdot 9} = \dfrac{36 + 35}{45} = \dfrac{71}{45}$

Will this definition satisfy all the needs of a society? Is it possible to add *any* two rational numbers and get a rational number? To show that the addition of two rational numbers always results in a rational number, it is necessary to show that, given any two elements of the set Q, their sum is also an element of the set. For example, suppose we are given two rational numbers x/y and w/z. This means that x and w are integers, and y and z are nonzero integers. We wish to prove that $(xz + wy)/yz$ is a rational number. Now, xz and wy are integers, since the set of integers is closed for multiplication. Also, $xz + wy$ is an integer, by closure for addition in the set of integers. Similarly, yz must be a nonzero integer; thus,

$$\frac{xz + wy}{yz}$$

is an integer divided by a nonzero integer and therefore a member of Q.

Notice that the definition of addition of rational numbers makes no mention of lowest common denominator. As a matter of fact, as we can see by the given examples, the concept of lowest common denominator is not necessary and need not be discussed. However, at times it can make our arithmetic simpler, as we will see in Chapter 6.

SUBTRACTION OF RATIONALS

Subtraction is similar to addition.

> If a/b and c/d are rational numbers, then
>
> $$\frac{a}{b} - \frac{c}{d} = \frac{ad - bc}{bd}$$

Subtraction of rational numbers.

We can repeat our preceding discussion for subtraction to show that the rationals are closed for subtraction. The details are left to the student (see Problem 32 in this section).

EXAMPLES:

1. $\dfrac{2}{3} - \dfrac{^-1}{2} = \dfrac{2 \cdot 2 - 3(^-1)}{3 \cdot 2} = \dfrac{4 + 3}{6} = \dfrac{7}{6}$

2. $\dfrac{4}{5} - \dfrac{7}{9} = \dfrac{36 - 35}{45} = \dfrac{1}{45}$

MULTIPLICATION OF RATIONALS

To motivate a definition of multiplication, consider the following argument. Certainly,

$$\frac{ac}{bd} = \frac{ac}{bd}, \text{ where } a,b,c, \text{ and } d \text{ are integers, } b \text{ and } d \neq 0$$

Then,

$$\frac{ac}{bd} = ac \cdot \frac{1}{bd}$$

$$= ac \cdot \frac{1}{bd} \cdot \left(b \cdot \frac{1}{b}\right) \cdot \left(d \cdot \frac{1}{d}\right)$$

$$= \left(a \cdot \frac{1}{b}\right) \cdot \left(c \cdot \frac{1}{d}\right) \cdot \frac{1}{bd} \cdot b \cdot d$$

$$= \left(a \cdot \frac{1}{b}\right) \cdot \left(c \cdot \frac{1}{d}\right) \cdot \left(\frac{1}{bd} \cdot bd\right)$$

$$= \frac{a}{b} \cdot \frac{c}{d} \cdot 1$$

$$= \frac{a}{b} \cdot \frac{c}{d}$$

This argument has proved nothing, but it suggests that the following definition of multiplication is reasonable.

Multiplication of rational numbers.

> *If a/b and c/d are rational numbers, then*
> $$\frac{a}{b} \cdot \frac{c}{d} = \frac{ac}{bd}$$

To multiply 4/5 by 7/9, we apply the definition:

$$\frac{4}{5} \cdot \frac{7}{9} = \frac{28}{45}$$

EXAMPLE: $\frac{^-3}{5} \cdot \frac{^-1}{2} = \frac{(^-3)(^-1)}{(5)(2)} = \frac{3}{10}$

Now we can see that the set of rational numbers is closed under the operation of multiplication. Suppose x/y and z/w are any two rational numbers. We wish to show that their product is also a rational number. By definition, $x/y \cdot z/w = xz/yw$. Now, since $x,z,y,$ and w are integers, with $y \neq 0$ and $w \neq 0$, we know that xz and yw must also be integers and $yw \neq 0$; by the definition of a rational number, xz/yw is a rational number.

Therefore, we know that the set is closed for the operation of multiplication.

DIVISION OF RATIONALS

Just as the integers gave rise to integers when added, multiplied, or subtracted, so all sums, products, and differences of rational numbers are rational numbers. But are the rationals a closed set for division?

Consider the problem

$$\frac{2}{3} \div \frac{4}{5} = \boxed{}$$

Before considering this problem, look at a more familiar problem,

$$\frac{12}{4} = *$$

By the definition of division, we must find some number for * so that

$$4 \cdot * = 12$$

Similarly, for

$$\frac{2}{3} \div \frac{4}{5} = \boxed{}$$

we must find some number for $\boxed{}$ so that

$$\frac{4}{5} \cdot \boxed{} = \frac{2}{3}$$

What do we put into the box to get the answer? Do it in two parts. First multiply 4/5 by 5/4 to obtain 1. *Then* multiply by 2/3 to obtain the result:

$$\frac{4}{5} \cdot \boxed{\frac{5}{4} \cdot \frac{2}{3}} = \frac{2}{3}$$

Thus,

$$\frac{2}{3} \div \frac{4}{5} = \frac{2}{3} \cdot \frac{5}{4}$$

which seems to suggest that we "invert the fraction we are dividing by, and then multiply." We can show this a little more abstractly by dividing a/b by c/d. Suppose b, c, and d are nonzero integers. Then,

$$\frac{\frac{a}{b}}{\frac{c}{d}} \cdot 1 = \frac{\frac{a}{b} \cdot \frac{d}{c}}{\frac{c}{d} \cdot \frac{d}{c}} = \frac{\frac{a}{b} \cdot \frac{d}{c}}{1} = \frac{a}{b} \cdot \frac{d}{c} = \frac{ad}{bc}$$

Thus, we have the following definition:

Division of rational numbers

Stated in words, we say that, to divide fractions, we "invert and multiply."

> *If a/b and c/d are rational numbers, c ≠ 0, then*
>
> $$\frac{\dfrac{a}{b}}{\dfrac{c}{d}} = \frac{ad}{bc}$$

Note the additional requirement that $c \neq 0$. If $c = 0$, then $c/d = 0$, and this is the same as our previous restriction for division—namely, *you can never divide by zero* (remember you are dividing by c/d). However, the rationals are closed for nonzero division, since ad/bc is a rational (can you explain why?).

EXAMPLES:

1. $\dfrac{^-3}{4} \div \dfrac{^-4}{7} = \dfrac{^-3}{4} \cdot \dfrac{7}{^-4} = \dfrac{^-21}{^-16} = \dfrac{21}{16}$

2. $\dfrac{4}{5} \div \dfrac{7}{9} = \dfrac{4}{5} \cdot \dfrac{9}{7} = \dfrac{36}{35}$

Having reached this stage in the development of numbers, you might ask if we need any other numbers. We have shown that the set of rationals is closed for addition, subtraction, multiplication, and nonzero division; thus, no new numbers could be formed by applying these operations. Could society advance without any new numbers? Is there a need to invent any more numbers to combine with the set of rationals?

PROBLEM SET 4.5

A Problems

Which of the fractions in Problems 1–6 are equal?

EXAMPLE: $\dfrac{14}{15} \overset{?}{=} \dfrac{42}{45}$

Apply the definition of equality: $(14)(45) \overset{?}{=} (15)(42)$

$$630 = 630$$

Thus they are equal.

1. $\dfrac{5}{6} \overset{?}{=} \dfrac{455}{546}$ 2. $\dfrac{23}{27} \overset{?}{=} \dfrac{92}{107}$ 3. $\dfrac{^-7}{9} \overset{?}{=} \dfrac{7}{^-9}$

4. $\dfrac{-5}{-8} \overset{?}{=} \dfrac{5}{8}$

5. $\dfrac{9}{15} \overset{?}{=} \dfrac{55.8}{93}$

6. $\dfrac{14}{71} \overset{?}{=} \dfrac{98}{497}$

Perform the indicated operations in Problems 7–30.

7. $\dfrac{2}{3} + \dfrac{7}{9}$

8. $\dfrac{-5}{7} + \dfrac{4}{3}$

9. $\dfrac{-12}{35} - \dfrac{8}{15}$

10. $2^{-1} + 3^{-1} + 5^{-1}$

11. $\left(\dfrac{4}{5} + 2\right) + \dfrac{-7}{25}$

12. $2 + 2^{-1}$

13. $\dfrac{7}{9} - \dfrac{2}{3}$

14. $\dfrac{4}{7} - \dfrac{-5}{9}$

15. $\dfrac{-3}{5} - \dfrac{-6}{9}$

16. $\left(\dfrac{3}{10} - \dfrac{7}{100}\right) - \dfrac{9}{1000}$

17. $\dfrac{3}{10} - \left(\dfrac{7}{100} - \dfrac{9}{1000}\right)$

18. $5 - 5^{-1}$

19. $\dfrac{2}{3} \cdot \dfrac{5}{7}$

20. $\dfrac{-1}{8} \cdot \dfrac{2}{-5}$

21. $5 \cdot \dfrac{6}{7}$

22. $\dfrac{47}{-5} \cdot \dfrac{-15}{26}$

23. $\left(\dfrac{-5}{7} \cdot \dfrac{-14}{25}\right) \cdot \dfrac{15}{-21}$

24. $7 \cdot 7^{-1}$

25. $\dfrac{4}{9} \div \dfrac{2}{3}$

26. $\dfrac{\frac{-2}{9}}{\frac{6}{7}}$

27. $\dfrac{105}{-11} \div \dfrac{-15}{33}$

28. $\left(\dfrac{5}{7} \div \dfrac{-14}{25}\right) \div \dfrac{15}{21}$

29. $\dfrac{5}{7} \div \left(\dfrac{-14}{25} \div \dfrac{15}{21}\right)$

30. $6 \div 6^{-1}$

31. Complete the following table. Use a ✔ to show that the number listed at the top of a column is a member of the set listed at the side. Use an X if the number is not a member of the set.

Sets	5	−7	0	$^4/_5$	$^{-1}/_2$
Natural numbers					
Whole numbers					
Integers					
Rational numbers					

B Problems

32. Show that the rationals are closed for subtraction.

33. Show that the rationals are closed for nonzero division.

34. Are the integers closed for addition, subtraction, multiplication, and non-zero division?

35. Which properties of addition are satisfied by the rationals? Support your answer.

36. Which properties of multiplication are satisfied by the rationals? Support your answer.

37. We have shown that if $a/b = c/d$, then $ad = bc$. Show that if $ad = bc$, then $a/b = c/d$ ($b \neq 0$, $d \neq 0$).

Mind Bogglers

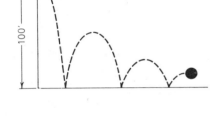

38. A rubber ball is known to rebound half the height it drops. If the ball is dropped from a height of 100 feet, how far will it have traveled by the time it hits the ground for

   a. the first time?
   b. the second time?
   c. the third time?
   d. the fourth time?
   e. the fifth time?
   f. Look for a pattern, and try to make a conjecture about the total distance traveled by the ball after it hits the ground for the nth time.
   g. Is there a maximum distance the ball will travel if we assume that it will bounce indefinitely?

39. *Computer Problem*. Write a BASIC program that will print out a table for Problem 38. That is, the output of the program should give the distance the ball travels by the time it hits the ground for the first, second, third, . . . time.

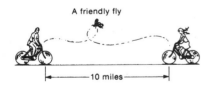

A friendly fly

10 miles

40. A boy cyclist and a girl cyclist are 10 miles apart and pedaling toward each other. The boy's rate is 6 miles per hour, and the girl's rate is 4 miles per hour. There is also a friendly fly zooming continuously back and forth from one bike to the other. If the fly's rate is 20 miles per hour, how far does the fly fly by the time the cyclists reach each other?

41. Answer the question in the following *Peanuts* cartoon strip.

4.6 REAL NUMBERS

IRRATIONALS

The type of society that we have been considering has used numbers only as they relate to practical problems. However, this need not be the case, since numbers can be appreciated for their beauty and interrelationships. The Pythagoreans, to our knowledge, were among the first to investigate numbers for their own sake. The Pythagoreans were a secret society founded in the sixth century B.C. They had their own philosophy, religion, and way of life. This group investigated music, astronomy, geometry, and number properties. Because of their strict secrecy, much of what we know about them is legend, and it is difficult to tell just what work can be attributed to Pythagoras himself.

Much of the Pythagoreans' life-style was embodied in their beliefs about numbers. They considered the number 1 the essence of reason; the number 2 was identified with opinion; and 4 was associated with justice because it is the first number that is the product of equals (the first perfect squared number other than 1). Of the numbers greater than 1, odd numbers were masculine and even numbers were feminine; thus, 5 represented marriage, since it was the union of the first masculine and feminine numbers ($2 + 3 = 5$).

The Pythagoreans were also interested in special types of numbers that had mystical meanings: perfect numbers, friendly numbers, deficient numbers, abundant numbers, prime numbers, triangular numbers, square numbers, and pentagonal numbers. Of these, the square numbers are probably the most important. They are called square numbers because they can be arranged into squares (see Figure 4.6).

The Pythagoreans discovered the famous property of square numbers that today bears Pythagoras' name. They found that if they constructed any right triangle and then constructed squares on each of the legs of the triangle, the area of the larger square was equal to the sum of the areas of the smaller squares (see Figure 4.7).

𝕳istorical 𝕹ote

It was considered impious for a member of the Pythagorean society to claim any discovery for himself. Instead, each new idea was attributed to their founder, Pythagoras.

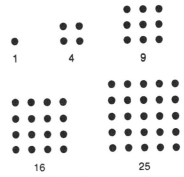

FIGURE 4.6 Square numbers

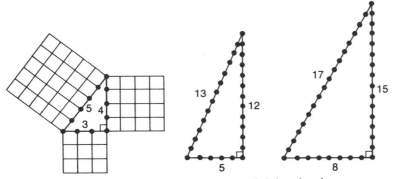

FIGURE 4.7 Relationships of sides of right triangles

𝕳istorical 𝕹ote

Every evening each member of the Pythagorean society had to answer three questions to himself: (1) What good have I done today? (2) What have I failed at today? (3) What have I not done today that I should have done?

Today we state the Pythagorean theorem algebraically by saying that, if a and b are the lengths of the legs of a right triangle and c is the length of the hypotenuse, then the square of the hypotenuse is equal to the sum of the squares of the other two sides.

BY PERMISSION OF JOHN HART AND FIELD ENTERPRISES, INC.

Pythagorean theorem

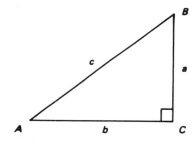

For a right triangle ABC, with sides a, b, and c,

$$a^2 + b^2 = c^2$$

where c is the side opposite the right angle.

The Pythagoreans were overjoyed with this discovery, but it led to a revolutionary idea in mathematics—one that caused the Pythagoreans many problems.

Legend tells us that one day while the Pythagoreans were at sea, one of their group came up with the following argument. Suppose each leg of a right triangle is 1. Then the hypotenuse must be

$$1^2 + 1^2 = c^2$$
$$2 = c^2$$

If we denote the number whose square is 2 by $\sqrt{2}$, we have $\sqrt{2} = c$. Now remember that, for the Pythagoreans, mathematics and religion were one; they asserted that all natural phenomena could be expressed by whole numbers or ratios of whole numbers. Thus they believed that $\sqrt{2}$ must be some whole number or fraction (ratio of two whole numbers). Suppose we try to find such a rational number:

$$\frac{7}{5} \cdot \frac{7}{5} = \frac{49}{25} = 1.96$$

We try again:

$$\frac{707}{500} \cdot \frac{707}{500} = \frac{499,849}{250,000}$$
$$= 1.999396$$

This time we might really get down to business and decide to use a calculator, which gives

$$1.414213562$$

If you square this number, do you obtain 2? Notice that the last digit of this multiplication will be 4; what should it be if it were the square root of 2?

There is an interesting argument to show that all such efforts are in vain. However, before we can follow this reasoning, we need to review some of the properties of even and odd numbers.

Recall that an integer is *even* if it is a multiple of 2—that is, if it may be expressed as $2k$, where k stands for an integer. Then the set of even integers is

$$\{\ldots, ^-6, ^-4, ^-2, 0, 2, 4, 6, \ldots\}$$

An integer that is not even is said to be *odd*. Each odd integer may be expressed in the form $2k + 1$, where k stands for an integer. Then the set of odd integers is

$$\{\ldots, ^-7, ^-5, ^-3, ^-1, 1, 3, 5, 7, \ldots\}$$

EXAMPLES:

1. Prove that the square of an even integer is an even integer.
   Proof: Any even integer may be expressed as $2k$, where k stands for an integer. Then the square of the integer may be expressed as $(2k)^2$.

   $$(2k)^2 = 4k^2 = 2 \cdot 2k^2 = 2(2k^2)$$

   which is even.
2. Prove that the square of an odd integer is an odd integer.
   Proof: Let $2k + 1$ be an odd integer. Show that $(2k + 1)^2$ is odd. The proof is left as a problem.
3. Prove that if a is an integer and a^2 is even, then a is even.
   Proof: Suppose a is odd. Then a^2 is odd by Example 2. This contradicts the fact that a^2 is even, and, since any integer is either even or odd, it follows that a is even.

Let's return to our argument to show that $\sqrt{2}$ can or cannot be represented as a fraction. (The following formal argument is generally attributed to Euclid.)

Suppose that $\sqrt{2}$ is a fraction a/b, where a and b are whole numbers. Moreover, to make matters simpler, let us suppose that any factors common to a and b are canceled. (That is, a/b is in reduced form.) We assume, then, that

$$\sqrt{2} = \frac{a}{b} \qquad (1)$$

Using a computer, I obtained the following possibility for $\sqrt{2}$: 1.41421356237309504880168872420969078569671875376 94

Historical Note

The Egyptians and the Chinese knew of the Pythagorean theorem before the Pythagoreans (but they didn't call it the Pythagorean theorem). An early example is shown below. It is attributed to Chou Pei, who was probably a contemporary of Pythagoras'.

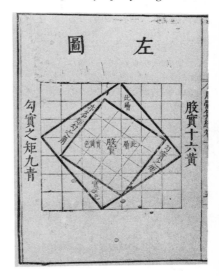

Shown below is a Moslem manuscript, written in 1258, which shows the Pythagorean theorem.

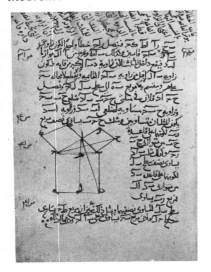

ℌistorical ℕote

The word root *is* radix *in Latin and it was represented by* r. *This led to* r, *then to* √, *and finally to* √ . *Students often say work with radicals is difficult. However, in the 16th century there were four symbols for indicating roots: the letters* R *and* ℓ *(both lower case and capital),* √, *and the fractional exponent. To complicate matters, these symbols were used not only for roots but for unknown quantities as well. Also, the radical signs for cube and fourth roots had quite different shapes and by the close of the 16th century there were 25 or more symbols for radicals.*

Square both sides:

$$2 = \frac{a^2}{b^2}$$

Multiply both sides by b^2:

$$2b^2 = a^2 \tag{2}$$

The left side of this equation is an even number because it contains a factor of 2. Hence the right side must also be an even number. But if a^2 is even, then, according to Example 3, a must be even. If a is even, it must contain 2 as a factor. That is,

$$a = 2d$$

where d is some integer. If we substitute this value of a into equation (2), we obtain

$$2b^2 = (2d)^2 = 2d \cdot 2d = 4d^2 \tag{3}$$

Then, since

$$2b^2 = 4d^2$$

we may divide both sides of this equation by 2 and obtain

$$b^2 = 2d^2 \tag{4}$$

We now see that b^2 is an even number; so, by again appealing to the result of Example 3, we find that b is an even number.

What we have shown in this argument is that if $\sqrt{2} = a/b$, then a and b must be even numbers. But at the very beginning we had canceled any common factor in a and b; yet we find that a and b still contain 2 as a common factor. This result contradicts the fact that a and b have no common factors.

Why do we arrive at a contradiction? Since our reasoning is correct, the only possibility is that the assumption that $\sqrt{2}$ equals a fraction is not correct. In other words, $\sqrt{2}$ cannot be a ratio of two whole numbers (that is, it is not rational).

This result shattered the Pythagorean philosophy. Clearly, they had to change their religion or deny logic. Legend has it that they took the latter course: they set the man who discovered it to sea alone in a small boat and pledged themselves to strict secrecy.

But secrets will get out (indeed, can there be any mathematical or scientific secrets?), and we see that we are faced with the need for still more numbers. We will invent a new set and call it the set of *irrational numbers*. We have seen that $\sqrt{2}$ is irrational, and it can also be shown that $\sqrt{3}$, $\sqrt{5}$, $\sqrt{7}$, and the square root of any number that is not a perfect square, the cube root of any number that is not a perfect cube, and so on are all irrational numbers. The number π, which is the ratio of the circumference of any circle to its diameter, is also not rational.

As before, we will consider the most general set of numbers we have, the rationals, and annex additional elements to that set in order to obtain a more complete set. This new set is called the set of *real numbers*.

> DEFINITION: The set of *real numbers*, denoted by R, is defined as the union of the set of rationals and the set of irrationals.

Definition of real numbers

From this definition, we see that the real numbers may be classified as either rational or irrational. The distinguishing characteristic is their decimal representation. Let's now consider some different ways of representing real numbers.

REPRESENTING REAL NUMBERS

Real numbers that are *rational* can be represented by *terminating decimals* or by *repeating decimals*.

$$\frac{1}{4} = .25 \qquad \frac{5}{8} = .625 \qquad \frac{2}{3} = .66\ldots \qquad \frac{1}{6} = .1666\ldots \qquad \frac{5}{11} = .4545\ldots$$

$$\frac{58}{10} = 5.8$$

Recall that the ellipses (...) indicate that the pattern is continuing in the same fashion.

The first two examples are terminating decimals, since the division ends.

```
      .25                  .625
  4)1.00              8)5.000
     8                   4 8
    ──                   ──
    20                    20
    20                    16
    ──                    ──
     0                    40
                          40
                          ──
                           0
```

The next three examples are repeating decimals, since the division never ends but repeats some sequence of numerals.

```
     .66..              .4545...            .1666...
  3)2.00            11)5.0000           6)1.0000
    1 8                 4 4                 6
    ──                  ──                  ──
    20                  60                  40
    18                  55                  36
    ──                  ──                  ──
     2...               50                  40
                        44                  36
                        ──                  ──
                        60                  40
                        55                  36
                        ──                  ──
                         5...                4...
```

Historical Note

History of the Decimal Point

Author	Time	Notation
Before Simon Stevin		$24\frac{375}{1000}$
Simon Stevin	1585	$24\ 3^{(1)}\ 7^{(2)}\ 5^{(3)}$
Franciscus Vieta	1600	$24|375$
John Kepler	1616	$24(375$
John Napier	1617	$24{:}3\ 7\ 5$
Henry Briggs	1624	24^{375}
William Oughtred	1631	$24|375$
Balam	1653	$24{:}375$
Ozanam	1691	$24 \cdot 3\ 7\ 5$
Modern		24.375

Real numbers that are *irrational* have decimal representations that are *nonterminating* and *nonrepeating*.

$$\sqrt{2} = 1.414214\ldots \qquad \pi = 3.1415926\ldots$$

In each of these examples, the numerals exhibit no repeating pattern and are irrational. Other decimals that do not terminate or repeat are also irrational:

$$0.12345678910111213\ldots \quad \text{and} \quad 0.10110111011110\ldots$$

We see that the real numbers may be classified in several ways. Any real number is:

1. positive, negative, or zero;
2. a rational number or an irrational number;
3. expressible as a terminating, a repeating, or a nonterminating and nonrepeating decimal:
   a. if it terminates, it is rational,
   b. if it eventually repeats, it is rational,
   c. if it does not terminate or repeat, it is irrational.

CHANGING FROM ONE REPRESENTATION TO ANOTHER

From Fractional Representation to Decimal Representation

Given a fraction a/b, we change it to a decimal by using the definition: a/b means a divided by b. (See Table 4.6.)

EXAMPLES:

1. Change $^4/_5$ to a decimal.

$$\begin{array}{r} .8 \\ 5\overline{)4.0} \\ \underline{4\,0} \\ 0 \end{array}$$

Thus, $^4/_5 = .8$.

2. Change $^1/_7$ to a decimal.

$$\begin{array}{r} .142857\ldots \\ 7\overline{)1.000000} \\ \underline{7} \\ 30 \\ \underline{28} \\ 20 \\ \underline{14} \\ 60 \\ \underline{56} \\ 40 \\ \underline{35} \\ 50 \\ \underline{49} \\ 1\ldots \end{array}$$

A mnemonic for remembering π:

"Yes, I have a number"

Notice that the number of letters in the words gives an approximation for π (the comma is the decimal point).

Thus, $1/7 = .142857\ldots$. We will also denote a repeating fraction by using a bar over the numerals that are to be repeated; thus, $1/7 = .\overline{142857}$.

3. Change $6^2/_9$ to a decimal. Now $6^2/_9 = 6 + {}^2/_9$; thus, we concentrate our attention on $^2/_9$.

$$
\begin{array}{r}
.2\ldots \\
9\overline{)2.0} \\
\underline{1\ 8} \\
2\ldots
\end{array}
$$

Thus, $6^2/_9 = 6 + .\overline{2} = 6.\overline{2}$.

TABLE 4.6   Decimal Representations of $1/n$

n	Decimal Representation of $1/n$
1	1.
2	0.5
3	0.333…
4	0.25
5	0.2
6	0.1666…
7	0.142857142857…
8	0.125
9	0.111…
10	0.1
11	0.0909…
12	0.08333…
13	0.076923076923…
14	0.0714285714285…
15	0.0666…
16	0.0625
17	0.05882352941176470588235294117647…
18	0.0555…
19	0.05263157894736842105263157894736842 10…
20	0.05
21	0.047619047619…
22	0.04545…
23	0.04347826086956521739130434782608695652173913…
24	0.041666…
25	0.04
26	0.038461538461538…
27	0.037037…
28	0.0357142857142857…
29	0.0344827586206896551724137931 0344827586206896…
30	0.0333…
31	0.032258064516129032258064516129032258 0645…
32	0.03125
33	0.0303…
34	0.0294117647058823529411764705882 35…
35	0.02857142857142857142857…

Historical Note

It is not at all obvious that π is an irrational number. In the Bible (I Kings 7:23) π is given as 3. In 1892 The New York Times printed an article showing π as 3.2. In 1897, House Bill No. 246 of the Indiana State Legislature defined π as 4; this definition was "offered as a contribution to education to be used only by the State of Indiana free of cost . . . ." Actually, the bill's author had 3.2 in mind but worded the bill so that it came out to be the value 4. The bill was never passed. In a book written in 1934 π was proposed to equal $3^{13}/_{81}$. The symbol itself was first published in 1706 but was not used widely until Euler adopted it in 1737. By 1873 William Shanks had calculated π to 700 decimal places. It took him 15 years to do so; however, computer techniques have shown that his last 100 or so places are incorrect.

From Decimal Representation to Fractional Representation

A. If Decimal Is Terminating:

Write in expanded notation. Recall that $10^{-1} = {}^1/_{10}$, $10^{-2} = {}^1/_{100}$, $10^{-3} = {}^1/_{1000}$, . . . $10^{-n} = {}^1/_{10^n}$. Thus, .5 means $5 \times 10^{-1} = 5 \cdot {}^1/_{10} = {}^5/_{10}$, and we have a fractional representation for .5. The remaining step is to reduce the fraction: $.5 = {}^5/_{10} = {}^1/_2$.

EXAMPLES:

1. .123 means $(1 \times 10^{-1}) + (2 \times 10^{-2}) + (3 \times 10^{-3})$

$$= \frac{1}{10} + \frac{2}{100} + \frac{3}{1000} = \frac{123}{1000}.$$

Thus, $.123 = \dfrac{123}{1000}.$

2. 6.28 means $(6 \times 10^0) + (2 \times 10^{-1}) + (8 \times 10^{-2})$

$$= 6 + \frac{2}{10} + \frac{8}{100} = \frac{628}{100} = \frac{157}{25}$$

Thus, $6.28 = \dfrac{157}{25}.$

3. .3479 means $(3 \times 10^{-1}) + (4 \times 10^{-2}) + (7 \times 10^{-3}) + (9 \times 10^{-4})$

$$= \frac{3}{10} + \frac{4}{100} + \frac{7}{1000} + \frac{9}{10,000} = \frac{3479}{10,000}.$$

Thus, $.3479 = \dfrac{3479}{10,000}.$

B. If Decimal Is Repeating:

We carry out the following procedure.

1. Let n = repeating decimal.
2. Multiply both sides of the equation by 10^k, where k is the number of digits repeating.
3. Subtract the original equation.
4. Solve the resulting equation for n.

EXAMPLES:

1. Change $.\overline{7}$ to fraction form a/b.

Let $n = .7777\ldots$

Multiply both sides of the equation by 10^1 (since there is one digit repeating).

Subtract original equation.

$$n = .7777\ldots$$
$$10n = 7.777\ldots$$
$$n = .7777\ldots$$
$$\overline{\qquad\qquad}$$
$$9n = 7$$

Solve for n.

$$n = \frac{7}{9}$$

Thus, $.\overline{7} = \dfrac{7}{9}.$

Historical Note

The Pythagoreans were not the only people to attach special significance to certain types of numbers. In Hong Kong certain car license numbers are considered lucky, and the government has had 27 auctions since 1973 to sell these lucky numbers. For example, the number 8888 is considered lucky because the word eight *in Cantonese* (baht) *also sounds like the word* prosperity. *In a recent auction, Cheng Yu-Ting paid the equivalent of $30,800 for this number.*

2. Change $.\overline{123}$ to fractional form.
Let n = the number.
Multiply both sides by 10^3 (since there are 3 repeating digits).
Subtract original equation.

$$n = \quad .123123\ldots$$
$$1000n = 123.123123\ldots$$
$$n = \quad .123123\ldots$$
$$\overline{}$$
$$999n = 123$$

Solve for n.

$$n = \frac{123}{999} = \frac{41}{333}$$

Thus, $.\overline{123} = \frac{41}{333}$.

3. Change $7.5\overline{64}$ to fractional form.
Let n = the number.
Multiply both sides by 10^2.
Subtract original equation.

$$n = \quad 7.5646464\ldots$$
$$100n = 756.464646\ldots$$
$$n = \quad 7.564646\ldots$$
$$\overline{}$$
$$99n = 748.9$$

Solve for n.

$$n = \frac{748.9}{99} \cdot \frac{10}{10}$$
$$= \frac{7489}{990}$$

Thus, $7.5\overline{64} = \frac{7489}{990}$.

REAL NUMBER LINE

If we once again consider a number line and associate points on the number line with rational numbers, it appears that the number line is just about "filled up" with the rationals. The reason for this feeling of "fullness" is that the rationals form a *dense* set. That is, between every two rationals we can find another rational (see Figure 4.8).

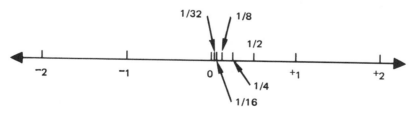

FIGURE 4.8 Real number line

The questions, then, are "Are there any 'holes'?" and "Is there any room left for the irrationals?"

We have shown that $\sqrt{2}$ is irrational. We can show that there is a place on the number line representing this length by using the *Pythagorean theorem* (see Figure 4.9).

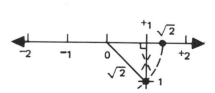

FIGURE 4.9 Construction of a segment of length $\sqrt{2}$

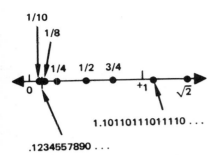

1.10110111011110 . . .

.1234557890 . . .

FIGURE 4.10 Real number line

Thus, we have a length equal to the $\sqrt{2}$ on the number line. We could show that other irrationals have their place on the number line (see Figure 4.10).

Before we consider the relative size of the set of rationals and the set of irrationals, let's review some preliminary concepts.

A. Infinite series

$$\frac{1}{2} + \frac{1}{4} + \frac{1}{8} + \frac{1}{16} + \frac{1}{32} + \cdots + \frac{1}{2^n} + \cdots$$

What is the sum of this series?

First term:
$$\frac{1}{2} = \frac{1}{2}$$

Sum of the first two terms:
$$\frac{1}{2} + \frac{1}{4} = \frac{3}{4}$$

Sum of the first three terms:
$$\frac{1}{2} + \frac{1}{4} + \frac{1}{8} = \frac{7}{8}$$

Sum of the first four terms:
$$\frac{1}{2} + \frac{1}{4} + \frac{1}{8} + \frac{1}{16} = \frac{15}{16}$$

Clearly, if we stop at some particular place in the series, the sum will be less than 1. Thus,

$$\frac{1}{2} + \frac{1}{4} + \frac{1}{8} + \frac{1}{16} + \frac{1}{32} + \cdots \leq 1$$

B. Consider the set

$$A = \left\{ \frac{1}{4}, \frac{1}{8}, \frac{1}{16}, \frac{1}{32}, \frac{1}{64}, \cdots \right\}$$

Now, what is the sum of the series of numbers in set A?

$$\frac{1}{4} = \frac{1}{4}$$

$$\frac{1}{4} + \frac{1}{8} = \frac{3}{8}$$

$$\frac{1}{4} + \frac{1}{8} + \frac{1}{16} = \frac{7}{16}$$

$$\frac{1}{4} + \frac{1}{8} + \frac{1}{16} + \frac{1}{32} = \frac{15}{32}$$

Thus,

$$\frac{1}{4} + \frac{1}{8} + \frac{1}{16} + \frac{1}{32} + \cdots \leq \frac{1}{2}$$

C. What is the sum of the series

$$\frac{1}{8} + \frac{1}{16} + \frac{1}{32} + \frac{1}{64} + \cdots ?$$

Answer:

$$\frac{1}{8} + \frac{1}{16} + \frac{1}{32} + \frac{1}{64} + \cdots \leq \frac{1}{4}$$

D. Let

$$S = \left\{ \frac{1}{2}, \frac{1}{3}, \frac{2}{3}, \frac{1}{4}, \frac{3}{4}, \frac{1}{5}, \frac{2}{5}, \frac{3}{5}, \frac{4}{5}, \frac{1}{6}, \cdots \right\}$$

That is, let S be the fractions between 0 and 1. Now,

$$A = \left\{ \frac{1}{4}, \frac{1}{8}, \frac{1}{16}, \frac{1}{32}, \frac{1}{64}, \cdots \right\}$$

and the sum of all the numbers in A is less than or equal to $1/2$. Show that the elements of A will cover the points in S, by using a number line.

EXAMPLES:

1. Locate $1/2$ on the number line (see Figure 4.11). Cover the point corresponding to $1/2$ with a segment $1/4$ unit long.
2. Cover $1/3$ with a segment $1/8$ unit long.
3. Cover $2/3$ with a segment $1/16$ unit long.
4. Cover $1/4$ with a segment $1/32$ unit long.

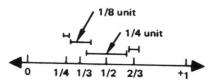

FIGURE 4.11 Covering the rational numbers

See the pattern? We are making the following correspondence:

$$S = \left\{ \frac{1}{2}, \frac{1}{3}, \frac{2}{3}, \frac{1}{4}, \frac{3}{4}, \frac{1}{5}, \frac{2}{5}, \frac{3}{5}, \frac{4}{5}, \frac{1}{6}, \cdots \right\}$$

$$\updownarrow \quad \updownarrow \quad \updownarrow \quad \updownarrow \quad \updownarrow \quad \updownarrow \quad \updownarrow \quad \updownarrow \quad \updownarrow \quad \updownarrow$$

$$A = \left\{ \frac{1}{4}, \frac{1}{8}, \frac{1}{16}, \frac{1}{32}, \frac{1}{64}, \frac{1}{128}, \frac{1}{256}, \frac{1}{512}, \frac{1}{1024}, \frac{1}{2048}, \cdots \right\}$$

Since there is a one-to-one correspondence between S and A, we see that every point in S has been covered. But how long is the cover? It is less than or equal to $\frac{1}{2}$. Thus some points on the number line must be uncovered, and all these remaining numbers are irrational. We have shown that the irrationals must take up *at least* one-half the space between 0 and 1 and therefore must take up at least one-half the space on the entire number line.

E. Repeat this argument for the set S and

$$B = \left\{ \frac{1}{8}, \frac{1}{16}, \frac{1}{32}, \frac{1}{64}, \ldots \right\}$$

Here the cover is less than or equal to $\frac{1}{4}$; therefore we can show that the rationals take up less than one-fourth the space on the number line.

F. Repeat for the set S and

$$C = \left\{ \frac{1}{16}, \frac{1}{32}, \frac{1}{64}, \ldots \right\}$$

Now we have covered the rationals with less than one-eighth the space on the number line. As a matter of fact, we can make the cover as small as we please! We see that almost all the space on the number line is taken up by the irrationals, and very little is used by the dense set we called the rationals.

RELATIONSHIPS BETWEEN SETS OF NUMBERS

The relationships among the various sets of numbers we have been discussing are shown in Figure 4.12.

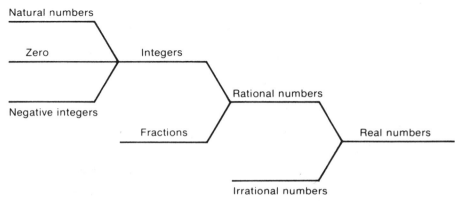

FIGURE 4.12 Sets of numbers

PROBLEM SET 4.6

A Problems

The Pythagorean theorem tells us that the sum of the squares on the legs of a right triangle is equal to the square on the hypotenuse. Verify that the theorem is true by tracing the squares in Problems 1 and 2 and fitting them onto the pattern shown in Figure 4.13. Next, cut the squares along the lines and rearrange the pieces so that they all fit into the large square in Figure 4.13.

1.

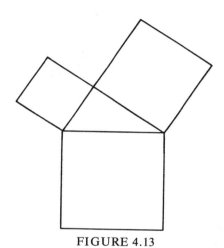

2.

FIGURE 4.13

3. With only a straightedge and compass, use a number line and the Pythagorean theorem to construct a segment whose length is $\sqrt{2}$. Measure the segment as accurately as possible, and write your answer in decimal form. *Do not consult any tables.*

4. What is the distinguishing characteristic between the rational and irrational numbers?

5. Draw a Venn diagram showing the natural numbers, integers, rationals, and irrationals where the universe is the set of reals.

6. Tell whether or not each number is an integer, a rational number, an irrational number, and a real number (since these sets are not all disjoint, you may have more than one answer).

   a. 7
   b. 4.93
   c. .656656665...
   d. 1.868686...
   e. $\sqrt{2}$
   f. 3.14159...
   g. $^{17}/_{43}$
   h. .002727
   i. $\sqrt{8}$
   j. $\sqrt{9}$

Two young braves and three squaws are sitting proudly side by side. The first squaw sits on a buffalo skin with her 50-pound son. The second squaw is on a deer skin with her 70-pound son. The third squaw, who weighs 120 pounds, is on a hippopotamus skin. Therefore, the squaw on the hippopotamus is equal to the sons of the squaws on the other two hides.

Historical Note

Carl Friedrich Gauss (1777–1855), along with Archimedes and Isaac Newton, is considered one of the three greatest mathematicians of all time. Earlier in the text we mentioned child prodigies, and Gauss was one of the greatest. When he was three years old, he corrected an error in his father's payroll calculations. However, unlike the other prodigies we mentioned, Gauss went on to become a great mathematician. By the time he was 21 he had contributed more to mathematics than most mathematicians do in a lifetime. Gauss kept a scientific diary containing 146 entries, some of which were independently discovered and published by others. On July 10, 1796, he wrote

EUREKA!

$$NUM = \triangle + \triangle + \triangle$$

What do you think this notation meant?

Express each of the numbers in Problems 7–14 as a decimal.

7. a. $\dfrac{3}{5}$ b. $\dfrac{27}{15}$ 8. a. $\dfrac{5}{6}$ b. $\dfrac{2}{17}$

9. a. $\dfrac{7}{8}$ b. $\dfrac{3}{7}$ 10. a. $\dfrac{3}{25}$ b. $2^{1}/_{6}$

11. a. $14^{2}/_{11}$ b. $52^{4}/_{9}$ 12. a. $\dfrac{-2}{3}$ b. $-2^{2}/_{13}$

13. a. $\dfrac{15}{3}$ b. $\dfrac{12}{11}$ 14. a. $\dfrac{-17}{6}$ b. $\dfrac{-14}{5}$

Express each of the numbers in Problems 15–22 as the quotient of two integers.

15. a. $0.\overline{45}$ b. $0.\overline{234}$ 16. a. $.1111$ b. $0.5\overline{22}$

17. a. $98.\overline{7}$ b. $.\overline{63}$ 18. a. $5.\overline{123}$ b. $.6\overline{2}$

19. a. $.555$ b. $.5\overline{55}$ 20. a. $.243\overline{33}$ b. 16.4752

21. a. $.\overline{4567}$ b. 3.872 22. a. $.3939$ b. $.\overline{3939}$

23. In the text we spoke of square numbers. Other kinds of numbers are also of interest.

   a. Continue the sequence of triangular numbers: $1,3,6,10,15,$ ——— , ——— , ——— .

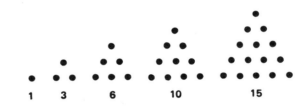

1   3   6   10   15

   b. Continue the sequence of tetrahedral numbers: $1,4,10,20,$ ——— , ——— , ——— .

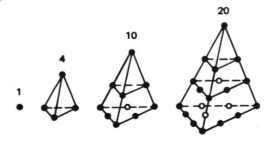

1   4   10   20

   c. Answer the question at the bottom of the historical note about Carl Gauss.

24. One of the most famous problems in mathematics is *Fermat's Four-Square Problem*. Fermat (see historical note at the end of this section) stated that any positive integer can be expressed as the sum of four squares. For example,

$$1 = 1^2 + 0^2 + 0^2 + 0^2 \qquad 6 = 2^2 + 1^2 + 1^2 + 0^2$$
$$2 = 1^2 + 1^2 + 0^2 + 0^2 \qquad 7 = 2^2 + 1^2 + 1^2 + 1^2$$
$$3 = 1^2 + 1^2 + 1^2 + 0^2 \qquad 8 = 2^2 + 2^2 + 0^2 + 0^2$$
$$4 = 2^2 + 0^2 + 0^2 + 0^2 \qquad 9 = 2^2 + 2^2 + 1^2 + 0^2$$
$$5 = 2^2 + 1^2 + 0^2 + 0^2 \qquad 10 = 2^2 + 2^2 + 1^2 + 1^2$$

Write the following numbers using exactly four squares.
a. 11   b. 39   c. 143   d. 1000   e. 1980

B Problems

25. Prove that the set of even integers is closed for multiplication (that is, prove that the product of two even numbers is even).
26. Prove that the set of odd integers is closed for multiplication.
27. Prove that the square of an odd integer is an odd integer.
28. a. Express .999 . . . as the quotient of two integers. Are you surprised at the result? Are there any mistakes in your arithmetic? If not, are there any mistakes in the method?
    b. Show that $1/3 = .333 \ldots$.
    c. Multiply both sides of the equation

$$\frac{1}{3} = .333\ldots$$

   by 3. Is your result consistent with part a?
   d. Look for a pattern in the following problem. Verify that:

$$\frac{1}{9} = .111\ldots$$

$$\frac{2}{9} = .222\ldots$$

$$\frac{3}{9} = .333\ldots$$

$$\frac{4}{9} = .444\ldots$$

$$\vdots$$

$$\frac{7}{9} = .777\ldots$$

$$\frac{8}{9} = .888\ldots$$

What is $9/9$ according to the pattern?

e. Check your answer to this problem by division.

$$1 = .999\ldots$$

$$\frac{9}{9} = .999\ldots$$

and

$$
\begin{array}{r}
.999\ldots \\
9\overline{)9.000\ldots} \\
8\,1 \\
\hline
90 \\
81 \\
\hline
90 \\
81 \\
\hline
9 \\
\vdots
\end{array}
$$

Is there anything "wrong" with this division? (Note: We usually require the remainder to be smaller than the number we're dividing, but this is a matter of definition and is not mandatory.)

29. In the text we stated that every rational number has a decimal representation that is either terminating or repeating. Study Table 4.6 on page 259 and make a conjecture about when a decimal will terminate and when it will repeat. Also, make a conjecture about the number of digits that repeat.

30. Show that $1/8 + 1/16 + 1/32 + 1/64 + \cdots \leq 1/4$.

We will talk about random numbers again in Chapter 9.

31. Let's choose a random number between 0 and 1.

a. Pick one from your head.
b. Is your choice rational or irrational?
c. Let's create a device for selecting our random number. Since the number is between 0 and 1, we must begin with a decimal point. Now *every* rational number can be written as a repeating decimal, since we can consider a terminating decimal such as

All real numbers between 0 and 1 have a decimal representation.

$$\frac{1}{2} = .5$$

$$= .5000\ldots$$

to end with the digit zero repeating. Build a spinner similar to the one shown in Figure 4.14. We select the digits by spinning the dial.

Remember that, for the number to be rational, the sequence must either repeat zeros for a terminating decimal or else repeat indefinitely some other sequence of digits.

i. *First spin:* Suppose it is a 4.   Our number is now .4
ii. *Second spin:* Suppose it is a 3.   .43
iii. *Third spin:* Suppose it is a 3.   .433
iv. *Fourth spin:* Suppose it is a 1.   .4331
$$\vdots \qquad\qquad\qquad\qquad \vdots$$

If we continue to spin, what do you suppose the chances are for obtaining a rational number? This finding is consistent with the demonstration in the text that almost all the space on the number line is taken up by the irrationals.

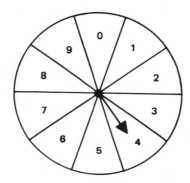

FIGURE 4.14   Random number spinner

32. a. i. Draw a circle.
    ii. Measure the diameter.
    iii. Use a string to approximate the circumference (it is *not* legal to use any formulas here). Measure the string.
    iv. Form the ratio of the diameter into the circumference of the circle. That is, divide the diameter into the circumference. This will serve as an approximation for π.

   b. Repeat these steps for a different circle and compare the results.

33. *Computer Problem.* In the text we considered the series $\frac{1}{2} + \frac{1}{4} + \frac{1}{8} + \frac{1}{16} + \cdots$. Write a BASIC program that will output the partial sums for this series.

34. *Computer Problem.* Repeat Problem 33 for the following series:

   a. $\dfrac{1}{4} + \dfrac{1}{8} + \dfrac{1}{16} + \cdots$ b. $\dfrac{1}{8} + \dfrac{1}{16} + \dfrac{1}{32} + \cdots$

   c. $\dfrac{1}{16} + \dfrac{1}{32} + \dfrac{1}{64} + \cdots$

35. *Census Taking for the Set of Rationals.* We've been writing the set of rationals by using the description methods. Devise a scheme that lists the set of rationals by roster. Limit your attention to those rationals between 0 and 1.

36. *Computer Problem.* Write a BASIC program to output a roster of the set of rational numbers. A simple version of the program would allow the same rational to appear more than once (for example, $\frac{1}{2}$ also appears as $\frac{2}{4}, \frac{3}{6}, \frac{4}{8}, \ldots$). A more complex version of the program would eliminate such repetitions.

Mind Bogglers

37. We have shown that $\sqrt{2}$ is irrational (review this proof). In a similar manner, show that $\frac{1}{2}$ is not an integer.

38. a. How are the square numbers embedded in Pascal's triangle?
    b. How are the triangular numbers embedded in Pascal's triangle?
    c. How are the tetrahedral numbers embedded in Pascal's triangle?

39. *The Extra Square Inch*. Here is a strange and interesting relationship. The square in Figure 4.15 has an area of 64 square units (8×8). When it is rearranged as indicated by the rectangle at the right, it appears to have an area of 65 square units. Where did the "extra" square unit come from?

    Construct your own square 8 inches on a side, and then cut it into the four pieces as illustrated. Place the four pieces together as shown. Be sure to do your measuring and cutting very carefully. Satisfy yourself that this "extra" square inch has appeared. Can you explain this?

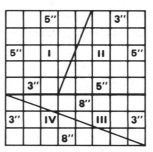

8″ × 8″ = 64 sq in.

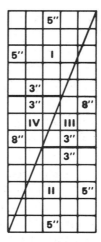

13″ × 5″ = 65 sq in.

FIGURE 4.15 Where did the extra square inch come from?

40. *Spider and Fly Problem.* A room is 80 feet long, 40 feet wide, and 20 feet high.

   a. A fly wishes to get from one lower corner of the room to the opposite upper corner by traveling the shortest possible distance. How far does the fly fly?
   b. A spider also wishes to get to the upper corner from the opposite lower corner (to eat the fly, no doubt). What is the shortest distance that the spider must travel?

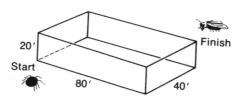

41. In 1907, the University of Göttingen offered the Wolfskehl Prize of 100,000 marks to anyone who could prove Fermat's Last Theorem, which seeks any replacements for *a, b,* and *c* such that $a^n + b^n = c^n$ (where *n* is greater than 2 and *a, b,* and *c* are counting numbers). In 1937, the mathematician Samuel Krieger announced that 1324, 731, and 1961 solved the equation. He would not reveal *n*—the power—but said that it was less than 20. That is,

$$1324^n + 731^n = 1961^n$$

However, it is easy to show that this cannot be a solution *for any n.* See if you can explain why by investigating some patterns for powers of numbers ending in 4 and 1.

Problems for Individual Study

42. Find any replacements for *a, b,* and *c* such that $a^n + b^n = c^n$, where *n* is greater than 2 and where *a, b,* and *c* are counting numbers. You might wish to do some research for this problem. Look for references to Fermat's Last Theorem. What can you find out about this problem?

43. *Pythagorean Relationship.* What are some unusual proofs of the Pythagorean theorem? What are some of the unusual relationships that exist between Pythagorean numbers? What models can be made to visualize and prove the relationship? How can this relationship be used to measure indirectly?
   Exhibit suggestions: models of cardboard, marbles, BB shot, or plastic to show the Pythagorean relationship, designs, tiles, cloth, or wallpaper based on this relationship, charts of unusual proofs, historical sidelights, mock-ups of applications in indirect measurement.

44. *Computer Problem.* Write a BASIC program that will output Pythagorean triplets (numbers that satisfy the relation $a^2 + b^2 = c^2$).

Historical Note

Pierre de Fermat (1601–1665) was a lawyer by profession, but he was an amateur mathematician in his spare time. He became Europe's finest mathematician, and he wrote well over 3000 mathematical papers and notes. However, he did not present them formally (he published only one) because he didn't need the status—he did them just for fun. Every theorem that Fermat said he proved has subsequently been verified—with one notable exception, Fermat's Last Theorem. *In the margin of a book he wrote:*

"To divide a cube into two cubes, a fourth power, or in general any power whatever above the second, into powers of the same denomination, is impossible, and I have assuredly found an admirable proof of this, but the margin is too narrow to contain it."

Many of the most prominent mathematicians since his time have tried to prove or disprove this conjecture, but the answer is still unresolved.

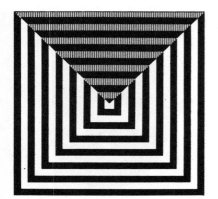

4.7 SUMMARY AND REVIEW

CHAPTER OUTLINE

I. Early Numeration Systems
   A. The difference between a number and a numeral
      1. A **number** is an idea or a concept.
      2. A **numeral** is the symbol or representation for a number.
   B. Types of numeration systems
      1. Simple grouping system, such as the Egyptian numeration system
      2. Positional numeration system
         a. Babylonian system
         b. Hindu-Arabic numeration system
      3. Additive and subtractive properties of a numeration system
   C. Expanded notation
      1. It is used to tell what a number means.
      2. We define

$$b^n = \underbrace{b \cdot b \cdot b \ldots b}_{n \text{ factors}}$$

$$b^{-n} = \frac{1}{b^n} \qquad (b \neq 0)$$

$$b^0 = 1 \qquad (b \neq 0)$$

(n is *any* counting number.)

II. Natural Numbers
   A. *Definition:* $N = \{1,2,3,4,\ldots\}$
   B. Operations
      1. Addition
      2. Multiplication: $a \times b$ means $\underbrace{a + a + \cdots + a}_{b \text{ addends}}$

      3. Subtraction: $a - b = x$ means $a = b + x$
      Division: $\dfrac{a}{b} = x$ means $a = b \cdot x$
      (never divide by zero: $b \neq 0$)
   C. Properties—Let a, b, and c be natural numbers
      1. Closure for addition: $a + b$ is a natural number
      2. Closure for multiplication: ab is a natural number
      3. Commutative for addition: $a + b = b + a$
      4. Commutative for multiplication: $ab = ba$
      5. Associative for addition: $(a + b) + c = a + (b + c)$
      6. Associative for multiplication: $(ab)c = a(bc)$
      7. Distributive: $a(b + c) = ab + ac$

III. Integers
   A. *Definition:* $Z = \{\ldots, ^-3, ^-2, ^-1, 0, 1, 2, 3, \ldots\}$
   B. **Absolute value** of x is the undirected distance of x from zero.
   C. Operations
      1. Addition
         a. same signs; add absolute values
         b. opposite signs; subtract absolute values
      2. Multiplication and division
         a. same signs; positive
         b. opposite signs; negative
      3. Subtraction; add the opposite

IV. Rationals
   A. *Definition:* $Q = \{\frac{p}{q} \mid p, q \in Z, q \neq 0\}$
   B. Operations—Let a/b and c/d be rational numbers

      1. Addition: $\dfrac{a}{b} + \dfrac{c}{d} = \dfrac{ad + bc}{bd}$

      2. Subtraction: $\dfrac{a}{b} - \dfrac{c}{d} = \dfrac{ad - bc}{bd}$

      3. Multiplication: $\dfrac{a}{b} \cdot \dfrac{c}{d} = \dfrac{ac}{bd}$

      4. Division: $\dfrac{\frac{a}{b}}{\frac{c}{d}} = \dfrac{ad}{bc} \qquad (c \neq 0)$

   C. Fractional and decimal representation of rational numbers
      1. Changing from fractional to decimal form: divide and obtain a terminating or repeating decimal
      2. Changing from decimal to fractional form:
         a. If decimal terminates: write in expanded notation.
         b. If decimal repeats:
            i. Let $n = $ repeating decimal.
            ii. Multiply both sides of the equation by 10^k, where k is the number of digits repeating.
            iii. Subtract the original equation.
            iv. Solve the resulting equation for n.

V. Real Numbers
   A. Irrational: A number whose decimal representation does not terminate and does not repeat

B. *Definition:* The set of real numbers, R, is defined by $R = Q \cup Q'$, where $Q = \{\text{rationals}\}$ and $Q' = \{\text{irrationals}\}$

C. Pythagorean theorem. In a right triangle ABC with hypotenuse c and sides a and b,

$$a^2 + b^2 = c^2$$

D. Real number line
1. Rational numbers are dense.
2. Rational numbers take up almost no space on the real number line.
3. There is a one-to-one correspondence between the real numbers and points on the line.

E. Relationship between sets of numbers; see Figure 4.12 on page 264.

REVIEW PROBLEMS

1. What do we mean when we say that a numeration system is positional? Give examples.

2. Is addition easier in a positional system or in a grouping system? Discuss and show examples.

3. What are some of the characteristics of the Hindu-Arabic numeration system?

4. Let n be any counting number. Define:
a. b^n b. b^{-n} c. b^0 $(b \neq 0)$

5. Write 436.20001 in expanded notation.

6. Write $(4 \times 10^6) + (2 \times 10^4) + (5 \times 10^0) + (6 \times 10^{-1}) + (2 \times 10^{-2})$ in decimal notation.

7. In your own words, explain the closure, commutative, and associative properties.

For Problems 8–10, let

$$A = \{5, 0.7, 0, {}^-6, {}^5/_7, \sqrt{2}, 192.1, 17, \sqrt{16}\}$$
$$R = \{\text{reals}\} \qquad\qquad N = \{\text{naturals}\}$$
$$Z = \{\text{integers}\} \qquad\qquad Q = \{\text{rationals}\}$$
$$Q' = \{\text{irrationals}\}$$

8. Draw a Venn diagram showing the relationships among the sets A, R, Z, N, Q, and Q'.

9. Find
a. $A \cap R$ b. $A \cap Z$ c. $A \cap Q$

10. Find
a. $A \cap N$ b. $Z \cap Q$ c. $A \cap Q'$

11. In the set of rational numbers, why is division by zero excluded?

Perform the indicated operations in each of Problems 12–15.

12. a. $8 + {}^-15$ b. $8 - {}^-15$
    c. $^-10 + {}^-7$ d. $^-10 - {}^-7$
    e. $^-14 + 6$ f. $^-14 - 6$

13. a. $7 - {}^-2$ b. $({}^-5)({}^-3)$ c. $^-5 + {}^-9$
    d. $\dfrac{52}{{}^-13}$ e. $({}^-5)(3) - ({}^-4)$ f. $5 - \dfrac{{}^-11}{5}$

14. a. $\dfrac{4}{7} + \dfrac{5}{9}$ b. $\dfrac{3}{4} - \dfrac{7}{9}$ c. $\dfrac{5}{3} \times 7$

    d. $\left(\dfrac{11}{12} + 2 \right) + \dfrac{{}^-11}{12}$ e. $\dfrac{\dfrac{7}{8}}{\dfrac{1}{2}}$

15. a. $\dfrac{4}{5} \times 3$ b. $\dfrac{2}{3} - \dfrac{3}{5}$

    c. $\dfrac{1}{2} + \dfrac{1}{3} + \dfrac{1}{5}$ d. $\dfrac{5/8}{1/3}$

    e. $\left(\dfrac{152}{313} + 7 \right) + \dfrac{{}^-152}{313}$

16. In your own words, state the Pythagorean theorem.

17. Express $^5/_8$ as a decimal.

18. Express $.\overline{72}$ as a quotient of two integers.

19. Express $.666$ as a quotient of two integers.

20. Find x:

$$\frac{3}{5}\left(\frac{25}{97}\right) + \frac{3}{5}\left(\frac{72}{97}\right) = x$$

CULTURAL

1215: Magna Carta

1250: Rise of European universities. Thomas Aquinas applies Aristotelian logic to the Christian faith

1270: Marco Polo begins his travels

Marco Polo

1291: Crusades end

1299: Florence bankers are forbidden to use Hindu numerals

1310: Dante's *Divine Comedy*

1340: The Chinese are printing with movable type (it won't be done in Europe for 100 years)

1349: To combat Black Death, physicians prescribe alcoholic drinks, which have just been invented by alchemists.

The stone masons of the Cathedrals personify geometry by creating extremely intricate designs

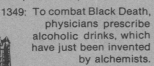

1390: Chaucer's *Canterbury Tales*

| 1200 |
| 1225 |
| 1250 |
| 1275 |
| 1300 |
| 1325 |
| 1350 |
| 1375 |

MATHEMATICAL

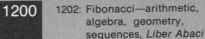

1202: Fibonacci—arithmetic, algebra, geometry, sequences, *Liber Abaci*

Fibonacci's *Liber Abaci* accelerated the adoption of Hindu-Arabic numerals in Europe and very ably summarizes Arabian algebra. His proof that
$x^3 + 2x^2 + 10 = 20$
cannot be solved by square roots and his accurate numerical solution were both far ahead of his time.

Fibonacci

1250: Sacrobosco—Hindu-Arabic numerals

1303: Chu Shí-Kié—algebra, solutions of equations, Pascal's triangle

1325: Thomas Bradwardine—arithmetic, geometry star polygons

1360: Nicole Oresme—coordinates, fractional exponents

5
THE NATURE OF
DAILY ARITHMETIC

There was an old man who said, "Do
Tell me how I should add two and two?
I think more and more
That it makes about four —
But I fear that is almost too few."

Anonymous

5.1 INTEREST

There are certain arithmetical skills that will enable us to make intelligent decisions about how we spend the money we earn. One of the most fundamental mathematical concepts that consumers, as well as business personnel, must understand is the notion of interest. Simply stated, interest is money paid for the use of money. That is, we receive interest when we let others use our money (when we deposit money in a savings account, for example), and we pay interest when we use the money of others (for example, borrowing from a bank).

The amount of the deposit or loan is called the *principal,* and the interest is stated as a percent of the principal, called the *interest rate.* The fundamental interest formula is

$$I = Prt$$

where I = amount of interest
P = principal
r = interest rate
t = time (in years)

Principal
The interest rate may be stated as a percent (8%) or a decimal (.08). Percent means hundredth, *so*

8 percent = 8 hundredths

= .08

In this book, interest rates are annual interest rates unless stated otherwise.

Look at the cartoon; suppose Blondie saves 20¢ per day, but only for a year. At the end of the year she will have saved $73. If she then puts the money into a savings account at 5½% interest, how much interest does the bank pay her after one year?

$$I = Prt$$

$$= 73(0.055)(1)$$

You can do this calculation on a calculator by pressing

$$\boxed{73} \quad \boxed{\times} \quad \boxed{.055} \quad =$$

$5\frac{1}{2}\% = 5.5\% = 0.055$

to obtain the result $4.0\ 15$. To the nearest cent, she earns $4.01.

EXAMPLE: How much interest does Blondie earn on her initial deposit of $73 in three years?

Solution: $I = Prt$

$= 73(0.055)(3)$

$= 12.045$

or $12.04.

On a deposit of $73, the total amount she would have in three years at simple interest is

$73 + $12.04 = $85.04.

This is an example of *simple interest*, but banks pay *compound interest* on savings accounts. Suppose Blondie's bank pays 5½% interest *compounded annually*. This means that, at the end of the first year that her money is in the bank, she has

principal + interest

$=$ $73.00 + 4.01$

$=$ 77.01

This amount becomes the principal for the second year:

$I = Prt$

$= 77.01(0.055)(1)$

$= 4.23555$

Notice that, for compound interest, interest is paid on the interest earned during the first period.

or $4.24. For the third year she has $77.01 + $4.24 = $81.25 on which she earns interest:

$I = Prt$

$= 81.25(0.055)(1)$

$= 4.46875$

or $4.47. The total amount she would have in three years compounded annually is $81.25 + $4.47 = $85.72.

The process described here is fairly easy to understand but rather tedious to apply, particularly if the number of years is very large. However, if we investigate matters a little more closely we can find a formula for the amount of money, A, present from compound interest.

Notice that this is 68¢ more than when the amount was compounded at simple interest.

After one year,

Amount present $=$ principal + interest

$A = P + Pr$

$= P(1 + r)$

$interest = principal \cdot rate \cdot time$

$= P \cdot r \cdot 1$

$= Pr$

> "The boy that by addition grows
> And suffers no subtraction
> Who multiplies the thing he knows
> And carries every fraction
> Who well divides the precious time
> The due proportion giving
> To sure success aloft will climb
> Interest compound receiving."

After two years,

A = amount after one year + interest for second year

$$= \qquad P(1 + r) \qquad + \qquad \underbrace{P(1 + r)}_{\text{principal for the first year}} \overset{\text{rate (time = 1)}}{r}$$

$$= P(1 + r)(1 + r)$$

$$= P(1 + r)^2$$

For three years,

A = amount from previous years + interest for third year

$$= P(1 + r)^2 \qquad\qquad + \qquad \underbrace{P(1 + r)^2}_{\text{principal for the previous year}} \overset{\text{rate (time = 1)}}{r}$$

$$= P(1 + r)^2(1 + r)$$

$$= P(1 + r)^3$$

The pattern indicates that, if we continue with this process, the amount P at r percent interest compounded annually for t years is

$$A = P(1 + r)^t.$$

We can verify this formula for the example of Blondie's interest after three years:

$$A = 73(1 + 0.055)^3$$

$$= 73(1.055)^3$$

$$= 73(1.055)(1.055)(1.055)$$

On a calculator,

$$\boxed{73} \quad \boxed{\times} \quad \boxed{1.055} \quad \boxed{\times} \quad \boxed{1.055} \quad \boxed{\times} \quad \boxed{1.055} \quad = $$

The result is ᴬ5.7 1962038 or $85.72.

At this point you might be saying to yourself that this looks like a considerable amount of arithmetic and you don't even own a calculator. Although the use of calculators to work problems such as these is almost universal, it is possible to use interest tables instead.

Table 5.1 shows compound interest for $1 for n periods. This means that, for the previous example, find the column headed 5½% and look in the row $n = 3$. The entry is 1.1742. Multiply this number by the principal (73) to obtain

$$85.7166 \quad \text{or} \quad \$85.72$$

EXAMPLE: You are considering a six-year $1000 certificate of deposit paying 8% compounded annually. How much money will you have at the end of six years?

Solution: By calculator,

$$A = 1000(1.08)^6$$

$\boxed{1000} \times \boxed{1.08} \times \boxed{1.08} \times \boxed{1.08} \times \boxed{1.08} \times \boxed{1.08} \times \boxed{1.08} =$

The result **1586.874323** to the nearest cent is $1586.87.

By Table 5.1, find the column headed 8% and the row headed $n = 6$ to find 1.5869. Multiply this by 1000 to find (to the nearest dime) $1586.90.

The difference in the answers is due to rounding in the table.

TABLE 5.1   Compounded Amount of $1.00 for n Periods

n	1%	2%	3%	4%	4½%	5%	5½%	6%	6½%	7%	7½%	8%	8½%	9%	9½%	10%
1	1.0100	1.0200	1.0300	1.0400	1.0450	1.0500	1.0550	1.0600	1.0650	1.0700	1.0750	1.0800	1.0850	1.0900	1.0950	1.1000
2	1.0201	1.0404	1.0609	1.0816	1.0920	1.1025	1.1130	1.1236	1.1342	1.1449	1.1556	1.1664	1.1772	1.1881	1.1990	1.2100
3	1.0303	1.0612	1.0927	1.1249	1.1412	1.1576	1.1742	1.1910	1.2079	1.2250	1.2423	1.2597	1.2773	1.2950	1.3129	1.3310
4	1.0406	1.0824	1.1255	1.1699	1.1925	1.2155	1.2388	1.2625	1.2865	1.3108	1.3355	1.3605	1.3859	1.4116	1.4377	1.4641
5	1.0510	1.1041	1.1593	1.2167	1.2462	1.2763	1.3070	1.3382	1.3701	1.4026	1.4356	1.4693	1.5037	1.5386	1.5742	1.6105
6	1.0615	1.1262	1.1941	1.2653	1.3023	1.3401	1.3788	1.4185	1.4591	1.5007	1.5433	1.5869	1.6315	1.6771	1.7238	1.7716
7	1.0721	1.1487	1.2299	1.3159	1.3609	1.4071	1.4547	1.5036	1.5540	1.6058	1.6590	1.7138	1.7701	1.8280	1.8876	1.9487
8	1.0829	1.1717	1.2668	1.3686	1.4221	1.4775	1.5347	1.5938	1.6550	1.7182	1.7835	1.8509	1.9206	1.9926	2.0669	2.1436
9	1.0937	1.1951	1.3048	1.4233	1.4861	1.5513	1.6191	1.6895	1.7626	1.8385	1.9172	1.9990	2.0839	2.1719	2.2632	2.3579
10	1.1046	1.2190	1.3439	1.4802	1.5530	1.6289	1.7081	1.7908	1.8771	1.9672	2.0610	2.1589	2.2610	2.3674	2.4782	2.5937
11	1.1157	1.2434	1.3842	1.5395	1.6229	1.7103	1.8021	1.8983	1.9992	2.1049	2.2156	2.3316	2.4532	2.5804	2.7137	2.8531
12	1.1268	1.2682	1.4258	1.6010	1.6959	1.7959	1.9012	2.0122	2.1291	2.2522	2.3818	2.5182	2.6617	2.8127	2.9715	3.1384
13	1.1381	1.2936	1.4685	1.6651	1.7722	1.8856	2.0058	2.1329	2.2675	2.4098	2.5604	2.7196	2.8879	3.0658	3.2537	3.4523
14	1.1495	1.3195	1.5126	1.7317	1.8519	1.9799	2.1161	2.2609	2.4149	2.5785	2.7524	2.9372	3.1334	3.3417	3.5629	3.7975
15	1.1610	1.3459	1.5580	1.8009	1.9353	2.0789	2.2325	2.3966	2.5718	2.7590	2.9589	3.1722	3.3997	3.6425	3.9013	4.1772
16	1.1726	1.3728	1.6047	1.8730	2.0224	2.1829	2.3553	2.5404	2.7390	2.9522	3.1808	3.4259	3.6887	3.9703	4.2719	4.5950
17	1.1843	1.4002	1.6528	1.9479	2.1134	2.2920	2.4848	2.6928	2.9170	3.1588	3.4194	3.7000	4.0023	4.3276	4.6778	5.0545
18	1.1961	1.4282	1.7024	2.0258	2.2085	2.4066	2.6215	2.8543	3.1067	3.3799	3.6758	3.9960	4.3425	4.7171	5.1222	5.5599
19	1.2081	1.4568	1.7535	2.1068	2.3079	2.5270	2.7656	3.0256	3.3086	3.6165	3.9515	4.3157	4.7116	5.1417	5.6088	6.1159
20	1.2202	1.4859	1.8061	2.1911	2.4117	2.6533	2.9178	3.2071	3.5236	3.8697	4.2479	4.6610	5.1120	5.6044	6.1416	6.7275
21	1.2324	1.5157	1.8603	2.2788	2.5202	2.7860	3.0782	3.3996	3.7527	4.1406	4.5664	5.0338	5.5466	6.1088	6.7251	7.4002
22	1.2447	1.5460	1.9161	2.3699	2.6337	2.9253	3.2475	3.6035	3.9966	4.4304	4.9089	5.4365	6.0180	6.6586	7.3639	8.1403
23	1.2572	1.5769	1.9736	2.4647	2.7522	3.0715	3.4262	3.8197	4.2564	4.7405	5.2771	5.8715	6.5296	7.2579	8.0635	8.9543
24	1.2697	1.6084	2.0328	2.5633	2.8760	3.2251	3.6146	4.0489	4.5330	5.0724	5.6729	6.3412	7.0846	7.9111	8.8296	9.8497
25	1.2824	1.6406	2.0938	2.6658	3.0054	3.3864	3.8134	4.2919	4.8277	5.4274	6.0983	6.8485	7.6868	8.6231	9.6684	10.8347
26	1.2953	1.6734	2.1566	2.7725	3.1407	3.5557	4.0231	4.5494	5.1415	5.8074	6.5557	7.3964	8.3401	9.3992	10.5869	11.9182
27	1.3082	1.7069	2.2213	2.8834	3.2820	3.7335	4.2444	4.8223	5.4757	6.2139	7.0474	7.9881	9.0490	10.2451	11.5926	13.1100
28	1.3213	1.7410	2.2879	2.9987	3.4297	3.9201	4.4778	5.1117	5.8316	6.6488	7.5759	8.6271	9.8182	11.1671	12.6993	14.4210
29	1.3345	1.7758	2.3566	3.1187	3.5840	4.1161	4.7241	5.4184	6.2107	7.1143	8.1441	9.3173	10.6528	12.1722	13.8998	15.8631
30	1.3478	1.8114	2.4273	3.2434	3.7453	4.3219	4.9840	5.7435	6.6144	7.6123	8.7550	10.0627	11.5583	13.2677	15.2203	17.4494
31	1.3613	1.8476	2.5001	3.3731	3.9139	4.5380	5.2581	6.0881	7.0443	8.1451	9.4116	10.8677	12.5407	14.4618	16.6662	19.1943
32	1.3749	1.8845	2.5751	3.5081	4.0900	4.7649	5.5473	6.4534	7.5022	8.7153	10.1174	11.7371	13.6067	15.7633	18.2495	21.1138
33	1.3887	1.9222	2.6523	3.6484	4.2740	5.0032	5.8524	6.8406	7.9898	9.3253	10.8763	12.6760	14.7632	17.1820	19.9832	23.2252
34	1.4026	1.9607	2.7319	3.7943	4.4664	5.2533	6.1742	7.2510	8.5092	9.9781	11.6920	13.6901	16.0181	18.7284	21.8816	25.5477
35	1.4166	1.9999	2.8139	3.9461	4.6673	5.5160	6.5138	7.6861	9.0623	10.6766	12.5689	14.7853	17.3796	20.4140	23.9604	28.1024
36	1.4308	2.0399	2.8983	4.1039	4.8774	5.7918	6.8721	8.1473	9.6513	11.4239	13.5115	15.9682	18.8569	22.2512	26.2366	30.9127
37	1.4451	2.0807	2.9852	4.2681	5.0969	6.0814	7.2500	8.6361	10.2786	12.2236	14.5249	17.2456	20.4597	24.2538	28.7291	34.0039
38	1.4595	2.1223	3.0748	4.4388	5.3262	6.3855	7.6488	9.1543	10.9467	13.0793	15.6143	18.6253	22.1988	26.4367	31.4584	37.4043
39	1.4741	2.1647	3.1670	4.6164	5.5659	6.7048	8.0695	9.7035	11.6583	13.9948	16.7853	20.1153	24.0857	28.8160	34.4469	41.1448
40	1.4889	2.2080	3.2620	4.8010	5.8164	7.0400	8.5133	10.2857	12.4161	14.9745	18.0442	21.7245	26.1330	31.4094	37.7194	45.2593
41	1.5038	2.2522	3.3599	4.9931	6.0781	7.3920	8.9815	10.9029	13.2231	16.0227	19.3976	23.4625	28.3543	34.2363	41.3027	49.7852
42	1.5188	2.2972	3.4607	5.1928	6.3516	7.7616	9.4755	11.5570	14.0826	17.1443	20.8524	25.3395	30.7644	37.3175	45.2265	54.7637
43	1.5340	2.3432	3.5645	5.4005	6.6374	8.1497	9.9967	12.2505	14.9980	18.3444	22.4163	27.3666	33.3794	40.6761	49.5230	60.2401
44	1.5493	2.3901	3.6715	5.6165	6.9361	8.5572	10.5465	12.9855	15.9729	19.6285	24.0975	29.5560	36.2167	44.3370	54.2277	66.2641
45	1.5648	2.4379	3.7816	5.8412	7.2482	8.9850	11.1266	13.7646	17.0111	21.0025	25.9048	31.9204	39.2951	48.3273	59.3793	72.8905
46	1.5805	2.4866	3.8950	6.0748	7.5744	9.4343	11.7385	14.5905	18.1168	22.4726	27.8477	34.4741	42.6352	52.6767	65.0204	80.1795
47	1.5963	2.5363	4.0119	6.3178	7.9153	9.9060	12.3841	15.4659	19.2944	24.0457	29.9363	37.2320	46.2592	57.4176	71.1973	88.1975
48	1.6122	2.5871	4.1323	6.5705	8.2715	10.4013	13.0653	16.3939	20.5485	25.7289	32.1815	40.2106	50.1912	62.5852	77.9611	97.0172
49	1.6283	2.6388	4.2562	6.8333	8.6437	10.9213	13.7838	17.3775	21.8842	27.5299	34.5951	43.4274	54.4574	68.2179	85.3674	106.7190
50	1.6446	2.6916	4.3839	7.1067	9.0326	11.4674	14.5420	18.4202	23.3067	29.4570	37.1897	46.9016	59.0863	74.3575	93.4773	117.3909

The disadvantages of working with tables are (1) a different table is needed for each different rate, (2) tables are not as readily available as calculators, and (3) even when using a table a final multiplication by the principal is necessary.

If you have deposited money or seen savings and loan advertisements lately, you know that most financial institutions will compound interest more frequently than annually. This is to your advantage, because the shorter the compounding period, the sooner you earn interest on your interest. If Blondie's money is deposited in a 6% account that compounds semiannually, then the period is one-half year and each period pays 3%. The advantage to Blondie is that she begins earning interest on the interest after one-half year instead of after one year. In general, if the money is compounded n times per year, then the rate per period is r/n and the interest formula becomes

$$A = P\left(1 + \frac{r}{n}\right)^{nt}$$

What difference does it make to Blondie how often the interest is compounded? The amounts are shown in the following table.

The differences in these amounts seem insignificant, but, if the amount invested was $73,000, the difference between them would be $70 for one year with accounts paying the same interest rate!

Length of Time (Years)	Compounded Annually	Compounded Semiannually
0	73.00	73.00
1	77.38	77.45
5	97.69	98.11
12	146.89	148.39
50	1344.67	1402.96

Many savings institutions compound interest quarterly ($n = 4$), monthly ($n = 12$), or daily ($n = 360$). (Note: Most savings institutions assume a 360-day year.) Amounts after one year are compared below for a deposit of $73 at 6%.

Method	Formula	Amount
Annually, $n = 1$	$73(1 + .06)^1$	$77.38
Semiannually, $n = 2$	$73\left(1 + \dfrac{.06}{2}\right)^2$	77.45
Quarterly, $n = 4$	$73\left(1 + \dfrac{.06}{4}\right)^4$	77.48
Monthly, $n = 12$	$73\left(1 + \dfrac{.06}{12}\right)^{12}$	77.50
Daily, $n = 360$	$73\left(1 + \dfrac{.06}{360}\right)^{360}$	77.51

Also NEW. Highest Interest Ever on Insured Savings.

8.33% annual yield on   8% interest compounded daily

$1,000 Minimum. 8-Year Term Certificate Account.

Annual yield based on daily compounding when funds and interest remain on deposit a year.
Note: Federal regulations require a substantial interest penalty for early withdrawal of principal from Certificate Accounts.

You have, no doubt, seen advertisements that say

8.00% = 8.33%*

*Effective annual yield when principal and interest are left in the account.

In order to understand effective annual yield, we can look at the earning of $1 at 8% compounded quarterly:

Annually

$$(1)(1 + .08) = 1.08$$

Quarterly

$$(1) \left(1 + \frac{.08}{4}\right)^4 = (1.02)^4$$

$$= 1.08243216$$

The $1 compounded quarterly at 8% is the same as $1 compounded annually at a rate of 8.243216%. Therefore, 8.24% is called the *effective yield* of 8% compounded quarterly.

Let's verify the bank's claim in the advertisement above. They compounded on a daily basis ($n = 360$), so $1 compounded daily is

$$(1) \left(1 + \frac{.08}{360}\right)^{360} \approx (1.00022222)^{360}$$

$$\approx 1.083277348$$

This means that if $1 was compounded annually at 8.3277348% the interest would be the same as $1 compounded daily at 8%. Is the bank justified in advertising 8% = 8.33%?

The same procedure that we use to calculate interest can also be used to calculate inflation rates. For rates of inflation between 1% and 9%, we can use Table 5.1. The government releases monthly or annual inflation

Can you check the claim that $1 compounded quarterly at 8% is the same as $1 compounded annually at 8.243216%?

This calculation was done on a calculator with an exponent key.

≈ means approximately equal to.

rates. For example, in August 1975 the inflation rate was about 9%. Suppose a person was earning $10,000 a year in August 1975. What would her salary need to be in 1981, if this same rate of inflation continued, to maintain her standard of living. Use a calculator or Table 5.1.

t = 6 for six years (from 1975 to 1981).

By calculator: $10,000(1 + 0.09)^6 \approx 16,771.0011$

By Table 5.1:   From the column headed 9% and the row headed $n = 6$, find 1.6771. Multiply this amount by $10,000.

Her salary would need to be about $16,771 to maintain her standard of living.

PROBLEM SET 5.1

A Problems

Use the formula

$$I = Prt$$

to find the simple interest for the information given in Problems 1–10.

1. $500 at 5½% for 1 year
2. $300 at 5½% for 5 years
3. $1200 at 8% for 3 years
4. $900 at 8% for 20 years
5. $1600 at 6% for 12 years
6. $400 at 6% for 7 years
7. $5000 at 8% for 6 years
8. $10,000 at 8% for 6 years
9. $1000 at 8% for 40 years
10. $10,000 at 9% for 40 years

Use a calculator or Table 5.1 to find the interest compounded annually for Problems 11–20.

To compare simple interest and compound interest for some particular examples, compare Problems 1–10 with Problems 11–20.

11. $500 at 5½% for 1 year
12. $300 at 5½% for 5 years
13. $1200 at 8% for 3 years
14. $900 at 8% for 20 years
15. $1600 at 6% for 12 years
16. $400 at 6% for 7 years

17. $5000 at 8% for 6 years

18. $10,000 at 8% for 6 years

19. $1000 at 8% for 40 years

20. $10,000 at 9% for 40 years

B Problems

The prices for certain items in 1978 are given in Problems 21–27. Calculate the expected price of each item in the year 2000, assuming a 6% rate of inflation.

21. Sunday paper, $.50

22. Big Mac, $.95

23. Monthly rent for a one-bedroom apartment, $175

24. Tuition at a private college, $2500

25. Small car, $4000

26. Yearly salary $15,000

27. Average house, $50,000

28. If $5000 is compounded annually at 5½% for 12 years, what is the total interest received at the end of that time?

29. If $10,000 is compounded annually at 8% for 18 years, what is the total interest received at the end of that time?

30. *Calculator Problem.* What is the effective yield of 6% compounded quarterly?

31. *Calculator Problem.* What is the effective yield of 6% compounded monthly?

32. *Calculator Problem.* What is the effective yield of 8% compounded monthly?

33. *Calculator Problem.* Is it better to deposit your money at 5½% compounded annually or 5¼% compounded daily?

To work Problems 33 and 34, you will need a calculator with an exponent key.

34. *Calculator Problem.* How much money would you have in six years if you purchased a $1000 six-year savings certificate that paid 8% compounded daily?

35. *Calculator Problem.* Blondie's $73 will grow to $77.38 at 6% interest if the interest is compounded annually, but it will grow to $77.51 at the *same* interest if the interest is compounded daily. This doesn't seem like a great deal of difference, but let's take the viewpoint of the savings and loan company paying this interest. One savings and loan advertises "Seven Billion Dollars Strong." If the company has $7 billion on deposit and it pays 6% interest compounded daily, how much more money will they pay out than if they were to pay interest compounded annually?

36. *Calculator Problem.* What is the difference in the interest paid in one year on $1 billion between an account at 5½% compounded annually and another account at 5¼% compounded daily?

37. *Calculator Problem.* Blondie's $73 grows to $77.51 in one year if 6% interest is compounded daily. What would she receive if she received interest compounded hourly?

Mind Bogglers

38. Some savings and loan companies advertise that they pay interest *continuously*. Explain what this means.

39. *Calculator Problem.* The problem we solved concerning the Blondie cartoon was simplified for ease of calculation. We assumed that she made a single $73 deposit at the end of a year of saving in a cookie jar. The cartoon, however, implies she will make a $73 deposit *every* year. That is, Blondie will make a series of equal payments over equal time intervals. This is called an *annuity, A,* where P dollars per year are invested at interest rate r compounded continuously over t years. The total amount at the end of t years is approximated by the formula

Annuity

$$A = \frac{P}{r}(2.718^{rt} - 1).$$

For example, if Blondie deposits $73 every year for ten years at 8%, the amount at the end of ten years would be

$$A = \frac{73}{0.08}(2.718^{0.08(10)} - 1)$$

$$\approx \$1118.14$$

a. For how many years would Blondie have to deposit $73 a year in order to save $3650 at 8% interest compounded continuously?
b. How much would she have actually deposited when her annuity reaches $3650?
c. If inflation was a constant 6% per year during the period Blondie was saving, how much would that same coat cost if the price rose to keep even with inflation?

Problems for Individual Study

40. Conduct a survey of banks, savings and loan companies, and credit unions in your area. Prepare a report on the different types of savings accounts available and the interest rates they pay. Include methods of payment as well as the interest rates.

41. Consult an almanac or governmental sources, and then write a report on the current inflation rate. Project some of these results to the year of your own expected retirement.

5.2 INSTALLMENT BUYING

In the last section we discussed interest in terms of savings accounts. In this section we'll discuss interest in terms of borrowing money. The next two sections will look at those items you are most likely to purchase on credit—an automobile and a home.

There are two types of consumer credit that allow you to make installment purchases. The first, called *closed-end*, is the traditional installment loan. An installment loan is an agreement to pay off a loan or a purchase by making equal payments at regular intervals (usually monthly) for some specified period of time. The loan is said to be *amortized* if it is completely paid off by these payments, and the payments are called *installments*. *If the loan is not amortized, then there will be a larger final payment, called a balloon payment*. With an *interest-only* loan, there is a monthly payment equal to the interest with a final payment equal to the amount received when the loan was obtained.

Closed-end loan

The second type of consumer credit is called an *open-end* loan, or credit card loan (Mastercharge, VISA, Sears, and so on). This type of loan allows for purchases or cash advances up to a specified "line of credit," and it has a flexible repayment schedule.

Open-end loan

CLOSED-END CREDIT

Suppose you want to purchase a refrigerator for $500 and you want to pay for it with installments over 2 years, or 24 months. When you ask what the interest rate will be, you are told that it is 9%. Then the salesperson carries out the following calculations:

$$I = Prt$$
$$= (500)(.09)(2)$$
$$= 90$$

Amount due: $500 + \text{interest} = 500 + 90$
$$= 590$$

Payments: $\dfrac{590}{24} = 24.5833\ldots$

You must pay $24.59 per month for 2 years. The interest, when calculated in this fashion, is called *add-on* interest. The important thing to notice is that you would be paying more than 9% interest for this loan. After the first month you will have repaid $24.59 and should no longer pay interest on that portion of the loan. By the time you make your last payment, you will owe the store only $24.59 but you will still be paying interest on the full $500.

*consumer spending and credit.
After wartime restrictions were
lifted, many plans were
reinstated, and the railroads and
airlines began to issue their own
travel cards. In 1949 the Diners
Club was established and was
followed a short time later by
American Express and Carte
Blanche.*

In 1969 a *Truth-in-Lending Act* was passed by Congress that requires all lenders to state the true annual interest rate, which is called the *annual percentage rate* (APR). Regardless of the rate quoted, when you ask the salesperson what the APR is, by law you must be told this rate. This regulation allows you to compare interest rates *before* you sign the contract, which must state the APR even if you have not asked for it.

Let's calculate the APR for the refrigerator purchase, which had a 9% add-on interest rate. To find the true annual interest, calculate the total interest and divide by the amount financed.

$$\frac{90}{500} = .18$$

Multiply by $100 and then use Table 5.2 at the top of page 289 (which was prepared by the Federal Reserve Bank).

$$.18 \times \$100 = \$18.00$$

Find the number of payments (24 for this example) and read across until you find the amount closest to the amount just calculated ($18.09 in this case). Then read up to find that the APR for this loan is just under $16\frac{1}{2}\%$.

EXAMPLE: Suppose you see an advertisement in the newspaper for a typewriter costing $240 with monthly payments of $21.89 for $12 months. What is the annual percentage rate?

Solution   total payments: $12 \times \$21.89 = \$262.68$

total interest: $\$262.68 - \$240 = \$22.68$

$$\frac{interest}{amount} = \frac{22.68}{240} = .0945$$

multiply by 100: $.0945 \times 100 = 9.45$

On Table 5.2 look across the row for 12 payments:
17% APR

On a calculator:

press	display
12	12.
×	12.
21.89	21.89
=	262.68
−	262.68
240	240.
=	22.68
÷	22.68
240	240.
=	.0945
×	.0945
100	100.
=	9.45

The interest rates for closed-end loans can vary greatly. In 1978 the author made a survey of businesses in his hometown and found annual percentage rates from 12% (credit union) to 30% (finance company). Smaller purchases will usually be charged higher percentage rates, so if at all possible it's better to pay cash for smaller items. However, if you must finance small purchases, it is generally better to shop around and arrange your own financing rather than to finance the purchase through the store selling the item. The store that has the best selling price may not have the best financing terms. Remember, your cost is the cost of the item plus the cost of financing.

TABLE 5.2 True Annual Interest Rate

Number of Pay- ments	10%	12%	14%	14$\frac{1}{2}$%	15%	15$\frac{1}{2}$%	16%	16$\frac{1}{2}$%	17%	18%	20%	25%	30%
6	$ 2.94	3.53	4.12	4.27	4.42	4.57	4.72	4.87	5.02	5.32	5.91	7.42	8.93
12	5.50	6.62	7.74	8.03	8.31	8.59	8.88	9.16	9.45	10.02	11.16	14.05	16.98
18	8.10	9.77	11.45	11.87	12.29	12.72	13.14	13.57	13.99	14.85	16.52	20.95	25.41
24	10.75	12.98	15.23	15.80	16.37	16.94	17.51	18.09	18.66	19.82	22.15	28.09	34.19
30	13.43	16.24	19.10	19.81	20.54	21.26	21.99	22.72	23.45	24.92	27.89	35.49	43.33
36	16.16	19.57	23.04	23.92	24.80	25.68	26.57	27.46	28.35	30.15	33.79	43.14	52.83
42	18.93	22.96	27.06	28.10	29.15	30.19	31.25	32.31	33.37	35.51	39.85	51.03	62.66
48	21.74	26.40	31.17	32.37	33.59	34.81	36.03	37.27	38.50	41.00	46.07	59.15	72.83

OPEN-END CREDIT

The most common types of open-end credit used today are through credit cards such as VISA, Mastercharge, department stores, and oil companies. Because it isn't necessary for you to apply for credit each time you want to charge an item, this type of credit is very convenient.

Some of these credit card companies charge no interest if the balance is paid within 30 days, while others use an *average daily balance* method, in which the finance charge is determined by dividing the sum of balances outstanding for each day of the billing period by the number of days in the period and then multiplying by the daily rate of interest. With open-end credit you can extend the time you take to make your payments. Most companies require a minimum payment (say $10) each month, but for balances over a specified amount (usually $200) the minimum amount to be paid each month varies from 2% to 5% of the unpaid balance.

Interest rates on credit card purchases vary from 12% to 18%. Remember that both VISA and Mastercharge cards are issued by different banks, and the terms can differ greatly, depending on the bank. For example, I have the following VISA credit cards:

Comparison of Interest Rates

Yearly Rate	Monthly Rate	Daily Rate
12%	1%	.03287%
18%	1.5%	.04932%

APR for Purchases

Rainier National Bank VISA	12%
Citibank VISA	18% on first $500, 12% over
BankAmericard VISA	18% on first $1500, 12% over

The rates for cash advances differ from the interest rates on purchases.

Some credit cards make no charge if the balance is completely paid within the first billing cycle; others charge interest even during this period.

The finance charges can vary greatly, *even on credit cards showing the same annual percentage rate*; depending on the method being used to calculate the balance. We've already mentioned the average daily balance method. This is one of the most costly methods (of course, that is why

The adjusted balance method is
sometimes called the closing
balance method.

Consumer Reports states that,
according to one accounting
study, the interest costs for
average daily balance and
previous balance methods are
about 16% higher than those for
the adjusted balance method.

Average daily balance method for
this example assumes that 10 days
are required for the receipt of the
payment.

The 4% mentioned here is in
addition to the finance charges.

1/1/79 represents the purchase;
statement on 2/1/79 is the first bill
received. It is assumed that this
credit card uses the best method
for consumers, the adjusted
balance method.

many banks use it). The other two methods are the *adjusted balance method* and the *previous balance method*. On the adjusted balance method the finance charge is calculated on the unpaid balance of the previous month, less credits and payments. The previous balance method is almost always the most costly to the consumer and, therefore, is the one preferred by many retailers. With this method, the finance charge is calculated on the amount owed on the final date of the previous month, and it does not take into account any payments made to reduce the balance.

EXAMPLE: Suppose we consider a $500 credit card purchase at 18% interest. When we receive the bill, we make a payment of $25. The interest charged by the different methods would be:

average daily balance, $7.25
adjusted balance, $7.13
previous balance, $7.50

However, suppose instead of a $25 payment, we make a $300 payment. The interest would be:

average daily balance, $4.50
adjusted balance, $3.00
previous balance, $7.50

Many credit cards have an additional charge for cash advances; these charges can be as high as 4%. Suppose you borrow $600 on a credit card showing 18% interest.

TABLE 5.3 Charges for $600 Cash Advance on a Credit Card Charging 18% Interest

Date	Balance	Interest	Payment	New Balance
1/1/79	$600.00	—	—	$600.00
2/1/79	$624.00	$24.00	$ 50	$574.00
3/1/79	$582.61	$ 8.61	$ 50	$532.61
4/1/79	$540.60	$ 7.99	$ 50	$490.60
5/1/79	$497.96	$ 7.36	$ 50	$447.96
6/1/79	$454.68	$ 6.72	$ 50	$404.68
7/1/79	$410.75	$ 6.07	$ 50	$360.75
8/1/79	$366.16	$ 5.41	$ 50	$316.16
9/1/79	$320.90	$ 4.74	$ 50	$270.90
10/1/79	$274.96	$ 4.06	$ 50	$224.96
11/1/79	$228.33	$ 3.37	$ 50	$178.33
12/1/79	$181.00	$ 2.67	$ 50	$131.00
1/1/80	$132.97	$ 1.97	$132.97	—

TOTAL PAYMENTS: $682.97

TRUE ANNUAL INTEREST: $(82.97/600)100 \approx 13.83$. From Table 5.2 we find that the annual percentage rate is a little under 25%!

Another thing you might notice from this example is that the interest is added to the balance; you are paying interest on the interest the following month. This was called an *effective interest rate* in the last section. Consider the following example.

EXAMPLE: What is the effective rate of annual interest for a credit card that charges $1\frac{1}{2}\%$ interest on your unpaid balance each month?

Solution: $1\frac{1}{2}\%$ per month is 18% per year.

$$\left(1 + \frac{0.18}{12}\right)^{12} = (1.015)^{12}$$

$$\approx 1.19561872$$

Therefore, the effective rate is almost 20% (19.56%)!

PROBLEM SET 5.2

A Problems

1. Briefly explain what the terms *closed-end loan* and *open-end loan* mean.

2. What does the term *fully amortized loan* mean?

3. What does *APR* mean, and why should you ask about it before buying on credit?

Find the APR for each of the loans in Problems 4–7, where the interest rate quoted is the simple add-on interest rate.

4. $500 at 9% for 3 years

5. $800 at 8% for 2 years

6. $4000 at 10% for 4 years

7. $5000 at 7% for 4 years

Find the APR for each of the loans in Problems 8–11.

8. $1000 with monthly payments of $35.65 for 36 months

9. $1500 with monthly payments of $93.93 for 18 months

10. $4500 with monthly payments of $128.69 for 48 months

11. $100 with monthly payments of $9.07 for 12 months

Use the loan chart in the margin to calculate the total finance charges for each of the loans given in Problems 12–15.

12. $800 for 24 months

13. $800 for 18 months

14. $2000 for 30 months

15. $2000 for 36 months

Loan Chart

The first payment is due one month from the date the loan check is issued. Subsequent payments are due on the same day at each month thereafter.

Amount of Loan	Monthly Payments	Annual Percentage Rate
36 Equal Payments		
$3,500	$126.53	18.0
3,200	115.68	18.0
3,000	108.45	18.0
2,500	93.29	20.2
2,200	82.95	21.0
2,000	75.98	21.6
1,800	68.92	22.1
1,600	61.70	22.7
30 Equal Payments		
$3,500	$145.73	18.0
3,200	133.24	18.0
3,000	124.91	18.0
2,500	107.04	20.3
2,000	86.95	21.6
1,800	78.78	22.2
1,500	66.26	23.0
24 Equal Payments		
$3,500	$174.73	18.0
3,200	159.75	18.0
3,000	149.77	18.0
2,500	127.78	20.4
2,000	103.51	21.7
1,800	93.67	22.3
1,500	78.66	23.1
1,200	63.54	24.1
1,000	53.40	25.0
800	43.17	26.1
18 Equal Payments		
$3,200	$204.17	18.0
2,500	162.51	20.5
2,000	131.27	21.8
1,800	118.64	22.3
1,200	80.17	24.2
1,000	67.25	25.1
800	54.23	26.2
700	47.66	26.8
500	34.31	27.9
12 Equal Payments		
$3,200	$293.37	18.0
2,500	232.21	20.5
2,000	187.02	21.8
1,500	141.28	23.2
1,000	95.12	25.1
800	76.52	26.2
500	48.23	27.9
400	38.72	28.6

Historical Note

Bank participation in charge-card credit plans didn't begin until the Franklin National Bank in New York issued charge cards in August of 1951. In the next two years almost 100 banks entered the credit card field, but about half of them discontinued the service in a short time. A major stimulus to bank cards came in 1958 when the Bank of America introduced its program. Chase Manhattan Bank and 40 others quickly followed suit, and there was a rapid increase in the number of bank-card plans. The third period of growth for the bank cards came in the late '60s when various banking groups agreed to interchange their cards. An example of one of these agreements is the Interbank group issuing the Mastercharge card. Current demand has induced almost every bank in the country to enter some type of bank-card plan.

B Problems

16. *Calculator Problem.* Table 5.3 outlines a cash advance of $600 on a credit card charging 18% interest using the adjusted balance method. Complete a similar table for a bank card using the previous balance method, and calculate the APR. Pay off the balance on 1/1/80.

17. *Calculator Problem.* If Table 5.3 had been constructed for a $600 purchase rather than for a $600 cash advance, the first bill on 2/1/79 would have shown no interest. Recalculate Table 5.3 for a $600 purchase using the adjusted balance method. Pay off the balance on 1/1/80, and calculate the APR.

18. Most credit cards provide for a minimum finance charge of 50¢ per month. Suppose you buy a $30 item and make five payments of $5 and then pay the remaining balance. What is the APR for this purchase?

FINANCE CHARGE is based upon account activity during the billing period preceding the current billing period, and is computed upon the "previous balance" ("new balance" outstanding at the end of the preceding billing period) before deducting payments and credits or adding purchases made during the current billing period. The FINANCE CHARGE will be the greater of 50¢ (applied to previous balances of $1.00 through $33.00) or an amount determined as follows:

PREVIOUS BALANCE	PERIODIC RATE	ANNUAL PERCENTAGE RATE
$1,000 or under	1.5% per month	18%
Excess over $1,000	1.0% per month	12%

Statement on a Sears Revolving Charge card. Why is the limitation on the 50¢ FINANCE CHARGE from $1.00 to $33.00?

19. Suppose you wish to purchase a home air conditioner from Sears that costs $800. You have three credit plans from which to choose:

1. Revolving: payments of $80 per month until paid. (APR 18%)
2. Easy Payment: payments of $29.02 per month for 36 months.
3. Modernizing Credit Plan: payments $38.69 per month for 24 months.

What is the APR for each of these choices, and which is the least expensive?

In order to calculate the monthly payment, m, for a loan of P dollars at an interest rate of r for t months, use the formula

$$m = \frac{PK^t(K - 1)}{K^t - 1}$$

where $K = 1 + r/12$.

Use this formula in Problems 20–24 to find the monthly payments for each of the indicated loans.

20. *Calculator Problem*. $500 for 12 months at 12%

21. *Calculator Problem*. $100 for 18 months at 18%

22. *Calculator Problem*. $2300 for 24 months at 15%

23. *Calculator Problem*. $3000 for 36 months at 17%

24. *Computer Problem*. Write a BASIC program to calculate the monthly payment. The program should ask the user to input the amount of loan, the interest rate, and the number of months.

Mind Bogglers

25. *Calculator Problem*. Suppose the following disclosure is made for a certain type of credit.

SUPPLEMENTAL DISCLOSURE FOR EXPRESS CREDIT CHECKS

The monthly periodic FINANCE CHARGE is assessed on the statement closing date by applying the appropriate monthly periodic FINANCE CHARGE rates to an "Average Daily Balance" which is the average of the outstanding balances on each of the calendar days in the billing period, including the amounts of any new Loan Advances Checks, and Credit Life Insurance charges posted during the billing period, but excluding the amounts of any unpaid periodic FINANCE CHARGE or Late Charges.

Monthly Periodic FINANCE CHARGE rates are applied to the Average Daily Balance as follows:

MONTHLY PERIODIC RATE	AVERAGE DAILY BALANCES	NOMINAL ANNUAL PERCENTAGE RATE
1.33%	the first $1000	16.0%
1.04%	the next $1500	12.5%
0.95%	over $2500	11.5%

What is the APR for each of the following 30-day loans?

a. $5,000
b. $10,000

26. *Computer Problem*. Write a computer program to calculate the APR for borrowing a given input amount according to the disclosure statement of Problem 25. Use your program to find the APR for borrowing the following amounts:

a. $800   b. $2500   c. $5000   d. $10,000   e. $12,500

Problems for Individual Study

27. Complete the following chart for sources around your own home.

Loan	Monthly rate	APR
credit card		
credit union—personal loan		
credit union—car loan		
bank—personal loan		
bank—home loan		
savings and loan—home loan		
car dealer—car loan		
life insurance loan		
finance company		

28. Interest rates are sometimes more costly if you pay off the loan early. The "Rule of 78" is a method of charging interest in which the consumer pays for the largest portion of the interest early in the loan. This means that, if it is paid off early, the APR rises. Do some research on the "Rule of 78," and prepare a report for the class. As part of your report, apply your findings to Problem 19 for the three types of loans (assume that you decide to make payments for 12 months and then pay off the balance).

5.3 BUYING AN AUTOMOBILE

One of the things you are most likely to buy in your lifetime is an automobile. Unless you are paying cash, there are two prices that must be added together to give the true cost of your car—the cost of the car and the cost of the financing. As we said in the previous section, the dealer with the best price for the car may not have the best price for financing, and you should shop for the two items separately.

PRICE OF THE AUTOMOBILE

You can do a great deal of "shopping around," but before you talk to a salesperson about price you should decide on the size, style, and accessories you want. An excellent source for helping you make these decisions is *Consumer Reports* (available in most libraries). Every April they devote an issue to the new cars and car buying. For example, in the April 1979 issue, there were articles on dealing with the dealers, choosing options, car loans, dealer cost versus list price, car ratings, specifications, frequency-of-repair records, and tips for buying used cars.

When you're buying a car from a dealer, it's important to remember that the salesperson is a professional and you are an amateur. The price on the window is not the price that is paid by most people buying the car. It is necessary to negotiate a fair price. Table 5.4 will help you approxi-

TABLE 5.4 Dealer's Cost for American Automobiles

Type of Car	Dealer's Cost
Subcompact	.86(.83 — .91)
Compact	.85(.76 — .88)
Large	.79(.75 — .83)
Luxury	.77(.77 — .80)

mate the price the dealer paid for the car. The information in "How to Read an Automobile Price Sticker" will help you to decide on the true value of the car. Most dealers will sell the car for 10% over dealer cost, but the higher volume dealers might settle for as little as 5% over their cost.

EXAMPLE: If the dealer cost on a car is $5120, make a 5% offer and a 10% offer.

Solution: 5% of $5120 is

$$.05(\$5120) = \$256$$
$$\text{A 5\% offer is } \$5120 + \$256 = \$5376.$$
$$\text{A 10\% offer is } \$5120 + \$512 = \$5632.$$

The price you actually pay for this car should be somewhere between $5376 and $5632.

RIDE-ENGINEERED

	MFR'S SUGGESTED RETAIL PRICE
CAPRI 3-DR SEDAN/61D	$ 4481 00
BRIGHT RED	
HIGH BACK BUCKET SEATS	NC
FULL INSTRUMENTATION WITH TACH	NC
SPORT STEERING WHEEL	NC
FRONT DISC BRAKES	NC
RACK AND PINION STEERING	NC
FRONT STABILIZER BAR	NC
ELECTRONIC IGNITION SYSTEM	NC
COLOR-KEYED FRONT/REAR BUMPER	NC
CIGAR LIGHTER	NC
CUT PILE CARPETING	NC
DAY/NIGHT MIRROR	NC
REMOTE CTL SAIL-MOUNTED MIRROR	NC
UNIQUE WIDE BODYSIDE MOLDING	NC
LIFTBACK THIRD DOOR	NC
2.8 LITRE V-6 ENGINE	273 00
CALIFORNIA EMISSIONS SYSTEM	76 00
RS OPTION	249 00
3-SPEED AUTOMATIC	307 00
BELTS - DELUXE COLOR-KEYED	-20 00
TRX 190/65RX390 RADIAL TIRES	116 00
CONSOLE	127 00
POWER DISC BRAKES - FRONT	70 00
RADIO - AM/FM STEREO	176 00
GLASS - TINTED COMPLETE	59 00
LIGHT GROUP	28 00
WHEELS - ALUMINUM TRX	240 00
TRANSPORTATION AND HANDLING	264 00
TOTAL	$ 6486 00

HOW TO READ AN AUTOMOBILE PRICE STICKER

Example for a 1979 Capri

Manufacturer's Suggested Retail Price. *This is the list price, or base price, for the car. To find the dealer's cost, multiply this number by a cost factor. Since this is a subcompact, Table 5.4 tells you that the dealer's cost will vary between*

$$(4481)(.83) = 3719$$

and

$$(4481)(.91) = 4078$$

Options. *The cost factor for the options is often different from the cost factor for the car (see Table 5.6). For this Capri they total $1741. This puts the dealer cost for these options between $1445 and $1584.*

Total of base price and options. *For this Capri the total is between $5164 and $5662.*

Add destination charge.

Dealer's cost *is between $5428 and $5926. You can use this as a basis for making the dealer an offer on this car.*

FINANCING THE AUTOMOBILE

The rules for determining the finance rate for a car are the same as the general rules we discussed in the last section. Many people are very careful about negotiating a good price for the car but then take the dealer's first offer on financing. The finance rates vary as much as the prices of the car; as you can see from Table 5.5, there are several sources for automobile loans. Even though these rates are subject to frequent change, you can see that they vary considerably. In addition, if you have the dealer arrange the loan through a bank or through a company like the General Motors Acceptance Corporation, the dealer will usually raise the rate to cover the costs of arranging the loan. The most economical sources are listed first on Table 5.5.

TABLE 5.5 Sources for New Car Loans*

Passbook Loans—generally 2% above the prevailing rate of the passbook account

Life-Insurance Loans—6%

Credit-Union Loans—12%

Commercial-Bank Loans—these can vary considerably depending on the area of the country; 9% to 13.38%

Dealer Loans—12% to 25%

Finance-Company Loans—15% to 25%

*Keep in mind that interest rates are constantly changing (on almost a daily basis). This table was complete in December 1977 and will not provide the most up-to-date interest-rate information. It will, however, show you the comparisons among various types of loans. That is, you would expect all of these interest rates to fluctuate, but the differences shown here should remain about the same.

As I was gathering information for Table 5.5, I commonly obtained quotes of lower rates from the same bank, dealer, or finance company just by asking if they had any other rates. Do not rely on their arithmetic. For example, when I called Beneficial Finance Company and asked for a $4000 36-month loan for a 1976 Granada, I was told that the payments would be $160.00 per month and that the interest rate was 18%. When I asked what the total interest for this loan would be, I was told $1370. Notice that

$$
\begin{array}{rl}
36 \times \$160 = & \$5760 \\
\text{less loan} & \underline{4000} \\
& \$1760
\end{array}
$$

When I asked about this discrepency, I was told that the interest rate was 18.05% (it raised 5/100%), but I still said that the total interest was $1760. Again I was told that the interest was $1370. I insisted that I be informed why the additional $390 was included. Then I was told it was for insurance.

I said I didn't want insurance, and then I was given a price of $145 per month. Even though this still did not equal $1370 interest, I questioned no

$$
\begin{array}{rl}
36 \times \$145 = & \$5220 \\
\textit{less loan} & \underline{4000} \\
& \$1220
\end{array}
$$

further and assumed that this "cut rate" was due to my questioning. The reason I relate this story is to give you the warning: BUYER BEWARE AND BE INFORMED— Don't hesitate to question.

PROBLEM SET 5.3

A Problems

Use Table 5.6 to determine the dealer's cost for the list price of each of the cars given in Problems 1–10.

1 Chevrolet Chevette

2. Datsun 210

3. Honda Civic CVCC

4. Buick Century

5. Dodge Magnum XE

6. Volkswagen Dasher

7. Buick LeSabre

8. Oldsmobile 98

9. Lincoln Versailles

10. Mercedes-Benz 280SE

Use Table 5.6 to determine the dealer's cost for the options of each of the cars named in Problems 11–20.

11. Chevrolet Chevette

12. Datsun 210

13. Honda Civic CVCC

14. Buick Century

15. Dodge Magnum XE

16. Volkswagen Dasher

17. Buick LeSabre

18. Oldsmobile 98

19. Lincoln Versailles

20. Mercedes-Benz 280SE

If you want insurance for a loan, buy it from an insurance company. A finance company selling insurance is like a car dealer selling financing. Separate these functions if you want the best price. For this example, the finance company was charging $540 for insurance that could be obtained from an insurance company for about $210. If you were interested in life insurance only, a 25-year-old male could buy a 10-year (not a 3-year) decreasing term insurance with a value of $5070 for only $2.95 a month. Insurance will be discussed in Section 5.5.

TABLE 5.6 Prices for Selected New 1979 Cars

Automobile	List	List Factor	Options	Option Factor	Destination Charges
Chevrolet Chevette 4-door hatchback	$ 3299	0.87	$ 868	0.83	$125
Datsun 210 2-door hatchback	$ 4809	0.86	$ 942	0.86	$125
Honda Civic CVCC 2-door	$ 3999	0.91	$ 150	0.91	$125
Buick Century 2-door	$ 4716	0.83	$1262	0.79	$150
Dodge Magnum XE	$ 5886	0.84	$1235	0.76	$200
Volkswagen Dasher 2-door hatchback	$ 6950	0.85	$ 770	0.85	$150
Buick LeSabre 4-door	$ 5888	0.80	$ 865	0.77	$200
Oldsmobile 98 4-door	$ 7919	0.78	$1565	0.77	$200
Lincoln Versailles	$13,466	0.77	$ 760	0.77	$200
Mercedes-Benz 280SE	$25,415	0.80	$1250	0.80	$200

B Problems

To determine a range of prices to pay for a car:
  a. find the dealer's cost for the car (Problems 1–10)
  b. find the dealer's cost for the options (Problems 11–20)
  c. total a and b and add 5% to the total
  d. total a and b and add 10% to the total
  e. add the destination charge (varies depending on where you live, but for these problems use the amount shown in Table 5.6) to your answers to parts c and d. Round your answers to the nearest dollar.

The results of part e give you a range of prices that you should reasonably expect to pay for the car. Carry out these steps in Problems 21–30 for each of the cars listed.

21. Chevrolet Chevette

22. Datsun 210

23. Honda Civic CVCC

24. Buick Century

25. Dodge Magnum XE

26. Volkswagen Dasher

27. Buick LeSabre

28. Oldsmobile 98

29. Lincoln Versailles

30. Mercedes-Benz 280SE

31. A newspaper advertisement offers a $4000 car for nothing down and 36 easy monthly payments of $141.62. What is the annual percentage rate?

32. A car dealer will sell you the $6798 car of your dreams for $798 down and payments of $168.51 per month for 48 months. What is the annual percentage rate?

33. A car dealer carries out the following calculations:

LIST PRICE	$5368
OPTIONS	$1625
DESTINATION CGS	$ 200
SUBTOTAL	$7193
TAX	$ 432
LESS TRADE-IN	$2932
AMOUNT TO BE FINANCED	$4693
8% interest for 48 months	$1501.76
TOTAL	$6194.76
MONTHLY PAYMENT	$ 129.06

What is the annual percentage rate charged?

Use the formula

$$m = \frac{PK^t(K - 1)}{K^t - 1}$$

where m = monthly payment
    P = amount of loan
    r = interest rate
    t = number of months
    K = 1 + r/12

to calculate the monthly payment for the cars indicated in Problems 34–37.

34. *Calculator Problem.* Chevrolet Chevette, 10% offer (Problem 21), 36 months at 11.5%

35. *Calculator Problem.* Datsun 210, full price (Table 5.6), 36 months at 14.5%

36. *Calculator Problem.* Honda Civic, 10% offer (Problem 23), 36 months at 12%

37. *Calculator Problem.* Lincoln Versailles, 5% offer (Problem 29), 36 months at 9.7%

Problem for Individual Study

38. Select a car of your choice, find the list price, and calculate a 5% and a 10% offer price. Check out available money sources in your community, and prepare a report showing the different costs for the same car.

5.4 BUYING A HOME

For many people, buying a home is the single most important financial step of a lifetime. There are three steps in buying and financing a home. The first is finding a home you would like to buy and then reaching an agreement with the seller on the price and terms. At this step you will need to negotiate a *sales contract*. The second step is finding a lender to finance the purchase. You will need to understand interest as you shop around to obtain the best terms. And finally, the third step is payment of certain *closing costs* in a process called *settlement*. For this step you will need to shop around and find a settlement agent who will provide the services you need at the best price.

The best source of information is a booklet for sale for 45¢ by the Superintendent of Documents (U.S. Government Printing Office, Washington, D.C. 20402), called *Settlement Costs*. This booklet is reprinted by many lending institutions and is available free for the asking; it is also available at most public libraries.

A sales contract is an agreement between a buyer and a seller stating the terms of the sale.

Closing costs are charges paid when the transfer of title takes place. Some closing costs are paid by the buyer and some by the seller.

NEGOTIATING A SALES CONTRACT

The first step in buying a home is finding a house you can afford and then coming to an agreement with the seller. A real estate agent can help you find the type of home you want that is consistent with the amount of money you can afford to pay. A great deal depends on the amount of down payment you can make, but a useful rule of thumb to determine the monthly payment you can afford is given below:

1. Subtract any monthly bills (not paid off in the next six months) from your gross monthly income.
2. Divide by 4.

Your house payment should not exceed this amount. Others say that the purchase price should not exceed four times your annual salary.

EXAMPLE: If your gross monthly income is $1500 and your current monthly payments on bills are $285, what is the maximum amount you should plan to spend for house payments?

Solution: 1. $1500 - 285 = 1215$
             2. $1215 \div 4 = 303.75$

The maximum house payment should be about $300.

After you have found a home that you can afford and that you would like to buy, you will be asked to sign a sales contract and make a deposit. This agreement should specify: (1) the purchase price, (2) the amount of down payment, (3) the method of financing (with a refund of deposit if you are unable to secure from a lending institution a mortgage with the interest rate, length of time, and terms that you specify in this contract), (4) whether the buyer or the seller will provide inspection reports (5) which items of personal property are to be included in the sale, and (6) that seller provides free title. You may negotiate with the seller about who pays various settlement fees and other charges. There are generally no fixed rules about which party pays which fee, but many are controlled by local custom.

This deposit is sometimes called earnest money, as evidence of good faith. The amount will vary depending on the price of the home.

The down payment can be as large as you wish, but most lenders have some minimum down payment requirements depending on the appraised value of the property. Although it is sometimes possible to buy a home with a 5% down payment, the amount usually runs between 10% and 25% of the appraised value.

EXAMPLES: Determine the down payment for the following homes.

	Purchase Price	Down Payment Required
1.	$36,000	10%
2.	$54,000	20%
3.	$62,300	15%

Solutions:

1. $(36,000)(.10) = \$3600$
2. $(54,000)(.20) = \$10,800$
3. $(62,300)(.15) = \$9345$

OBTAINING A MORTGAGE

A mortgage is a loan contract. A lender agrees to provide the money you need to buy a specific home. You, in turn, promise to repay the money based on terms set forth in the agreement. The contract should state the amount of the loan, the interest rate, the size of the payment, and the frequency of payments. The contract also may include other provisions, such as penalties and prepayment privileges and any special conditions agreed upon by the lender and you.

As the borrower, you pledge your home as security. It remains pledged until the loan is paid off. If you fail to meet the terms of the contract, the lender has the right to *foreclose*.

There are three types of mortgage loans: (1) conventional loans made between you and a private lender, (2) VA loans made to eligible veterans (these are guaranteed by the Veteran's Administration so they cost less than the other types of loans), and (3) FHA loans made by private lenders and insured by the Federal Housing Administration.

Mortgage

Foreclosure is the obtaining of possession of the property.

The government does not make loans. All loans are with conventional lenders. VA and FHA loans are simply guaranteed loans. These insured loans have certain regulations, such as the ceiling rates shown in Table 5.7.

TABLE 5.7 Rates for FHA, VA, and Conventional Loans

January 1 Year	FHA	VA	Conventional
1970	8.5%	8.5%	8.20%
1971	7.5%	7.5%	7.54%
1972	7%	7%	7.38%
1973	7%	7%	7.86%
1974	8.25%	8.25%	8.51%
1975	9%	9%	10%
1976	9%	9%	9%
1977	8%	8%	8.75%
1978	8.5%	8.5%	9.5%
1979	9.5%	9.5%	10.75%

After you select your home, the most important consideration is obtaining the best possible loan. If you pay off your loan, you will be paying more in interest charges than you paid for the home, so it is very important to compare lender costs. By lender costs we mean all the charges required by a lender, including *loan origination fee, discount points,* and other *one-time settlement charges*. Of course the most important lender cost is the *interest rate*.

The *origination fee* is a one-time charge to cover the lender's administrative costs in processing the loan. It may be a flat $100-to-$300 fee, or it may be expressed as a percentage of the loan.

We will define and discuss each of these terms in this section.

The *discount points,* usually called simply *points,* are a one-time charge used to adjust the yield on the loan to what the market conditions demand. It is used to offset constraints placed on the yield by state or federal regulators. Each point is equal to 1% of the amount of the loan.

EXAMPLE: If you are obtaining a $45,000 loan and the bank charges $3^{1}/_{2}$ points, what is the fee charged?

Solution: $(45,000)(.035) = 1575$
The fee is $1575.

There is a rule of thumb that you can use to calculate the combined effects of the interest rate on your loan and the one-time settlement charges you pay for points. While not perfectly accurate, it is usually close enough for meaningful comparisons between lenders. The rule states that one-time settlement charges equaling 1% of the loan amount increase the interest charge by $^{1}/_{8}$ of 1%.

EXAMPLE: Suppose you wish to borrow $45,000. Lender A quotes 10.5% + 3 pts + $150, and lender B quotes 11% + 2 pts + $250. What are the actual charges for each lender, and which one is giving you the better offer?

Solution:	Lender A	Lender B
points:	3%	2%
origination fee	.3%	.6%
	3.3%	2.6%
	$^{1}/_{8}(3.3\%) \approx .41\%$	$^{1}/_{8}(2.6\%) \approx .32\%$
comparable	10.5 %	11.00%
interest	.41%	.32%
rate	10.91%	11.32%

Now that you can find the comparable interest rates, the next step is to determine the amount of your mortgage payments. Most lenders assume a 30-year period, but you might be able to afford a 20- or 25-year loan, which would greatly reduce the total finance charge.

The amount of money you borrow, of course, depends on the cost of the home and the size of the down payment. Some financial counselors suggest making as large a down payment as you can afford, and others suggest making as small a down payment as is allowed. There are many factors, such as your tax bracket and your investment potential, that you will have to assess in order to determine how large a down payment you want to make. Table 5.8 may help you with this decision.

This means that if loans are hard to find or money is scarce the loan points will go up. If the money supply is plentiful, then there may be no point charges.

The $^{1}/_{8}$ factor corresponds to a pay-back period of approximately 15 years. If you intend instead to hold the property for only five years and pay off the loan at that time, the factor increases to $^{1}/_{4}$.

The origination fee is found as a percentage of the total loan: $150/45,000 \approx .3\%$.

If a lender charges other fees, they would be included here in the same way—as a percentage of the loan.

At this step it helps to have some idea of the period of time you plan to own the house before you pay off the mortgage. Remember, you are paying off the mortgage if you sell your home to someone else.

TABLE 5.8 Effect of Down Payment on the Cost of a $60,000 Home, with Interest at 9%

Down Payment	Monthly Payment (Principal and Interest)			Total Interest		
	20 Years	25 Years	30 Years	20 Years	25 Years	30 Years
$ 0	$540	$504	$483	$69,600	$91,200	$113,880
1,500	528	492	471	66,720	87,600	109,560
3,000	513	480	459	63,120	84,000	105,240
6,000	486	453	435	56,640	75,900	96,600
9,000	459	429	411	50,160	68,700	87,960
12,000	432	402	387	43,680	60,600	79,320
15,000	405	378	363	37,200	53,400	70,680

Note: Monthly payments are rounded to the nearest $1 and total interest to the nearest $10.

Now that you know the amount to be financed, you can calculate your monthly payments using the formula given in the previous sections. But for most purposes an estimation will work just as well. Table 5.9 gives the cost to finance $1000, as well as the monthly payments, for selected years and rates of interest.

Some calculators have a single button that will calculate the monthly payment on a loan.

TABLE 5.9 Cost to Finance $1000 for Selected Years and Rates of Interest

Rate of Interest	Years Financed					
	20 Years		25 Years		30 Years	
	Monthly Cost	Total Cost	Monthly Cost	Total Cost	Monthly Cost	Total Cost
6 %	$ 7.17	$1721	$ 6.45	$1935	$ 6.00	$2160
7 %	7.76	1862	7.07	2121	6.66	2398
8 %	8.37	2009	7.72	2316	7.34	2642
9 %	9.00	2160	8.40	2520	8.05	2898
10 %	9.66	2318	9.09	2727	8.78	3161
10½%	9.98	2397	9.44	2833	9.15	3294
11 %	10.32	2477	9.80	2940	9.52	3427
11½%	10.66	2559	10.16	3049	9.90	3564
12 %	11.01	2643	10.53	3160	10.29	3704
15 %	13.17	3161	12.81	3843	12.64	4550
18 %	15.43	3703	15.17	4551	15.07	5425

EXAMPLE: The home you select costs $68,500, and you pay 20% down. Estimate your monthly payments for a 20-year, a 25-year, and a 30-year mortgage for a 9% interest rate.

Solution: $(68,500)(.2)$ $= 13,700$ down payment
              $68,500 - 13,700 = 54,800$ amount financed

From Table 5.9,   *Total cost per $1000*
    20 years at 9%: $9.00 mo   *$2160*
    25 years at 9%: $8.40 mo   *$2520*
    30 years at 9%: $8.05 mo   *$2898*

                                                          Total cost of loan
Multiply each of these by 54.8 (for $54.8 \times 1000 = $54,800)
    20 years at 9%: 9.00(54.8) = $493.20 monthly payment   *$118,368*
    25 years at 9%: 8.40(54.8) = $460.32 monthly payment   *$138,096*
    30 years at 9%: 8.05(54.8) = $441.14 monthly payment   *$158,810*

Notice that if you pay off this loan in 30 years you will be paying back $158,810 for a house that cost $68,500. The interest charge for this house is $104,010.

CLOSING COSTS

Buying a home and getting a loan to finance it involve filling out a number of papers before the property officially becomes yours.

Between the signing of the sales contract and the closing of the loan, three things usually need to be done: (1) the property will have to be appraised, (2) evidence of title will have to be obtained, and (3) a survey of the land will need to be made. At the time the loan is closed, the note and mortgage will need to be signed and the deed transferring the title to the buyer will need to be executed and then recorded.

In order to carry out these tasks, you will need to select a settlement agent. Settlement practices vary from locality to locality and even within the same county or city. In various areas settlements are conducted by lending institutions, title insurance companies, escrow companies, real estate brokers, or attorneys for the buyer or seller. By investigating and comparing practices and rates, you may find that the first suggested settlement agent may not be the least expensive. You might save money by taking the initiative in arranging for settlement and selecting the firm and location that meets your needs.

Figure 5.1 shows the form developed by HUD (U.S. Department of Housing and Urban Development) that will be filled out by the person who will conduct the settlement meeting. This statement must be delivered or mailed to you before settlement, unless you waive that right.

A sample worksheet for a family purchasing a $35,000 house and obtaining a $30,000 loan is shown below. Line 103 assumes that their total settlement charges are $1000 and is merely illustrative. The amount may be higher in some areas and for some types of transactions and lower for others.

J.	SUMMARY OF BORROWER'S TRANSACTION	
100.	GROSS AMOUNT DUE FROM BORROWER:	
101.	Contract sales price	35,000.00
102.	Personal property	200.00
103.	Settlement charges to borrower *(line 1400)*	1,000.00
104.		
105.		
Adjustments for items paid by seller in advance		
106.	City/town taxes to	
107.	County taxes to	
108.	Assessments 6/30 to 7/31 (owner's assn).	20.00
109.	Fuel oil 25 to gal. @ 50/gal	12.50
110.		
111.		
112.		
120.	GROSS AMOUNT DUE FROM BORROWER	36,232.50
200.	AMOUNTS PAID BY OR IN BEHALF OF BORROWER:	
201.	Deposit or earnest money	1,000.00
202.	Principal amount of new loan(s)	30,000.00
203.	Existing loan(s) taken subject to	
204.		
205.		
206.		
207.		
208.		
209.		
Adjustments for items unpaid by seller		
210.	City/town taxes to	
211.	County taxes 1/1 to 6/30 @$600/yr	300.00
212.	Assessments 1/1 to 6/30 @$100/yr	50.00
213.		
214.		
215.		
216.		
217.		
218.		
219.		
220.	TOTAL PAID BY/FOR BORROWER	31,350.00
300.	CASH AT SETTLEMENT FROM/TO BORROWER	
301.	Gross amount due from borrower *(line 120)*	36,232.50
302.	Less amounts paid by/for borrower *(line 220)*	31,350.00
303.	CASH (☒ FROM) (☐ TO) BORROWER	4,882.50

HUD-1 REV. 5/76

A.

U.S. DEPARTMENT OF HOUSING AND URBAN DEVELOPMENT

SETTLEMENT STATEMENT

FORM APPROVED
OMB NO. 63-R-1501

B. TYPE OF LOAN

1. ☐ FHA 2. ☐ FmHA 3. ☐ CONV. UNINS
4. ☐ VA 5. ☐ CONV. INS.

6. FILE NUMBER: 7. LOAN NUMBER:

8. MORTGAGE INSURANCE CASE NUMBER:

C. NOTE: *This form is furnished to give you a statement of actual settlement costs. Amounts paid to and by the settlement agent are shown. Items marked "(p.o.c.)" were paid outside the closing; they are shown here for informational purposes and are not included in the totals.*

D. NAME OF BORROWER:

E. NAME OF SELLER:

F. NAME OF LENDER:

G. PROPERTY LOCATION:

H. SETTLEMENT AGENT:

PLACE OF SETTLEMENT:

I. SETTLEMENT DATE:

J. SUMMARY OF BORROWER'S TRANSACTION		K. SUMMARY OF SELLER'S TRANSACTION	
100. GROSS AMOUNT DUE FROM BORROWER:		400. GROSS AMOUNT DUE TO SELLER:	
101. Contract sales price		401. Contract sales price	
102. Personal property		402. Personal property	
103. Settlement charges to borrower (line 1400)		403.	
104.		404.	
105.		405.	
Adjustments for items paid by seller in advance		*Adjustments for items paid by seller in advance*	
106. City/town taxes to		406. City/town taxes to	
107. County taxes to		407. County taxes to	
108. Assessments to		408. Assessments to	
109.		409.	
110.		410.	
111.		411.	
112.		412.	
120. GROSS AMOUNT DUE FROM BORROWER		420. GROSS AMOUNT DUE TO SELLER	
200. AMOUNTS PAID BY OR IN BEHALF OF BORROWER:		500. REDUCTIONS IN AMOUNT DUE TO SELLER	
201. Deposit or earnest money		501. Excess deposit (see instructions)	
202. Principal amount of new loan(s)		502. Settlement charges to seller (line 1400)	
203. Existing loan(s) taken subject to		503. Existing loan(s) taken subject to	
204.		504. Payoff of first mortgage loan	
205.		505. Payoff of second mortgage loan	
206.		506.	
207.		507.	
208.		508.	
209.		509.	
Adjustments for items unpaid by seller		*Adjustments for items unpaid by seller*	
210. City/town taxes to		510. City/town taxes to	
211. County taxes to		511. County taxes to	
212. Assessments to		512. Assessments to	
213.		513.	
214.		514.	
215.		515.	
216.		516.	
217.		517.	
218.		518.	
219.		519.	
220. TOTAL PAID BY/FOR BORROWER		520. TOTAL REDUCTION AMOUNT DUE SELLER	
300. CASH AT SETTLEMENT FROM/TO BORROWER		600. CASH AT SETTLEMENT TO/FROM SELLER	
301. Gross amount due from borrower (line 120)		601. Gross amount due to seller (line 420)	
302. Less amounts paid by/for borrower (line 220) ( )		602. Less reductions in amount due seller (line 520) ( )	
303. CASH (☐ FROM) (☐ TO) BORROWER		603. CASH (☐ TO) (☐ FROM) SELLER	

L.	SETTLEMENT CHARGES	PAID FROM BORROWER'S FUNDS AT SETTLEMENT	PAID FROM SELLER'S FUNDS AT SETTLEMENT
700. TOTAL SALES/BROKER'S COMMISSION based on price $ @ % =			
Division of Commission (line 700) as follows:			
701. $ to			
702. $ to			
703. Commission paid at Settlement			
704.			
800. ITEMS PAYABLE IN CONNECTION WITH LOAN			
801. Loan Origination Fee %			
802. Loan Discount %			
803. Appraisal Fee to			
804. Credit Report to			
805. Lender's Inspection Fee			
806. Mortgage Insurance Application Fee to			
807. Assumption Fee			
808.			
809.			
810.			
811.			
900. ITEMS REQUIRED BY LENDER TO BE PAID IN ADVANCE			
901. Interest from to @ $ /day			
902. Mortgage Insurance Premium for months to			
903. Hazard Insurance Premium for years to			
904. years to			
905.			
1000. RESERVES DEPOSITED WITH LENDER			
1001. Hazard insurance months @ $ per month			
1002. Mortgage insurance months @ $ per month			
1003. City property taxes months @ $ per month			
1004. County property taxes months @ $ per month			
1005. Annual assessments months @ $ per month			
1006. months @ $ per month			
1007. months @ $ per month			
1008. months @ $ per month			
1100. TITLE CHARGES			
1101. Settlement or closing fee to			
1102. Abstract or title search to			
1103. Title examination to			
1104. Title insurance binder to			
1105. Document preparation to			
1106. Notary fees to			
1107. Attorney's fees to			
(includes above items numbers;)			
1108. Title insurance to			
(includes above items numbers;)			
1109. Lender's coverage $			
1110. Owner's coverage $			
1111.			
1112.			
1113.			
1200. GOVERNMENT RECORDING AND TRANSFER CHARGES			
1201. Recording fees: Deed $; Mortgage $; Releases $			
1202. City/county tax/stamps: Deed $; Mortgage $			
1203. State tax/stamps: Deed $; Mortgage $			
1204.			
1205.			
1300. ADDITIONAL SETTLEMENT CHARGES			
1301. Survey to			
1302. Pest inspection to			
1303.			
1304.			
1305.			
1400. TOTAL SETTLEMENT CHARGES (enter on lines 103, Section J and 502, Section K)			

HUD-1 REV. 5/76

FIGURE 5.1 HUD Settlement Statement

PROBLEM SET 5.4

A Problems

In your own words explain each of the terms given in Problems 1–6.

1. mortgage

2. origination fee

3. sales contract

4. earnest money

5. points

6. closing costs

Determine each maximum monthly payment for a house, given the information in Problems 7–10.

7. Gross monthly income $985, current monthly payments $147

8. Gross monthly income $1240, current monthly payments $215

9. Gross monthly income $1480, current monthly payments $520

10. Gross monthly income $2300, current monthly payments $350

Determine the down payment for each of the homes in Problems 11–14.

	Purchase Price	Down Payment Required
11.	$48,500	5%
12.	$69,900	20%
13.	$85,000	20%
14.	$112,000	30%

B Problems

Round your answers to the nearest hundredth percent in Problems 15–20.

For Problems 15–20, determine the comparable interest rate for a $50,000 loan, when the quoted information is as given.

15. 11.5 % + 1 pt + $150

16. 10.25% + 3 pts + $250

17. 10.75% + 2 pts + $200

18. 11.75% + 1 pt + $100

19. 11.25% + 1 pt + $350

20. 11.00% + 2 pts + $350

*Use Table 5.9 to estimate the monthly payments for each of the loans indi-
cated in Problems 21–32. Also state the total amount of interest paid if the loan
is carried to full term.*

	Purchase Price	Down Payment	Number of Years	Interest Rate
21.	$48,500	5%	20	12%
22.	$48,500	5%	25	12%
23.	$48,500	5%	30	12%
24.	$69,900	20%	20	$11^{1}/_{2}$%
25.	$69,900	20%	25	$11^{1}/_{2}$%
26.	$69,900	20%	30	$11^{1}/_{2}$%
27.	$85,000	20%	20	11%
28.	$85,000	20%	25	11%
29.	$85,000	20%	30	11%
30.	$112,000	30%	20	$10^{1}/_{2}$%
31.	$112,000	30%	25	$10^{1}/_{2}$%
32.	$112,000	30%	30	$10^{1}/_{2}$%

Problem for Individual Study

33. Do some research and write a report about buying a home in your community. Interview a real estate agent, various lenders, and some escrow
agents. Be sure to include specific information about the local customs and
interest rates.

5.5 BUYING INSURANCE

In the last two sections we discussed buying an automobile and buying
a home. After you buy them you will have to insure them. In addition,
you may want to consider buying some life insurance, which would pay
off the loans if you were to die before they were repaid.

AUTOMOBILE INSURANCE

Buying insurance is like buying any other consumer item, and, if you
are willing to do some preliminary work, you can generally find the kind
of auto insurance you need at a substantial savings over that recommended by the car dealer. The first step is to decide what coverage you
want. The main types of automobile coverage are liability, medical, collision, comprehensive, uninsured motorist, and wage loss.

Liability insurance pays others for damage that was your fault. It is usually quoted using a numerical shorthand such as 10/20/5 or 100/300/25. These numbers represent amounts in thousands of dollars, and the first two numbers state the maximum amount that can be paid for *bodily injury* that you cause. The first represents the maximum amount the company will pay for any one person who is injured or killed; the second number represents the maximum amount the company will pay for any one accident to all the victims or their survivors. The third number represents the maximum amount the company will pay for *property damage*.

EXAMPLE: If you have 10/20/5 coverage and there is a judgment against you for $15,000 for bodily injury to another person and $4000 for damage to that person's car, how much would your insurance pay and how much would you have to pay?

Solution: Your insurance will pay $10,000 for bodily injury, so you would be liable for $5000. Since your coverage will pay up to $5000 for property damage, you would not need to pay anything for property damage.

One of the difficulties with liability insurance is that it only pays a claim to someone who can prove someone else's negligence. This process results in long delays and large legal and administrative expenses. To avoid this difficulty, many states have passed *no-fault* laws. The purpose of no-fault insurance is to provide quick and adequate compensation to everyone who has a major loss, injury, or death in an automobile accident. Unfortunately, there are many political aspects to the passing of no-fault laws, and much work in this area is still needed.

In 1979 there were 24 states with some form of no-fault coverage. See Table 5.10.

Collision coverage pays for the damage you do to your own car; it requires that you pay something before you can collect from the company. The amount you have to pay is called the *deductible*. The higher the deductible, the lower the cost of the insurance. For example, the following rates are based on a basic coverage of $100 deductible.

Deductible

For an excellent discussion of automobile insurance in general and no-fault insurance in particular, see Consumer Reports, *June, July, and August 1977, for a three-part series on automobile insurance.*

$150 deductible; 80%
$250 deductible; 60%
$500 deductible; 45%
$1000 deductible; 25%

EXAMPLE: If your collision insurance costs $145 for $100 deductible, how much would you expect to pay for $250 deductible?

Solution: 60% of $145 = .60 × $145
= $87

TABLE 5.10 States Having No-Fault Legislation in 1979
The amounts shown are the minimum benefits mandated for loss of earnings.

Arkansas.	70% up to $140 per week in 8 days, $7280 limit.
Colorado.	100% up to $125 per week, $6500 limit.
Connecticut.	85% up to $200 per week, $5000 limit.
Delaware.	100% up to 10/20 minimum; deductibles of $250 to $10,000 available.
Florida.	60% up to $10,000 limit; deductibles of $250 to $8000 available.
Georgia.	85% up to $200 per week, $5000 limit.
Hawaii.	100% up to $800 per month, $15,000 limit; deductibles of $100 to $500 available.
Kansas.	85% up to $650 per month, $7800 limit.
Kentucky.	100% up to $200 per week, $10,000 limit; deductibles of $250 to $1000 available.
Maryland.	100% up to $2500 limit.
Massachusetts.	75% up to $2000 limit; deductibles of $250 to $2000 available.
Michigan.	100% up to $1000 per month, $36,000 limit; deductibles up to $300 available.
Minnesota.	85% up to $200 per week, $10,000 limit; deductible of $200 available.
New Jersey.	100% up to $100 per week, $5200 limit.
New York.	80% up to $1000 per month, $36,000 limit; deductible of $200 available.
North Dakota.	85% up to $150 per week, $15,000 limit.
Oregon.	70% up to $750 per month, $9000 limit.
Pennsylvania.	100% up to $15,000 limit; deductible of $100 available.
Puerto Rico.	1st year 59%, $50 per week limit; 2nd year 25%, $25 per week limit.
South Carolina.	100% up to $1000 limit.
South Dakota.	100% up to $60 per week in 14 days, $3120 limit. Purchase is optional.
Texas.	80% up to $2500 per person limit.
Utah.	85% up to $150 per week, $7800 limit; deductibles up to $500 available.
Virginia.	100% up to $100 per week, $5200 limit. Purchase is optional.

The last major type of automobile coverage, called *comprehensive,* pays for loss or damage to your car due to something other than collision, such as damage done during a storm.

After you decide what kinds of coverage you want, check with three or four companies and make your selection. You can use Table 5.11 to help you with this comparison.

TABLE 5.11 Comparison of Automobile Insurance

	Coverage Desired	Company A Annual Rate	Company B Annual Rate	Company C Annual Rate
Liability	_____	_____	_____	_____
Medical	_____	_____	_____	_____
Collision	_____	_____	_____	_____
	deductible			
Comprehensive	_____	_____	_____	_____
	deductible			
Uninsured motorist	_____	_____	_____	_____
Towing	_____	_____	_____	_____
Wage loss	_____	_____	_____	_____
Other fees	_____	_____	_____	_____
TOTAL		_____	_____	_____

Fill in the "Coverage Desired" column first; then obtain quotations from three companies. Be sure to compare annual rates since some companies make quotations for less than one year.

HOME INSURANCE

Protection for your home and personal property is available in various kinds of insurance policies. To help you decide on the kind and amount of coverage you need for your home and its contents, discuss your situation with a qualified insurance agent in your area. Be sure to ask about the cost of different policies. Then you can select the policy that best suits your needs at the least cost to you.

A *standard fire insurance* policy protects you against losses from damage caused by fire and lightning. Fire insurance can be extended to protect your house against losses caused by hailstorm, tornado, wind, explosion, riot, aircraft damage, vehicle damage, and smoke damage. This is known as *extended coverage*. The extra protection of the extended coverage is well worth the small additional cost.

A *personal liability insurance* policy may be desirable, in addition to the fire and extended coverage insurance. With a liability policy, you are protected in the event that someone is injured on your property. Injuries or damage resulting from activities of members of your family are also covered. Some policies have special provisions to pay medical costs within certain limits, regardless of your liability.

A *theft insurance* policy protects your personal property against robbery, burglary, and larceny. Property-theft policies vary widely, so it's important to select a policy carefully. You can get a policy in a broad form that protects your possessions both in your home and away from home, but it costs more than the more limited form.

A *homeowners'* policy provides insurance on the house, garage, and other buildings, insurance on personal property, and coverage for personal liability. Policies for renters are also available. You can buy a homeowners' policy, with its combined coverage, for less money than if you bought these coverages separately. Insurance companies usually have more than one homeowners' policy from which to choose. The policies differ in the extent of coverage, and some package policies may include more protection than you need.

LIFE INSURANCE

There are two basic types of life insurance, *term* and *whole life*. With term insurance you are buying life insurance only, and there is no cash-value buildup. The premium stays the same for the period of the term. You can continue your insurance without the need for a new medical exam if you have a *renewable* term policy. However, the premium goes up after each term, as you can see by looking at Table 5.12. A term policy is said to be *convertible* if you can convert it to whole life insurance.

Whole life insurance is sometimes called straight life *or* ordinary life.

Term insurance is commonly issued for one-year or five-year terms, but other terms are available.

TABLE 5.12 Comparison of Annual Premium Rates of Life Insurance for Term, Whole Life, and Endowment Policies per $1000 of Insurance

Age	5-Year Term	10-Year Term	Whole Life	20-Year Endowment 20 Pay Life at Age 65*
20	$ 3.39	$ 4.90	$ 9.68	$26.07
25	3.49	5.36	11.44	28.63
30	3.71	6.09	13.69	31.60
35	4.26	7.22	16.64	35.13
40	5.55	9.27	20.49	39.34
45	7.56	12.61	25.51	44.41
50	12.06	17.88	31.71	50.70
55	17.95	26.01	40.05	58.85
60	26.95	n.a.	51.36	69.85

The ages given are for males. For females, subtract three years. For example, rates for a 23-year-old female would be the same as the 20-year-old male rates from this table. Women pay lower premiums than men of the same age because they live longer. The figures given in this table are for a participating policy and may vary among companies. This table is given for illustrative purposes only, and you will need to consult your local agent for a quotation.

*This means that you pay the premiums for 20 years and then receive the endowment at age 65.

Instead of increasing the payments to maintain constant coverage, you can keep the same payments while the amount of the insurance decreases. Such a policy is called *decreasing term insurance*. It can be used to protect the mortgage on your home, car, or other installment purchase. As we noted in Section 5.2, it is usually cheaper to arrange your own insurance than to pay for it as part of the item you are buying.

An example of decreasing term insurance is a policy that costs about $12 per month and provides the following coverage:

Age	Amount of Insurance
under 30	$40,000
30–34	35,000
35–39	30,000
40–44	25,000
45–49	18,000
50–54	11,000
55–59	7000
60–66	4000

For younger people, term insurance may be the best insurance buy. But as you can see from Table 5.12, the premiums for term insurance become very expensive as you grow older. At some point you will probably want to purchase a *whole life insurance policy*. Be aware of the fact that insurance agents make a larger commission on whole life policies than on other types of insurance.

Whole life insurance provides insurance plus savings.

A *whole life* policy is characterized by a constant payment throughout the life of the policy, until your death or retirement. The premium is determined by your age when you take out the policy, but it does not become greater as you grow older. It also builds up a *cash value,* which is a form of forced savings or investment program.

The cash value is the amount you would receive if you cancelled the policy.

The extreme step in a forced savings plan with insurance is the *endowment policy*. This type of policy guarantees payment at the end of some period if you live, as well as payment if you die during the interim. These policies are far more expensive than the other types of insurance, and you would probably be better off buying a term or whole life policy while investing the difference in premium rates directly into the bank or other investment program. A new type of policy, called *adjustable life,* is also available, which provides a combination of term insurance and whole life insurance. This type of insurance may be the best type to consider, depending on your own situation.

Endowment insurance policies pay whether you live or die.

As we've said before in this chapter, it is generally more expensive to buy from a company that is not in that particular business. Just as a finance company charges more for insurance than an insurance company does, an insurance company doesn't pay you as well for your investment as a bank or other type of savings program.

EXAMPLE: If a $10,000 10-year term policy is purchased at age 45 and renewed at age 55, for a total of 20 years, what is the total cost?

Solution: 10 years at $126.10 per year (from Table 5.12), and
10 years at $260.10 per year = 10(126.10) + 10(260.10)
= $3862.00

EXAMPLE: If a $10,000 20-year endowment is purchased at age 45, what is the total cost?

See Table 5.12 for these figures.

Solution: 20 years at $444.10 per year = 20(444.10)
= $8882.

At age 65 the amount returned is $10,000, so the net result is a gain of $1118. Remember that all during that 20 years you had $10,000 insurance coverage.

The savings that result from buying term insurance rather than an endowment policy are
first ten years
  444.10 − 126.10 = 318 per year

second ten years
  444.10 − 260.10 = 184 per year

From the examples above, it might appear that a 20-year endowment is a better buy, but let's look at it from another viewpoint. Suppose you are 45 years old and want a 20-year endowment of $10,000. You also want $10,000 of life insurance protection. However, instead of purchasing an endowment policy, you decide to purchase term insurance and put the difference in the premium prices into your own savings account. At the end of 20 years you would have $11,927 in the bank (as shown by the calculation in the margin) instead of only $10,000 from the endowment policy. By buying term insurance and saving the difference, you will have $1927 more than if you had bought an endowment policy.

All life insurance policies are either *participating* or *nonparticipating*. Participating policies pay dividends to the policyholder, while nonparticipating policies do not. Although it is very difficult to compare different policies, *Consumer Reports* did an excellent comparison of five-year term policies (January 1974), one-year term policies (February 1974), and whole life policies (March 1974). These issues should be available in your local library.

EXAMPLE: Compare the costs for a female purchasing a $10,000 10-year term policy at age 53 or two consecutive 5-year term policies at ages 53 and 58.

Solution: 10-year term: 10(178.80) = 1788

first 5-year term:	5(120.60) =	603.00
second 5-year term:	5(179.50) =	897.50
	total =	1500.50

This person would save $287.50 if she purchased two consecutive five-year term policies.

In Problem 39 of Problem Set 5.1, the idea of an annuity was discussed. An annuity for ten years at 6% with an annual deposit of $318.00 would grow to $4357. If this amount were then deposited in a 10-year time certificate paying 8% it would become $9406. The second ten years would be a 10-year annuity at 6% with an annual deposit of $184, which would grow to $2521. Therefore, the grand total saved is $11,927.

Use the figure for 50-year-old males from Table 5.12.

PROBLEM SET 5.5

A Problems

In your own words, explain each of the terms given in Problems 1–8.

1. deductible

2. convertible life

3. cash value

4. no-fault insurance

5. comprehensive

6. bodily injury insurance

7. extended coverage

8. property damage insurance

9. In your own words, compare and contrast term, whole life, and endowment policies.

10. If you have 20/30/10 automobile coverage and there is a judgment against you for $120,000 for bodily injury to another person and $6000 damage to that person's car, how much would your insurance pay and how much would you have to pay?

11. Explain what each of the following sets of numbers means with regard to your automobile insurance.

   a. 40/80/10   b. 100/200/20

In Problems 12–15 state the premium cost for a collision insurance policy with a $100 deductible. For each given deductible, estimate the new cost of the policy.

12. $195; $250 deductible

13. $155; $500 deductible

14. $175; $150 deductible

15. $175; $1000 deductible

B Problems

Use Table 5.12 to determine the annual cost for each of the policies indicated in Problems 16–21.

16. $50,000 10-year term for a 35-year-old male

17. $20,000 20 year endowment for a 45-year-old male

18. $10,000 whole life for a 28-year-old female

19. $15,000 5-year term for a 33-year-old female

20. $60,000 whole life for a 28-year-old female

21. $100,000 5-year term for a 40-year-old male

22. If a man purchases a $50,000 10-year term at age 35, renews it at age 45, and renews it again at age 55, for a total of 30 years, what is the total cost?

23. If a $50,000 20-year endowment is purchased by a male at age 35, paid for 20 years, what is the total cost?

24. If a $50,000 whole life is purchased by a male at age 35 and kept for 30 years, what is the total cost?

25. If a $20,000 10-year term is purchased by a male at age 45 and renewed at age 55, for a total of 20 years, what is the total cost?

26. If a $20,000 20-year endowment is purchased by a male at age 45 and paid for 20 years, what is the total cost?

27. If a $20,000 whole life is purchased by a male at age 45 and kept for 20 years, what is the total cost?

28. If a $10,000 10-year term is purchased by a male at age 25 and renewed at ages 35, 45, and 55, for a total of 40 years, what is the total cost?

29. If a $10,000 20-year endowment is purchased by a male at age 25 and paid for 20 years, what is the total cost?

30. If a $10,000 whole life is purchased by a male at age 25 and kept for 40 years, what is the total cost?

Mind Boggler

For Problem 31, see Problem 39 in Section 5.1 about annuities.

31. Compare the costs of the policies in Problems 22–24 by investing the difference in premiums in an 8% savings account and then cashing in at age 65.

5.6 SUMMARY AND REVIEW

Chapter Outline

I. Interest
   A. Meaning of percent
   B. Simple interest formula
$$I = Prt$$
   C. Compount interest formula
$$A = P(1 + r)^t$$
   D. If money is compounded *n* times per year, the compound interest formula is
$$A = P(1 + r/n)^{nt}$$

For these interest formulas,
I = amount of interest
P = principal
r = annual interest rate
t = number of years
A = amount present
n = number of times per year money is compounded

II. Installment Buying
   A. Closed-end loan
      1. Annual percentage rate (APR)
      2. To find the APR
         a. Divide the total interest by the amount financed
         b. Multiply by 100
         c. Use Table 5.2 to find APR
   B. Open-end loan
      1. Average daily balance method
      2. Adjusted balance method
      3. Previous balance method

III. Automobile Buying
   A. Price
   B. Financing

IV. Home Buying
   A. Sales contract
   B. Obtaining a mortgage
   C. Closing costs

V. Buying Insurance
   A. Automobile insurance
      1. Liability
         a. Bodily injury—one person
         b. Bodily injury—more than one person
         c. Property damage
      2. No-fault insurance
      3. Collision
      4. Comprehensive
   B. Home insurance
      1. Fire
      2. Liability
      3. Theft
      4. Homeowners

C. Life insurance
1. Term
2. Whole life
3. Endowment

REVIEW PROBLEMS

1. What is the simple interest for $1000 at 8% for 20 years?

2. Use a calculator or Table 5.1 on page 281 to find the compound interest for $1000 at 8% for 20 years, compounded annually.

3. What is the effective yield of 8% compounded quarterly?

4. Explain the difference between closed-end and open-end credit.

5. What is APR, and why it is useful?

6. If the simple add-on interest rate for a $1000 loan is quoted at 9% for 3 years, use Table 5.2 on page 289 to find the APR for monthly payments.

7. If a $4195 automobile is advertised at $121.42 per month for 48 months with no down payment, what is the total amount of interest paid for the car?

8. *Calculator Problem*. Use Table 5.2 to determine the APR for the automobile described in Problem 7.

9. The Ford Mustang advertised in Problem 7 has a list price of $4316 with a cost factor of 0.87. Is the dealer's price a good price? How does it compare with a 5% offer? A 10% offer?

10. Suppose you wanted the following options on the Ford Mustang advertised in Problem 7:

    automatic transmission: $368
    power steering $140
    air conditioning $465

    Assume the list price is $4316 (cost factor 0.87), and the option factor is 0.85.

    a. Find the list price of the car with the options you want.
    b. Make a 5% offer on the car with options.
    c. Make a 10% offer on the car with options.

11. Suppose you purchase a $550 TV set on a VISA credit card and receive the bill. You then make a $50 payment; the APR is 18%. Show the total interest charged on your next statement if that credit card uses the following methods of calculating interest.

    a. average daily balance (allow 10 days for payment)
    b. adjusted balance
    c. previous balance

12. Repeat Problem 11 assuming you make a $500 payment.

13. If the average rent for a two-bedroom apartment today is $250 per month, how much would it be in 20 years if there is a 9% inflation rate?

For Problems 14–18 assume that you select a home that costs $65,000 and you will finance the home for 30 years.

14. Determine the following down payments.

    a. 10% down
    b. 20% down
    c. 25% down

15. Use Table 5.9 on page 303 and Problem 14 to estimate the monthly payments for each 30-year loan described.

    a. 10% down, 12% interest rate
    b. 20% down, 11% interest rate
    c. 25% down, 11% interest rate

16. If your gross monthly income is $1800 with monthly payments of $335, estimate the maximum monthly house payment you could afford. What down payment would be necessary for you to afford the payments at 11% (see Problem 15).

17. Suppose you decide to pay 30% down. Lender A quotes 12% + 2 pts + $200, and Lender B quotes $12\frac{1}{4}\%$ + 1 pt + $300. Which lender is giving you the better offer, and by how much?

18. Suppose you decide to pay 30% down and can obtain a 9% loan. What is the total interest paid if you pay off the loan in

    a. 20 years?
    b. 25 years?
    c. 30 years?

19. What is the total cost of a $50,000 10-year term policy purchased by a male at age 45 and renewed at age 55 for a total of 20 years?

20. What is the total cost of a $50,000 20-year endowment policy purchased by a male at age 45?

Use Table 5.12 on page 311 for Problems 19 and 20.

CULTURAL   **1400**   MATHEMATICAL

In Florence, the commercial activity has produced
several books on mercantile arithmetic

1425

1431: Joan of Arc burned at the stake

1435: Ulugh Beg—trig tables
1436: Regiomontanus—established trigonometry

Leonardo da Vinci

1450: Printing from movable type

1450

1453: Constantinople falls
1454: Gutenberg prints Bible

1460: Georg von Peurbach—
        arithmetic, table of
        sines

1470: First printed
        arithmetic book

1475

1482: First printing of
        Euclid's *Elements*

1489: Johann Widmann—first use of + and - signs
1492: Pellos—use of decimal point

1492: Columbus discovers America

1500: Michelangelo's *Pieta*

1500

1505: Leonardo da Vinci—geometry, art, optics
1506: Scipione dal Ferro—cubic equations

1507: da Vinci's *Mona Lisa*
1509: Erasmus: *In Praise of Folly*

1510: Albrecht Dürer—perspective, polyhedra, curves

1513: *The Prince* by Machiavelli
1516: Thomas More's *Utopia*

1520: Luther excommunicated
Martin Luther   1521: Aztec Empire falls to Cortez

6

THE NATURE OF ALGEBRA

Proof requires a person who can give and a person who can receive . . . .

A blind man said, As to the Sun,
I'll take by Bible oath there's none;
For if there had been one to show
They would have known it long ago.
How came he such a goose to be?
Did he not know he couldn't see?
Not he.

Augustus De Morgan

ℌistorical Note

Algebra and algebraic ideas date back 4000 years to the Babylonians and the Egyptians; the Hindus and the Greeks also solved algebraic problems. The Arab mathematician al-Khowârizmî's text (about 825 A.D.*) entitled* Hisâb al-jabr w'almugâbalah *is the origin of the word* algebra. *The book was widely known in Europe through Latin translations. The word* al-jabr *or al-ge-bra became synonymous with equation solving. Interestingly enough, the Arabic word* al-jabr *was also used in connection with medieval barbers. The barber, who also set bones and let blood in those times, usually called himself an* algebrista. *In the 19th century there was a significant change in the prevalent attitude toward mathematics. Up to that time, mathematics was expected to have immediate and direct applications, but now mathematicians became concerned with the structure of their subject and with the* validity, *rather than the practicality, of their conclusions. Thus there was a move toward* pure mathematics *and away from* applied mathematics.

6.1 MATHEMATICAL SYSTEMS

Many people think of algebra as simply a high school mathematics course in which variables are manipulated. It is not the intent of this chapter to duplicate the material of such a course but rather to give you some insight into the nature of algebra. In mathematics, the word *algebra* refers to a structure, or a set of axioms, which form the basis for what is accepted and what is not when manipulating the symbols of that system. There are many different algebras, most of them too advanced to discuss in this book, but they are all studied and defined according to some basic properties. In Chapter 4 we saw some of these properties: closure, commutative properties, associative properties, and a distributive property.

The basic understanding of an algebra is related to what is called a mathematical system.

> A *mathematical system* is a set with at least one defined operation and some developed properties.

An operation is the process of carrying out rules of procedure, such as addition, subtraction, multiplication, or division. However, the word *operation* is not limited to these four fundamental operations. For example, taking square roots is an operation; other abstract operations were discussed in Chapter 4.

As an example of a mathematical system, let's consider the set of real numbers with the operations of addition and multiplication. As we saw in Chapter 4, certain properties are satisfied in the set of real numbers.

ADDITION	MULTIPLICATION
closure: $(a + b)$ is a real number	closure: ab is a real number
associative: $(a + b) + c = a + (b + c)$	associative: $(ab)c = a(bc)$
commutative: $a + b = b + a$	commutative: $ab = ba$

Distributive for multiplication over additon:
$$a(b + c) = ab + ac$$

There are two additional properties that are important in this set of properties, the identity and the inverse properties.

IDENTITY PROPERTY

In Chapter 4 we saw that the development of the concept of zero and a numeral for it did not take place at the same time as the development of the counting numbers. The Greeks were using the letter ''oh'' (omicron) for zero as early as 150 A.D., but the predominant system in Europe was the Roman numeration system, which did not include zero. It was not until the 15th century, when the Hindu-Arabic numeration system finally replaced the Roman system, that the zero symbol came into common usage.

The number 0 (zero) has a special property for addition that allows it to be added to any real number without changing the value of that number. This property is called the identity property for addition of real numbers.

IDENTITY FOR ADDITION: There exists a real number 0, called zero, so that

$$0 + a = a + 0 = a$$

for any real number a.

Remember, when you are studying algebra, you're studying ideas and not just rules about specific numbers. A mathematician would attempt to isolate the *concept* of an identity. First, does an identity property apply to other operations?

Multiplication	*Subtraction*	*Division*
$\square \cdot a = a \cdot \square = a$	$\triangle - a = a - \triangle = a$	$\bigcirc \div a = a \div \bigcirc = a$

Is there a real number that will satisfy any of the blanks for multiplication, subtraction, or division?

IDENTITY FOR MULTIPLICATION: There exists a real number, 1, called one, so that

$$1 \cdot a = a \cdot 1 = a$$

for any real number a.

Notice that there are no real numbers satisfying either the subtraction or division identity properties. There may be identities for other operations or for sets other than the set of real numbers.

EXAMPLE: Is there an identity set for the operation of union?

Solution: We wish to find a single set $\square$ so that

$$\square \cup X = X \cup \square = X$$

holds for every set X. We see that the empty set satisfies this condition; that is,

$$\varnothing \cup X = X \cup \varnothing = X$$

Thus, $\varnothing$ is the identity set for the operation of union.

Historical Note

The Egyptians did not have a zero symbol, but their numeration system did not require such a symbol. On the other hand, the Babylonians, with their positional system, had a need for a zero symbol but did not really use one until around 150 A.D. The Mayan Indians' numeration system was one of the first to use a zero symbol, not only as a placeholder but as a number zero. The exact time of its use is not known, but the symbol was noted by the early 16th-century Spanish expeditions into Yucatan. Evidently, the Mayans were using the zero long before Columbus arrived in America.

Identity property for multiplication

INVERSE PROPERTY

In Chapter 4 we spoke of opposites when we were adding and subtracting integers. Recall the property of opposites:

$$5 + (^-5) = 0$$
$$^-128 + 128 = 0$$
$$a + (^-a) = 0$$

When opposites are added, the result is zero, the identity for addition. This idea, which can be generalized, is called the inverse property.

Inverse property for addition

> INVERSE PROPERTY FOR ADDITION: For each real number a, there is a unique real number ^-a, called the opposite of a, so that
>
> $$a + (^-a) = ^-a + a = 0$$

For multiplication, we substitute the identity element for addition with the identity element for multiplication and check:

1. $5 \cdot \square = \square \cdot 5 = 1$
2. $^-128 \cdot \triangle = \triangle \cdot (^-128) = 1$

.
.
.

For 1. $5 \cdot \dfrac{1}{5} = \dfrac{1}{5} \cdot 5 = 1$, and 1/5 is a real number

2. $^-128 \cdot \dfrac{1}{^-128} = \dfrac{1}{^-128} \cdot (^-128) = 1$, and $1/(^-128)$ is a real number.

To show the inverse property for multiplication, we seek to find a replacement for the box for each and every real number a:

$$a \cdot \square = \square \cdot a = 1$$

Does the inverse property for multiplication hold? No, because if $a = 0$, then

$$0 \cdot \square = \square \cdot 0 = 1$$

does not have a replacement for the box in the set of real numbers. However, the inverse property for multiplication holds for all *nonzero* replacements of a.

INVERSE PROPERTY FOR MULTIPLICATION: For each real number a, $a \neq 0$, there exists a number, $1/a$, called the reciprocal of a, so that

$$a \cdot \frac{1}{a} = \frac{1}{a} \cdot a = 1$$

Inverse property for multiplication

Notice that every nonzero real number has a reciprocal.

EXAMPLE: Given the set $A = \{^-1, 0, 1\}$, the operation of multiplication has the identity 1 and satisfies the inverse property:

$1 \cdot 1 = 1$, so the inverse of 1 is 1
$^-1 \cdot {}^-1 = 1$, so the inverse of $^-1$ is $^-1$

The only element in the set that does not have an inverse is 0, so we say that the inverse property is satisfied.

FIELD

We have now introduced 11 properties satisfied by the set of real numbers. *Any* set satisfying these properties is called a *field*.

DEFINITION: A *field* is a set, F, with two operations, $+$ and $\cdot$, satisfying the following properties for any a, b, and c elements of F.
1. *Closure for addition:* $(a + b)$ is an element of F
2. *Closure for multiplication:* ab is an element of F
3. *Commutative for addition:* $a + b = b + a$
4. *Commutative for multiplication:* $ab = ba$
5. *Associative for addition:* $(a + b) + c = a + (b + c)$
6. *Associative for multiplication:* $(ab)c = a(bc)$
7. *Identity for addition:* There exists a real number 0 so that $0 + a = a + 0 = a$ for any element a in F.
8. *Identity for multiplication:* There exists a real number 1 such that $1 \cdot a = a \cdot 1 = a$ for any element a in F.
9. *Inverse for addition:* For each number a in F there is a unique number ^-a in F so that

$$a + (^-a) = (^-a) + a = 0$$

10. *Inverse for multiplication:* For each nonzero a in F there is a unique number $1/a$ in F so that

$$a \cdot \frac{1}{a} = \frac{1}{a} \cdot a = 1$$

11. *Distributive for multiplication over addition:*

$$a(b + c) = ab + ac$$

Field

In this book, a · b is often shortened to ab.

The set of real numbers is a field, but there are other fields. In our definition of a field we used the operations of addition and multiplication for the sake of understanding, but, for the mathematician, a field is defined as a set with *any two* operations satisfying the 11 stated properties.

PROBLEM SET 6.1

A Problems

In Problems 1–6, explain, in your own words, the meaning of each property, and justify your explanations by giving examples.

1. Closure

2. Associative

3. Identity

4. Inverse

5. Commutative

6. Distributive

7. Give the identity (if any) for the set {1,4,7,9} with operation * defined by Table 6.1.

8. Does the set {1,4,7,9} with operation * defined by Table 6.1 satisfy the inverse property?

Identify each of the properties in Problems 9–18.

9. $5 + 7 = 7 + 5$

10. $3(4 + 8) = 3 \cdot 4 + 3 \cdot 8$

11. mustard + catsup = catsup + mustard

12. $5 \cdot 1 = 5$

13. $5 \cdot 1/5 = 1$

14. $a + (10 + b) = (a + 10) + b$

15. $a + (10 + b) = (10 + b) + a$

16. (red + blue) + yellow = red + (blue + yellow)

17. $15 + (a + {}^-a) = 15 + 0$

18. $A \cap (B \cup C) = (A \cap B) \cup (A \cap C)$

TABLE 6.1

*	1	4	7	9
1	9	7	1	4
4	7	9	4	1
7	1	4	7	9
9	4	1	9	7

B Problems

In Problems 19–23, you are asked to check various field properties for the sets N (natural numbers), W (whole numbers), Z (integers), Q (rationals), and R (reals). Fill in the table with yes or no answers, but be prepared to justify your answers. For these exercises division means nonzero division.

	19.				20.				21.				22.				23.			
Set	*N*				*W*				*Z*				*Q*				*R*			
Operation	+	×	−	÷	+	×	−	÷	+	×	−	÷	+	×	−	÷	+	×	−	÷
Closure																				
Associative																				
Identity																				
Inverse																				
Commutative																				
Distributive																				

24. *Computer Problem.* In Problems 19–23, it was necessary to check the properties by working specific examples. We can write a BASIC program to help us do this. It will not answer the questions, but it will help find counterexamples and lead us to the correct conclusions. Write a program that will check the field properties for three given numbers. Here is an example of what your output might look like:

It is important to note that trying examples does not constitute proof.

```
RUN
WHAT THREE NUMBERS WOULD YOU LIKE ME TO CHECK? 2,5,9

COMMUTATIVE PROPERTY
A+B=            7
B+A=            7

A−B=            −3
B−A=            3

A*B=            1Ø
B*A=            1Ø

A/B=            .4
B/A=            2.5


ASSOCIATIVE PROPERTY
(A+B)+C=        16
A+(B+C)=        16

(A−B)−C=        −12
A−(B−C)=        6

(A*B)*C=        9Ø
A*(B*C)=        9Ø

(A/B)/C=        .Ø4444444
A/(B/C)=        3.6


DISTRIBUTIVE PROPERTY
A*(B+C)=        28
(A*B)+(A*C)=    28

A+(B*C)=        47
(A+B)*(A+C)=    77
```

```
THERE ARE OTHER POSSIBILITIES THAT I HAVE NOT TRIED, BUT I AM NOT
PROGRAMMED TO CHECK THEM.

THE IDENTITY FOR ADDITION AND SUBTRACTION IS Ø.
THE IDENTITY FOR MULTIPLICATION AND DIVISION IS 1.


THE ADDITIVE INVERSE OF A IS                    -2
THE ADDITIVE INVERSE OF B IS                    -5
THE ADDITIVE INVERSE OF C IS                    -9


THE MULTIPLICATIVE INVERSE OF A IS .5
THE MULTIPLICATIVE INVERSE OF B IS .2
THE MULTIPLICATIVE INVERSE OF C IS .1111111

READY
```

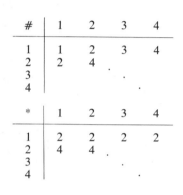

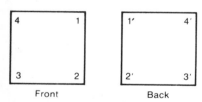

FIGURE 6.1 Square construction

25. a. Given the set {1,2,3,4} and the operation # defined by

$$a \mathbin{\#} b = ab,$$

construct a table for # showing all possible answers for the numbers in the set. That is, complete the table in the margin.

b. Given the set {1,2,3,4} and the operation * defined by

$$a * b = 2a,$$

construct a table for * showing all possible answers for the numbers in the set. That is, complete the table in the margin.

c. Verify the field properties for the operations of # and *.

26. *Symmetries of a Square.* Cut out a small square and label it as shown in Figure 6.1. Be sure that 1 is in front of 1′, 2 is in front of 2′, 3 is in front of 3′, and 4 is in front of 4′. We will study certain *symmetries* of this square—that is, the results that are obtained when the square is moved around according to certain rules that we will establish.

Hold the square with the front facing you and the 1 in the top right-hand corner as shown. This is called the *basic position*.

Now rotate the square 90° clockwise so that 1 moves into the position formerly held by 2 and so that 4 and 3 end up on top. We use the letter *A* to denote this rotation of the square. That is, *A* indicates a clockwise rotation of the square through 90°.

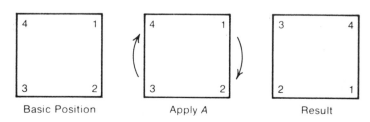

Other symmetries can be obtained similarly according to Table 6.2.

You should be able to tell how each of the results in the table was found. Do this before continuing with the problem.

TABLE 6.2 Symmetries of the Square

Letter	Description	Result
A	90° clockwise rotation	3 4 / 2 1
B	180° clockwise rotation	2 3 / 1 4
C	270° clockwise rotation	1 2 / 4 3
D	360° clockwise rotation	4 1 / 3 2
E	flip about a horizontal line through the middle of the square	3' 2' / 4' 1'
F	flip about a vertical line through the middle of the square	1' 4' / 2' 3'
G	flip along a line drawn from upper left to lower right	4' 3' / 1' 2'
H	flip along a line drawn from lower left to upper right	2' 1 / 3' 4'

We now have part of a mathematical system: a set of elements $\{A,B,C,D,E,F,G,H\}$. We must define an operation that combines a pair of these symmetries. Define an operation ☆ which means the following:

A ☆ B is defined as "A followed by B"; that is, begin with the square in basic position, apply A, and then apply B to get the result as shown.

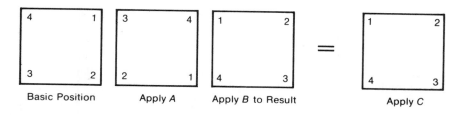

Basic Position   Apply A   Apply B to Result   =   Apply C

Thus we say $A \star B = C$, meaning "A followed by B is equivalent to the single element C."

Complete a table for the operation $\star$ and the set $\{A,B,C,D,E,F,G,H\}$.

27. Is the set $\{A,B,C,D,E,F,G,H\}$ for the operation "followed by" of Problem 26:
    a. closed?
    b. associative?
    c. commutative?

28. Does the set $\{A,B,C,D,E,F,G,H\}$ for the operation "followed by" of Problem 26 satisfy the:
    a. identity property?
    b. inverse property?

29. *A Geoboard Activity.* A geoboard is a device used by teachers for illustrating a wide variety of concepts. It consists of a board with pegs upon which rubber bands can be stretched. Let's consider a 3×3 geoboard, as shown in Figure 6.2.

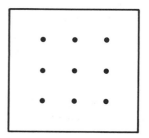

FIGURE 6.2 A 3×3 geoboard

We now stretch three rubber bands between two adjacent rows. There are six possible ways of doing this, which we label as follows:

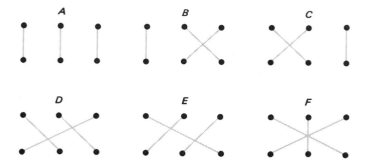

Next we combine these geoboard arrangements according to an operation Ω, which we call "snap." Snap is accomplished as shown in the following example.

EXAMPLE: Form $A \, \Omega \, B$.

Step 1: Form A.

Step 2: Form B by stretching the same rubber bands to the third row.

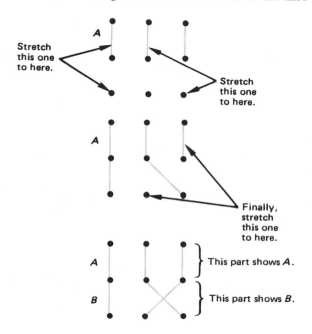

Step 3: Perform the operation of "snap" by releasing the bands from the middle row.

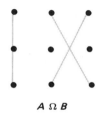

$A \, \Omega \, B$

The result is now one of the six original positions (in this case B), so we say

$$A \, \Omega \, B = B.$$

EXAMPLE: Find $E \, \Omega \, D$.

Step 1:

Step 2:

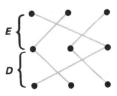

Step 3:

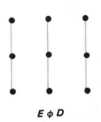

$E \, \phi \, D$

We see that $E \, \Omega \, D = A$.
Complete the following table.

Ω	A	B	C	D	E	F
A		B				
B						
C						
D						
E					A	
F						

30. Is the set $\{A,B,C,D,E,F\}$ for the operation ''snap'' of Problem 29:
    a. closed?
    b. associative?
    c. commutative?

31. Does the set $\{A,B,C,D,E,F\}$ for the operation ''snap'' of Problem 29 satisfy the:
    a. identity property?
    b. inverse property?

Mind Bogglers

32. *Symmetries of a Cube.* Consider a cube labeled as in Figure 6.3. List all the possible symmetries of this cube (patterning your work after that done for the symmetries of a square). Determine which of the following properties are satisfied: closure, associative, identity, inverse, and commutative.

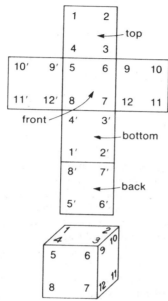

FIGURE 6.3 A cube

33. Suppose we use a placeholder ∘ to stand for an operation (addition, subtraction, multiplication, or division). See if you can determine what the operation ∘ stands for in each part.

a. $5 \circ 3 = 8$
   $7 \circ 2 = 9$
   $9 \circ 1 = 10$
   $8 \circ 2 = 10$

b. $5 \circ 3 = 15$
   $7 \circ 2 = 14$
   $9 \circ 1 = 9$
   $8 \circ 2 = 16$

c. $5 \circ 3 = 2$
   $7 \circ 2 = 5$
   $9 \circ 1 = 8$
   $8 \circ 2 = 6$

34. Using the idea presented in Problem 33, we can let ∘ represent operations other than the ordinary ones, as shown by the examples in the margin. Describe the operation ∘ for each part. Also, fill in the blanks by using the same operational rule.

a. $1 \circ 9 = 11$
   $2 \circ 7 = 10$
   $9 \circ 0 = 10$
   $9 \circ 8 = 18$
   $4 \circ 4 = \underline{\quad}$
   $4 \circ 8 = \underline{\quad}$
   $8 \circ 6 = \underline{\quad}$

b. $8 \circ 0 = 1$
   $5 \circ 4 = 21$
   $1 \circ 0 = 1$
   $5 \circ 6 = 31$
   $6 \circ 2 = \underline{\quad}$
   $2 \circ 3 = \underline{\quad}$
   $4 \circ 7 = \underline{\quad}$

c. $4 \circ 6 = 20$
   $8 \circ 2 = 20$
   $7 \circ 9 = 32$
   $6 \circ 8 = 28$
   $5 \circ 2 = \underline{\quad}$
   $4 \circ 3 = \underline{\quad}$
   $8 \circ 3 = \underline{\quad}$

35. Describe the operation ∘ and fill in the blanks. (See Problems 33 and 34.)

a. 4 ∘ 7 = 1   b. 4 ∘ 7 = 17   c. 5 ∘ 9 = 7

$$4 \circ 7 = 1 \qquad 4 \circ 7 = 17 \qquad 5 \circ 9 = 7$$
$$4 \circ 5 = 3 \qquad 5 \circ 6 = 26 \qquad 6 \circ 9 = 18$$
$$7 \circ 3 = 11 \qquad 6 \circ 4 = 37 \qquad 6 \circ 7 = 22$$
$$12 \circ 9 = 15 \qquad 2 \circ 8 = 5 \qquad 3 \circ 0 = 9$$
$$7 \circ 7 = \underline{\quad} \qquad 0 \circ 3 = \underline{\quad} \qquad 6 \circ 2 = \underline{\quad}$$
$$9 \circ 5 = \underline{\quad} \qquad 2 \circ 7 = \underline{\quad} \qquad 8 \circ 6 = \underline{\quad}$$
$$6 \circ 9 = \underline{\quad} \qquad 7 \circ 9 = \underline{\quad} \qquad 4 \circ 5 = \underline{\quad}$$

36. Describe the operation ∘ and fill in the blanks. (See Problems 33 and 34.)

$$\text{a.} \quad 4 \circ 5 = 1 \qquad \text{b.} \quad 5 \circ 8 = 5 \qquad \text{c.} \quad 0 \circ 8 = 64$$
$$3 \circ 5 = 2 \qquad 2 \circ 5 = 11 \qquad 2 \circ 6 = 52$$
$$5 \circ 2 = 3 \qquad 9 \circ 7 = 2 \qquad 4 \circ 8 = 112$$
$$7 \circ 0 = 3 \qquad 9 \circ 4 = 5 \qquad 9 \circ 7 = 193$$
$$3 \circ 6 = \underline{\quad} \qquad 3 \circ 3 = \underline{\quad} \qquad 2 \circ 7 = \underline{\quad}$$
$$5 \circ 5 = \underline{\quad} \qquad 3 \circ 2 = \underline{\quad} \qquad 6 \circ 4 = \underline{\quad}$$
$$1 \circ 0 = \underline{\quad} \qquad 0 \circ 8 = \underline{\quad} \qquad 4 \circ 4 = \underline{\quad}$$

37. *Pretty Kitty Kelly.* By looking for a pattern, formulate a rule that tells what Pretty Kitty Kelly likes. After you find the rule, don't divulge it—simply add more examples to the list. You'll know when you have it!

Pretty Kitty Kelly likes school but doesn't like homework;
> she likes green but doesn't like blue;
> she likes apples but doesn't like oranges;
> she likes happy but doesn't like sad;
> she likes good but doesn't like bad.

Pretty Kitty Kelly likes the bedroom but doesn't like the kitchen;
> she likes trees but doesn't like leaves;
> she likes puppies but doesn't like dogs;
> she likes three but doesn't like four;
> she likes little but doesn't like more.

6.2 SOLVING LINEAR EQUATIONS

By "ordinary algebra" we mean the algebra that is taught in high school.

The field properties of the last section cannot be proved and must be assumed as postulates of the set of real numbers. If they are all assumed, then the resulting mathematical system leads to ordinary algebra. Even though there are many aspects of ordinary algebra that are important to the scientist and mathematician, the ability to solve simple equations is important to the layperson and can be used in a variety of everyday applications.

A variable is a symbol that is used to represent an unspecified member of some set. A variable is a "placeholder" for the name of some member of a set.

An *equation* is a statement of equality that may be true or false or may depend on the values of the variable. If it is always true, as in

$$2 + 3 = 5$$

then it is called an *identity*. If it is always false, as in

$$2 + 3 = 15$$

then it is called a *contradiction*. If it depends on the values of the variable, as in

$$2 + x = 15$$

then it is called a *conditional equation*. The values that make the conditional equation true are said to *satisfy* the equation and are called the *solutions* or *roots* of the equation. Generally, when we speak of equations we mean conditional equations. Our concern when solving equations is to find the numbers that satisfy a given equation, so we look for things to do to equations to make the solutions or roots more obvious. Two equations with the same solutions are called *equivalent equations*. An equivalent equation may be easier to solve than the original equation, so we try to get successively simpler equivalent equations until the solution is obvious. There are certain procedures you can use to create an equivalent equation. The following properties specify the operations that are permissible.

Some values for the variable make the equation true (as with x = 13), and some values make the equation false (as with x = 10).

Given an equation, the following operations result in an equivalent equation:
1. *Addition property:* adding the same number to both sides;
2. *Subtraction property:* subtracting the same number from both sides;
3. *Multiplication property:* multiplying both sides by the same nonzero number;
4. *Division property:* dividing both sides by the same nonzero number.

Properties for solving equations

When these properties are used to obtain equivalent equations, the goal is to *isolate* the variable on one side of the equation, as illustrated in the examples that follow. You can always check the solution to see if it is correct; substituting the solution into the original equation will verify that it satisfies the equation. Notice how the field properties are used when solving these equations.

EXAMPLES: Solve the given equations.

1.
$$x + 15 = 25$$
$$x + 15 - \mathbf{15} = 25 - \mathbf{15} \qquad \text{Subtract 15 from both sides}$$
$$x = 10$$

The root of the simpler equivalent equation is 10. We often display the answer in the form of an equation, $x = 10$, whose root is obvious.

2.
$$x - 36 = 42$$
$$x - 36 + \mathbf{36} = 42 + \mathbf{36} \qquad \text{Add 36 to both sides}$$
$$x = 78$$

The root is 78; you may leave your answer in the form $x = 78$.

3.
$$52 = 14 + x$$
$$52 - \mathbf{14} = 14 + x - \mathbf{14}$$
$$38 = x$$

4.
$$15 - x = 0$$
$$15 - x + \mathbf{x} = 0 + \mathbf{x}$$
$$15 = x$$

The goal is to isolate the variable on one side of the equals sign. Perform the appropriate opposite operations to find a simpler equivalent equation.

Actually do the arithmetic here. Don't just put down the answer you expect to get. If you always do the arithmetic, you will seldom make a mistake when solving equations.

15 = x is the same as x = 15. This is a general property of equality called the symmetric property of equality: *if a = b. then b = a.*

Sometimes it's necessary to *combine similar terms* when solving equations. This is done by using the distributive property. For example,

$$5x + 3x = (5 + 3)x \qquad \text{By the distributive property}$$
$$= 8x$$

EXAMPLES: Simplify the given expressions by combining similar terms.

1. $\qquad\qquad 3a + 7a = (3 + 7)a$
$$= 10a$$

2. $\qquad 5a + 4b + 3a = (5a + 3a) + 4b \qquad$ Associative and commutative properties
$$= 8a + 4b$$

8a + 4b cannot be simplified further.

EXAMPLES: Find the root of the given equations.

1. $\qquad 5x + 3 - 4x + 5 = 6 + 9$
$$(5x - 4x) + (3 + 5) = 15$$
$$x + 8 = 15$$
$$x + 8 - \mathbf{8} = 15 - \mathbf{8}$$
$$x = 7$$

Combine similar terms.
Subtract 8 from both sides.

2. $6x + 3 - 5x + 7 = 11 + 2$
$$x + 10 = 13$$
$$x = 3$$

Combine similar terms first.
Mentally *subtract 10 from both sides.*

3. $\qquad 5x + 2 = 4x - 7$
$$5x + 2 - \mathbf{4x} = 4x - 7 - \mathbf{4x}$$
$$x + 2 = {}^-7$$
$$x + 2 - 2 = {}^-7 - \mathbf{2}$$
$$x = {}^-9$$

Remember that the goal is to isolate the variable on one side. Use the addition property for solving equations to obtain an equivalent equation that has all the variables on one side.

Once we have a root, we might want to check our answer. Remember that a root is a value that satisfies the equation. If we substitute the root into the original equation, we should obtain an equality. For example, we can check the solution of Example 3 above:

$$5x + 2 = 4x - 7 \qquad \text{Check for } x = {}^-9$$
$$5({}^-9) + 2 \stackrel{?}{=} 4({}^-9) - 7$$
$$^-45 + 2 \stackrel{?}{=} {}^-36 - 7$$
$$^-43 = {}^-43$$

Since we are checking the equality, we write $\stackrel{?}{=}$ until we verify the equality by doing the arithmetic.

EXAMPLES: Solve the following equations.

1. $4x + x = 20$

$$5x = 20 \qquad \text{Combine similar terms}$$

$$\frac{5x}{5} = \frac{20}{5} \qquad \text{Divide both sides by 5}$$

$$x = 4 \qquad \text{Simplify}$$

2. $3(m + 4) + 5 = 5(m - 1) - 2$
   $3m + 12 + 5 = 5m - 5 - 2$ Eliminate parentheses
       $3m + 17 = 5m - 7$ Combine similar terms
            $24 = 2m$ Add 7 to both sides; subtract $3m$
                        from both sides
            $12 = m$ Divide both sides by 2

Parentheses are eliminated by using the distributive property.

3. $5(y + 2) + 6 = 3(2y) + 5(3y)$
   $5y + 10 + 6 = 6y + 15y$
       $5y + 16 = 21y$
           $16 = 16y$
            $1 = y$

See if you can follow what was done at each step.

PROBLEM SET 6.2

A Problems

Solve the equations in Problems 1–15.

1. $A + 13 = 18$

2. $5 = 3 + B$

3. $6 = C - 4$

4. $2D + 2 = 10$

5. $15E - 5 = 0$

6. $16F - 5 = 11$

7. $6 = 5G - 24$

8. $4(H + 1) = 4$

9. $5(I - 7) = 0$

10. $\dfrac{J}{5} = 3$

11. $\dfrac{2K}{3} = 6$

12. $\dfrac{3L}{4} = 5$

13. $\dfrac{2M}{3} + 7 = 1$

14. $\dfrac{2N}{3} + 11 = 7$

15. $2(\emptyset + 5) = {}^{-}10$ (Note: O is the 15th letter of the alphabet—the letter "oh")

B Problems

Find the roots for each of the equations in Problems 16–26.

16. ${}^{-}5 = \dfrac{2P + 1}{3}$

17. $\dfrac{2 - 5Q}{3} = 4$

18. $\dfrac{5R - 1}{2} = 5$

Historical Note

Our modern approach to algebra can be traced to the work of Niels Abel (1802–1829) and Évariste Galois (1811–1832). Abel, now considered to be Norway's greatest mathematician, was, for the most part, ignored during his lifetime. He investigated fifth-degree equations, infinite series, functions, and integral calculus, as well as solutions of the general class of equations now called "Abelian." Although he died at the age of 26 "of poverty, disappointment, malnutrition, and chest complications," he left mathematicians enough material to keep them busy for several hundreds of years.

Galois also was not recognized during his lifetime. He had trouble with his teachers in school because of his ability to do a great amount of work in his head; his teachers always wanted him to "show all of his work." At 16 he wrote up several of his discoveries and sent them to a famous mathematician of the time, Augustin-Louis Cauchy, who promised to present Galois' results. However, Cauchy lost Galois' work. Galois tried again by sending his work to another mathematician, Joseph Fourier. This time Galois' work was lost because Fourier died before he could read it. When Galois was 20, he had an affair with a

19. $4(6S - 81) = {}^-3(4 + 5S)$

20. $5T + 3(T + 2) + (T + 4) + 17 = 0$

21. $3(U - 3) - 2(U - 12) = 18$

22. $6(V - 2) - 4(V + 3) = 10 - 42$

23. $5(W + 3) - 6(W + 5) = 0$

24. $4(X + 2) + 7(X + 1) = 10X - 7$

25. $6(Y + 2) = 4 + 5(Y - 4)$

26. $5(Z - 2) - 3(Z + 3) = 9$

Mind Bogglers

27. In solving Problems 1–26, you found a numerical value for each letter of the alphabet. Replace each number in parentheses with the letter that had that value. The result will be a paraphrase of a proverb; state the proverb.

$(3)(2)(7)({}^-2)(3)(7)({}^-3)({}^-10)(3)(8)$ $({}^{20}/_3)(7)({}^-2)(3)(7)(4)$ $(2)(3)({}^-3)$
$({}^-6)({}^-10)({}^-6)(^1/_3)$ $({}^-3)({}^-10)$ $(7)({}^-6)(6)(3)(^{11}/_5)(6)(7)({}^-3)(5)({}^-3)(^1/_3)$.

28. An elephant and a little bird want to play on a teeter-totter.

The bird says that it's impossible, but the elephant assures the little bird that it will work out and that she will prove it, since she has had a little algebra. She presents the following argument:

Let E = the weight of the elephant;
  b = the weight of the bird.
Now there must be some weight, w (probably very large), so that

$$E = b + w$$

Multiply both sides by $E - b$:

$$E(E - b) = (b + w)(E - b)$$

Using the distributive property:

$$E^2 - Eb = bE + wE - b^2 - wb$$

Subtract wE from both sides:

$$E^2 - Eb - wE = bE - b^2 - wb$$

Use the distributive property to factor both sides:

$$E(E - b - w) = b(E - b - w)$$

Divide both sides by $E - b - w$:

$$E = b$$

Thus, the weight of the elephant is the same as the weight of the bird. "Now," says the elephant, "since our weights are the same, we'll have no problem on the teeter-totter."

"Wait!" hollers the bird. "Obviously this is false." But where is the error in the reasoning?

6.3 PRIME NUMBERS

A set of numbers that is important, not only in algebra but in all of mathematics, is the set of prime numbers. In order to understand prime numbers, you first must understand the idea of divisibility, along with some new terminology and notation.

The counting number 10 is divisible by 2, since there is a counting number 5 so that $10 = 2 \cdot 5$; it is not divisible by 3, since there is no counting number k such that $10 = 3 \cdot k$. This leads us to the following definition.

DEFINITION: If m and d are natural numbers, and if there is a natural number k so that $m = d \cdot k$, we say that:

d is a *divisor* of m;
d is a *factor* of m;
d *divides* m;
m is a *multiple* of d.

Divisibility

Let's denote this relationship as $d\,|\,m$. That is, $5\,|\,30$ means that 5 divides 30 (because there exists some counting number k—namely 6—so that $30 = 5 \cdot k$).

EXAMPLES: Which of the following are true?
1. $7\,|\,63$
2. $8\,|\,104$
3. $14\,|\,2$
4. $6\,|\,15$

Solutions:

1. $7|63$ says "7 divides 63," which is true, since we can find a natural number k—namely 9—such that $63 = 7 \cdot k$.
2. $8|104$ is true, since $104 = 8 \cdot 13$.

Do not confuse the notation 14|2
with the notation 14/2 = 7 or ¹⁴/₂ =
7 used for fractions.

3. 14|2 is false. We write 14∤2 to say that 14 does not divide 2.
4. 6|15 is false. That is, we can find no natural number k so that $15 = 6 \cdot k$.

It is easy to see that 1 divides every counting number m, since

$$m = 1 \cdot m$$

Also, we see that every counting number m divides itself, since

$$m = m \cdot 1$$

We have proved the following theorem.

Distinct means *different.*

> THEOREM: Every counting number greater than 1 has at least two distinct divisors, itself and 1.

Since every counting number greater than 1 has at least two divisors, can these numbers have more than two?

Checking: 2 has exactly two divisors: 1,2
           3 has exactly two divisors: 1,3
           4 has more than two divisors: 1,2, and 4

Thus some numbers (such as 2 and 3) have exactly two divisors, and some (such as 4 and 6) have more than two divisors. Do any counting numbers have fewer than two divisors?

We now state a definition that classifies the counting numbers according to the number of divisors of each.

*Definition of a prime number
The Latin word* primus *means
"first in importance," and the
primes lead us to the important
properties of numbers.*

> DEFINITION: A *prime number* is a counting number that has exactly two divisors. The counting numbers that have more than two divisors are called *composite numbers.*

Thus 2 is prime, 3 is prime, 4 is composite (since it is divisible by three counting numbers), 5 is prime, and 6 is composite (since it is divisible by 1,2,3, and 6). Note that every counting number greater than 1 is either prime or composite. The number 1 is neither prime nor composite.

One method for finding primes smaller than some given number was first used by a Greek mathematician named Eratosthenes more than 2000 years ago. The technique is known as the *Sieve of Eratosthenes.* Suppose we wish to find the primes less than 100. First, we prepare a table of the counting numbers through 100, as is shown in Table 6.3.

TABLE 6.3 Finding Primes Using the Sieve of Eratosthenes

1̸	②	③	4̸	⑤	6̸	⑦	8̸	9̸	1̸0̸
⑪	1̸2̸	⑬	1̸4̸	1̸5̸	1̸6̸	⑰	1̸8̸	⑲	2̸0̸
2̸1̸	2̸2̸	㉓	2̸4̸	2̸5̸	2̸6̸	2̸7̸	2̸8̸	㉙	3̸0̸
㉛	3̸2̸	3̸3̸	3̸4̸	3̸5̸	3̸6̸	㊲	3̸8̸	3̸9̸	4̸0̸
㊶	4̸2̸	㊸	4̸4̸	4̸5̸	4̸6̸	㊼	4̸8̸	4̸9̸	5̸0̸
5̸1̸	5̸2̸	㊾③	5̸4̸	5̸5̸	5̸6̸	5̸7̸	5̸8̸	㊾	6̸0̸
㊶①	6̸2̸	6̸3̸	6̸4̸	6̸5̸	6̸6̸	㊷	6̸8̸	6̸9̸	7̸0̸
㊀①	7̸2̸	㊺	7̸4̸	7̸5̸	7̸6̸	7̸7̸	7̸8̸	㊴	8̸0̸
8̸1̸	8̸2̸	㊳	8̸4̸	8̸5̸	8̸6̸	8̸7̸	8̸8̸	㊨	9̸0̸
9̸1̸	9̸2̸	9̸3̸	9̸4̸	9̸5̸	9̸6̸	㊾⑦	9̸8̸	9̸9̸	1̸0̸0̸

Cross out 1, since it is not classified as a prime number.

Draw a circle around 2, the smallest prime number. Then cross out every following multiple of 2, since each one is divisible by 2 and thus is not prime.

Draw a circle around 3, the next prime number. Then cross out each succeeding multiple of 3. Some of these numbers, such as 6 and 12, will already have been crossed out because they are also multiples of 2.

Circle the next open number, 5, and cross out all subsequent multiples of 5.

The next open number is 7; circle 7 and cross out multiples of 7. Since 7 is the largest prime less than $\sqrt{100} = 10$, we will now show that all of the remaining numbers are prime.

The process is a simple one, since you do not have to cross out the multiples of 3 (for example) by checking for divisibility by 3 but can simply cross out every third number. Thus anyone who can count can find primes by this method. Also, notice that, in finding the primes under 100, we had crossed out all the composite numbers by the time we crossed out the multiples of 7. That is, to find all primes less than 100: (1) find the largest prime smaller than or equal to $\sqrt{100} = 10$ (7 in this case); (2) cross out multiples of primes up to and including 7; and (3) all the remaining numbers in the chart are primes.

This result generalizes. If we wish to find all primes smaller than n:

1. Find the *largest* prime smaller than or equal to $\sqrt{n}$;
2. Cross out the multiples of primes less than or equal to $\sqrt{n}$;
3. All the remaining numbers in the chart are primes.

Phrasing this another way, if n is composite, then one of its factors must be less than or equal to $\sqrt{n}$. That is, if $n = ab$, then it can't be true that *both* a and b are greater than $\sqrt{n}$ (otherwise $ab > \sqrt{n}\sqrt{n} = n = ab$,

𝕳istorical 𝕹ote

Eratosthenes (275–194 B.C.) was a highly talented and versatile person; athlete, astronomer, geographer, historian, poet, and mathematician. In addition to his sieve for prime numbers, he invented a mechanical "mean finder" for duplicating the cube, and he made some remarkably accurate measurements of the size of the earth. Unfortunately, most of his mathematical contributions are lost. It is said that he committed suicide by starving himself to death upon discovering that he had ophthalmia.

Sieve for numbers

so $ab > ab$ is a contradiction). Thus one of the factors must be less than or equal to $\sqrt{n}$.

This is a complicated way of stating a well-known fact. Suppose you wish to determine the divisors of 100. You begin checking:

$$1 \cdot 100$$
$$2 \cdot 50$$
$$3 \cdot \text{none}$$
$$4 \cdot 25$$
$$5 \cdot 20$$
$$6 \cdot \text{none}$$
$$7 \cdot \text{none}$$
$$8 \cdot \text{none}$$
$$9 \cdot \text{none}$$
$$10 \cdot 10$$

Do you need to check further? The answer is no, since you can now list all divisors (*both* smaller and larger than 10): 1, 2, 4, 5, 10, 20, 25, 50, and 100.

We use this property to tell us when we have excluded all composite numbers from a set. In the set of numbers $\{1,2,3, \ldots ,100\}$, we have considered the primes 2, 3, 5, and 7. The next prime is 11. Since $11^2 = 121$, and since we have 100 numbers in our table, we know we are finished and all remaining numbers are primes. Hence the last prime we need to consider is 7.

This method of Eratosthenes' is not very satisfactory to use if we wish to determine whether or not a given number n is a prime. For centuries, mathematicians have tried to find a formula that would yield *every* prime—or even a formula that would yield only primes. Let's try to find a formula that results in giving only primes. A possible candidate is

$$n^2 - n + 41$$

If we try this formula for $n = 1$, we obtain $1^2 - 1 + 41 = 41$. For $n = 2: 2^2 - 2 + 41 = 43$. For $n = 3: 3^2 - 3 + 41 = 47$.

So far, so good. That is, we are obtaining only primes. Continuing, we have Table 6.4.

Notice that all the numbers in the right columns are primes. If we reason inductively and conclude from the pattern that we have a formula for primes, our conjecture would be shattered by the next case, $n = 41$.

$$41^2 - 41 + 41$$
$$1681 - 41 + 41 = 1681$$

Now, 1681 is divisible by 1, 1681, and 41 and therefore is not a prime.

Historical Note

In 1900 the mathematician David Hilbert delivered an address before the International Congress of Mathematicians in Paris. Instead of solving a problem Hilbert presented a list of 23 problems. The mathematical world went to work on these problems and, even today, is still working on some of them. In the process of solving these problems, entire new frontiers of mathematics were opened. Hilbert's tenth problem was solved in 1970 by Yuri Matyasievich, who built upon the previous work of Martin Davis, Hilary Putnam, and Julia Robinson. A consequence of Hilbert's tenth problem is the fact that there must exist a polynomial with integer coefficients that will produce only prime numbers.

TABLE 6.4 Evaluation of the Formula $n^2 - n + 41$

Value of n	Value of $n^2 - n + 41$	Value of n	Value of $n^2 - n + 41$
4	53	23	547
5	61	24	593
6	71	25	641
7	83	26	691
8	97	27	743
9	113	28	797
10	131	29	853
11	151	30	911
12	173	31	971
13	197	32	1033
14	223	33	1097
15	251	34	1163
16	281	35	1231
17	313	36	1301
18	347	37	1373
19	383	38	1447
20	421	39	1523
21	461	40	1601
22	503		

A more serious attempt to find a prime-number formula was made by Fermat, who tried

$$2^{2^n} + 1$$

For $n = 1$, $2^{2^1} + 1 = 5$, and 5 is a prime.
For $n = 2$, $2^{2^2} + 1 = 2^4 + 1 = 17$, and 17 is a prime.
For $n = 3$, $2^{2^3} + 1 = 2^8 + 1 = 257$, and 257 is a prime.

Now the numbers get large rather quickly:

For $n = 4$, $2^{2^4} + 1 = 2^{16} + 1 = 65{,}537$

Checking this number with a computer, we find that it, too, is a prime number. But, alas, this formula fails for $n = 5$:

$$2^{2^5} + 1 = 2^{32} + 1 = 4{,}294{,}967{,}297$$

It turns out that this number is not prime! It is divisible by 641. Whether this formula generates any other primes is still an unsolved problem.

Historical Note

Fermat was not able to show that 4,294,967,297 was not a prime. Euler later discovered that it is divisible by 641. In the 1800s there was a young American named Colburn who had a great capacity to do things in his head. He was shown this number and asked if it was a prime. He said "No, because 641 divides it." When asked how he knew that, he replied "Oh, I just felt it." He was unable to explain this great gift—he just had it.

In 1970 a young Russian named Matyasievich discovered several explicit polynomials of this sort that generate only prime numbers, but all of those he discovered are too complicated to reproduce here. The largest prime number known at that time was

$$2^{11,213} - 1$$

and was discovered at the University of Illinois through the use of number-theoretic techniques and computers. The mathematicians were so proud of this discovery that the following ad was used on the university's postage meter:

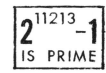 

Another large prime is

$$2^{19,937} - 1$$

which was found by Bryant Tuckerman on March 4, 1971, at the IBM Research Center. It is more than

$$1.6 \times 10^{2617}$$

times as great as the University of Illinois' number.

A larger prime number wasn't found until 1978 when it was discovered by students Laura Nickel and Curt Noll. They began their work in 1975 as high school students and on October 30, 1978, found the prime

$$2^{21,701} - 1$$

by using the computer at Cal State Hayward. This prime number is shown in Table 6.5. A few months later Curt Noll found a larger prime

$$2^{23,209} - 1$$

and in May 1979 Harry Nelson of the Livermore Lab found a prime almost twice as large:

$$2^{44,497} - 1$$

Historical Note

Laura Nickel and Curt Noll, two 18-year-old students at Cal State Hayward, spent three years finding the then largest known prime number:

$$2^{21,701} - 1$$

At the time of this writing they are at work trying to find an even larger prime. As you can see from Table 6.5, they used a computer to help them find this number, and they spent over 1900 hours on the problem. The number shown in Table 6.5 was discovered at 9 P.M. on October 30, 1978.

TABLE 6.5 The largest known prime in 1978. This *single* number has no divisors other than itself and 1.

```
                          THE LARGEST KNOWN PRIME
                        2 TO THE 21701TH POWER MINUS 1

                                                      44867916611904333479495141036159 1
7787272090237293886130103648044751278560915805363716201839592018310368914961397303553362113455167471528788000713434534719468102573
2056939825423723521750452980127215084299527266875706892007262798468825185681532142985720637290299313726344463257416449344509835102
4588167890163949458936967051685024361802322595515762603295399186364437045681350697590862198047118901225352609650331560624641680529 3
6095027632251954119937878161188790368067065457094570602703933944508795918017973611326798274338403964825448745270434393258865935 31
1826270281291301767537362073506047170679601808698618819230547730826314303368930940240111602318421873986179317338167429362403908014 6
4465331678931424034416262439863548298681529638568560976130813700085429090217885276018126751691666618919665288927240545754801572 75
8322393638969062394826852973840659534253688170621194098884299042222510205198907427651269590866489020849331593183544730378436867361
9905441075619665362090487700417583472204239335761771260615063239687136058218014855376128809648109025328444692060981149171853122 623
2144903151992703476798402506504466243611586327126073718204410761290405760761564139819464433841749359455575892695505513844442496417
3858932643106235024425974988495427561113138125373546749571089212977055534105306317668036525450115619994704703080975001831039096 32
8170193148671686659220369588362291591980763340251182094830945643052580945876198697254235294718583716788165976228557375284260972377
346244685323620326927812128869122804644917707943185384476057711810558228653120572801809938250702311793201429792131750687465985 253
4317769457992501015403459055756096150131796369342571712916973547166474698856592132612717244424968586154241059761035472955012 53R058
449710905646399196440357223209008827554713220022843977757307555156586140292296094752220794567511997522268424383084190316769975 828
96940816151292973116618877463197394261300350532803052843592986194302881466472243934096043321299157094797164800017511529010477300220
9922649639362263334540500936114495516338936448113785644513097993390353792975161976866955505630024237016161236691536299577305442 11605
591046268240572830493302269563487603354791811717670443165137272975361148520795728309760775693316925291386128772683383000075590 7168
49637364320188662722605918682498704615895029012148031010622082176086512056292049756568081153177263527091318013801387158672559 8473961
011785486552173088901587886092432642791917812170832604266985767528270606097792953278073806000017055574112385644959778735492110 13123
9116174462947424279402757853073650206058072098130908133815521220237989176959032725532187981806390188976611966692338378008058 49571180
30412795376342687245889400317644073344813078432076085825035772016369730815188469726917304803914235111646000934113546568255979 54514
0706575716471268280269434403999279968415403789437874392343874032494256225397330000397729149794558036848864231378472855395291175411510838
529005486556378452316132591114338403795092561182541488447059744712363488652928979195070962647432495742758096709852132553243 687088
002636637356883646720928138446466834966147353947767611471395239057428387113864360892768954627026543622707236136155696932626 9725551
377958061315120783963923811176302148831371912295396724184966500356865168976353282982941355943827107890964801254681815583067064973
1025924561052079518365511691315910359549578098542021350858601956949766303513660742781294511017929376440855043952755985238806 4138
6024393469009569031381270272833205694814223411334423569759378875423953534853015462848353277020068736893276116304379664911400779519
4015497259024341651238058760990011147082576731067693944194282683442761299775689234066920340874365922446663372321259665041724 973082
52151183035770112727658129337136614839277853233573127542905472394542069919985611732624482347646980930805958600708197626100309 7963
19881288287374262137746392666878649820492244780213560739855252676622720512431000805663291936355319496671705670944669080905362 50206
5721841983654762461859537535203901870544902429903708892328245566751153611632372977920734703117075921951295707710873733317931 43
3110923531405719416365240907518118794096775462324423434250371711420947553751555904657975806450898706151111865933553368007417 6
4997430690178998890828912250160625775343227050906397483026836459676407507010511274491368155114450216775684148263040093943008 537622
40103073543150732971096790978525911737213760340840390379732608426485090962292426511418406848508069512389853069176379179345 8820
6083928373369371925855787898368194089900915303776031168373058432938426749218951088422975303906809888393523259437606901944039 838823
31522548871187135553033467187970328749361217101889072004779121998300657569055022754324745901876793980240930546444765551 54058
6061556618233956250052608068535805591607847721844153592871626756041813259510251752292898135383307406724693311157087329533181550172
57112413836980510540228678649224861387416522711795380646257788142883734020101220075694246933111361280630641004515 77701609
67151754387744251152420312129302326059837561013693228792446847369626893292863959906611559181514200293625478634965230187 6009932048
60736747921056014359652196606969325320755082334969013066924068368550517444551573915101718569640158204816271224232 9264120
89910712229530383437665419250549717375916989292375645057151727543470106206531135735154136115144300015873080448570355596137 321380
0627488813320314721428230249495100224093701777129942352897027904900255826941441774321681772576645929282506199883549556147 63200793811 7
264594610448118080133362151215059862559419546016019333689732902990369357171414581341823547346460807428977
20332813488432789225262118290180024463097969378133742810292359981284016339591088387881070079633919558475740204934996444756554 5558
196048749757202301277031727328614477186919611098975942332752633167063512960112700154075209382155519550328268932942978 93465031763 5
10075286098417437360840424080940185091096451848280564960260939346916500676021037081798285049955718351590057099179191497337 35278914364
020463842602726374574875276434149756920202000079346259656661516651525829193931343391222614621242014153365037286833 6629211862904235477
896637837854678930126380410821437895846874998049417994854048338667781255945413472465246231194884140131607162842728171304 224786918
563120019233369896693354436162939131104173095650169466275455887564434519126927960069355180927195645026429409285741082 8353511882751
```

ON 10/30/78, CURT NOLL AND LAURA NICKEL OF CAL STATE HAYWARD PROVED THE ABOVE NUMBER TO BE PRIME, USING AN IMPLEMENTATION OF THE LUCAS-LEHMER TEST ON A CDC 174. THE PROOF OF PRIMALITY TOOK 7 HOURS, 8 MINUTES, AND 20.023 SECONDS.

PROBLEM SET 6.3

A Problems

1. What is a prime number?

2. Which of the following numbers are primes?
   a. 59   b. 57   c. 1   d. 97   e. 79

3. Which of the following numbers are primes?
   a. 63   b. 73   c. 37   d. 43   e. 87

4. a. Do any counting numbers have fewer than two distinct divisors?
   b. Are all counting numbers either prime or composite?

5. Are the following true or false?
   a. $8 \mid 48$ b. $6 \nmid 39$ c. $15 \mid 5$ d. $2 \mid 628,174$ e. $10 \mid 148,729,320$

6. Are the following true or false?
   a. $7 \mid 65$ b. $12 \mid 156$ c. $16 \nmid 576$ d. $5 \mid 817,543$
   e. $10 \nmid 518,728,641$

7. a. Is every odd number a prime number?
   b. Is every prime number greater than 2 an odd number?

8. Find a pair of prime numbers that differ by 1, and show that there is only one such pair possible.

9. Find all the prime numbers less than or equal to 300.

B Problems

10. What is the largest prime you need to consider to be sure that you have excluded, in the Sieve of Eratosthenes, all primes less than or equal to:
    a. 200?   b. 500?   c. 1000?

11. We used the Sieve of Eratosthenes in Table 6.3 by arranging the first 100 numbers into 10 rows and 10 columns. Repeat the sieve process for the first 100 numbers by arranging the numbers in the following patterns.

    a. by 6:   1   2   3   4   5   6
                 7   8   9   10   11   12
                 13   14   15
                                .
                                .
                                .

    b. by 7:   1   2   3   4   5   6   7
                 8   9   10   11   12   13   14
                 15
                                .
                                .
                                .

    c. by 21.

    As you are using the above sieves, look for patterns. Name some of the patterns you notice. Do you think that these sieves are better than the one shown in Table 6.3? Why or why not?

12. Using the sieves of Problem 11, make a conjecture about primes and multiples of 6.

If you have access to a computer, Problem 19 may be of some help with Problem 13.

13. Pairs of consecutive odd numbers that are primes are called prime twins. For example, 3 and 5, 11 and 13, and 41 and 43 are prime twins. Can you find any others?

14. For what values of n is $11 \cdot 14^n + 1$ a prime? (Hint: Consider n even and then consider n odd.)

15. Suppose three new states were added to the United States. Then the number of states would be 53, a prime. Could we still arrange the stars into rows and columns so that the design would be symmetric?

16. Three consecutive odd numbers that are primes are called *prime triplets*. It is easy to show that 3, 5, and 7 are the only prime triplets. Can you explain why this is true?

17. *Computer Problem*. The following BASIC program will find primes less than or equal to a given number *n*.

```
10   PRINT "PROGRAM TO FIND PRIMES LESS ";
11   PRINT "THAN OR EQUAL TO N."
12   PRINT "WHAT IS N";
13   INPUT N
15   IF INT(N) <> N GOTO 145
18   IF N < = 0 GOTO 145
20   IF N = 1 GOTO 140
25   PRINT
30   PRINT "PRIMES < = N ARE"
40   PRINT 2
50   LET A = 1
60   LET A = A + 2
70   LET B = 3
80   IF A > N GOTO 150
90   IF B < = SQR(A) GOTO 110
105  PRINT A
108  GOTO 60
110  IF A/B < = INT(A/B) GOTO 60
120  LET B = B + 2
130  GOTO 90
140  PRINT "THERE ARE NO PRIMES LESS THAN 1."
142  GOTO 150
145  PRINT "N MUST BE A POSITIVE INTEGER."
148  GOTO 12
150  END
```

Use this program to find all primes less than or equal to 500.

18. *Computer Problem*. Write a BASIC program to output the values found in Table 6.4.

19. *Computer Problem*. Use the following BASIC program to help you answer Problem 13.

```
 5   PRINT
 6   PRINT
10   PRINT "THIS PROGRAM WILL FIND THE TWIN PRIMES LESS THAN";
12   PRINT " OR EQUAL TO N"
13   PRINT
15   PRINT "WHAT IS N ";
20   INPUT N
22   PRINT
25   PRINT "TWIN PRIMES LESS THAN OR EQUAL TO ";N;" ARE:"
26   IF N < = 1 THEN 98
27   IF INT(N) <> N THEN 98
```

```
30   LET T = 1
32   LET L = 2
34   GOTO 40
37   LET L = T
40   LET T = T + 2
42   LET C = 3
44   IF T > N THEN 90
46   LET Z = SQR(T)
48   IF C < = Z THEN 80
50   LET R = ABS(C − SQR(T)) − .001
52   IF R < = 0 THEN 80
54   LET G = L + 2 − T
56   IF G < > 0 THEN 37
60   PRINT L,T
62   GOTO 37
80   LET P = T/C
82   LET Q = INT(T/C)
84   IF P < = Q THEN 40
86   LET C = C + 2
88   GOTO 46
90   PRINT
92   PRINT
95   STOP
98   PRINT "TRY AGAIN !!"
99   GOTO 15
100  END
```

20. *Computer Problem.* Use a computer to check the formula given in Problem 14. Check it for several different values of *n*.

Mind Bogglers

21. A man goes to a well with two cans whose capacities are 3 gallons and 5 gallons. Explain how he can obtain exactly 4 gallons of water from the well.

22. Some primes are 1 more than a square. For example, $5 = 2^2 + 1$. Can you find any other primes p so that $p = n^2 + 1$?

23. Some primes are 1 less than a square. For example, $3 = 2^2 - 1$. Can you find any other primes p so that $p = n^2 - 1$?

24. *Calculator Problem.* In the text we tried some formulas that might have generated only primes, but, alas, they failed. Below are some other formulas. Show that these, too, do not generate only primes.

   a. $n^2 + n + 41$
   b. $n^2 - 79n + 1601$
   c. $2n^2 + 29$
   d. $9n^2 - 489n + 6683$
   e. $n^2 + 1$, *n* an even integer

Historical Note

It has been proved that, of the first 2398 numbers generated by the formula of Problem 24a, exactly one-half are prime. After listing all numbers below 10 million, the mathematicians Ulam, Stein, and Wells found the proportion to be .475. . . .

Problems for Individual Study

25. *Infinitude of Primes.* A proof that there are infinitely many primes using $p! + 1$ is attributed to Euclid. Consult some elementary number-theory textbook, and then write a convincing argument showing that there are infinitely many prime numbers.

26. *Paradoxes and Fallacies.* What fallacies result when a number is "divided" by zero? What fallacies occur from incorrect constructions in geometry? What fallacies can be explained by limits? What fallacies depend on false probabilities?

    Exhibit suggestions: charts and illustrations of famous paradoxes; experiments with coins or dice to illustrate false probabilities.

    References: Bakst, Aaron, *Mathematics, Its Magic and Mastery* (2nd ed.) (New York: Van Nostrand, 1952).

    Kasner, Edwin, and James Newman, *Mathematics and the Imagination,* Chap. 6 (New York: Simon and Schuster, 1940).

    Kline, Morris, *Mathematics in Western Culture,* Chap. XXV (New York: Oxford University Press, 1953).

27. *Infinity.* What is infinity? Is there more than one infinity? How can you compute with infinite numbers?

    Exhibit suggestions: illustrations and comparison of infinite amounts, paradoxes of the infinite.

    References: Gamow, George, *One, Two, Three, Infinity,* Chap. 1 (New York: Viking Press, 1947).

    Newman, James, *The World of Mathematics,* Vol. 3, Part X (New York: Simon and Schuster, 1956). See also the references for Problem 26.

28. *Primes.* Investigate some of the properties of primes not discussed in the text. Why do primes hold such fascination for mathematicians? Why are primes important in mathematics? What are some of the important theorems concerning primes?

    Reference: Gardner, Martin, "The Remarkable Lore of the Prime Number," *Scientific American*, March 1964 (Vol. 210, No. 3).

6.4 PRIME FACTORIZATION

Figure 6.4 shows the frequency of primes among the first 10,000 or the first 65,000 integers. But is there a largest prime? In this section we will answer this question conclusively after showing that every counting number greater than 1 is either a prime or a product of primes.

𝕳istorical 𝕹ote

In 1978 Hugh C. Williams of the University of Manitoba discovered a new prime number:
11,111,111,111,111,111,111,111,-
111,111,111,111,111,111,111,111,
111,111,111,111,111,111,111,111,
111,111,111,111,111,111,111,111,
111,111,111,111,111,111,111,111,
111,111,111,111,111,111,111,111,
111,111,111,111,111,111,111,111,
111,111,111,111,111,111,111,111,
111,111,111,111,111,111,111,111,
111,111,111,111,111,111,111,111,
111,111,111,111,111,111,111,111,
111,111,111,111,111,111,111,111,
111,111,111,111,111,111,111,111,
111,111

He called this number R_{317} for short. Do some research and see if you can find some newly discovered prime numbers (see Problem 28).

𝔥istorical 𝔑ote

In 1964, Stanislaw Ulam of the Los Alamos Scientific Laboratory was passing the time by arranging the counting numbers into a grid in a spiral pattern as shown:

17	16	15	14	13
	5	4	3	12
	6	1	2	11
	7	8	9	10

Next, he crossed out the primes and noticed that the primes seemed to congregate into straight lines:

100	99	98	97	96	95	94	93	92	91
65	64	63	62	61	60	59	58	57	90
66	37	36	35	34	33	32	31	56	89
67	38	17	16	15	14	13	30	55	88
68	39	18	5	4	3	12	29	54	87
69	40	19	6	1	2	11	28	53	86
70	41	20	7	8	9	10	27	52	85
71	42	21	22	23	24	25	26	51	84
72	43	44	45	46	47	48	49	50	83
73	74	75	76	77	78	79	80	81	82

How would the grid look, Ulam wondered, if this process were extended to thousands of primes? The result is shown in Figure 6.4.

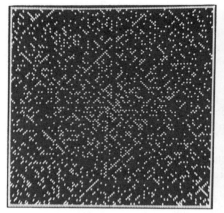 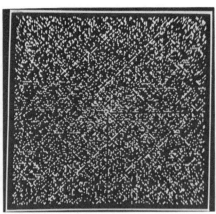

FIGURE 6.4 Photographs of a computer grid showing primes as a spiral of integers from 1 to about 10,000 (left) and from 1 to about 65,000 (right). For a description, see the historical note in the margin.

The operation of factoring is the reverse of the operation of multiplying. Multiplying 3 by 6 yields

$$3 \cdot 6 = 18$$

and the answer is unique (only one answer is possible). The reverse process is called *factoring*. Suppose we are given the number 18 and are asked for numbers that can multiplied together to give 18. Then we can list the different possible factorizations of 18:

$$
\begin{aligned}
18 &= 1 \cdot 18 \\
&= 18 \cdot 1 \\
&= 2 \cdot 9 \\
&= 1 \cdot 1 \cdot 2 \cdot 9 \\
&= 3 \cdot 6 \\
&= 2 \cdot 3 \cdot 3 = 2 \cdot 3^2 \\
& \quad \vdots
\end{aligned}
$$

In this problem we see that the answer is certainly not unique; in fact, there are infinitely many possibilities. Some agreements are in order.

1. We will not consider the order in which the factors are listed as important. That is, $2 \cdot 9$ and $9 \cdot 2$ will be considered the same factorization.
2. We will not consider 1 as a factor when writing out any factorizations. That is, prime numbers will not have factorizations.
3. Recall that we are working in the set of counting numbers; thus, $18 = 36 \cdot \frac{1}{2}$ and $18 = (^-2)(^-9)$ are not considered factorizations of 18.

With these agreements, we have greatly reduced the possibilities.

$$18 = 2 \cdot 9$$
$$= 3 \cdot 6$$
$$= 2 \cdot 3^2$$

are the only possible factorizations. Notice that the last factorization contains only prime factors; thus it is called the *prime factorization* of 18.

It should be clear that, if a number is composite, it can be factored into two counting numbers greater than 1. These two numbers themselves will be prime or composite. If they are prime, then we have a prime factorization. If one or more is composite, we repeat the process. This continues until we have written the original number as a product of primes. It is also true that this representation is unique. This is one of the most important results in arithmetic, and we state it formally:

> *Fundamental Theorem of Arithmetic:* Every counting number greater than 1 is either a prime or a product of primes, and the factorization is unique (except for the order in which the factors appear).

Fundamental Theorem of Arithmetic

EXAMPLE: Find the prime factorization of 385. One of the easiest ways of finding the prime factors of a number is to try division by each of the prime numbers in order:

$$2,3,5,7,11,13,17, \text{ and } 19$$

If none of these primes divides 385, then 385 must be prime. We see by inspection that 385 is not divisible by 2 or 3. It is divisible by 5, so

$$385 = 5 \cdot 77$$

Recall that we need only check primes less than or equal to $\sqrt{385}$, since the next prime is 23 and $23^2 = 529$.

Since 77 is a composite number that is divisible by 7 and 11, we write

$$385 = 5 \cdot 7 \cdot 11$$

Some people prefer to write these steps using a "factor tree."

A number is divisible by
  2 if the last digit is even;
  3 if the sum of the digits is divisible by 3;
  5 if the last digit is 0 or 5.

EXAMPLE: Find the prime factorization of 1400. Using the factor tree, we may find *any* factors of 1400:

Now find any factors of 10 and 140:

Continue finding factors of nonprimes:

Write the factorization using exponents:

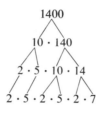

$$2^3 \cdot 5^2 \cdot 7$$

This is called the canonical representation.

EXAMPLE: Find the prime factorization of 3465 using a factor tree.

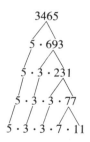

The prime factorization is $3^2 \cdot 5 \cdot 7 \cdot 11$.

n	% of primes $\leq$ n
10	40
100	25
500	19
1,000	16.8
10,000	12.37
1,000,000	6*

*approximate

By the Fundamental Theorem of Arithmetic this is the only possible prime factorization of 3465. Regardless of how we begin the factor tree, the end result will always be the same.

You might wonder how many primes exist. If we count the primes less than or equal to 10, we have 2, 3, 5, and 7 (or 40% of the numbers). If we do this for other numbers, as shown by the table in the margin, we find that the primes are "thinning out" and, indeed, may run out. In fact, we can create a sequence of consecutive nonprime numbers that is as large as we please. Consider

$$3 \cdot 2 \cdot 1 + 1$$

The next two consecutive numbers are

$$3 \cdot 2 \cdot 1 + 2 \qquad \text{Divisible by 2}$$
$$3 \cdot 2 \cdot 1 + 3 \qquad \text{Divisible by 3}$$

This is a sequence of two consecutive nonprimes.

and these are not prime. Consider $19 \cdot 18 \cdot 17 \cdots 3 \cdot 2 \cdot 1 + 1$ along with the next 18 consecutive numbers:

$$19 \cdot 18 \cdot 17 \cdots 3 \cdot 2 \cdot 1 + 2 \qquad \text{Divisible by 2}$$
$$19 \cdot 18 \cdot 17 \cdots 3 \cdot 2 \cdot 1 + 3 \qquad \text{Divisible by 3}$$
$$19 \cdot 18 \cdot 17 \cdots 3 \cdot 2 \cdot 1 + 4 \qquad \text{Divisible by 4}$$

This is a sequence of 18 consecutive nonprimes.

$$\vdots \qquad\qquad \vdots$$

$$19 \cdot 18 \cdot 17 \cdots 3 \cdot 2 \cdot 1 + 18 \qquad \text{Divisible by 18}$$
$$19 \cdot 18 \cdot 17 \cdots 3 \cdot 2 \cdot 1 + 19 \qquad \text{Divisible by 19}$$

Notice that $1001 \cdot 1000 \cdot 999 \cdots 3 \cdot 2 \cdot 1 + 1$ is followed by 1000 consecutive nonprimes. You can thus show that there exist a million consecutive nonprimes, or a billion, or any number you wish.

Indeed, the primes do seem to be thinning out. However, using the Fundamental Theorem, we can present an argument that will lead us to conclude that there must be infinitely many prime numbers. Let's begin by considering a proof that 19 is not the largest prime. There are two ways to proceed. The obvious way is not much fun. That is, we could simply find a larger prime—say, 23—and be finished.

Let's proceed differently. We will prove that 19 is not the largest prime by contradiction. That is, assume that there are only finitely many primes, ending with 19. *We are assuming that 19 is the largest prime.*

Consider the number $2 \cdot 3 \cdot 5 \cdot 7 \cdot 11 \cdot 13 \cdot 17 \cdot 19$. Certainly this number is larger than 19, but it is not a prime (why?). So we have no contradiction yet.

Consider the number

$$M = [2 \cdot 3 \cdot 5 \cdot 7 \cdot 11 \cdot 13 \cdot 17 \cdot 19] + 1$$

This number is larger than 19. Is it a prime? According to our assumption, it must be composite, since it is larger than 19. But if it is composite, it has a prime divisor (Fundamental Theorem). Check all primes: 2 does not divide M, since 2 divides $2 \cdot 3 \cdot 5 \cdot 7 \cdot 11 \cdot 13 \cdot 17$, and it cannot divide that number plus 1; 3 does not divide M for the same reason; but then 5, 7, 11, 13, 17, and 19 do not divide M. Thus, if M is not divisible by any prime, then either it must be prime or there is a prime divisor larger than 19. In either case, we have found a prime larger than 19.

Now if *anyone* claims to be in possession of the largest prime, we need only carry out an argument like the above to find a larger prime. Thus we are saying that there are infinitely many primes, since it is impossible to have a largest prime.

PROBLEM SET 6.4

A Problems

Write the prime factorization for each of the numbers in Problems 1–10.

1. a. 24	b. 30	2. a. 300	b. 144
3. a. 28	b. 76	4. a. 215	b. 125
5. a. 108	b. 740	6. a. 699	b. 123
7. a. 120	b. 90	8. a. 75	b. 975
9. a. 490	b. 4752	10. a. 143	b. 51

Indicate which of the numbers in Problems 11–22 are prime. If the number is not prime, write its prime factorization.

11. 83	12. 97
13. 127	14. 113
15. 377	16. 151
17. 105	18. 187
19. 67	20. 229
21. 315	22. 111

Historical Note

A statement equivalent to the Fundamental Theorem of Arithmetic is found in Book IX of Euclid's Elements *(Proposition IX 14). Very little is known about Euclid except through these books, called the* Elements. *This work is not only the earliest known major Greek mathematical book, but it is also the most influential textbook of all time. It was composed around 300* B.C. *and first printed in 1482. Except for the Bible, no other book has been through so many editions and printings. Even though we do not know the year of Euclid's birth or death, we do know he lived around 300* B.C. *because he was the first professor of mathematics at the Museum of Alexandria, which opened in 300* B.C. *He was also the author of at least ten other books, five of which have been lost to us. However, Euclid's 13-volume* Elements *is a masterpiece of mathematical thinking. It was described by Augustus de Morgan (1806–1871) with the statement: "The thirteen books of Euclid must have been a tremendous advance, probably even greater than that contained in the* Principia *of Newton."*

23. In your own words, state the Fundamental Theorem of Arithmetic.

24. *Calculator Problem.* Find the prime factorization of 101,101.

25. *Calculator Problem.* Find the prime factorization of 514,080.

B Problems

26. Use an argument similar to the one in the text to show that 23 is not the largest prime.

Find the prime factorization for each of the numbers in Problems 27–30.

27. 567   28. 568   29. 2869   30. 793

31. *Computer Problem.* It would be helpful to write a program to compute the prime factorization of a number, particularly if the number to be factored is quite large. The following BASIC program shows four runs of such a program, for the numbers 35, 467, 923, and 1968.

```
10   PRINT "WHAT IS N";
20   INPUT N
25   IF N = 0 GOTO 260
26   IF INT(N) <> N GOTO 260
30   PRINT "PRIME FACTORS OF N ARE:"
32   IF N = 1 GOTO 250
34   IF N < 0 GOTO 254
40   LET I = 2
50   GOSUB 200
60   FOR I = 3 TO N STEP 2
70   GOSUB 200
75   NEXT I
80   GOTO 300
200  IF INT(N/I) <> N/I GOTO 230
210  LET N = N/I
220  PRINT I
225  GOTO 200
230  IF N = 1 GOTO 300
240  RETURN
250  PRINT "1 HAS NO PRIME FACTORS."
252  GOTO 300
254  PRINT "−1"
256  GOTO 40
260  PRINT "N MUST BE A NON-ZERO INTEGER."
262  PRINT "PLEASE TRY AGAIN. NOW, ";
264  GOTO 10
300  END
```

```
RUN
WHAT IS N? 35
PRIME FACTORS OF N ARE:
 5
 7

READY

RUN
WHAT IS N? 467
PRIME FACTORS OF N ARE:
 467

READY

RUN
WHAT IS N? 923
PRIME FACTORS OF N ARE:
 13
 71

READY

RUN
WHAT IS N? 1968
PRIME FACTORS OF N ARE:
 2
 2
 2
 2
 3
 41

READY
```

This program finds prime factors. Write a program that will find *all* the factors of a given number N.

32. As seen in the preceding problem, 1968 is not a prime number; there are only seven prime-number years in the period 1950–2000. What are the prime-number years from 1970 to 1979?

33. *Computer Problem.*   Using the program for prime factorizations given in Problem 31, find the prime factorization of each of the following.

   a. 1763   b. 15,049   c. 128,137   d. 614,777

34. In the text, we showed that 19 was not the largest prime by considering

$$M = (2 \cdot 3 \cdot 5 \cdot 7 \cdot 13 \cdot 17 \cdot 19) + 1$$

Now, M is either prime or composite. If it is prime, then it is larger than 19, and we have a prime larger than 19. If it is composite, it has a prime divisor larger than 19. In either case, we find a prime larger than 19. Show that this number M does not always generate primes. That is,

$$2 + 1 = 3, \text{ a prime;}$$
$$2 \cdot 3 + 1 = 7, \text{ a prime;}$$
$$2 \cdot 3 \cdot 5 + 1 = 31, \text{ a prime.}$$

Find an example in which the product of consecutive primes plus 1 does not yield a prime.

35. We showed that 19 was not the largest prime. Generalize the argument in the text to show that there is *no* largest prime.

36. *Calculator Problem.* Find the one composite number in the following set:

   31
   331
   3331
   33331
   333331
   3333331
   33333331
   333333331

Hint for Problem 36: The composite number has a prime factor under 19.

Mind Bogglers

37. The Pythagoreans studied numbers in order to find certain mystical properties in them. We mentioned that they had masculine and feminine numbers, amicable numbers, deficient numbers, and so on.

   A *perfect number* is a counting number that is equal to the sum of all its divisors that are less than itself. These divisors less than the number itself are called *proper divisors*. The proper divisors of 6 are {1,2,3} and $1 + 2 + 3 = 6$, so 6 is the smallest perfect number.

   On the other hand, 24 is not perfect, since its proper divisors, {1,2,3,4,6,8,12}, have the sum

$$1 + 2 + 3 + 4 + 6 + 8 + 12 = 36$$

All even perfect numbers are of the form $2^{p-1}(2^p - 1)$, where p and $(2^p - 1)$ are prime numbers.

   Find the perfect number furnished by the formula

$$2^{p-1}(2^p - 1) \text{ when } p = 5$$

Show that it is perfect.

A number is perfect if it is the sum of its proper divisors, deficient if it exceeds the sum of its proper divisors, and abundant if it is less than the sum of its proper divisors. God created the world in six days, a perfect number. The whole human race is descended from the eight persons of Noah's ark, and this number is deficient. The smallest odd abundant number is 945.

𝕳istorical 𝕹ote

The Pythagoreans discovered the first four perfect numbers. Fourteen centuries later, the fifth perfect number was discovered. There are only 27 known perfect numbers. The 27th perfect number is

$$2^{44,496}(2^{44,497} - 1)$$

It is so large that it would take over 25,000 digits to write it out (a more than 200-foot-wide piece of paper would be needed to write it on an ordinary typewriter).

𝕳istorical 𝕹ote

The Pythagoreans discovered that 220 and 284 are friendly. The next pair of friendly numbers was found by Pierre de Fermat: 17,296 and 18,416. In 1638 the French mathematician René Descartes found a third pair, and the Swiss mathematician Leonhard Euler found more than 60 pairs. Another strange development was the discovery in 1866 of another pair of friendly numbers: 1184 and 1210. The discoverer, Nicolo Pagonini, was a 16-year-old Italian schoolboy. Evidently the great mathematicians had overlooked this simple pair. Today there are more than 600 known pairs of friendly numbers.

38. *Computer Problem.* Use the program you wrote in Problem 31 to help you find the next perfect number after 6.

39. Two numbers are called *amicable,* or *friendly,* if each is the sum of the proper divisors of the other (a proper divisor includes the number 1 but not the number itself). For example, the numbers 220 and 284 are friendly. The proper divisors of 220 are {1,2,4,5,10,11,20,22,44,55,110}, and

$$1 + 2 + 4 + 5 + 10 + 11 + 20 + 22 + 44 + 55 + 110 = 284$$

The proper divisors of 284 are {1,2,4,71,142}, and

$$1 + 2 + 4 + 71 + 142 = 220$$

Show that 1184 and 1210 are friendly.

Problems for Individual Study

40. *Computer Problem.* Do some research to find additional perfect numbers. Then use a computer to help show that each of them is a perfect number. Hint: There are no odd perfect numbers less than or equal to 10^{200}— according to Bryant Tuckerman, IBM Research Paper RC-1925.
    Reference: Prielipp, Robert, "Perfect Numbers, Abundant Numbers, and Deficient Numbers," *Mathematics Teacher,* December 1970 (Vol. LXIII, No. 8), pp. 692–696.

41. Write out a proof of the Fundamental Theorem of Arithmetic. (Hint: You must show uniqueness.)

6.5 APPLICATIONS OF PRIME FACTORIZATION

We can make use of unique prime factorization in many situations. Recall that if $m = d \cdot k$, then m is a multiple of d and k, and d and k are factors of m.

GREATEST COMMON FACTOR

To find the set of *common* factors of 24 and 30, we can find the set of all factors of each.

$$A = \text{set of factors of 24} = \{1,2,3,4,6,8,12,24\}$$
$$B = \text{set of factors of 30} = \{1,2,3,5,6,10,15,30\}$$

The intersection of these two sets is the set of *common factors* of 24 and 30:

$$A \cap B = \{1,2,3,6\}$$

The number 1 is the least element in this set, and 6 is the greatest. Now, 1 will always be a common factor for any pair of counting numbers, but the greatest of the common factors depends on the numbers chosen.

Thus we will be concerned with finding the *greatest common factor* (g.c.f), and we will say that 6 is the g.c.f. of 24 and 30. That is, 6 is the *largest* number to divide *both* 24 and 30.

Definition of greatest common factor

The method just used is lengthy. We would like to find an algorithm that will give us the g.c.f. directly.

We find the g.c.f. by first finding the prime factorization of each number:

$$24 = 2^3 \cdot 3$$
$$30 = 2 \cdot 3 \cdot 5$$

We write these factorizations in *canonical form* as follows:

$$24 = 2^3 \cdot 3^1 \cdot 5^0 \qquad \text{(Recall that } b^0 = 1.\text{)}$$
$$30 = 2^1 \cdot 3^1 \cdot 5^1$$

Writing the canonical form of a number means writing its prime factorization using exponential notation with the factors arranged in order of increasing magnitude.

Now we select one representative from each of the columns in the factorization. The representative we select when finding the g.c.f. will be the one with the smallest exponent:

$$24 = 2^3 \cdot 3^1 \cdot 5^0$$
$$30 = 2^1 \cdot 3^1 \cdot 5^1$$
$$\text{g.c.f.} = 2^1 \cdot 3^1 \cdot 5^0$$

Finally, we take the product of these representatives to find the g.c.f. (See also Figure 6.5.)

$$\text{g.c.f.} = 2^1 \cdot 3^1 \cdot 5^0 = 2 \cdot 3 \cdot 1 = 6$$

EXAMPLE: Find the g.c.f. of 300, 144, and 108.

Solution:

$$300 = 2^2 \cdot 3^1 \cdot 5^2$$
$$144 = 2^4 \cdot 3^2 \cdot 5^0$$
$$108 = 2^2 \cdot 3^3 \cdot 5^0$$
$$\text{g.c.f.} = 2^2 \cdot 3^1 \cdot 5^0 = 4 \cdot 3 \cdot 1 = 12$$

EXAMPLE: Find the g.c.f. of 15 and 28.

Solution:

$$15 = 2^0 \cdot 3^1 \cdot 5^1 \cdot 7^0$$
$$28 = 2^2 \cdot 3^0 \cdot 5^0 \cdot 7^1$$
$$\text{g.c.f.} = 2^0 \cdot 3^0 \cdot 5^0 \cdot 7^0 = 1$$

If the g.c.f. of two numbers is 1, we say that the numbers are *relatively prime* (notice that relatively prime numbers are not necessarily prime numbers).

Definition of relatively prime numbers

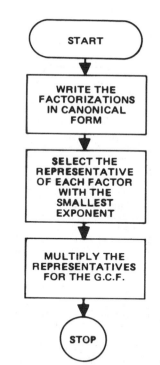

FIGURE 6.5 Flow chart for finding the greatest common factor

EXAMPLES:

1. 15 and 33 are not relatively prime, since their g.c.f. is not 1 (it is 3).
2. 15 and 28 are relatively prime, since their g.c.f. is 1. Notice that these are not prime numbers—they are simply *prime to each other*.

LEAST COMMON MULTIPLE

Consider the set of multiples of 24 and 30:

C = set of multiples of 24: $\{24,48,72,96,120,144,168,192,216,240, \ldots\}$
D = set of multiples of 30: $\{30,60,90,120,150,180,210,240,270,300, \ldots\}$

The set of common multiples of 24 and 30 is infinite:

$$C \cap D = \{120,240,360, \ldots\}$$

Definition of least common multiple.

There is no largest member in this set, so we focus attention on the smallest member, 120, which is called the *least common multiple* (l.c.m.). In general, the l.c.m. of two or more counting numbers is the smallest number into which the counting numbers divide. Notice that the l.c.m. is never smaller than the given numbers.

Again we use an algorithm to find the l.c.m. We proceed as we did with the g.c.f. to find the canonical representations of the numbers involved.

$$24 = 2^3 \cdot 3^1 \cdot 5^0$$
$$30 = 2^1 \cdot 3^1 \cdot 5^1$$

Now, for the l.c.m., we choose the representative of each factor with the largest exponent. The l.c.m. is the product of these representatives. (See also Figure 6.6.)

$$\text{l.c.m.} = 2^3 \cdot 3^1 \cdot 5^1 = 120$$

EXAMPLE: Find the l.c.m. of 300, 144, and 108.

Solution:

$$300 = 2^2 \cdot 3^1 \cdot 5^2$$
$$144 = 2^4 \cdot 3^2 \cdot 5^0$$
$$108 = 2^2 \cdot 3^3 \cdot 5^0$$
$$\text{l.c.m.} = 2^4 \cdot 3^3 \cdot 5^2 = 16 \cdot 27 \cdot 25 = 10{,}800$$

This means that the *smallest* number that 300, 144, and 108 *all* divide into is $2^4 \cdot 3^3 \cdot 5^2$ or 10,800.

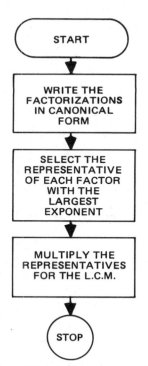

FIGURE 6.6 Flow chart for finding the least common multiple

SUMMARY

To Find the Greatest Common Factor:	To Find the Least Common Multiple:
1. Find the prime factorization.	1. Find the prime factorization.
2. Write in canonical form.	2. Write in canonical form.
3. Choose the representative of each factor with the *smallest exponent*.	3. Choose the representative of each factor with the *largest exponent*.
4. Take the product of the representatives.	4. Take the product of the representatives.

The greatest common factor will be less than or equal to the numbers involved. The least common multiple will be greater than or equal to the numbers involved.

RATIONAL NUMBERS REVISITED

We may use the ideas of g.c.f. and l.c.m. when working with rational numbers. The idea of greatest common factor, for example, is useful when reducing or simplifying rational expressions. If the g.c.f. of the numerator and denominator is 1, then we say that the fraction is in lowest terms, or *reduced*.

A fraction is reduced if the numerator and denominator are relatively prime.

EXAMPLE: Reduce 24/30.

Solution: Note: $24 = 2^3 \cdot 3^1$
$30 = 2^1 \cdot 3^1 \cdot 5^1$ The g.c.f. is 6.

Thus,

$$\frac{24}{30} = \frac{(2 \cdot 3) \cdot 2^2}{(2 \cdot 3) \cdot 5} = \frac{2^2}{5} = \frac{4}{5}$$

After finding the g.c.f. (in parentheses), we wish to simplify the fraction (see arrow). But how? We will need one of the most fundamental ideas of arithmetic, and that is the changing of a rational number from one form to another. It is well known that one element in the set of rational numbers has several representations. For example,

$$\frac{1}{2} = \frac{2}{4} = \frac{19}{38} = \frac{17,475}{34,950} = \cdots$$

The reason why we know these are all equal is summarized with the following property:

Fundamental Property of Fractions: If a/b is any rational number and x is any nonzero integer, then

$$\frac{a \cdot x}{b \cdot x} = \frac{x \cdot a}{x \cdot b} = \frac{a}{b}$$

Fundamental Property of Fractions

That is, given some fraction to simplify:

1. Find the g.c.f. of the numerator and denominator (this is the x of the Fundamental Property).
2. Use the Fundamental Property of Fractions to simplify the fraction. Notice that this property works only for *factors*.

EXAMPLE: Reduce 300/144.

Solution:

$$\frac{300}{144} = \frac{(2^2 \cdot 3^1) \cdot 5^2}{(2^2 \cdot 3^1) \cdot (2^2 \cdot 3^1)}$$

Since the g.c.f. is 12, or $(2^2 \cdot 3)$, we write:

$$\frac{300}{144} = \frac{12 \cdot 5^2}{12 \cdot 2^2 \cdot 3}$$

$$= \frac{25}{12}$$

"Nonsense, Miss Murgle! Their eager, little minds are ready for fractions. There must be something wrong with you as a teacher."

We see that 25/12 is reduced, since the g.c.f. of 25 and 12 is 1.

As we pointed out in Section 4.5, the definitions for addition and subtraction do not require that we first obtain a least common denominator. However, it is sometimes convenient to find one before adding. We can use the least common multiple of the denominators of the two fractions when we add or subtract fractions. For example, the least common multiple of 24 and 30 is 120. Thus,

We are simply multiplying each fraction by the identity, 1, since ⁵/₅ and ⁴/₄ are both equal to 1.

$$\frac{5}{24} = \frac{5}{24} \cdot \frac{5}{5} = \frac{25}{120}$$

$$+ \frac{7}{30} = \frac{7}{30} \cdot \frac{4}{4} = \frac{28}{120}$$

$$\frac{53}{120}$$

The answer is in reduced form, since 53 and 120 are relatively prime.

EXAMPLE: Add: $\dfrac{19}{300} + \dfrac{55}{144} + \dfrac{25}{108}$

Solution: On page 356 we found the l.c.m. of 300, 144, and 108 to be $2^4 \cdot 3^3 \cdot 5^2$.
Thus we write:

$$\frac{19}{300} = \frac{19}{300} \cdot \frac{2^2 \cdot 3^2}{2^2 \cdot 3^2} = \frac{19 \cdot 2^2 \cdot 3^2}{2^4 \cdot 3^3 \cdot 5^2}$$

This is the step during which we multiply each fraction by 1.

$$\frac{55}{144} = \frac{5 \cdot 11}{144} \cdot \frac{3 \cdot 5^2}{3 \cdot 5^2} = \frac{3 \cdot 5^3 \cdot 11}{2^4 \cdot 3^3 \cdot 5^2}$$

$$+ \frac{25}{108} = \frac{5^2}{108} \cdot \frac{2^2 \cdot 5^2}{2^2 \cdot 5^2} = \frac{2^2 \cdot 5^4}{2^4 \cdot 3^3 \cdot 5^2}$$

Adding:

$$\frac{19 \cdot 2^2 \cdot 3^2}{2^4 \cdot 3^3 \cdot 5^2} + \frac{3 \cdot 5^3 \cdot 11}{2^4 \cdot 3^3 \cdot 5^2} + \frac{2^2 \cdot 5^4}{2^4 \cdot 3^3 \cdot 5^2}$$

$$= \frac{684}{2^4 \cdot 3^3 \cdot 5^2} + \frac{4125}{2^4 \cdot 3^3 \cdot 5^2} + \frac{2500}{2^4 \cdot 3^3 \cdot 5^2} = \frac{7309}{2^4 \cdot 3^3 \cdot 5^2}$$

Now, 7309 is not divisible by 2, 3, or 5 (see the rules of divisibility in Section 6.3);
thus, 7309 and $2^4 \cdot 3^3 \cdot 5^2$ are relatively prime, and the solution is complete:

$$\frac{7309}{2^4 \cdot 3^3 \cdot 5^2} \quad \text{or} \quad \frac{7309}{10,800}$$

PROBLEM SET 6.5

A Problems

1. What do we mean when we say that numbers are relatively prime?

2. Find the prime factorization of each of the following, if possible:

   a. 60   b. 72   c. 95   d. 1425

3. Find the prime factorization of each of the following, if possible:

   a. 12   b. 54   c. 171   d. 53

Find the g.c.f. of the sets of numbers in Problems 4–12.

4. $\{60, 72\}$	5. $\{95, 1425\}$	6. $\{12, 54, 171\}$
7. $\{11, 13, 23\}$	8. $\{12, 20, 36\}$	9. $\{6, 8, 10\}$
10. $\{9, 12, 14\}$	11. $\{3, 6, 15, 54\}$	12. $\{90, 75, 120\}$

EXAMPLE: $\{12, 90, 336\}$

Solution:

Find the prime factorization of each:

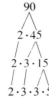

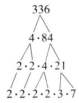

Write in canonical form:

$12 = 2^2 \cdot 3^1 \cdot 5^0 \cdot 7^0$
$90 = 2^1 \cdot 3^2 \cdot 5^1 \cdot 7^0$
$336 = 2^4 \cdot 3^1 \cdot 5^0 \cdot 7^1$

To find the g.c.f. choose the representative from each column with the *smallest* exponent:

$2^1 \cdot 3^1 \cdot 5^0 \cdot 7^0$

Take the product:

g.c.f. $= 2 \cdot 3 \cdot 1 \cdot 1$
    $= 6$

To find the l.c.m. choose the representative from each column with the *largest* exponent:

$2^4 \cdot 3^2 \cdot 5^1 \cdot 7^1$

Take the product:

l.c.m. $= 16 \cdot 9 \cdot 5 \cdot 7$
    $= 5040$

Find the l.c.m. of the sets of numbers in Problems 13–21.

13. $\{60, 72\}$ 14. $\{95, 1425\}$ 15. $\{12, 54, 171\}$

16. $\{11, 13, 23\}$ 17. $\{12, 20, 36\}$ 18. $\{6, 8, 10\}$

19. $\{9, 12, 14\}$ 20. $\{3, 6, 15, 54\}$ 21. $\{90, 75, 120\}$

22. a. Find the g.c.f. of 6 and 15.
    b. Find the l.c.m. of 6 and 15.
    c. Find the product of your answers to parts a and b.
    d. Compare your answer to part c with the product of 6 and 15.
    e. Repeat parts a–d for 8 and 12.
    f. Repeat parts a–d for 5 and 10.
    g. Repeat parts a–d for 5 and 12.

23. Make a conjecture based on your findings in Problem 22.

B Problems

For Problems 24–27 consider the set $\{1,2,3,5,6,10,15,30\}$. Let us define an operation $\varnothing$, meaning the greatest common factor. For example,

$$5 \varnothing 10 = 5;$$
$$5 \varnothing 6 = 1;$$
$$10 \varnothing 30 = 10.$$

Let us also define a second operation M, meaning the least common multiple. For example,

$$5 M 10 = 10;$$
$$5 M 6 = 30;$$
$$10 M 30 = 30.$$

24. Is the set closed for the operations of $\varnothing$ and M?

25. Are the operations associative?

26. The identity for $\varnothing$ is 30 and for M is 1. Does the set have the inverse property for $\varnothing$ and M?

27. Does the set form a field for the operations of $\varnothing$ and M?

28. Find the g.c.f. and the l.c.m. of $\{120, 75, 975\}$.

29. Reduce:

a. $\dfrac{78}{455}$ b. $\dfrac{75}{500}$ c. $\dfrac{240}{672}$ d. $\dfrac{5670}{12,150}$

Perform the indicated operations in Problems 30–53.

30. $\dfrac{\frac{1}{2} + \frac{-2}{3}}{\frac{5}{6} - \frac{-3}{5}}$ 31. $\dfrac{2^{-1} + 3^{-2}}{2^{-1} + 3^{-1}}$

32. $\dfrac{^-3}{4} \cdot \dfrac{119}{200} + \dfrac{^-3}{4} \cdot \dfrac{81}{200}$

33. $\dfrac{^-3}{4} \cdot \left(\dfrac{119}{200} + \dfrac{81}{200}\right)$

34. $\dfrac{2}{3} - \dfrac{7}{12}$

35. $\dfrac{^-7}{24} + \dfrac{^-13}{16}$

36. $\dfrac{28}{9} - \dfrac{4}{27}$

37. $\dfrac{1}{-8} + \dfrac{1}{-7}$

38. $^-.3 + .16 + .476$

39. $5^{-1} + 5^{-2} + 5$

40. $^-5 + (^-5)^2$

41. $^-3 - \dfrac{5}{29}$

42. $\dfrac{^-2}{15} + \dfrac{3}{5} + \dfrac{7}{12}$

43. $^-2^3/_5 + 4^1/_8 - 7^1/_{10}$

44. $\dfrac{1}{10} \cdot \dfrac{^-2}{5}$

45. $\left(\dfrac{3}{7} \cdot \dfrac{3}{5}\right) \div \dfrac{1}{2}$

46. $\dfrac{2}{^-3} \cdot \dfrac{^-2}{15}$

47. $\dfrac{4}{5}\left(\dfrac{17}{95}\right) + \dfrac{4}{5}\left(\dfrac{78}{95}\right)$

48. $\dfrac{^-7}{8} \cdot \dfrac{131}{147} + \dfrac{^-7}{8} \cdot \dfrac{16}{147}$

49. $\dfrac{\left(\dfrac{3}{5} \cdot \dfrac{1}{7}\right) + \dfrac{1}{3}}{2}$

50. $\dfrac{\dfrac{1}{3} - \dfrac{^-1}{4}}{\dfrac{7}{8} - \dfrac{3}{16}}$

51. $\dfrac{^-3}{8} + \dfrac{3}{4} - \dfrac{9}{10}$

52. $\dfrac{\dfrac{15}{16}}{\dfrac{3}{8}}$

53. $\dfrac{\dfrac{4}{5} + \dfrac{^-2}{3}}{\dfrac{1}{5}} + 2$

54. *Euclidean Algorithm.* The g.c.f. of two numbers can be found using another method, which is attributed to Euclid. It is called the Euclidean Algorithm and is based on repeated division. For example, find the g.c.f. of 108 and 300.

Procedure: (1) Divide 300 by 108. List the quotient and remainder.

$$\begin{array}{r} 2 \\ 108\overline{)300} \\ 216 \\ \hline 84 \end{array} \qquad 300 = 2 \cdot 108 + 84$$

(2) Divide 108 by remainder of Step 1.

$$84\overline{)108} \qquad 108 = 1 \cdot 84 + 24$$
$$\underline{84}$$
$$24$$

(3) Repeat the process.

$$24\overline{)84} \qquad 84 = 3 \cdot 24 + 12$$
$$\underline{72}$$
$$12$$

(4) Repeat until the remainder is 0.

$$12\overline{)24} \qquad \text{The remainder is 0, so we stop.}$$
$$\underline{24}$$
$$0$$

(5) The last divisor (in this case, 12) is the g.c.f.

Find the g.c.f. using the Euclidean Algorithm:

a. {91, 107}   b. {126, 180}   c. {51, 1995}

EXAMPLE: {357, 629}

$$357\overline{)629} \qquad\qquad 272\overline{)357}$$
$$\underline{357} \qquad\qquad\qquad \underline{272}$$
$$272 \quad r = 272 \qquad 85 \quad r = 85$$

$$85\overline{)272} \qquad\qquad 17\overline{)85}$$
$$\underline{255} \qquad\qquad\qquad \underline{85}$$
$$17 \quad r = 17 \qquad\quad 0$$

r = 0; thus
we stop.

The last divisor is the g.c.f.
Therefore the g.c.f. of 357 and 629 is 17.

55. *Computer Problem*. The g.c.f. of two numbers can easily be programmed using BASIC as follows:

```
10   PRINT "WHAT ARE THE TWO NUMBERS";
20   INPUT A,B
25   IF A < = 0 GOTO 98
26   IF B < = 0 GOTO 98
27   IF INT(A) <> A GOTO 98
28   IF INT(B) <> B GOTO 98
30   IF A > = B GOTO 60
40   LET C = A
45   LET A = B
50   LET B = C
60   LET D = A − B * INT(A/B)
65   LET A = B
70   LET B = D
80   IF D <> 0 GOTO 60
90   PRINT "THE GREATEST COMMON FACTOR IS ";
95   PRINT A
96   GOTO 105
98   PRINT "PLEASE INPUT POSITIVE INTEGERS ONLY."
99   PRINT
100  GOTO 10
105  END
```

Notice that three trial runs are shown for the sets {45, 100}, {544, 5321}, and {31,75}. Use this program to find the g.c.f. of the following numbers:

a. {7957, 11,023}   b. {18,847, 23,641}
c. {44,329, 140,299}

```
RUN
WHAT ARE THE TWO NUMBERS? 45,100
THE GREATEST COMMON FACTOR IS  5

READY

RUN
WHAT ARE THE TWO NUMBERS? 544,5321
THE GREATEST COMMON FACTOR IS  17

READY

RUN
WHAT ARE THE TWO NUMBERS? 31,75
THE GREATEST COMMON FACTOR IS  1

READY
```

56. *Computer Problem*. The l.c.m. of two numbers can be programmed using BASIC as follows:

```
10   PRINT "WHAT ARE THE TWO NUMBERS";
20   INPUT A,B
23   LET N = A*B
25   IF A < = 0 GOTO 98
26   IF B < = 0 GOTO 98
27   IF INT(A) <> A GOTO 98
28   IF INT(B) <> B GOTO 98
30   IF A > = B GOTO 60
40   LET C = A
45   LET A = B
50   LET B = C
60   LET D = A − B*INT(A/B)
65   LET A = B
70   LET B = D
80   IF D <> 0 GOTO 60
90   PRINT "THE LEAST COMMON MULTIPLE IS";
95   PRINT N/A
96   GOTO 105
98   PRINT "PLEASE INPUT POSITIVE INTEGERS ONLY."
99   PRINT
100  GOTO 10
105  END
```

```
RUN
WHAT ARE THE TWO NUMBERS? 45,100
THE LEAST COMMON MULTIPLE IS  900

READY

RUN
WHAT ARE THE TWO NUMBERS? 544,5321
THE LEAST COMMON MULTIPLE IS  170272

READY

RUN
WHAT ARE THE TWO NUMBERS? 31,75
THE LEAST COMMON MULTIPLE IS  2325

READY
```

Notice that three trial runs are shown. Use this program to find the l.c.m. of the following numbers:

a. {127, 809}   b. {2201, 3317}   c. {109, 1135}

57. *Computer Problem*. Study the programs given in Problems 55 and 56, and then try to write a program that will find both the g.c.f. and the l.c.m. of two given numbers.

Mind Boggler

58. The sum and difference of two squares may be primes. For example,

$$9 − 4 = 5 \quad \text{and} \quad 9 + 4 = 13$$

Can the sum and difference of two primes be square? Can you find more than one example?

Problem for Individual Study

59. *A Fraction Machine*. Take two sheets of lined paper. Label one *numerator* and one *denominator*, and number the lines consecutively from 0 as shown in Figure 6.7. Fasten the pages together at 0 so that the top page pivots at this point (or else always be sure that the 0 lines coincide).

Suppose you wish to find a fraction equal to $5/6$ using a denominator of 18. Slide the top page down until the lines 5 (numerator) and 6

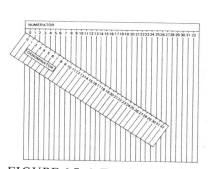

FIGURE 6.7 A Fraction Machine

(denominator) coincide. Now you can read other equivalent fractions from the Fraction Machine by noticing those places where the lines coincide. From Figure 6.7 we see that

$$\frac{5}{6} = \frac{15}{18}$$

Also notice from the figure that $^5/_6 = {}^{10}/_{12}$ and $^5/_6 = {}^{20}/_{24}$. Build a Fraction Machine and fill in the following blanks.

a. $\dfrac{5}{7} = \dfrac{?}{21}$ b. $\dfrac{4}{9} = \dfrac{?}{18}$

c. $\dfrac{5}{8} = \dfrac{?}{4}$ d. $\dfrac{3}{4} = \dfrac{1}{?}$

Demonstrate a Fraction Machine to the class. Can you think of any other uses for a Fraction Machine?*

Reference: "Reducing Fractions Can Be Easy, Maybe Even Fun," by M. Wassmansdorf, *Arithmetic Teacher*, February 1974, pp. 99–102.

6.6 A FINITE ALGEBRA

At the beginning of this section we mentioned that there are many types of algebra. We then looked at the algebraic system taught in high school and solved some equations with that algebra. In this section we'll look at an algebra based on a finite set rather than on the infinite set of real numbers.

A FINITE SYSTEM: THE 12-HOUR CLOCK

We now consider a mathematical system based on the 12-hour clock. We'll need to define some operations for this set of numbers, and we'll use the way we tell time as a guide to our definitions. For example, if you have an appointment at 4:00 P.M. and you are two hours late, you arrive at 6:00 P.M. On the other hand, if your appointment was at 11:00 A.M. and you are two hours late, you arrive at 1:00 P.M. That is,

$$4 + 2 = 6 \quad \text{and} \quad 11 + 2 = 1 \quad \text{on a 12-hour clock}$$

Using the clock as a guide, we define an operation called "clock addition" (which is different from ordinary addition) according to Table 6.6.

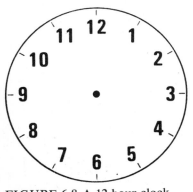

FIGURE 6.8 A 12-hour clock

*My thanks to Mark Wassmansdorf of the Philadelphia School District for this problem.

TABLE 6.6 Addition on a 12-Hour Clock

+	1	2	3	4	5	6	7	8	9	10	11	12
1	2	3	4	5	6	7	8	9	10	11	12	1
2	3	4	5	6	7	8	9	10	11	12	1	2
3	4	5	6	7	8	9	10	11	12	1	2	3
4	5	6	7	8	9	10	11	12	1	2	3	4
5	6	7	8	9	10	11	12	1	2	3	4	5
6	7	8	9	10	11	12	1	2	3	4	5	6
7	8	9	10	11	12	1	2	3	4	5	6	7
8	9	10	11	12	1	2	3	4	5	6	7	8
9	10	11	12	1	2	3	4	5	6	7	8	9
10	11	12	1	2	3	4	5	6	7	8	9	10
11	12	1	2	3	4	5	6	7	8	9	10	11
12	1	2	3	4	5	6	7	8	9	10	11	12

EXAMPLES:

1. $7 + 5 = 12$ 2. $9 + 5 = 2$ 3. $4 + 11 = 3$

Subtraction, multiplication, and division may be defined as they are in ordinary arithmetic:

Subtraction is defined as the inverse or opposite operation of addition: $a - b = x$ means $a = b + x$.

Multiplication is defined as repeated addition ($b \neq 0$):

$$a \times b \text{ means } \underbrace{a + a + a + \cdots + a}_{b \text{ addends}}$$

If $b = 0$, then $a \cdot 0 = 0$.

Division is defined as the inverse or opposite of multiplication ($b \neq 0$):

$$\frac{a}{b} = x \text{ means } a = b \cdot x$$

EXAMPLES:

1. Find $4 - 9$. This means find the number that, when added to 9, produces 4. Thus, solve $4 = 9 + t$. From the table, $t = 7$.
2. Find 4×9. This means find $9 + 9 + 9 + 9 = 12$.
3. Find $4 \div 7$ or $4/7 = t$. This means find the number that, when multiplied by 7, produces 4. Thus, solve $4 = 7 \times t$. Since $7 \times 4 = 4$, we see that $4/7 = 4$.
4. Find $4/9 = t$. This means find the number that, when multiplied by 9, produces 4. There is no such number.

Since the addition and multiplication tables for the 12-hour clock are rather big, we shorten our arithmetic system by considering a clock with fewer than 12 hours.

A FINITE SYSTEM: THE 5-HOUR CLOCK

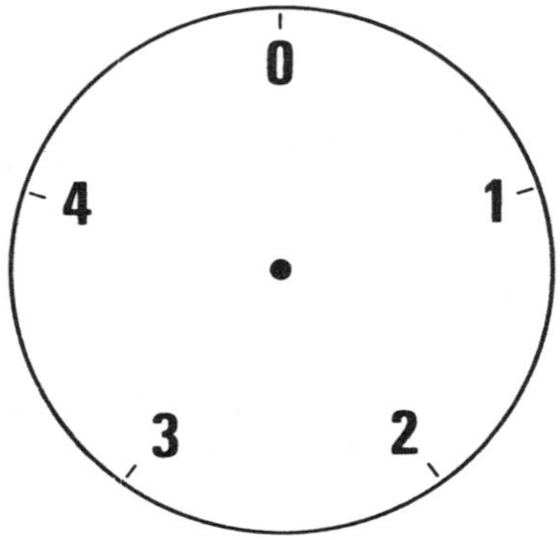

FIGURE 6.9 A 5-hour clock

Consider a mathematical system based on a clock for five hours, numbered 0,1,2,3, and 4, as shown in Figure 6.9. Clock addition on the 5-hour clock is the same as on an ordinary clock, except that the only numerals are 0,1,2,3, and 4 (notice that 5 has been replaced by 0, and we agree that 0 and 5 have the same meaning in this system).

We define the operations of multiplication, subtraction, and division as before, and thus we can complete Table 6.7 (since subtraction and division are the opposite operations of addition and multiplication, we need only two tables).

TABLE 6.7 Addition and Multiplication on a 5-Hour Clock

+	0	1	2	3	4
0	0	1	2	3	4
1	1	2	3	4	0
2	2	3	4	0	1
3	3	4	0	1	2
4	4	0	1	2	3

×	0	1	2	3	4
0	0	0	0	0	0
1	0	1	2	3	4
2	0	2	4	1	3
3	0	3	1	4	2
4	0	4	3	2	1

OTHER FINITE SYSTEMS

Instead of speaking about "arithmetic on the 5-hour clock," mathematicians usually speak of "modulo 5 arithmetic." The set {0,1,2,3,4}, together with the operations defined in Table 6.7, is called a *modulo 5* or *mod 5* system.

Notice that in a mod 5 system (or on a 5-hour clock):

$$4 + 9 = 3 \quad \text{and} \quad 2 + 1 = 3$$

Thus, $4 + 9 = 2 + 1$.

Since we do not wish to confuse this with ordinary arithmetic, in which

$$4 + 9 \neq 2 + 1$$

we invent the following notation:

$$4 + 9 \equiv 2 + 1 \ (\text{mod } 5)$$

where $\equiv$ is read

"4 + 9 is congruent to 2 + 1 in modulo 5."

𝕳istorical 𝕹ote

Carl Friedrich Gauss (see historical note on page 266) was the first mathematician to work with these modular systems. Gauss called mathematics the Queen of the Sciences. When he was 20 he wrote that he was so overwhelmed with mathematical ideas that there was not enough time to record more than a small fraction of them. An excellent article about Gauss was published in honor of the 200th anniversary of his birth in the July 1977 issue of Scientific American *(pp. 122–131).*

We define congruence as follows.

DEFINITION: a and b are *congruent modulo m,* written $a \equiv b \pmod{m}$, if a and b differ by a multiple of m.

Definition of congruence mod m

EXAMPLES:

1. $3 \equiv 8 \pmod 5$, since $8 - 3 = 5$, and 5 is a multiple of 5.
2. $3 \equiv 53 \pmod 5$, since $53 - 3 = 50$, and 50 is a multiple of 5.
3. $3 \not\equiv 19 \pmod 5$, since $19 - 3 = 16$ and 16 is not a multiple of 5.

Another way of determining two congruent numbers mod m is to divide each by m and check the remainders. If the remainders are the same, then the numbers are congruent mod m. For example, $3 \div 5$ gives a remainder 3, and $53 \div 5$ gives a remainder 3; thus $3 \equiv 53 \pmod 5$.

EXAMPLES:

1. $4 + \quad 9 \equiv 2 + 1 \pmod 5$, since
   $$13 \equiv 3 \pmod 5.$$
2. $15 + \quad 92 \equiv x \pmod 5$
   $$107 \equiv x \pmod 5$$
   $$2 \equiv x \pmod 5$$

Notice that every whole number is congruent modulo 5 to exactly one element in the set $I = \{0,1,2,3,4\}$. This set I contains all possible remainders when dividing by 5. Since the number of elements in I is rather small, we can easily solve equations by trying the numbers 0,1,2,3, and 4.

EXAMPLE: $2 + 4 \equiv x \pmod 5$

Method 1: $2 + 4 = 6$; so
$$6 \equiv x \pmod 5$$
$$1 \equiv x \pmod 5 \longleftarrow$$

We found this number by putting down the remainder obtained when we divided 6 by the modulus 5.

Method 2:   Check all possibilities for x, and determine if any make the sentence true.

$$x \equiv 0? \quad 2 + 4 \equiv 0 \quad \text{no}$$
$$x \equiv 1? \quad 2 + 4 \equiv 1 \quad yes$$
$$x \equiv 2? \quad 2 + 4 \equiv 2 \quad \text{no}$$
$$x \equiv 3? \quad 2 + 4 \equiv 3 \quad \text{no}$$
$$x \equiv 4? \quad 2 + 4 \equiv 4 \quad \text{no}$$

You must try all possibilities, since, for mod m, there may be more than one solution. Be careful not to assume that you can "solve" these equations using ordinary rules of algebra. This is *not* ordinary algebra—it is modular arithmetic. For example, in mod 6 you cannot divide both sides of the equivalence relation by 3.

EXAMPLES:

1. $2 - 4 \equiv x$ (mod 5)

    Now, $2 - 4 \equiv x$ means $2 \equiv 4 + x$. Directly, or by substitution, we find that $x \equiv 3$ (mod 5).

2. $2 \times 4 \equiv x$ (mod 5)

    $8 \equiv x$ (mod 5)

    $3 \equiv x$ (mod 5)

3. The manager of a T.V. station hired a college student, U. R. Stuck, as an election-eve runner. The request for reimbursement turned in is shown. The station manager refused to pay the bill, since the mileage was not honestly recorded. How did he know? (Note: It has nothing to do with the illegible digits in the mileage report.)

    Solution: Let x = the number of miles in one round trip. From the reimbursement request we see there are 8 trips, so the total mileage is $8x$, and since the last digits are legible we see that the difference is 5 (mod 10). Thus,

    $$8x \equiv 5 \quad (\text{mod } 10)$$

    Solving:

    $x = 0:\ 8(0) \equiv 0$ (mod 10)
    $x = 1:\ 8(1) \equiv 8$ (mod 10)
    $x = 2:\ 8(2) \equiv 6$ (mod 10)
    $x = 3:\ 8(3) \equiv 4$ (mod 10)
    $x = 4:\ 8(4) \equiv 2$ (mod 10)
    $x = 5:\ 8(5) \equiv 0$ (mod 10)
    $x = 6:\ 8(6) \equiv 8$ (mod 10)
    $x = 7:\ 8(7) \equiv 6$ (mod 10)

    These are the only possible values for x (mod 10), and none gives a mileage reading of 5 (mod 10) in the units digit. Therefore, Stuck did not report the mileage honestly.

4. Suppose you are planning to buy paper to cover some shelves. You need to cover between 30 and 50 11-inch shelves and one shelf 100 inches long. The paper can be purchased in multiples of 36 inches. How much paper should you buy to eliminate as much waste as possible?

    Solution: Since the paper comes in multiples of 36 inches, you must buy $36x$ inches of paper, where x represents the number of multiples you buy. Also, you will cover k shelves at 11 inches and one shelf at 100 inches for a total of $11k + 100$ inches. We can write this as a congruence:

    $$36x \equiv 100 \ (\text{mod } 11)$$
    $$\text{or}$$
    $$3x \equiv 1 \ (\text{mod } 11)$$

REQUEST FOR REIMBURSEMENT
FOR MILEAGE

NAME: _U. R. Stuck_

ADDRESS: _1234 Fifth st._

S.S. NO.: _576-38-4459_

ENDING MILEAGE: _14,128_

BEGINNING MILEAGE: _14,6i3_

NO. OF TRIPS _8_

This congruence has a solution of $x \equiv 4 \pmod{11}$, which means that we can buy 4, 15, 26, . . . multiples of the 36-inch paper. If $x = 15$, then we buy

$$36 \cdot 15 = 540$$

inches of paper. This amount allows us to cover the 100-inch board and 40 11-inch boards.

5. If $[x]$ denotes the greatest integer less than or equal to x, then you can use this idea along with modular arithmetic to find the dates of Friday the 13th in any given year by using Table 6.8.

Notice that $36 \equiv 3$ and $100 \equiv 1$ (mod 11). We found this solution by considering:

$x = 0$: $3(0) \equiv 0 \pmod{11}$
$x = 1$: $3(1) \equiv 3 \pmod{11}$
$x = 2$: $3(2) \equiv 6 \pmod{11}$
$x = 3$: $3(3) \equiv 9 \pmod{11}$
$x = 4$: $3(4) \equiv 1 \pmod{11}$
$x = 5$: $3(5) \equiv 4 \pmod{11}$
$x = 6$: $3(6) \equiv 7 \pmod{11}$

TABLE 6.8 Months with Friday the 13th*

Day #	Day	Non-Leap Year	Leap-Year
0	Sunday	January, October	January, April, July
1	Monday	April, July	September, December
2	Tuesday	September, December	June
3	Wednesday	June	March, November
4	Thursday	February, March, November	February, August
5	Friday	August	May
6	Saturday	May	October

If there were no leap years, the first day of each year would move up by one day, but because of leap years the problem becomes more complicated. Let y = year in question.

$$y + \left[\frac{y-1}{4}\right] - \left[\frac{y-1}{100}\right] + \left[\frac{y-1}{400}\right] = x$$

      leap   leap   leap
      years   centuries quadracenturies

Find x mod 7 and use Table 6.8 to find the months with a Friday the 13th. For example, in 1979:

$$
\begin{aligned}
&1979 + [1978/4] - [1978/100] + [1978/400] \\
= &1979 + \quad 494 \quad - \quad 19 \quad + \quad 4 \\
= &2458 \\
\equiv &1 \pmod{7}
\end{aligned}
$$

The Gregorian calendar has a leap year every four years, except century years. In addition, every fourth century is a leap year.

The calculation $[1978/4]$ is easily done on a calculator:

$$\boxed{1978} \;\boxed{\div}\; \boxed{4} \;\boxed{=}$$

The result is 494.5; disregard the decimal part for this calculation since you are considering the greatest integer part.

Thus, from Table 6.8, the first day of the year is a Monday, and it is a non-leap year, so there will be a Friday the 13th in April and in July. Notice that if you happen to know the day of the week of January 1st, you need no calculation, but can go directly to Table 6.8.

*This table, along with the justification for the formula, is found in "An Aid to the Superstitious," by G. L. Ritter, S. R. Lowry, H. B. Woodruff, and T. L. Isenhour, *The Mathematics Teacher*, May 1977, pp. 456–457.

6. $^2/_4 \equiv x (\bmod 5)$
   This means $2 \equiv 4 \cdot x$
   Checking all possibilities:

$$4 \cdot \mathbf{0} = 0 \not\equiv 2 \;(\bmod 5)$$
$$4 \cdot \mathbf{1} = 4 \not\equiv 2 \;(\bmod 5)$$
$$4 \cdot \mathbf{2} = 8 \not\equiv 2 \;(\bmod 5)$$
$$4 \cdot \mathbf{3} = 12 \equiv 2 \;(\bmod 5)$$
$$4 \cdot \mathbf{4} = 16 \not\equiv 2 \;(\bmod 5)$$

Thus, there is one solution, $x \equiv 3 \;(\bmod 5)$.

Can you say $^2/_4 \equiv ^1/_2 \;(\bmod 5)$? You must be careful! In this case, we see that

$$\frac{2}{4} \equiv 3 \;(\bmod 5) \qquad \text{and} \qquad \frac{1}{2} \equiv 3 \;(\bmod 5).$$

Thus, $^2/_4 \equiv ^1/_2 \;(\bmod 5)$. Consider the following example in modulo 6.

EXAMPLE: $^2/_4 \equiv x \;(\bmod 6)$
This means $2 \equiv 4 \cdot x \;(\bmod 6)$.

$$4 \cdot \mathbf{0} = 0 \not\equiv 2 \;(\bmod 6)$$
$$4 \cdot \mathbf{1} = 4 \not\equiv 2 \;(\bmod 6)$$
$$4 \cdot \mathbf{2} = 8 \equiv 2 \;(\bmod 6)$$
$$4 \cdot \mathbf{3} = 12 \not\equiv 2 \;(\bmod 6)$$
$$4 \cdot \mathbf{4} = 16 \not\equiv 2 \;(\bmod 6)$$
$$4 \cdot \mathbf{5} = 20 \equiv 2 \;(\bmod 6)$$

We see that it has the solutions $x \equiv 2$ and $x \equiv 5 \;(\bmod 6)$. Also, $^1/_2 \equiv x \;(\bmod 6)$ has no solution (check all possibilities). Thus, $^2/_4 \not\equiv ^1/_2 (\bmod 6)$.

A FINITE ALGEBRA

It appears that some mod systems "behave" like ordinary algebra and others do not. The distinction is found by determining which systems form a field.

Let's explore the field properties for the set $I = \{0,1,2,3,4\}$ and the operations of addition and multiplication as defined by Table 6.7.

1. *Closure for addition.* The set I of elements in arithmetic modulo 5 is closed with respect to addition. That is, for any pair of elements, there is a unique element that represents their sum and that is also a member of the original set.
2. *Closure for multiplication.* The set is closed for multiplication as you can see from Table 6.7.
3. *Commutative for addition.* Addition in arithmetic modulo 5 satisfies the commutative property. That is,

$$a + b \equiv b + a$$

where a and b are any elements of the set I. Specifically, we see that the entries in Table 6.7 are symmetric with respect to the principal diagonal.

4. *Commutative for multiplication.* The set is commutative for multiplication as you can see from Table 6.7.

5. *Associative for addition.* Addition of elements in arithmetic modulo 5 satisfies the associative property. That is,

$$(a + b) + c \equiv a + (b + c)$$

for all elements a, b, and c of the set I.

As a specific example, we evaluate $2 + 3 + 4$ in two ways:

$$(2 + 3) + 4 \quad \text{and} \quad 2 + (3 + 4) \quad \text{(mod 5)}$$
$$0 + 4 \equiv 2 + 2 \quad \text{(mod 5)}$$

6. *Associative for multiplication.* This property is satisfied, and the details are left for you.

7. *Identity for addition.* The set I of elements modulo 5 includes an identity element for addition. That is, the set contains an element zero such that the sum of any given element and zero is the given element. That is,

$$0 + 0 \equiv 0, \quad 1 + 0 \equiv 1, \quad 2 + 0 \equiv 2, \quad 3 + 0 \equiv 3, \quad 4 + 0 \equiv 4.$$

8. *Identity for multiplication.* The identity element for multiplication is 1, since

$$0 \cdot 1 \equiv 0, \quad 1 \cdot 1 \equiv 1, \quad 2 \cdot 1 \equiv 2, \quad 3 \cdot 1 \equiv 3, \quad 4 \cdot 1 \equiv 4.$$

9. *Inverse for addition.* Each element in arithmetic modulo 5 has an inverse with respect to addition. That is, for each element a of set I, there exists a unique element a' of I such that $a + a' = 0$, the identity element. The element a' is said to be the inverse of a. Specifically, we have the following:

The inverse of 0 is 0;	$0 + 0 \equiv 0$ (mod 5).
The inverse of 1 is 4;	$1 + 4 \equiv 0$ (mod 5).
The inverse of 2 is 3;	$2 + 3 \equiv 0$ (mod 5).
The inverse of 3 is 2;	$3 + 2 \equiv 0$ (mod 5).
The inverse of 4 is 1;	$4 + 1 \equiv 0$ (mod 5).

elements ↑ ↑ │ ↑ identity
inverses

10. *Inverse for multiplication.* The inverse can be checked by finding the following:

There is no inverse for 0;	$0 \cdot ? \equiv 1$ (mod 5).
The inverse of 1 is 1;	$1 \cdot 1 \equiv 1$ (mod 5).
The inverse of 2 is 3;	$2 \cdot 3 \equiv 1$ (mod 5).
The inverse of 3 is 2;	$3 \cdot 2 \equiv 1$ (mod 5).
The inverse of 4 is 4;	$4 \cdot 4 \equiv 1$ (mod 5).

elements ↑ ↑ │ ↑ identity
inverses

Since every nonzero element has an inverse for multiplication, we say the inverse property for multiplication is satisfied.

the changing and the infinite rather than by the static and the finite. As an example of his tremendous ability, when Newton was 74 years old the mathematician Leibniz (see Historical Note on page 63) posed a challenge problem to all the mathematicians in Europe. Newton received the problem after a day's work at the mint (remember he was 74!) and solved the problem that evening. His intellect was monumental.

11. *Distributive for multiplication over addition.* We need to check
$$a(b + c) \equiv ab + ac$$

For example,
$$2(3 + 4) \equiv 2(2) \quad (\text{mod } 5)$$
$$\equiv 4 \quad (\text{mod } 5)$$

and
$$2 \cdot 3 + 2 \cdot 4 \equiv 1 + 3 \quad (\text{mod } 5)$$
$$\equiv 4 \quad (\text{mod } 5)$$

Thus, for this example, the distributive property holds. It would be possible (but very tedious) to check all possibilities, since it is a finite system, and find that this property is satisfied.

Therefore, modulo 5 arithmetic with addition and multiplication forms a field. You should expect that the solving of equations in mod 5 is consistent with ordinary algebra.

On the other hand, modulo 6 arithmetic does not form a field for addition and multiplication, since it does not satisfy the inverse property for multiplication (you are asked to show this in the problem set). You should not expect that the solving of equations in mod 6 is consistent with ordinary algebra. For example, consider
$$4x \equiv 2 \quad (\text{mod } 6)$$
In ordinary algebra, you would write
$$4x = 2$$
$$x = \frac{2}{4}$$
$$x = \frac{1}{2}$$

Remember $\frac{1}{2}$ means 1 divided by 2.

In mod 5, you could write
$$4x \equiv 2$$
$$x \equiv \frac{2}{4} \quad (\text{mod } 5)$$
$$\equiv \frac{1}{2} \quad (\text{mod } 5)$$
$$\equiv 3 \quad (\text{mod } 5)$$

But in mod 6, which is not a field, you would need to check each possibility:
$$4x \equiv 2$$
$$x \equiv 2 \quad \text{and} \quad x \equiv 5$$

This problem was solved on page 370.

Notice that in ordinary algebra a first-degree equation has one root. In mod 5, which is a field, there is one root. However, in mod 6, there may be more than one solution to a first-degree equation.

PROBLEM SET 6.6

A Problems

Perform the indicated operations in Problems 1–5 using arithmetic for a 12-hour clock.

1. a. $9 + 6$ b. $5 - 7$ c. 5×3 d. $2 \cdot 7$

2. a. $7 + 10$ b. $4 - 8$ c. 6×7 d. $1/5$

3. a. $5 + 7$ b. $7 - 9$ c. $2 - 6$ d. $9 \cdot 3$

4. a. $4 \cdot 8$ b. $2 \cdot 3$ c. $3 \cdot 5 - 7$ d. $5 \cdot 2 - 11$

5. a. $\dfrac{1}{12}$ b. $5 \cdot 8 + 5 \cdot 4$

> There was a young fellow named Ben
> Who could only count modulo ten
> He said when I go
> Past my last little toe
> I shall have to start over again.

6. a. Define precisely the concept of "congruence modulo m."
   b. Discuss the meaning of this definition in your own words.

7. Which of the following are true?
   a. $5 + 8 \equiv 1 \pmod 6$
   b. $4 + 5 \equiv 1 \pmod 7$
   c. $5 \equiv 53 \pmod 8$
   d. $102 \equiv 1 \pmod 2$

8. Which of the following are true?
   a. $47 \equiv 2 \pmod 5$
   b. $108 \equiv 12 \pmod 8$
   c. $5670 \equiv 270 \pmod{365}$
   d. $2001 \equiv 30 \pmod{73}$

9. Which of the following are true?
   a. $1975 \equiv 0 \pmod{1975}$
   b. $246 \equiv 150 \pmod 6$
   c. $126 \equiv 1 \pmod 7$
   d. $144 \equiv 12 \pmod{144}$

Perform the indicated operations in Problems 10–13.

10. a. $9 + 6 \pmod 5$ b. $7 - 11 \pmod{12}$

    c. $4 \cdot 3 \pmod 5$ d. $\dfrac{1}{2} \pmod 5$

11. a. $5 + 2 \pmod 4$ b. $2 - 4 \pmod 5$

    c. $6 \cdot 6 \pmod 8$ d. $5 \div 7 \pmod 9$

12. a. $4 + 3 \pmod 5$ b. $6 - 12 \pmod 8$

    c. $121 \cdot 47 \pmod{121}$ d. $\dfrac{2}{3} \pmod 7$

13. a. $7 + 41 \pmod 5$ b. $5 \cdot 4 \pmod{11}$

    c. $62 \cdot 4 \pmod 2$ d. $\dfrac{7}{12} \pmod{13}$

B Problems

Solve for x in Problems 14–18.

14. a. $x + 3 \equiv 0 \pmod 7$ b. $4x \equiv 1 \pmod 5$

    c. $x \cdot x \equiv 1 \pmod 4$ d. $\dfrac{x}{4} \equiv 5 \pmod 9$

EXAMPLE: $\dfrac{x}{3} \equiv 4 \pmod 5$

Solution: Check all possibilities:

$x \equiv 0; \dfrac{0}{3} \not\equiv 4 \pmod 5$

$x \equiv 1; \dfrac{1}{3} \not\equiv 4 \pmod 5$

$x \equiv 2; \dfrac{2}{3} \equiv 4 \pmod 5$

$x \equiv 3; \dfrac{3}{3} \not\equiv 4 \pmod 5$

$x \equiv 4; \dfrac{4}{3} \not\equiv 4 \pmod 5$

Therefore, there is one solution—namely $x \equiv 2 \pmod 5$.

Alternate solution: Use the definition of division to write

$$\frac{x}{3} \equiv 4 \pmod 5$$

as

$x \equiv 3 \cdot 4$
$x \equiv 12$
$x \equiv 2 \pmod 5$

15. a. $x + 5 \equiv 2 \pmod 9$ b. $4x \equiv 1 \pmod 6$
    c. $x^2 \equiv 1 \pmod 5$ d. $\dfrac{4}{6} \equiv x \pmod{13}$

16. a. $5x \equiv 2 \pmod 7$ b. $7x + 1 \equiv 3 \pmod{11}$
    c. $\dfrac{x}{3} \equiv 4 \pmod 8$ d. $4k + 2 \equiv x \pmod 4$, where k is any counting number.

17. a. $x - 2 \equiv 3 \pmod 6$ b. $3x \equiv 2 \pmod 7$
    c. $4k \equiv x \pmod 4$, where k is any counting number. d. $4k + 3 \equiv x \pmod 4$, where k is any counting number.

18. a. $2x^2 - 1 \equiv 3 \pmod 7$ b. $5x^3 - 3x^2 + 70 \equiv 0 \pmod 2$

19. Your doctor tells you to take a certain medication every 8 hours. If you begin at 8:00 A.M., show that you will not have to take the medication between midnight and 7 A.M.

20. a. Suppose you make six round trips to visit a sick aunt and wish to record your mileage. You forget the original odometer reading, but you do remember that the units digit has increased by 8 miles. What are the possible distances between your house and your aunt's house?
    b. If you know that your aunt lives somewhere between 10 and 15 miles from your house, how far exactly is her house, given the information in part a?

21. Suppose you are planning to purchase some rope. You need between 15 and 20 pieces that are 7 inches long and one piece that is 80 inches long. The rope can be purchased in multiples of 12 inches. How much rope should you buy to eliminate as much waste as possible?

Find the months containing a Friday the 13th for each of the years indicated in Problems 22–24.

22. 1984   23. 2001   24. 1943

25. Consider the set of numbers used in 12-hour clock arithmetic:
    a. Is the set closed with respect to addition?
    b. Does the set of numbers have an identity for addition?
    c. Does the set of numbers on the clock contain inverse elements with respect to addition?

26. Investigate the following addition properties of a modulo 6 system.
    a. closure
    b. commutative
    c. identity
    d. inverse

27. Investigate the following multiplication properties of a modulo 6 system.
    a. closure
    b. commutative
    c. identity
    d. inverse

28. *Computer Problem*. We can easily write a BASIC program that will allow us to input a positive integer and a mod m and have the computer convert the integer into a mod m numeral.

```
10   PRINT "WHAT IS THE INTEGER";
20   INPUT I
30   PRINT "WHAT IS THE MOD";
40   INPUT M
50   LET A = I − INT(I/M)*M
60   PRINT "THE INTEGER IN MOD";
70   PRINT M;
80   PRINT "IS";
90   PRINT A
100  END
```

Use this program to find x:

a. $5872 \equiv x \pmod{119}$
c. $5,827,300 \equiv x \pmod{365}$
b. $4,872,879 \equiv x \pmod{365}$
d. $5,710,000 \equiv x \pmod{4380}$

29. *Computer Problem*. Tables for the arithmetic operation in a given modulus can easily be generated with a short BASIC program. The following program asks for the modulus and then types out the addition table for the given modulus. (If you are running the program on a computer with a Teletype output, the entered modulus must be a positive integer less than or equal to 12 so that the table will fit on one page.)

```
10   PRINT "WHAT IS THE MOD";
20   INPUT M
30   PRINT
35   PRINT "      *  ";
40   FOR C = 0 TO M − 1
45   PRINT C;
47   PRINT " ";
50   NEXT C
55   PRINT
60   FOR I = 1 TO 4*M + 8
65   PRINT "*";
70   NEXT I
75   PRINT
80   PRINT "       *  ";
85   FOR R = 0 TO M − 1
90   PRINT R;
95   PRINT "  *  ";
100  FOR C = 0 TO M − 1
105  PRINT C + R − INT((C + R)/M)*M;
110  PRINT " ";
115  NEXT C
118  PRINT
119  PRINT "       *"
120  NEXT R
125  END
```

Some sample runs are shown. Using this program as a guide, write a program that will give you a multiplication table for any mod (less than 13).

Notice that we have used a new command, INT(X), *called the greatest integer, denoted by* [x] *in this section. It gives the greatest number not exceeding the one in the brackets, or parenthesis in* INT(X). *For example,*

$$INT(46.2) = 46$$
$$INT(19.9) = 19$$
$$INT(108) = 108$$
$$INT(\pi) = 3$$

Some sample runs of the program given in Problem 28 are shown below:

```
RUN
WHAT IS THE INTEGER? 23
WHAT IS THE MOD? 5
THE INTEGER IN MOD 5 IS 3

READY

RUN
WHAT IS THE INTEGER? 108
WHAT IS THE MOD? 13
THE INTEGER IN MOD 13 IS 4

READY

RUN
WHAT IS THE INTEGER? 23453
WHAT IS THE MOD? 61
THE INTEGER IN MOD 61 IS 29

READY

RUN
WHAT IS THE MOD? 5
```

*	0	1	2	3	4
0	0	1	2	3	4
1	1	2	3	4	0
2	2	3	4	0	1
3	3	4	0	1	2
4	4	0	1	2	3

```
RUN
WHAT IS THE MOD? 6

        *   Ø   1   2   3   4   5
**********************************
        *
  Ø     *   Ø   1   2   3   4   5
        *
  1     *   1   2   3   4   5   Ø
        *
  2     *   2   3   4   5   Ø   1
        *
  3     *   3   4   5   Ø   1   2
        *
  4     *   4   5   Ø   1   2   3
        *
  5     *   5   Ø   1   2   3   4
        *
```

30. Make a table for the addition and multiplication facts, mod 11.

31. Using the addition and multiplication tables for mod 11 found in Problem 30, answer the following questions:
    a. Is the set closed for the operations of addition and multiplication?
    b. Is the set commutative for these operations?
    c. Is the set associative for these operations?

32. Using the addition and multiplication tables for mod 11 found in Problem 30, answer the following questions:
    a. Does the set have an identity for addition? For multiplication?
    b. What is the additive inverse of each element?
    c. What is the multiplicative inverse of each element?

33. *Computer Problem.* Suppose we are interested in finding the multiplicative inverses in some modulus, as in Problem 32. We notice that for a particular mod, there may or may not be an inverse for each element in the set. Let's consider a BASIC program for finding inverses:

```
10   PRINT "THIS PROGRAM WILL FIND ALL OF THE MULTIPLICATIVE";
20   PRINT "INVERSES MOD M."
25   PRINT "IF A PARTICULAR NUMBER IS NOT LISTED, THEN IT DOES NOT";
26   PRINT "HAVE AN"
27   PRINT "INVERSE FOR MULTIPLICATION."
30   PRINT "WHAT IS M";
35   INPUT M
40   PRINT "     NUMBER          MULTIPLICATIVE"
50   PRINT "                          INVERSE"
60   FOR N = Ø TO M − 1
70   FOR R = Ø TO M − 1
80   GOSUB 100
95   NEXT R
96   NEXT N
98   GOTO 150
100  LET A = N*R − INT((N*R)/M)*M
110  IF A <> 1 GOTO 130
115  PRINT "      ";
120  PRINT N,"    ",R
130  RETURN
150  END
```

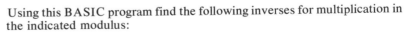

Using this BASIC program find the following inverses for multiplication in the indicated modulus:

a. mod 5   b. mod 6   c. mod 17

34. Find the multiplicative inverses modulo 7 if the identity is 1.

35. Find the multiplicative inverses modulo 10 if the identity is 1.

36. *Modular Codes.* In Arthur Conan Doyle's stories, Sherlock Holmes is often pursuing his archenemy, Professor Moriarty. It is imperative that Professor Moriarty be able to communicate with his accomplices by means of a secret message. Since the Professor has had some experience with

modular arithmetic, he sets up the code shown in Figure 6.10, based on a mod 29 system. For example, if he were alive today, he might send the message

THE FBI HAS BED BUGS.

First, he would select some encoding key—say, multiply by 2. Then he could use the mod 29 table shown in Figure 6.10.

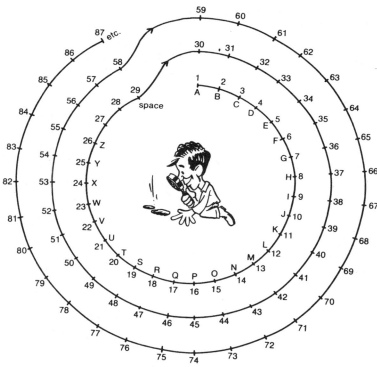

FIGURE 6.10 Mod 29 code

Message: T H E   F B I   H A S   B E D   B U G S
            20 8 5 29 6   2 9 29 8   1 19 29   2 5 4 29 2 21   7 19 28

Encode: 40-16-10-58-12- 4-18-58-16- 2-38-58- 4-10- 8-58- 4-42-14-38-56

Modulate: 11-16-10-29-12- 4-18-29-16- 2- 9-29- 4-10- 8-29- 4-13-14- 9-27

Translate: K P J   L D R   P B I   D J H   D M N I
CODED MESSAGE: KPJ LDR PBI DJH DMNI.

Encode *means multiply by the encoding key.*

Modulate *means find the number in mod 29.*

Decode the following message for which the encoding key is multiply by 2: NBXXMC RI B CAXX FBK,

37. Use Problem 36 to decode the following message for which the encoding key is multiply by 3: XEJSBQ LDJSBQ ,. C RCGG UEQZ

Hint for Problem 46: We want to find x so that

$$x \equiv 3 \ (mod \ 17)$$
$$x \equiv 4 \ (mod \ 11)$$
$$x \equiv 5 \ (mod \ \ 6).$$

We know that the value of x we are looking for is the smallest number in all three of the following sets:

$$\{3,20,37,54,71,88, \ldots\}$$
$$\{4,15,26,37,48,59, \ldots\}$$
$$\{5,11,17,23,29,36, \ldots\}$$

To find this number, write a BASIC program that follows the flow chart shown in Figure 6.11.

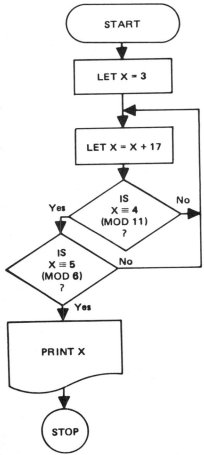

FIGURE 6.11 Flow chart for Pirate Problem (46)

38. Use Problem 36 to decode the following message for which the encoding key is multiply by 2 and add 10: LJMQKVTSSKQJ. SJKITJ,ZKJULECSJ.ISKGTKITJTESTSJSETTM

39. Use Problem 36 to decode the following message for which the encoding key is multiply by 2 and subtract 7: SKAXBVIXAVXVGKNDV.WSYQCLT

40. Use Problem 36 to encode the following message with the encoding key multiply by 2 and subtract 11: ALL MEN BY NATURE DESIRE KNOWLEDGE.

41. Use Problem 36 to encode the following message with the encoding key multiply by 3 and add 5: SHOW ME A DROPOUT FROM A DATA PROCESSING SCHOOL AND I WILL SHOW YOU A NINCOM-PUTER.

Mind Bogglers

42. If it is now 2 P.M., what time will it be 99,999,999,999 hours from now?

43. Write a schedule for 12 teams so that each team will play every other team once and no team will be idle (Hint: Consider mod 11 arithmetic.)

44. Write a BASIC program to determine the months in a given year that have a Friday the 13th.

45. A man buys 100 birds for $100. A rooster is worth $10, a hen is worth $3, and chicks are worth $1 a pair. How many roosters, hens, and chicks did he buy if he bought at least one of each type. (This is the Chinese problem of "One Hundred Fowls.")

46. *Computer Problem*. A band of 17 pirates decided to divide their doubloons into equal portions. When they found that they had 3 coins remaining, they agreed to give them to their Chinese cook, Wun Tu. But 6 of the pirates were killed in a fight. Now when the treasure was divided equally among them, there were 4 coins left that they considered giving to Wun Tu. Before they could divide the coins, there was a shipwreck and only 6 pirates, the coins, and the cook were saved. This time equal division left a remainder of 5 coins for the cook. Now Wun Tu took advantage of his culinary position to concoct a poison mushroom stew so that the entire fortune in doubloons became his own. What is the smallest number of coins that the cook would have finally received?

47. What is the next smallest number of coins that would satisfy the conditions of Problem 46?

48. Decode the messages in Figure 6.12. (Be careful. This is *not* the same code discussed in Problem 36.)*

*My thanks to Harold Andersen for Problems 36–41 and 48.

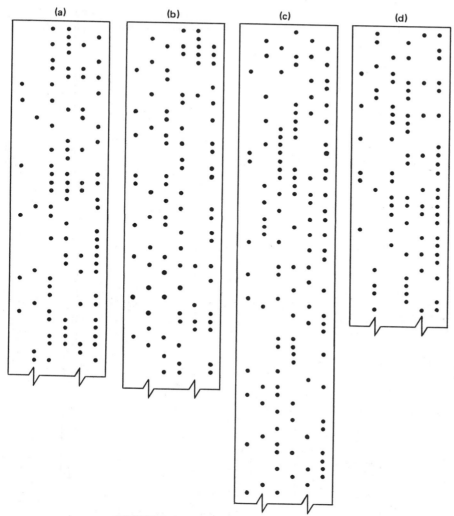

FIGURE 6.12 Code for Problem 48

49. Does the set of mod 8 numbers together with the operations of addition and multiplication form a field?

Problems for Individual Study

50. What is a Diophantine equation?
     References: "Puzzle Problems and Diophantine Equations," by Warren J. Himmelberger, *Mathematics Teacher,* February 1973, pp. 136–138.

     For a more complete reference, see Chapter 8 of *Number Theory and Its History,* by Oystein Ore (New York: McGraw-Hill, 1948).

51. *Modular Art*. When constructing addition and multiplication tables for the various modular systems, you probably noticed some patterns. Consider, for example, the multiplication tables for mod 19 and mod 21 (see Tables 6.9 and 6.10). The patterns from these tables can be translated into interesting figures.

TABLE 6.9 Mod 19

×	0	1	2	3	4	5	6	7	8	9	10	11	12	13	14	15	16	17	18
0	0	0	0	0	0	0	0	0	0	0	0	0	0	0	0	0	0	0	0
1	0	1	2	3	4	5	6	7	8	9	10	11	12	13	14	15	16	17	18
2	0	2	4	6	8	10	12	14	16	18	1	3	5	7	9	11	13	15	17
3	0	3	6	9	12	15	18	2	5	8	11	14	17	1	4	7	10	13	16
4	0	4	8	12	16	1	5	9	13	17	2	6	10	14	18	3	7	11	15
5	0	5	10	15	1	6	11	16	2	7	12	17	3	8	13	18	4	9	14
6	0	6	12	18	5	11	17	4	10	16	3	9	15	2	8	14	1	7	13
7	0	7	14	2	9	16	4	11	18	6	13	1	8	15	3	10	17	5	12
8	0	8	16	5	13	2	10	18	7	15	4	12	1	9	17	6	14	3	11
9	0	9	18	8	17	7	16	6	15	5	14	4	13	3	12	2	11	1	10
10	0	10	1	11	2	12	3	13	4	14	5	15	6	16	7	17	8	18	9
11	0	11	3	14	6	17	9	1	12	4	15	7	18	10	2	13	5	16	8
12	0	12	5	17	10	3	15	8	1	13	6	18	11	4	16	9	2	14	7
13	0	13	7	1	14	8	2	15	9	3	16	10	4	17	11	5	18	12	6
14	0	14	9	4	18	13	8	3	17	12	7	2	16	11	6	1	15	10	5
15	0	15	11	7	3	18	14	10	6	2	17	13	9	5	1	16	12	8	4
16	0	16	13	10	7	4	1	17	14	11	8	5	2	18	15	12	9	6	3
17	0	17	15	13	11	9	7	5	3	1	18	16	14	12	10	8	6	4	2
18	0	18	17	16	15	14	13	12	11	10	9	8	7	6	5	4	3	2	1

TABLE 6.10 Mod 21

×	0	1	2	3	4	5	6	7	8	9	10	11	12	13	14	15	16	17	18	19	20
0	0	0	0	0	0	0	0	0	0	0	0	0	0	0	0	0	0	0	0	0	0
1	0	1	2	3	4	5	6	7	8	9	10	11	12	13	14	15	16	17	18	19	20
2	0	2	4	6	8	10	12	14	16	18	20	1	3	5	7	9	11	13	15	17	19
3	0	3	6	9	12	15	18	0	3	6	9	12	15	18	0	3	6	9	12	15	18
4	0	4	8	12	16	20	3	7	11	15	19	2	6	10	14	18	1	5	9	13	17
5	0	5	10	15	20	4	9	14	19	3	8	13	18	2	7	12	17	1	6	11	16
6	0	6	12	18	3	9	15	0	6	12	18	3	9	15	0	6	12	18	3	9	15
7	0	7	14	0	7	14	0	7	14	0	7	14	0	7	14	0	7	14	0	7	14
8	0	8	16	3	11	19	6	14	1	9	17	4	12	20	7	15	2	10	18	5	13
9	0	9	18	6	15	3	12	0	9	18	6	15	3	12	0	9	18	6	15	3	12
10	0	10	20	9	19	8	18	7	17	6	16	5	15	4	14	3	13	2	12	1	11
11	0	11	1	12	2	13	3	14	4	15	5	16	6	17	7	18	8	19	9	20	10
12	0	12	3	15	6	18	9	0	12	3	15	6	18	9	0	12	3	15	6	18	9
13	0	13	5	18	10	2	15	7	20	12	4	17	9	1	14	6	19	11	3	16	8
14	0	14	7	0	14	7	0	14	7	0	14	7	0	14	7	0	14	7	0	14	7
15	0	15	9	3	18	12	6	0	15	9	3	18	12	6	0	15	9	3	18	12	6
16	0	16	11	6	1	17	12	7	2	18	13	8	3	19	14	9	4	20	15	10	5
17	0	17	13	9	5	1	18	14	10	6	2	19	15	11	7	3	20	16	12	8	4
18	0	18	15	12	9	6	3	0	18	15	12	9	6	3	0	18	15	12	9	6	3
19	0	19	17	15	13	11	9	7	5	3	1	20	18	16	14	12	10	8	6	4	2
20	0	20	19	18	17	16	15	14	13	12	11	10	9	8	7	6	5	4	3	2	1

Divide a circle into $n - 1$ parts, where n is the given modulus. For example, with mod 19 we divide the circle into 18 parts (for our purposes, we disregard the 0 element). Next, select some nonzero number in the table—say, 9. That is, we consider:

	1	2	3	4	5	6	7	8	...
9	9	18	8	17	7	16	6	15	...

We now connect the points 1 and 9, 2 and 18, 3 and 8, 4 and 17, and so on. The result is shown in Figure 6.13.

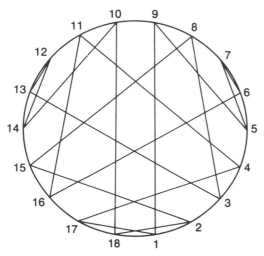

FIGURE 6.13 The (19,9) design

The (19, 2) design

We call this the (19,9) design. By shading in alternate regions, we obtain the pattern shown in Figure 6.14. Create some of your own modular art patterns, and make a presentation to the class.*

The (19, 18) design

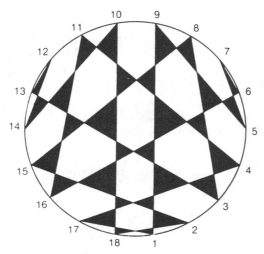

FIGURE 6.14 The (19, 9) design with shading

*My thanks to Phil Locke for this problem.

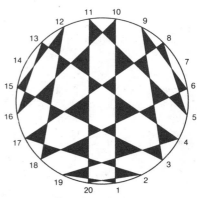

The (21, 10) design

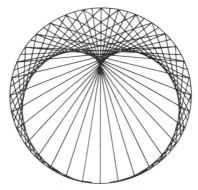

The (65, 2) design

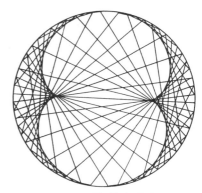

The (65, 3) design

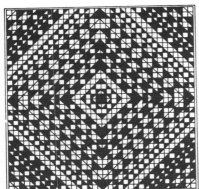

52. *Modular Arithmetic.* Many interesting designs can be created using patterns based on modular arithmetic (see Figure 6.15). Prepare a paper for class presentation based on the article "Using Mathematical Structures to Generate Artistic Designs" by Sonia Forseth and Andria Price Troutman, *Mathematics Teacher,* May 1974, pp. 393–398.

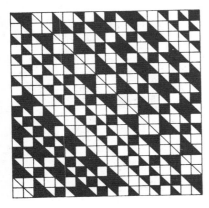

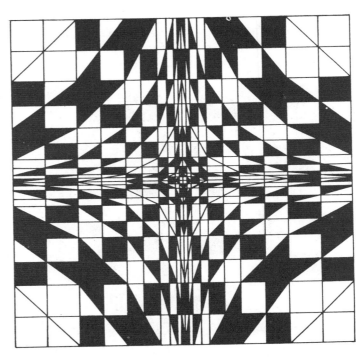

FIGURE 6.15 Design based on modular arithmetic

6.7 SUMMARY AND REVIEW

CHAPTER OUTLINE

I. Mathematical Systems
   A. A *field* is a set, F, with two operations, $+$ and $\cdot$, satisfying the following properties for any a, b, and c elements of F.
      1. *Closure for addition:* $(a + b)$ is an element of F
      2. *Closure for multiplication:* ab is an element of F
      3. *Commutative for addition:* $a + b = b + a$
      4. *Commutative for multiplication:* $ab = ba$
      5. *Associative for addition:* $(a + b) + c = a + (b + c)$
      6. *Associative for multiplication:* $(ab)c = a(bc)$
      7. *Identity for addition:* There exists a real number 0 so that $0 + a = a + 0 = a$ for any element a in F.
      8. *Identity for multiplication:* There exists a real number 1 such that $1 \cdot a = a \cdot 1 = a$ for any element a in F.
      9. *Inverse for addition:* For each number a in F there is a unique number ^-a in F so that

$$a + (^-a) = (^-a) + a = 0.$$

      10. *Inverse for multiplication:* For each nonzero a in F there is a unique number $1/a$ in F so that

$$a \cdot \frac{1}{a} = \frac{1}{a} \cdot a = 1$$

      11. *Distributive for multiplication over addition:*

$$a(b + c) = ab + ac$$

   B. Ordinary algebra
      1. The set of real numbers with addition and multiplication is a field
      2. Solving equations. Given an equation, the following operations result in an equivalent equation:
         a. *Addition property:* add the same number to both sides;
         b. *Subtraction property:* subtract the same number from both sides;
         c. *Multiplication property:* multiply both sides by the same nonzero number;
         d. *Division property:* divide both sides by the same nonzero number.

II. Prime Numbers
   A. *Definition:* A **prime number** is a counting number that has exactly two divisors. If it has more than two divisors, it is called composite.

B. Finding primes: *Sieve of Eratosthenes*.

C. Divisibility: If m, d, and k are natural numbers and $m = d \cdot k$, we say that d is a **divisor** of m; d is a **factor** of m; d **divides** m; and m is a **multiple** of d. We denote this relationship by $d \mid m$.

III. Prime Factorization
   A. Definitions
      1. A *set of factors* of a natural number M is a listing of *all* divisors of M.
      2. A *factorization* of a number M is a listing of a set of numbers whose product is M.
      3. A *prime factorization* of a number M is a listing of a set of prime numbers whose product is M.
      4. The *canonical representation* of a number is the representation of that number as a product of primes using exponential notation with the factors arranged in order of increasing magnitude.
   B. *Fundamental Theorem of Arithmetic:* Every counting number greater than 1 is either prime or a product of primes, and the factorization is unique (except for the order in which the factors appear).
   C. Use of a "factor tree" to find the prime factorization of a number.
   D. There is no largest prime.

IV. Applications of Prime Factorization
   A. Greatest common factor (g.c.f.)
      1. *Definition:* The **greatest common factor** of a set of numbers is the largest number that divides evenly into each member of the set.
      2. Two numbers are **relatively prime** if their greatest common factor is 1.
      3. To find the greatest common factor:
         a. Find the prime factorization.
         b. Write in canonical form.
         c. Choose the representative of each factor with the *smallest exponent* (be sure to include representatives with zero exponents).
         d. Take the product of the representatives.
   B. Least common multiple (l.c.m.)
      1. *Definition:* The **least common multiple** of a set of numbers is the smallest number that each member of the set divides into.
      2. To find the least common multiple:
         a. Find the prime factorization.
         b. Write in canonical form.
         c. Choose the representative of each factor with the *largest exponent*.
         d. Take the product of the representatives.
   C. Working with rational numbers
      1. A rational expression is *simplified* or *reduced* if the numerator and denominator are relatively prime.
      2. *Fundamental Property of Fractions:* If a/b is any rational number and x is any nonzero integer, then

$$\frac{a \cdot x}{b \cdot x} = \frac{x \cdot a}{x \cdot b} = \frac{a}{b}$$

    3. Use the least common multiple to find the common denominator when adding or subtracting fractions.

    4. Use the greatest common factor to reduce fractions.

V. Modular Arithmetic

  A. Operations

    1. **Addition:** Defined by table according to the "addition on a clock."

    2. **Subtraction:** $a - b = x$ means $a = b + x$.

    3. **Multiplication:** $a \times b$ means $\underbrace{a + a + \ldots + a + a}_{b \text{ addends}}$.

    4. **Division:** $\dfrac{a}{b} = x$ means $a = b \cdot x$.

  B. *Definition:* a and b are **congruent mod m** if and only if a and b differ only by a multiple of m. That is, if m is divided into a and also into b and yields the same remainder results, then $a \equiv b \pmod{m}$.

  C. Arithmetic on a clock with an arbitrary number of hours

  D. Solving equations in a modular system

    1. Use the definitions for the operations.

    2. Solve the equation by direct substitution of all the possibilities.

  E. Given a mod m system, check the field properties.

REVIEW PROBLEMS

1. *In your own words,* discuss each of the following properties: closure, associative, identity, inverse, commutative, and distributive.

2. Discuss what we mean when we write $a \mid b$.

3. Which of the following are true? If true, use the definition to prove it is true.

  a. $6 \mid 714$ b. $8 \mid 735{,}812$

  c. $11 \mid 6{,}825$ d. $9 \mid 598{,}364$

4. Consider arithmetic on a 4-hour clock. Construct the addition and multiplication tables.

5. Is the set $\{0,1,2,3\}$ a field for the operations of addition and multiplication (mod 4)? Why or why not? Discuss each of the properties involved.

6. Find the prime factorization of 6825.

7. Which of the following are prime numbers?

  a. 89   b. 1   c. 349

8. Discuss the *Fundamental Theorem of Arithmetic*.

9. What do we mean when we speak about the "greatest common factor" of a set of numbers?

10. What is the canonical representation of a number?

11. Find the greatest common factor of the following sets of numbers.

    a. $\{30, 42, 99\}$ b. $\{49, 1001, 2401\}$

12. Find the least common multiple of the sets of numbers in Problem 11.

13. Reduce $\dfrac{3536}{3952}$.

14. Perform the indicated operations.

    a. $\dfrac{3^{-1} + 4^{-1}}{6}$ b. $\dfrac{-7}{9} \cdot \dfrac{99}{174} + \dfrac{-7}{9} \cdot \dfrac{75}{174}$

15. Perform indicated operations.

    a. $\dfrac{7}{30} + \dfrac{5}{42} + \dfrac{5}{99}$ b. $\dfrac{5}{49} - \dfrac{1}{1001}$

16. Solve on a 5-hour clock:

    a. $4 \cdot 3$ b. $1 - 4$ c. $\dfrac{1}{3}$ d. $\dfrac{2}{3}$ e. $\dfrac{4}{7}$

17. Which of the following are true?

    a. $4 + 9 \equiv 1 \pmod{12}$ b. $2 \equiv 2001 \pmod 2$

    c. $7 \cdot 9 \equiv 22 \pmod 8$ d. $4 - 10 \equiv 6 \pmod{12}$

    e. $5 - 7 \equiv 24 \pmod{11}$

18. Solve for x.

    a. $\dfrac{x}{5} \equiv 2 \pmod 8$ b. $2x^2 + 7x + 1 \equiv 0 \pmod 2$

    c. $2x \equiv 3 \pmod 7$ d. $\dfrac{x}{3} \equiv 5 \pmod 6$

    e. $5x \equiv 6 \pmod 7$

19. Solve for x.

    a. $x + 11 = 14$ b. $2x + 5 = 13$

c. $3x + 2 = x + 6$ d. $3x + 1 = 7x$

e. $\dfrac{2x}{3} = 6$

20. Solve for x.

   a. $2x - 7 = 5x$ b. $14 = 5x - 1$

   c. $3(2x + 3) = 12$ d. $\dfrac{2x - 1}{2} = 5$

   e. $\dfrac{3(x - 1)}{2} = 7$

1534: Henry VIII becomes head of the Church of England

1543: The Copernican concept that the sun is in the center and the planets revolve about it formed the basis for modern astronomy and completely revolutionized current thinking

1558: Elizabeth I crowned Queen of England

1567: Monteverdi born

Nicolaus Copernicus

1582: Pope Gregory XIII created a new calendar recommended by a mathematical commission—it wasn't adopted worldwide until 1918

1588: England defeats Spanish Armada

1600: Shakespeare's *Hamlet*
1602: Dutch East Indian Company formed

1608: Telescope invented
1609: Kepler's Laws

1614: Cervantes' *Don Quixote*

1620: Pilgrims land at Plymouth

1632: Harvard University founded—first American university

1637: Descarte's *Discourse on Method*

1646: Building of the Taj Mahal

1649: Cromwell abolishes English monarchy

1525
1550
1575
1600
1625

1525: Stifel—number mysticism MATHEMATICAL
Rudolff—algebra, decimals
1530: Copernicus—astronomy, trigonometry

1545: Tartaglia—cubic equations, arithmetic
Ferrari—quartic equations
1550: Cardano, Scheubel—algebra

1557: Robert Recorde—arithmetic, algebra, first use of = sign

1572: Bombelli—algebra, cubic equations

1580: Vièta—algebra, geometry, trignometry, much modern notation
1583: Clavius—arithmetic, algebra, geometry

1593: Adrianus Romanus—value of π

Galileo

1600: Galileo—physics, astronomy, projectiles

1610: Kepler—astronomy, continuity

1614: Napier—logarithms, Napier's rods

Kepler

1630: Mersenne—number theory
Oughtred—first table of natural logs
1635: Cavalieri—number theory
Fermat—number theory, analytic geometry
1637: Descartes—analytic geometry
1640: Desargues—projective geometry

7
THE NATURE OF GEOMETRY

No power on earth, however great,
Can pull a string, however fine,
Into a horizontal line
That shall be absolutely straight.

William Whewell

COPERNICANVM
Systema
TIVS CREATI
THESI
CANA IN
EXHIBITVM

Historical Note

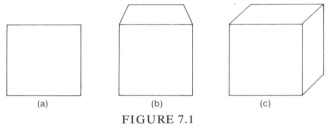

THE MEANING OF LIFE

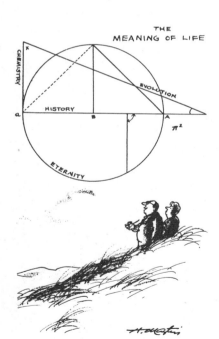

"*I wish I could see the expression on the faces of my students who said there was no value in studying geometry.*"

7.1 THE NATURE OF GEOMETRY

Geometry, or "earth measure," was one of the first branches of mathematics. Both the Egyptians and the Babylonians needed geometry for construction, for land measurement, and for commerce. As we've seen, they both discovered the Pythagorean theorem, although it was not proved until the Greeks developed geometry formally. This formal development utilizes the logic in Chapter 2 by beginning with certain assumptions, called *axioms*. Historically, the axioms that were assumed were those that seemed to conform to the physical world. Euclid's *Elements* collected all the material known about geometry and organized it into a logical, deductive system.

Geometry involves *points* as well as sets of points called *lines, planes,* and *surfaces*. We often draw physical models or pictures to represent these concepts; however, we must be careful not to try to prove assertions by looking at pictures, since a picture may contain hidden assumptions or ambiguities. For example, consider Figure 7.1a.

| (a) | (b) | (c) |

FIGURE 7.1

What do you see? A square? Perhaps. But what if we take a slightly different view of the same object, as shown in Figure 7.1b. What do you see now? A square and a trapezoid? Perhaps. But what if we have in mind a cube, as shown in Figure 7.1c? Even if you view this object as a cube, do you see the same cube as everyone else? Is the fly in Figure 7.2 inside or outside the cube? Or is it perhaps on an edge of the cube? Thus, although we may use a figure to help us understand a problem, we cannot prove results by this technique.

FIGURE 7.2 Fly on a cube

Geometry can be separated into two categories: (1) traditional (which is like high school geometry), and (2) transformational (which is more algebraic than the traditional approach). Traditional geometry begins with certain undefined terms. For example, what is a line? You might say ''I know what a line is!'' But try to define a line. Is it a set of points? Any set of points? What is a point?

1. A point is something that has no length, width, or thickness.
2. A point is a location in space.

Certainly these are not satisfactory definitions, because they involve other terms that are not defined. We take, therefore, the terms *point, line,* and *plane* as *undefined.* Next, certain properties associated with these terms are assumed as axioms. By accepting different sets of axioms, we can develop different geometries. The axioms you accepted in high school seemed to correspond to the world around us and led to a geometry that we call *Euclidean geometry,* an abstract deductive system that uses logic to deduce further results, called *theorems*.

An axiom *is a statement that is accepted without proof.*

A theorem *is a statement that can be proved from accepted axioms or previously proved results.*

Definition of a transformation

The transformational approach is quite different. It begins with the idea of a *transformation,* which is a function from one set onto itself. For example, one way to transform one geometric figure into another is by a *reflection.* Given a line L and a point P as shown in Figure 7.3, we call the point P' the *reflection* of P about the line L if line PP' is perpendicular to L and is also bisected by L. Each point in the plane has exactly one reflection point corresponding to a given line L. This means that there is a one-to-one correspondence between the set of points of a plane and their image points with respect to a given line L. *Any operation,* such as a reflection, *in which there is a one-to-one correspondence between the points of the plane and their image points is called a transformation,* and the line of reflection is called the line of symmetry. (See Figure 7.4 on page 392 for examples of symmetry in nature and in design.)

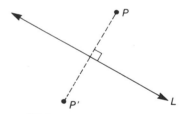

FIGURE 7.3 A reflection

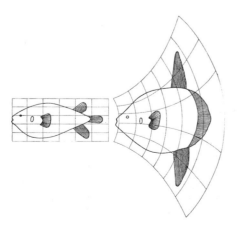

We can find an excellent example of a grid transformation by looking at a scheme devised by the Scottish biologist D'Arcy Wentworth Thompson. The illustration to the left is reproduced from his 1917 work On Growth and Form. *His method is useful in understanding the forces that shape sheets of cells into the tissue layers of the embryo.*

FIGURE 7.4 Symmetry in nature is illustrated by snowflakes. Symmetry in design is illustrated by the Apollo space capsule.

Other transformations include *translations, rotations, dilations,* and *contractions,* which are illustrated in Figure 7.5, and *inversions,* which are illustrated in Figure 7.6.

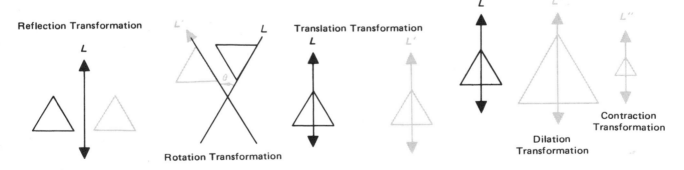

Reflection Transformation

Rotation Transformation

Translation Transformation

Dilation Transformation

Contraction Transformation

FIGURE 7.5 Transformations of an equilateral triangle

Congruence is a transformation that preserves distances between points.

Similarity is a transformation that preserves ratios of distances between pairs of points.

Geometry is concerned with a study of the relationships between geometric figures. A primary relationship is that of *congruence.* Two figures are said to be congruent if one can be made to coincide with the other by a rigid motion. A second relationship is called *similarity.* Two figures are said to be *similar* if they have the same shape, although not necessarily the same size.

This chapter is intended to provide you with a feeling for the nature of geometry, not a formal course in the subject. It assumes that you have had some exposure to Euclidean geometry in elementary and high school. In the next section, we'll consider some of the ideas of projective and non-Euclidean geometry. Then we'll consider some newer aspects of geometry, topology, and networks.

FIGURE 7.6 An example of an inversion transformation. Good music is not only beautiful to hear but often interesting to see as well. Bach, in his great *Art of the Fugue,* made use of mirror reflections in contrapuntal passages between the upper treble and lower bass figures. Melodic inversion is achieved by simply turning a page of music upside down. This example, usually attributed to Mozart, is a unique piece for violin and can be played simultaneously by two musicians facing each other and reading the music, laid flat between them, in opposite directions. The two parts, though different, will be in perfect harmony, without violating a single rule of classical composition.

If we were to develop the course formally, we would have to be very careful about the statement of our postulates. The notion of a mathematical proof requires a very precise formulation of all postulates and theorems. For example, Euclid based many proofs about congruence on a postulate that said "Things that coincide with one another are equal to one another (Common Notions, Book I, Elements)." This statement is not formulated precisely enough to be used as a justification for some of the things Euclid did with congruence. Even the explanation we gave for congruence is not precise enough to be used in a formal course. However, since we are not formally developing geometry, we simply accept the general drift of the statement. Remember, though, that you cannot base a mathematical proof on general drift.

PROBLEM SET 7.1

A Problems

1. What is the difference between an axiom and a theorem?

2. The first illustration below shows a cube with the top cut off. Use solid lines and shading to depict seven other different views of a cube with one side cut off.

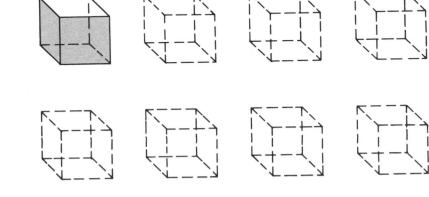

FIGURE 7.7 A young woman or an old woman?

3. Is the woman in Figure 7.7 a young woman or an old woman?

4. What are the two main approaches to the study of geometry? Briefly characterize each.

5. Which of the following pictures illustrate(s) line symmetry?

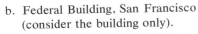

   a. An apple.

   b. Federal Building, San Francisco (consider the building only).

c. *Allegorical Garden of
Geometry*. Engraving by
Francesco Curti.

d. *Winged Lion*. China—Mid-T'ang
Dynasty, 7th/8th century A.D.

e. Human circulatory system.

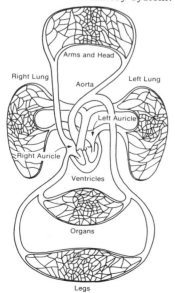

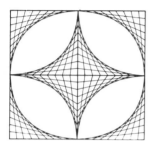

FIGURE 7.8 Aestheometry

B Problems

6. Below are two sets of three intersecting planes. Use solid lines and shading to illustrate different ways of viewing the planes. Can you find more than two ways?

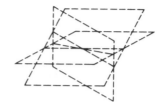 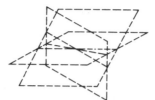

7. *Aestheometry*. Many curves can be illustrated by using only straight line segments. The result is called *aestheometry* and is depicted in Figure 7.8. The basic design is the parabolic curve, which is drawn by starting with an angle. (See Figure 7.9.)

   (1) Draw an angle with two sides of equal length.
   (2) Mark off equally distant units on both rays.
   (3) Connect #1 on the one ray to #1 on the other ray, #2 on one ray to #2 on the other, #3 to #3, . . .

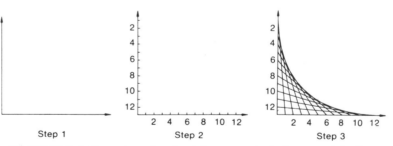

FIGURE 7.9 Procedure for obtaining a parabola from straight line segments

You can vary the unit's length, make the units unequal in length along different rays, or invent designs with several angles of different measures. Construct a design illustrating various curves using only straight line segments and one or more angles as a starting place.

8. *Aestheometry* (continued). The second basic design begins with a circle. (See Figure 7.10).

   (1) Draw a circle and mark off any number of equally spaced points.
   (2) Choose two points and connect them.
   (3) Connect succeeding points.

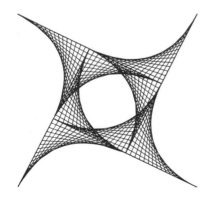

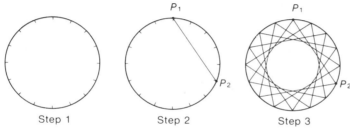

FIGURE 7.10 Procedure for obtaining a circle from straight line segments

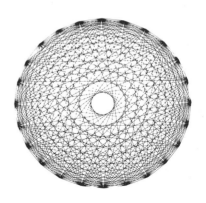

Construct a design illustrating various curves using only straight line segments and circles as a starting place.

9. a. Draw a square $ABCD$.
   b. Divide the square into two parts by completing the following steps:

      i. Call the midpoint of BC point E.
      ii. Call the midoint of AD point F.
      iii. Draw line segment EF.

   c. Draw a circle with center at F and radius FC.
   d. Extend the base line AD to intersect the circle at point G.
   e. This line forms the base of a rectangle $ABHG$ where H is the point on the extension of BC such that HG forms a right angle with AG.
   f. Measure the sides of rectangle $ABHG$.
   g. Divide the width by the length.

A compass is needed for this problem.

Problems 9–11 all build on the previous one of the sequence. Compare the results of these problems with the results obtained in Section 1.3.

10. a. Remove the square $ABCD$ from the rectangle $ABHG$ of Problem 9. The result is a rectangle $DCHG$.
    b. Measure the sides of rectangle $DCHG$.
    c. Divide the width by the length.
    d. Compare your answers to Problems 9g and 10c.

11. a. Draw a rectangle that is aesthetically pleasing to you.
    b. Measure the sides.
    c. Divide the width by the length.
    d. Compare your answer to part (c) with your answers to Problems 9g and 10c.

Mind Bogglers

12. Solve the equation

$$\frac{a + b}{a} = \frac{a}{b}$$

where $a = 1$. Compare this answer with your answer to Problem 11.

13. Connect the nine dots shown in the margin with four straight lines, but don't lift your pencil from the paper.

14. *Santa Rosa Street Problem.* On Saturday evenings a favorite pastime of the high school students in Santa Rosa, California, is to cruise certain streets. The selected routes are shown on the map in Figure 7.11. Is it possible to choose a route so that all of the permitted streets are traveled exactly once?

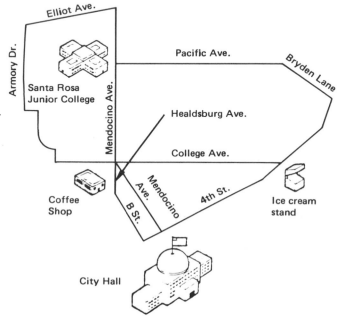

FIGURE 7.11 Santa Rosa Street Problem

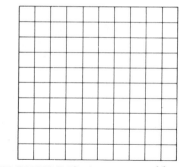

FIGURE 7.12 A square problem

15. How many squares are in Figure 7.12?

16. Generalize from Problem 15 to tell how many squares are in a general $n \times n$ square.

17. Generalize from Problem 15 to tell how many squares are in a general $m \times n$ rectangle.

18. Generalize from Problem 15 to tell how many squares are in a general $m \times n$ rectangle with a hole of size $p \times q$ cut in the middle.

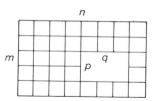

19. *Computer Problem.* Write a BASIC program to solve Problems 15, 16, and 17. (Or, if you really feel like a challenge, write a program for Problem 18.)

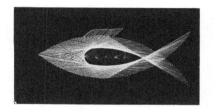

Problems for Individual Study

20. *Curve Stitching.* Use the ideas of aestheometry in Problems 7 and 8, some thread or string, and your own imagination to create your own curve-stitching design. For more specific information, you can check the references given below.

    References: Anderson, James, *Space Concepts through Aestheometry* (New Mexico: Aestheometry, Inc., 1965).

    Cundy, H. Martyn, and A. P. Rollett, *Mathematical Models* (2nd ed.) (Oxford: Clarendon Press, 1961).

    Seymore, Dale, and Joyce Snider, *Line Designs* (Palo Alto, Calif.: Creative Publications, 1968).

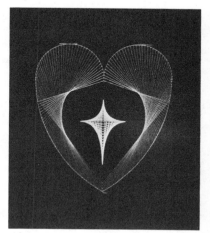

21. *Optical Illusions.* What are optical illusions? Why do these patterns cause illusions? Are these illusions still apparent with three-dimensional representations? What tricks depend on these illusions? How are these illusions used in advertising, dress designing, and architecture?

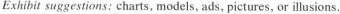

Exhibit suggestions: charts, models, ads, pictures, or illusions.

References: Bakst, Aaron, *Mathematics, Its Magic and Mastery,* (2nd ed.), pp. 469–478 (New York: Van Nostrand, 1952).

Gardner, Martin, "Mathematical Games," *Scientific American,* May 1970.

Gregory, Richard, "Visual Illusions," *Scientific American,* November 1968.

Penrose, Lionel, "Impossible Objects: A Special Type of Visual Illusion," *The British Journal of Psychology,* February 1958.

22. a. What is a regular polyhedron?
    b. How many regular polyhedra are possible? Name them.
    c. Construct models for the regular polyhedra.
       References: Coxeter, H. S. M., *Introduction to Geometry* (New York: Wiley, 1961).

       Cundy, H. Martyn, and A. P. Rollett, *Mathematical Models* (2nd ed.) (Oxford: Clarendon Press, 1961).

       Hogben, Lancelot, *Mathematics in the Making* (Garden City, N.Y.: Doubleday, 1960).

       Lines, L., *Solid Geometry* (New York: Dover Publications, 1965).

       Trigg, Charles W., "A Collapsible Model of a Tetrahedron," *The Mathematics Student Journal* (Vol. 2), February 1955, p. 1.

       ____, "Folding Tetrahedra," American Mathematical Monthly, January 1951 (Vol. 58), 39–40.

       ____, "Collapsible Models of the Regular Octrahedron," *The Mathematics Teacher,* October 1972, pp. 530–533.

       ____, "Geometry of Paper Folding," *School Science and Mathematics,* December 1954, (Vol. 54), pp. 683–689.

7.2 EUCLID DOES NOT STAND ALONE

PROJECTIVE GEOMETRY

Mighty is geometry; joined with art, resistless.— Euripides

As Europe passed out of the Middle Ages and into the Renaissance, artists were at the forefront of the intellectual revolution. No longer satisfied with flat-looking scenes, they wanted to portray people and objects as they looked in real life. The artists' problem was one of dimension. How could a flat surface be made to look three dimensional? Some of these artists tried to solve the problem by studying mathematics; however, the Euclidean geometry of the day did not provide all the answers they needed.

One of the first (but rather unsuccessful) attempts at portraying depth in a painting is shown in Figure 7.13, Duccio's *Last Supper*. Notice that the figures are in a boxed-in room. This technique is characteristic of the period and made the perspective easier to define.

Artists finally solved the problem of perspective by considering the surface of the picture to be a window through which to view the object to be painted. This technique was pioneered by Paolo Uccello (1397–1475), Piero della Francesca (1416–1492), Leonardo da Vinci (1452–1519), and Albrecht Dürer (1471–1528). As the lines of vision from the object converge at the eye, the picture captures a cross-section of them, as shown in Figure 7.14.

FIGURE 7.13 Duccio's *Last Supper*. Notice the receding wall and ceiling lines. However, the perspective is not complete. The table and the room are not seen from the same vantage point. The table seems to slant toward the front.

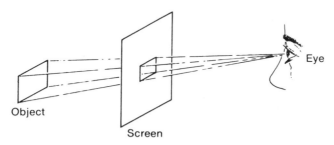

FIGURE 7.14 An attempt to understand perspective

The next step was the execution of this idea. Figure 7.15 shows an artist painting a picture using a grid system and a fixed viewpoint. As he changes his line of vision, the points projected onto the screen change. However, the method illustrated in this figure certainly has its limitations. Thus artists continued to seek mathematical solutions to their problems.

FIGURE 7.15 Albert Dürer's *Designer of the Lying Woman* shows how the problem of perspective can be overcome. The point in front of the artist's eye fixes the point of viewing the painting. The grid on the window corresponds to the grid on the artist's canvas.

In response to the needs of these artists, a geometry quite different from Euclidean geometry developed. This new geometry is called *projective geometry* and is studied in more advanced courses. It is a generalization of the Euclidean geometry you learned in high school.

NON-EUCLIDEAN GEOMETRIES

When Euclid formalized geometry, he used axioms and postulates that appeared to conform to the world in which we live. The so-called fifth postulate caused problems from the time it was stated. It somehow doesn't seem like the other postulates but, rather, like a theorem that should be proved. In fact, this postulate even bothered Euclid himself, since he didn't use it until he had proved his 29th theorem. Many mathematicians tried to find a proof for this postulate.

Historical Note

There are many ways of stating Euclid's fifth postulate, and several historical ones are repeated here.
1. **Poseidonius** *(about 135–51 B.C.): Two parallel lines are equidistant.*
2. **Proclus** *(410–485 A.D.): If a line intersects one of two parallel lines, then it also intersects the other.*
3. **Saccheri** *(1667–1733): The sum of the interior angles of a triangle is two right angles.*
4. **Legendre** *(1752–1833): A line through a point in the interior of an angle other than a straight angle intersects at least one of the arms of the angle.*
5. **Bolyai** *(1775–1856): There is a circle through every set of three noncollinear points.*
6. **John Playfair** *(1748–1819): Given a straight line and any*

EUCLID'S POSTULATES

1. A straight line can be drawn from any point to any other point in only one way.
2. A finite straight line can be drawn continuously in a straight line.
3. A circle can be described with any point as center and with a radius equal to any finite straight line drawn from the center.
4. All right angles are equal to each other.
5. If a straight line falling on two straight lines makes the interior angles on the same side less than two right angles, the two straight lines, if produced infinitely, meet on that side on which the angles are less than the two right angles.

One of the first serious attempts to prove Euclid's fifth postulate was made by Girolamo Saccheri (1667—1733), an Italian Jesuit. He constructed a quadrilateral with base angles A and B right angles and with sides AC and BD congruent (see Figure 7.16).

FIGURE 7.16 A Saccheri quadrilateral

As you probably know from high school geometry, the summit angles C and D are also right angles. However, this result used the fifth postulate. Now, it is also true that, *if* the summit angles are right angles, *then* Euclid's fifth postulate holds. The problem, then, was to establish the fact that angles C and D were right angles. First, Saccheri assumed the angles to be obtuse, and he didn't have too much difficulty in arriving at a contradiction. Next, he assumed the angles to be acute. No contradiction arose, and finally he gave up the quest because it "led to results that were repugnant to the nature of a straight line."

Saccheri never realized the significance of what he had started, and his work was forgotten until 1889. However, in the meantime, Johann Lambert (1728–1777) and Adrien-Marie Legendre (1752–1833) similarly investigated the possibility of eliminating Euclid's fifth postulate by proving it from the other postulates.

By the early years of the 19th century, three accomplished mathematicians began to suspect that the parallel postulate was independent and could not be eliminated by deducting it from the others. The great mathematician Carl Gauss, whom we've mentioned before, was the first to reach this conclusion, but, since he didn't publish this finding of his, the credit goes to two others. In 1811, an 18-year-old Russian named Nikolai Lobachevski pondered the possibility of a "non-Euclidean" geometry—that is, a geometry that did not assume Euclid's fifth postulate. In 1829, he published his ideas in an article entitled "Geometrical Researches on the Theory of Parallels." The postulate he used instead was subsequently named after him:

> THE LOBACHEVSKIAN POSTULATE: The summit angles of a Saccheri quadrilateral are acute.

point not on this line, there is one and only one line through that point that is parallel to the given line.

Playfair's statement is the one most often used in high school geometry textbooks. Remember that all of these statements are equivalent, *and, if you accept any one of them, you must accept them all.*

Saccheri's plan was a simple one:

1. *Assume that the angles are obtuse and deduce a contradiction.*
2. *Assume that the angles are acute and deduce a contradiction.*
3. *Therefore, by the first two steps, the angles must be right angles.*
4. *From Step 3, Euclid's fifth postulate can be deduced.*

𝕳istorical 𝕹ote

For centuries, we've assumed that the physical space in our universe is Euclidean; the notion that it wasn't was repugnant to Saccheri. However, Einstein's theory of relativity is based on a non-Euclidean geometry. This fact leads us to believe that, at least on a cosmic scale, Euclidean geometry is not sufficient to describe our universe.

Historical Note

Another mathematician, Janos Bolyai (1802–1860), was working on this same idea at about the same time as Lobachevski. (Clearly the world was nearly ready to accept the idea that a geometry need not be based on our physical experience.) Because of his work with this geometry, it is sometimes called Bolyai-Lobachevski *geometry.*

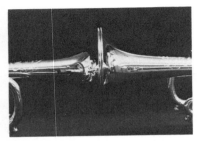

Two trumpets serve as a model of a tractrix rotated about a line.

Riemannian geometry is sometimes called spherical geometry *or* elliptic geometry.

This axiom, in place of Euclid's fifth postulate, leads to a geometry that we call *hyperbolic geometry*.

If we use the plane as a model for Euclidean geometry, what model will serve for hyperbolic geometry? A rough model for this geometry can be seen by placing two trumpet bells together. It is called a *pseudosphere* and is generated by a curve called a *tractrix* (as shown in Figure 7.17). The tractrix is rotated about the line *AB*. It has the property that, through a point not on a line, there are many lines parallel to the given one.

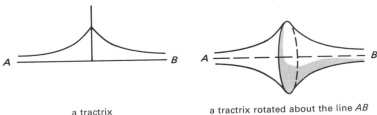

a tractrix   a tractrix rotated about the line *AB*

FIGURE 7.17

Georg Bernhard Riemann (1826–1866), who also worked in this area, pointed out that, although a straight line may be extended indefinitely, it need not have infinite length. It could instead be similar to the arc of a circle, which eventually begins to retrace itself. Such a line is called *re-entrant*. An example of a re-entrant line is found by considering a great circle on a sphere. A *great circle* is a circle on the sphere with a diameter equal to the diameter of the sphere. Taking this as a model, a Saccheri quadrilateral is constructed on a sphere with the summit angles obtuse. The resulting geometry is called *Riemannian geometry*. The shortest path between any two points on a sphere is an arc of the great circle through those points; thus these arcs correspond to line segments in Euclidean geometry. In 1854 Riemann showed that, with some other slight adjustments in the remaining postulates, another consistent non-Euclidean geometry can be developed. Notice that the fifth, or parallel, postulate fails to hold because any two great circles on a sphere must intersect at two points (see Figure 7.18).

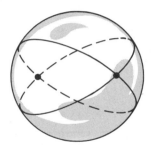

FIGURE 7.18 A sphere showing the intersection of two great circles.

We have not, by any means, discussed all possible geometries. We have merely shown that the Euclidean geometry you learned in high school is not the only model. A comparison of some of the properties of these geometries is shown in Table 7.1.

TABLE 7.1 Comparison of Major Two-Dimensional Geometries

Euclidean Geometry	Hyperbolic Geometry	Elliptic Geometry
Euclid	Gauss, Bolyai, Lobachevski	Riemann
Parabola	Hyperbola	Ellipse
Given a point not on a line, there is one and only one line through the point and parallel to the given line.	Given a point not on a line, there are an infinite number of lines through the point that do not intersect the given line.	There are no parallels.
The sum of the angles of a triangle is 180°.	The sum of the angles of a triangle is less than 180°.	The sum of the angles of a triangle is more than 180°.
Lines are infinitely long.	Lines are infinitely long.	Lines are finite in length.
$D = 90°$ $AB = CD$	$D < 90°$ $AB = CD$	$D > 90°$ $AB = CD$

PROBLEM SET 7.2

A Problems

Which of the figures in Problems 1–4 are Saccheri quadrilaterals?

1.

2.

3.

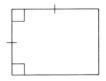

4.

5. Let A be any point on a sphere, and let ℓ be a line passing through A. How many lines can you draw through A that are perpendicular to ℓ?

6. Let P be a point at the pole, and let m be a line passing through P. How many lines can you draw through P that are perpendicular to m?

7. Consider Figure 7.19.
   a. Why do the lines appear to be parallel?
   b. In Table 7.1 we said that there are no parallels on a circle. What's wrong with our reasoning in part (a)?

Remember that a line on a sphere is a great circle.

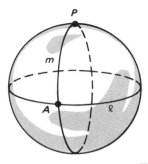

FIGURE 7.19

8. On a globe, locate San Francisco, Miami, and Detroit. Connect these cities with the shortest paths to form a triangle. Next, use a protractor to measure the angles. What is their sum?

9. Repeat Problem 8 for the cities of Tokyo, Seattle, and Honolulu.

10. Why do you think Euclidean geometry remains so prevalent today even though we know there are other valid geometries?

B Problems

11. Walking north or south on the earth is defined as walking along meridians, and walking east or west is defined as walking along parallels. There is a point on the surface of the earth from which it is possible to walk a mile south, then a mile east, then a mile north and be right back where you started. Find one such point.

12. Find two additional, different points satisfying the conditions of Problem 11.

13. Let's consider a new model. First draw a circle.

Now draw a "perpendicular circle," which is a circle whose tangents to the original circle at the points of intersection are perpendicular. The two circles are called *orthogonal circles.*

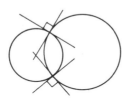

Next we set up another model of Lobachevskian geometry. Points of the plane are points inside the circle. Lines are both diameters of the circle and arcs of circles perpendicular to it and inside the original circle.

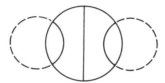

Draw a circle, and then draw several lines for this model. The drawing by Escher in Figure 7.20 is an example that uses this model. The white arcs through the backbones of the fish are lines in this model.

FIGURE 7.20   Escher print: *Circle Limit III*

"*Although I am absolutely innocent of training or knowledge in the exact sciences, I often seem to have more in common with mathematicians than with my fellow artists.*"
M. C. Escher

14. The Escher print in Figure 7.20 is an example of a *tessellation*. By skillfully altering a basic polygon, such as a triangle or hexagon, Escher was able to produce this interesting and artistic tessellation. The figure used in the problem is based on the one in Figure 7.20.

Step 1: Start with equilateral triangle ABC. Mark off the same curve on sides AB and AC as shown in Figure 7.21. Mark off another curve on side BC that is symmetric about the midpoint P. If you choose the curves carefully, as Escher did, an interesting figure suitable for tessellating will be formed.

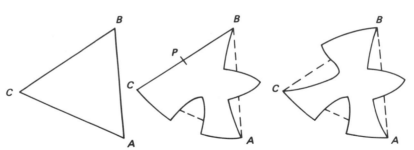

FIGURE 7.21 Tessellation pattern

Step 2: Six of these figures accurately fit together around a point forming a hexagonal array. Trace and cut one of the basic figures, and show how it can be used to continue the tessellation over the entire sheet.

15. Construct a tessellation using rectangles.
16. Construct a tessellation using hexagons.

Problems for Individual Study

17. Create some tessellations (see Problem 14) of your own similar to the one in Figure 7.22, which was drawn by Joseph L. Teeters. A method for drawing these is described in the article "How to Draw Tessellations of the Escher Type," in *The Mathematics Teacher,* April 1974, pp. 307–310.

18. The discovery and acceptance of non-Euclidean geometries had an impact on all of our thinking about the nature of scientific truth. Can we ever know truth in general? Write a paper on the nature of scientific laws, the nature of an axiomatic system, and the implications of non-Euclidean geometries.

19. Euclid clearly made a distinction between the definition of a figure and the proof that such a figure could be constructed. Two very famous problems in mathematics focus on this distinction:

    1. *Trisecting an angle:* Using only a straightedge and compass, trisect a given angle.
    2. *Squaring a circle:* Using only a straightedge and compass, construct a square with an area equal to the area of a given circle.

    These problems have been *proved* to be impossible (as compared with unsolved problems that *might* be possible). Write a paper discussing the nature of an unsolved problem as compared with an impossible problem.

20. *Non-Euclidean Geometry.* Build some models of spheres and pseudo-spheres with geometric figures drawn on the surfaces. Make some charts comparing theorems in different geometries. Make a report to the class.
    References: Kasner, Edward, and James Newman, *Mathematics and the Imagination,* pp. 135–153 (New York: Simon and Schuster, 1940).
    Kline, Morris, *Mathematics and Western Culture,* Chap. 26 (New York: Oxford University Press, 1953).
    Newman, James, *The World of Mathematics* (New York: Simon and Schuster, 1956). See especially the following articles:

FIGURE 7.22

If someone claims to be able to trisect an angle with a straightedge and compass, a mathematician, without ever looking at the construction, *will respond by saying that a mistake was made. This can be frustrating to someone who believes that he or she has accomplished the impossible. In fact, it was so frustrating to Daniel Wade Arthur that he was motivated to take out a paid advertisement in the* Los Angeles Times, *which is reproduced in Figure 7.23 on page 410.*

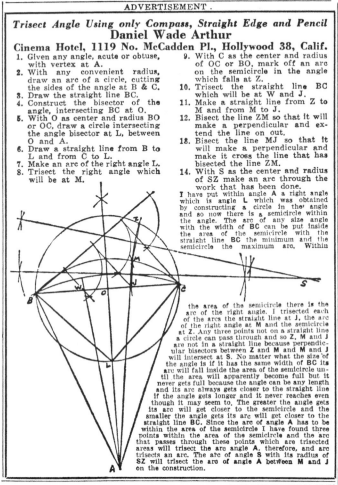

FIGURE 7.23

Hempel, Carl, "Geometry and Empirical Science," pp. 1635–1646.

Von Mises, Richard, "Mathematical Postulates and Human Understanding," pp. 1723–1754.

Wolfe, Harold, *Non-Euclidean Geometry* (New York: Holt, Rinehart, and Winston, 1966).

Encyclopedia entries: Nikolai Lobachevski, George Bernhard Riemann, Henri Poincaré, Non-Euclidean geometry.

21. Find some examples of art that illustrate true perspective and some that illustrate false perspective.

   Reference: H. W. Janson, *History of Art* (Rev. ed.) (New York: Abrams, 1969).

22. *Perspective and Projective Geometry.* How are three-dimensional objects represented in two dimensions? What are different ways of projecting lines or surfaces on a plane?
   Exhibit suggestions: models of perspective; examples of how projective geometry and art are related.
   References: Courant, Richard, *What is Mathematics?* (London: Oxford University Press, 1941).
   Kline, Morris, *Mathematics, a Cultural Approach* Chaps. 10–11 (Reading, Mass.: Addison-Wesley, 1962).
   Ogilvy, C. Stanley, *Excursions in Geometry* Chap. 7 (New York: Oxford University Press, 1969).

7.3 TOPOLOGY

A brief look at the history of geometry illustrates, in a very graphical way, the historical evolution of many mathematical ideas and the nature of the changes in mathematical thought. The geometry of the Greeks included very concrete notions of space and geometry. They considered space to be a locus in which objects could move freely about, and their geometry was a geometry of congruence. This geometry was tied to the physical world we observe and was described in Section 7.2 as Euclidean geometry.

In the 17th century, space came to be conceptualized as a set of points, and, with the non-Euclidean geometries of the 19th century, mathematicians gave up the notion that geometry had to describe the physical universe. The existence of multiple geometries was accepted, but space was still thought of as a locus with a geometry of congruence. The emphasis shifted to sets, and geometry was studied as a mathematical system with properties and operations called *congruent transformations*.

> *Euclid alone has looked on*
>   *Beauty bare.*
> *Let all who prate on*
>   *Beauty hold their peace*
> *And lay them prone upon*
>   *the earth and cease*
> *To ponder on themselves,*
>   *the while they stare*
> *At nothing, intricately*
>   *drawn nowhere*
> *In shape of shifting*
>   *lineage; let geese*
> *Gabble and hiss, but*
>   *heroes seek release*
> *From dusty bondage into*
>   *luminous air.*
> *O binding hour, O holy,*
>   *terrible day,*
> *When first the shaft into his*
>   *vision shone*
> *Of light anatomized. Euclid*
>   *alone*
> *Has looked on Beauty*
>   *bare.*
>   *Fortunate they*
> *Who, though once only and*
>   *then from far away,*
> *Have heard her massive*
>   *sandal set on stone.*
>   *Edna St. Vincent Millay*

Consider a triangle that is rotated or flipped to form a "new" triangle; these triangles are called *congruent*, but the angles and lengths of the sides have not changed, as illustrated in Figure 7.24.

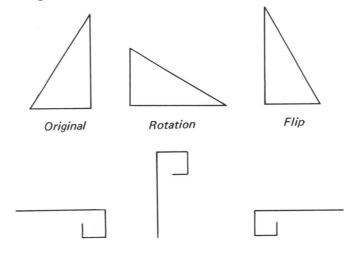

FIGURE 7.24

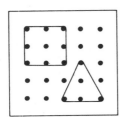

FIGURE 7.25 Geoboard with a square and a triangle

By the end of the 19th century, the idea that a geometry could be described as an abstract set of axioms along with a body of deduced theorems was generally accepted. Space was thought of simply as a set of points together with an abstract set of relations in which these points are involved. The time was right for geometry to be considered as the theory of such a space, and in 1895 Henri Poincaré published a book using this notion of space and geometry in a systematic development. This book was called *Vorstudien zer Topologie* (*Introductory Studies in Topology*). However, topology was not the invention of any one person, and the names of Cantor, Euler, Fréchet, Hausdorff, Möbius, and Riemann are associated with the origins of topology. Today it is a broad and fundamental branch of mathematics.

To obtain an idea of the nature of topology, consider a *geoboard*. You will recall that geoboards are used to illustrate a wide variety of geometric ideas, especially in elementary schools. Suppose we stretch one rubber band over the nails to form a square and another to form a triangle, as shown in Figure 7.25.

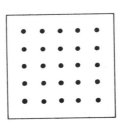

In the geometries we've been considering, in which we've been concerned with congruence, similarity, angles, number of sides, and so on, these figures are seen as different. However, in topology, we are concerned with the *mathematics of distortion*. The general idea is one called *elastic motion*, which includes bending, stretching, shrinking, or distorting the figure in any way that allows the points to remain distinct. It does not include cutting a figure unless we "sew up" the cut *exactly* as it was before.

> DEFINITION: Two geometric figures are said to be *topologically equivalent* if one figure can be elastically twisted and stretched into the same shape as the other.

Definition of topological equivalence.

Rubber bands can be stretched into a wide variety of shapes. All the forms in Figure 7.26 are topologically equivalent. These curves are *simple closed curves*. A curve is *closed* if it divides the plane into three disjoint

FIGURE 7.26 Topologically equivalent curves

subsets: the set of points on the curve itself, the set of points *interior* to the curve, and the set of points *exterior* to the curve. It is *simple* if it has only one interior. Notice that, to pass from a point in the interior to a point in the exterior, it is necessary to cross over the given curve at least once. This property remains the same for any distortion and is therefore called an *invariant*.

In topology, it is necessary to group together objects of the same dimension. Objects of different dimensions are never topologically equivalent. For the remainder of this chapter we'll restrict our attention to two-dimensional surfaces in a three-dimensional space. This means that when considering a doughnut or a coffee cup, as in Figure 7.27, we are considering the surfaces of these objects. Figure 7.28 shows four rows of figures. Each figure is equivalent to each other figure in its row.

The children and their distorted images are topologically equivalent.

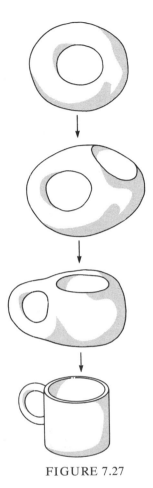

FIGURE 7.27

FIGURE 7.28 Topologically equivalent curves

Two-dimensional surfaces in a three-dimensional space are classified according to the number of cuts possible without slicing it into two pieces. For example, no closed cut can be made around a sphere without cutting it into two pieces, so its genus is 0. Roughly speaking, the genus of a surface without edges is the number of holes the object has. A doughnut has genus 1 since only one closed cut can be made without dividing it into two pieces. All figures with the same genus are topological equivalents. Figure 7.27 shows that a doughnut and a coffee cup are topologically equivalent. Figure 7.29 shows objects of genus 0, genus 1, and genus 2.

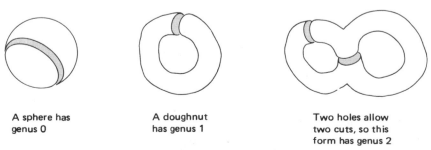

A sphere has genus 0

A doughnut has genus 1

Two holes allow two cuts, so this form has genus 2

FIGURE 7.29 Genus of a sphere, a doughnut, and a doughnut with two holes

PROBLEM SET 7.3

A Problems

Group Problems 1–15 into classes so that all the elements within each class are topologically equivalent and no elements from different classes are topologically equivalent.

1.

2.

3.

4.

5.

6.

7.

8.

9.

10.
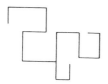

11.

12.

13.

14.

15.

16. Which of the figures of Problems 1–15 are simple closed figures?

TABLE 7.2 Genus of the Surfaces of Everyday Objects

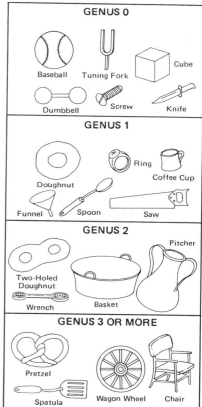

> *The golden age of mathematics—that was not the age of Euclid, it is ours.*
>
> C. J. Keyser

Group the objects in Problems 17–31 into classes so that all the elements within each class are topologically equivalent and no elements from different classes are topologically equivalent.

17. a glass   18. a bolt   19. a straw

20. a bowling ball   21. a record

22. a sheet of two-ring-binder paper

23. a sheet of typing paper

24. a sphere   25. a horseshoe

26. a ruler   27. a sewing needle

28. a funnel with a handle   29. a brick

30. a banana   31. a pencil

In Problems 32–36, determine whether each of the points A, B, and C is inside or outside the simple closed curve.

32.
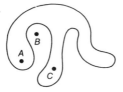

33.
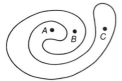

34.

35.
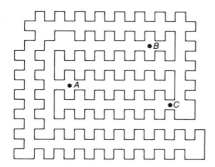

36.
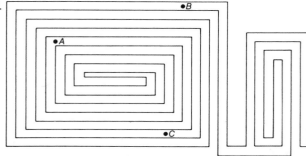

37. Let X be a point obviously outside the figure given in Problem 32. Draw AX. How many times does it cross the curve? Repeat for BX and CX.

38. Repeat Problem 37 for the figure given in Problem 33.

39. Repeat Problem 37 for the figure given in Problem 34.

B Problems

40. Make a conjecture based on Problems 37–39. Use your conjecture for the figure given in Problems 35 and 36.

41. Group the letters of the alphabet into classes so that all the elements within each class are topologically equivalent and no elements from different classes are topologically equivalent.

A B C D E F G H I J K L M
N O P Q R S T U V W X Y Z

This conjecture involves a theorem called the Jordan Curve *theorem. It says that a simple closed curve divides the plane into two regions, an inside and an outside. Sometimes a simple closed curve is called a* Jordan *curve.*

A Jordan curve has also been described as a Protestant curve: it doesn't cross itself.

42. *He killed the noble Mudjokivis.*
    Of the skin he made him mittens,
    Made them with the fur side inside
    Made them with the skin side outside.
    He, to get the warm inside,
    Put the skin side outside;
    He, to get the cold side outside,
    Put the warm side fur side inside.
    That's why he put the fur side inside,
    Why he put the skin side outside,
    Why he turned them inside outside.—Author unknown

    a. If a right-handed mitten is turned inside out, as is suggested in the poem, will it still fit the right hand?
    b. Is the right-handed mitten topologically equivalent to the left-handed mitten?

Mind Bogglers

43. *Two-Person Problem.* Tie a piece of cord to each of your wrists. Then tie a second piece of cord to the wrists of a second person and through your cord, as shown. Now see if you can separate yourselves without cutting the cord, untying the knots, or taking the string off your wrists. Since the figure in (a) is topologically equivalent to the figure in (b), the problem can be solved.

(a)   (b)

44. Is is possible to take an entire trip through the houses whose floor plans are shown in Figure 7.30 and pass through each door once and only once?

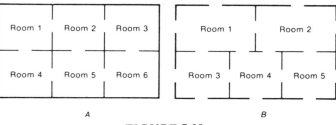

A   B

FIGURE 7.30

FOUR COLOR PROOF

URBANA, IL.—Kenneth Appel and Wolfgang Haken, professors of mathematics at University of Illinois, Urbana Champaign, have announced a proof of the four-color theorem.

Since the theorem was first stated formally in 1852, many unsuccessful attempts have been made to prove it. The first published "proof" is due to Kempe who enumerated four cases and disposed of each. However, in 1890 Heawood found an error in one case. Appel and Haken have broken Kempe's difficult case into about 1,930 cases. To each of these they have applied reasoning due to G. D. Birkhoff.

High-speed computers checked every possible arrangement of four colors for each case and a ring of up to 14 neighboring regions sometimes nearly 200,000 different combinations. Some 1,200 hours of computer time were used. The proof would have been impossible without the computers. Appel and Haken said.

45. *A five-color map?* One of the earliest and most famous problems in topology is the four-color problem. It was first stated in 1850 by the English mathematician Francis Guthrie. It states that any map on a plane or a sphere can be colored with at most four colors so that any two countries that share a common boundary are colored differently. All attempts to prove this conjecture had failed until Kenneth Appel and Wolfgang Haken of the University of Illinois announced their proof in 1976. The university honored their discovery by using the following postmark:

FOUR COLORS

SUFFICE

Some mathematicians were reluctant to accept this proof because of the necessity of computer verification. The proof was not "elegant" in the sense that it required the computer analysis of a large number of cases. Study the map in Figure 7.31 and determine for yourself if it is the *first five-color* map, providing a counterexample for the computerized "proof." Notice from Table 7.3 that the region in question is bounded by four colors.

TABLE 7.3

Region	Blue	Yellow	Green	Red
1	✔	✔	✔	
2	✔	✔	✔	✔
3	✔	✔	✔	✔
4		✔	✔	✔
5	✔		✔	✔
6		✔	✔	✔
7	✔		✔	✔
?	✔	✔	✔	✔

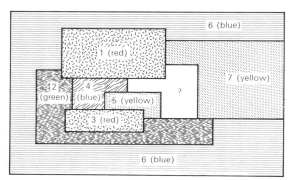

FIGURE 7.31 Is this the world's first five-color map?

Problems for Individual Study

46. Make drawings of geometric figures on a piece of rubber inner tube. Demonstrate to the class various ways in which these figures can be distorted.

47. Use modeling clay to show some of the transformations illustrated in the text. Make a presentation to the class.

7.4 KÖNIGSBERG BRIDGE PROBLEM

In the 18th century, in the German town of Königsberg (now the Russian city of Kaliningrad), a popular pastime was to walk along the bank of the Pregel River and cross over some of the seven bridges that connected two islands, as is shown in Figure 7.32.

FIGURE 7.32 Königsberg bridges

Compare the experience of the people of Königsberg with your findings in the Santa Rosa Street Problem page 398).

One day a native asked a neighbor this question: "How can you take a walk so that you cross each of our seven bridges once and only once?" The problem intrigued the neighbor and soon caught the interest of many other people of Königsberg as well. Whenever people tried it, they ended up either not crossing a bridge at all or else crossing one bridge twice.

This problem was brought to the attention of the Swiss mathematician Leonhard Euler, who was serving at the court of the Russian empress Catherine the Great in St. Petersburg. His study of this problem laid the groundwork for topology.

Euler began by drawing a diagram of the problem. With this diagram, the land became points and the bridges became lines connecting these points, as shown in Figure 7.33a. Now we can work the problem as Euler did. We begin by trying to trace a diagram like the one in Figure 7.33b. In order to solve the bridge problem, we need to draw the figure without lifting our pencil from the paper. Figures similar to the one in Figure 7.33b are called *networks*.

(a)

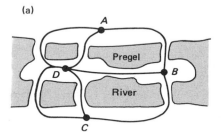

(b)

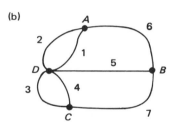

FIGURE 7.33 Networks for the Königsberg Bridge Problem

In a network, the points where the lines cross are called *vertices*, and the lines representing bridges are called *arcs*. A network is *traversed* by passing through all the arcs exactly once. You may pass through the vertices any number of times. In the network of the Königsberg bridges, the vertices are A, B, C, and D. The number of arcs to vertex A is 3, so the vertex at A is called an *odd vertex*. In the same way, D is an odd vertex, since 5 arcs go to it. Euler discovered that there must be a certain number of odd vertices in any network if you are to travel it in one journey without retracing any arcs. You may start at any vertex and end at any other vertex, as long as you travel the entire network. Also, the network must connect each point (this is called a *closed network*).

Let's examine the network more carefully and look for a pattern.

Number of Arcs Connecting a Vertex	Possibilities	
1	One departure; starting point	
	one arrival; ending point	
2	one arrival/one departure	
3	one arrival/two departures; starting point	
	two arrivals/one departure; ending point	
4	two arrivals/two departures	
5	two arrivals/three departures; starting point	
	three arrivals/two departures; ending point	

We see that, if the vertex is odd, then it must be a starting point or an ending point. What is the largest number of starting and ending points in any network? Note the following results:

1. If a network has more than two odd vertices, it is not traversible.
2. If a network has two odd vertices, it is traversible; one odd vertex must be a starting point and the other odd vertex must be an ending point.
3. If all vertices are even, then it is traversible.

EXAMPLES: Are the following networks traversible?

1.

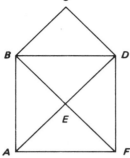

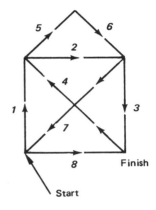

You may remember seeing this network before. It has four even vertices (*B*, *C*, *D*, and *E*) and two odd vertices (*A* and *F*) and is therefore traversible. In order to traverse it, we must start at *A* or *F* (at an odd vertex).

2.

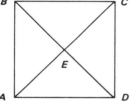

This network has one even vertex and four odd vertices, and it is therefore not traversible.

We can now solve the Königsberg Bridge Problem. Referring to Figure 7.33a, we see there are four odd vertices and the network is not traversible.

Another example of a network problem was given by Problem 44 on page 418. We will call this the *floor-plan problem*, which involves taking a trip through all the rooms and passing through each door only once. Let's label the rooms in Figure 7.34a *A, B, C, D, E*, and *F*. Rooms *A, C, D*, and *F* have two doors; rooms *B* and *E* have three doors. Make a conjecture about which room we should begin in if we wish to solve this problem.

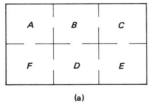   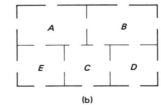

FIGURE 7.34 Floor-plan problem

How about the floor plan in Figure 7.34b? Rooms *A*, *B*, and *D* have an odd number of doors; Rooms *C* and *E* have an even number of doors. It is not possible to choose a route to traverse the floor plan of this house. Why not? If we barricade one of the doors in room A, the problem can then be solved as shown in Figure 7.35.

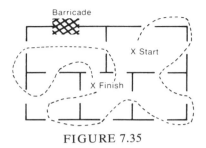

FIGURE 7.35

PROBLEM SET 7.4

A Problems

Which of the networks in Problems 1–10 are traversible? If a network can be traversed, show how.

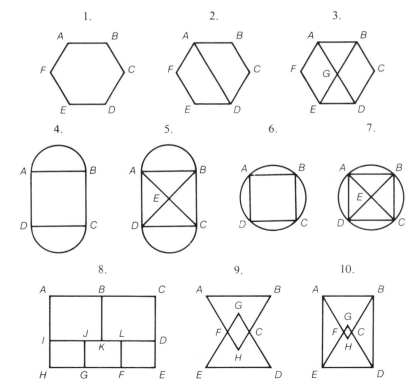

11. Present a complete solution to the Santa Rosa Street Problem originally posed in Problem 14, page 398.

12. The edges of a cube form a three-dimensional network. Are the edges of a cube traversible?

13. After Euler solved the Königsberg Bridge Problem, an eighth bridge was built as shown in Figure 7.36. Is this network traversible? If so, show how.

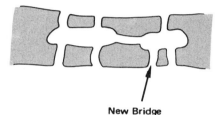

FIGURE 7.36 Königsberg with eight bridges

For which of the floor plans in Problems 14–17 can you pass through all the rooms while going through each door only once?

14.
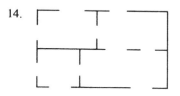

15.
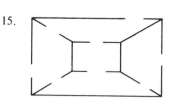

16.
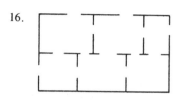

17.
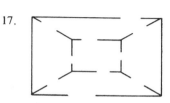

18. Figure 7.37 shows a portion of London's Underground transit system. Is it possible to travel the entire system and visit each station while taking each route exactly once?

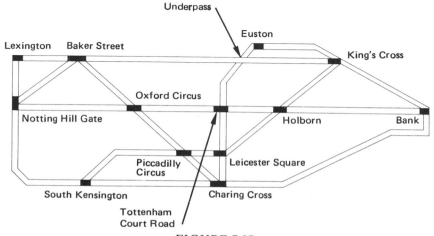

FIGURE 7.37

B Problems

19. Make a conjecture about when the floor plans of Problems 14–17 can be solved.

20. Make a conjecture about which room to begin with to solve a floor-plan problem.

21. Count the number of vertices, arcs, and regions for each of Problems 1–10.
    Let V = the number of vertices,
        A = the number of arcs,
        R = the number of regions.
    Compare $V + R$ with A. Make a conjecture relating V, R, and A. This finding is called *Euler's Formula for Networks*.

Euler's Formula

22. About a century ago, August Möbius made the discovery that, if you take a strip of paper and give it a single half-twist and paste the ends together, you will have a piece of paper with only one side! Construct a Möbius strip, and verify that it has only one side. How many edges does it have?

23. Cut your Möbius strip from Problem 22 in half down the center. What is the result?

24. Cut your resulting band from Problem 23 in half again. What is the result?

25. Construct a Möbius strip, and cut along a path that is about one-third the distance from the edge. What is the result?

26. Construct a Möbius strip, and mark a point A on it. Draw an arc from A around the strip until you return to the point A. Do you think you could connect *any* two points on the sheet of paper without lifting your pencil?

27. Take a strip of paper $11'' \times 1''$, and give it three half-twists; join the ends together. How many edges and sides does this band have? What happens if you cut down the center of this piece?

Mind Bogglers

28. *The Paper-Boot Puzzle.* The boot puzzle consists of three pieces, all of which are cut out of stiff paper. One piece is shaped like a pair of boots joined together at the top, as in Figure 7.38a. The remaining pieces are shaped as shown in Figures 7.38b and 7.38c.

Escher print: *Möbius Strip II*

FIGURE 7.38 Paper-Boot Puzzle

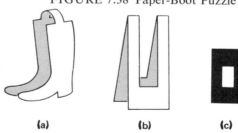

(a)   (b)   (c)

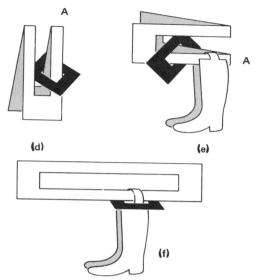

To assemble the puzzle, fold the large rectangular piece as in Figure 7.38b and slip the smaller piece over one of its arms, as in Figure 7.38d. Then hang the boots over part of the same arm, as in Figure 7.38e. Pull the small piece to the right and over the end of the arm at A. Then unfold the large piece, and the puzzle will be assembled as in Figure 7.38f.

The problem is to remove the boots without tearing the paper.

29. Three houses, A, B, and C, as shown in Figure 7.39, must connect to the water main, W, the gas main, G, and the electric line, E.

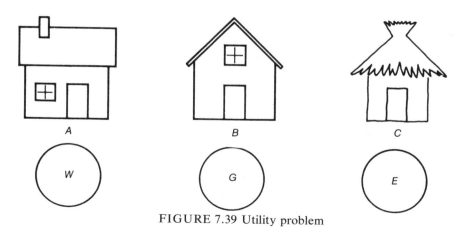

FIGURE 7.39 Utility problem

Is is possible to make these connections:

a. if the power lines are in the same plane?
b. if we do not require all nine power lines to be in the same plane?

Problem for Individual Study

30. What is a Klein bottle?

Hint: The figure and limerick should give you some clues for Problem 30.

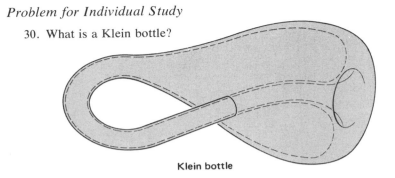

Klein bottle

A mathematician named
    Klein
Thought the Möbius strip
    was divine.
    Said he, "If you glue
    The edges of two,
You'll get a wierd bottle
    like mine."
—Anonymous

7.5 SUMMARY AND REVIEW

CHAPTER OUTLINE

I. The Nature of Geometry
   A. Geometry as a deductive system
      1. Undefined terms
         a. Point
         b. Line
         c. Plane
      2. **Axiom:** A statement that is accepted without proof
      3. **Theorem:** Results that are proved from previous terms, axioms, and theorems by using the rules of logic
   B. Geometry as a set of transformations
      1. A **transformation** is a function from one set onto itself. In particular, it is a one-to-one correspondence between the points of the plane and their image points.
      2. Some types of transformations
         a. Reflections
         b. Other types, including translations, rotations, inversions, dilations, and contractions
   C. Relationships of geometry
      1. Equality
      2. Congruence
      3. Similarity
      4. Topological equivalence
II. Different Kinds of Geometry
   A. Projective geometry; perspective in art is achieved as a result of the study of projective geometry.
   B. The Saccheri quadrilateral
      1. A quadrilateral $ABCD$ with base angles A and B right angles and with sides AC and BD equal is a Saccheri quadrilateral.
      2. The angles C and D of a Saccheri quadrilateral might be acute, right, or obtuse depending on additional assumptions that are made.

3. The assumptions we accept in determining the size of angles C and D of a Saccheri quadrilateral give rise to three geometries: Euclidean, hyperbolic, and elliptic.

C. Euclidean geometry
1. Probably the geometry you studied in high school; it conforms to the way we view the physical world.
2. It is based on five postulates (see page 402).
3. The plane serves as a model.
4. Important results
    a. Given a point not on a line, there is one and only one line through the point and parallel to the given line (Euclid's fifth postulate).
    b. The sum of the angles of a triangle is 180°.
    c. Lines are infinitely long.

D. Hyperbolic geometry
1. Lobachevski made the assumption that the summit angles of a Saccheri quadrilateral are acute.
2. The pseudosphere serves as a model.
3. Important results
    a. Given a point not on a line, there are an infinite number of lines through the point that do not intersect the given line.
    b. The sum of the angles of a triangle is less than 180°.
    c. Lines are infinitely long.

E. Elliptic geometry
1. Riemann considered *re-entrant* lines and thus concluded that the base angles of a Saccheri quadrilateral are obtuse.
2. The sphere serves as a model.
3. Important results
    a. There are no parallels.
    b. The sum of the angles of a triangle is more than 180°.
    c. Lines are finite in length.

III. Topology
A. Topology is the mathematics of distortion; it considers figures that are topologically equivalent.
1. Two geometric figures are topologically equivalent if one figure can be elastically twisted or stretched (without tearing) into the same shape as the other.
2. Simple closed curves (Jordan curves)
3. The inside and the ouside of a simple closed curve
B. The genus of a surface without any edges is the number of holes the object has.
C. Networks
1. Vertices and arcs
2. Traversible networks
3. Odd vertex/even vertex
4. Königsberg Bridge Problem
5. Floor-plan problem
6. Euler's Formula for Networks
D. Möbius strip

REVIEW PROBLEMS

1. What is a Saccheri quadrilateral? Discuss why this quadrilateral leads to different kinds of geometrics.

2. Briefly discuss Euclidean, hyperbolic, and elliptic geometries.

3. Group the following figures into classes so that all the elements within each class are topologically equivalent.

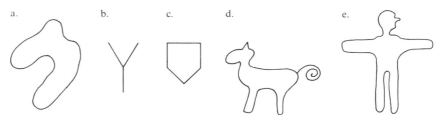

a.   b.   c.   d.   e.

4. Which of the following objects have the same genus?
   a. a garden hose
   b. a pullover sweater
   c. a sheet of three-ring-binder paper
   d. the letter *P*
   e. a pitchfork

5. Classify the surfaces of the following objects as having genus 0, 1, 2, 3, or more than 3.

a.   b.   c.   d.   e.

6. Determine whether the point *X* is inside or outside the simple closed curves.

a.   b.

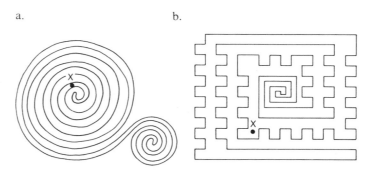

Indicate which of the networks in Problems 7 and 8 are traversible. If the network is traversible, show how.

7.
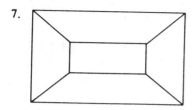

8.
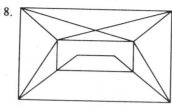

Indicate whether, for the floor plans in Problems 9 and 10, you can pass through all the rooms while going through each door only once.

9.
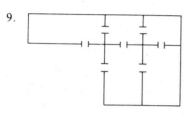

10.
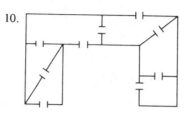

11. Draw a map with seven regions that can be colored with three colors, no two bordering regions having the same color.

12. Draw a map similar to the one in Problem 11 but that requires four colors.

13. Take a strip of paper 11″ × 1″, and give it four half-twists; join the edges together. How many edges and sides does this band have? Cut the band down the center. What is the result?

14. Count the number of faces (F), edges (E), and vertices (V) for each of the polyhedra in Figure 7.40.

15. Make a conjecture about the relationship among faces, edges, and vertices by comparing $V + F$ and E in Problem 14.

Tetrahedron

Hexahedron

Octahedron

Dodecahedron

Icosahedron

FIGURE 7.40 Polyhedra

CULTURAL

1650: Palace of Versailles built by Louis XIV, the Sun King

1677: John Locke writes his *Essay Concerning Human Understanding*

1684: Newton's encounter with the apple
1685: Bach and Handel born

Sir Isaac Newton

1698: The Glorious Revolution

1708: Charles XII invades Russia

1722: Peter the Great attacks Persia

1740: Accession of Frederick the Great
1742: Handel writes *The Messiah*

1650

1675

1700

1725

MATHEMATICAL

1650: Pascal—conics, probability, computing machines, Pascal's triangle
John Wallis—algebra, imaginary numbers

Blaise Pascal

1670: Sir Christopher Wren—architecture, imaginary numbers

1680: Sir Isaac Newton—calculus, gravitation, series, hydrodynamics
1682: Gottfried Leibniz—calculus, determinants, symbolic logic, notation, computing machines

1690: Nicolaus Bernoulli—probability, curves

Gottfried Leibniz

1700: Johann Bernoulli—applied calculus

1715: Brook Taylor—series, geometry

1720: Abraham De Moivre—probability, calculus, complex numbers

1731: Alexis Clairaut—solid analytic geometry
1733: Saccheri—beginnings of analytic geometry

1740: Colin Maclaurin—series, physics, higher-plane curves

8

THE NATURE OF COUNTING

"Threes (To Be Sung by Niels Bohr)"

I think that I shall never c
A # lovelier than 3;
For 3 < 6 or 4,
And than 1 it's slightly more.

All things in nature come in 3s,
Like ∴s, trios, Q.E.D.'s;
While $s gain more dignity
If augmented 3 × 3—

A 3 whose slender curves are pressed
By banks, for compound interest;
Oh, would that, paying loans or rent,
My rates were only 3%!

3^2 expands with rapture free,
And reaches outward ∞;
3 complements each x and y,
And intimately lives with π.

A ☉'s # of °
Are best ÷ up by 3s,
But wrapped in dim obscurity
Is the $\sqrt{-3}$.

Atoms are split by men like me,
But only God is 1 in 3.

John Atherton

8.1 COUNTING PROBLEMS

In this chapter we'll investigate the nature of counting. It may seem like a simple topic for a math book, since everyone knows how to count. However, there are easier ways to count than that shown in the cartoon (see Problem 18). In this first section, we'll pose some counting problems and introduce the Fundamental Counting Principle.

"Five trillion, four hundred eighty billion, five hundred twenty-three million, two hundred ninety-seven thousand, one hundred and sixty-two . . ."

ELECTION PROBLEM

Consider a club with five members:

{Alfie, Bogie, Calvin, Doug, Ernie}

In how many ways could they elect a president and secretary? There are several ways to solve this problem. The first, and perhaps the easiest, is by making a tree diagram (see Figure 8.1). We see that there are 20 possibilities. This method is effective for "small" tree diagrams, but the technique quickly gets out of hand. For example, if we wished to see how many ways the club could elect a president, secretary, and treasurer, this technique would be very lengthy.

A second method of solution is by using boxes or "pigeon holes" representing each choice separately. Here we determine the number of ways of choosing a president and the number of ways of choosing a secretary.

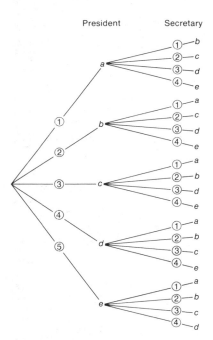

FIGURE 8.1 Tree diagram for election problem. a = Alfie, b = Bogie, c = Calvin, d = Doug, and e = Ernie

Ways of choosing a president	Ways of choosing a secretary
5	4

↑
Since we have chosen a president, only 4 members remain.

If we multiply the numbers in the pigeon holes,

$$5 \cdot 4 = 20$$

and we see that the result is the same as that from the tree diagram. This method is much more satisfying and is an example of a general result called the *Fundamental Counting Principle*.

> FUNDAMENTAL COUNTING PRINCIPLE: If task A can be performed in m ways, and, after task A is performed, a second task, B, can be performed in n ways, then task A followed by task B can be performed in $m \cdot n$ ways.

Definition of the Fundamental Counting Principle

EXAMPLES:

1. How many skirt-blouse outfits can a woman wear if she has three skirts and five blouses?

    Solution: There are two "tasks": (i) choosing a skirt, and (ii) choosing a blouse. We illustrate with two pigeon holes:

Task A	Task B
3	5
↑	↑
Skirts	Blouses

There are $3 \times 5 = 15$ skirt-blouse outfits.

2. In how many ways could the given club choose a president, secretary, and treasurer?

    Solution:

5	4	3	= 60
↑	↑	↑	
Ways of choosing a president	Ways of choosing a secretary	Ways of choosing a treasurer	

3. In how many ways could this club choose a president, vice-president, secretary, and treasurer?

    Solution:

$$5 \cdot 4 \cdot 3 \cdot 2 = 120$$

LICENSE-PLATE PROBLEM

The State of California "ran out" of codes that could be issued on license plates. The scheme in California is practiced in many states; that is, license plates consist of three letters followed by three numerals. For example, CWB 072 is a license-plate code. When the state ran out of available new numbers, a change had to take place. The decision was

made to leave the old numbers in circulation and to issue new plates in the order of three numerals followed by three letters, such as

The problem is to determine how many plates were available before the switch.* The solution to this problem is found by using the Fundamental Counting Principle.

C	W	B	0	7	2

↑   ↑   ↑   ↑   ↑   ↑

There are 26 possibilities   There are 10 possibilities
for each of these letters.   for each of these numerals.

Thus the total number of possible license plates that could be issued is:

$$26 \cdot 26 \cdot 26 \cdot 10 \cdot 10 \cdot 10 = 17,576,000$$

EXAMPLE: How many license plates can be formed if repetitions of letters or digits are not allowed?

Solution: The result is given by

$$26 \cdot 25 \cdot 24 \cdot 10 \cdot 9 \cdot 8 = 11,232,000$$

COMMITTEE PROBLEM

Consider the same club:

{Alfie, Bogie, Calvin, Doug, Ernie}

In how many ways could they elect a committee of two members?

One method of solution is easy (but tedious) and involves the enumeration of all possibilities.

---

*Certain plates are never issued because the combinations of letters produce obscene or confusing words, such as CHP, which might be misinterpreted as a car of the California Highway Patrol. But in our discussion here, we will assume that all possible license plates were issued.

$$\{a,b\} \qquad \{b,c\} \qquad \{c,d\} \qquad \{d,e\}$$
$$\{a,c\} \qquad \{b,d\} \qquad \{c,e\}$$
$$\{a,d\} \qquad \{b,e\}$$
$$\{a,e\}$$

We see that there are 10 possible two-member committees.

EXAMPLE: In how many ways can this club elect a committee of three members?

Solution: This is the same problem as the preceding one. Notice that every time we select a two-member committee, we also have a three-member committee (those left out). Hence the solution is 10.

If we like, we can also list the 10 possibilities:

$$\{c,d,e\} \qquad \{a,d,e\} \qquad \{a,b,e\} \qquad \{a,b,c\}$$
$$\{b,d,e\} \qquad \{a,c,e\} \qquad \{a,b,d\}$$
$$\{b,c,e\} \qquad \{a,c,d\}$$
$$\{b,c,d\}$$

Do you see why we can't use the Fundamental Counting Principle for the committee problem in the same way we did for the election problem?

We have presented two different counting problems, the election problem and the committee problem. For the election problem, we found an easy numerical method of counting, using the Fundamental Counting Principle. For the committee problem, we did not; further investigation is necessary. Later sections of this chapter will illustrate a numerical method of counting for the committee problem and apply this method to still other counting problems. The task of this chapter, then, is to develop a method of "counting without counting."

PROBLEM SET 8.1

A Problems

1. State the Fundamental Counting Principle, and explain in your own words what it says.

2. If a state issued license plates using the scheme of one letter followed by five digits, how many plates could it issue?

3. Answer Problem 2 if repetitions are not allowed.

4. Count the number of cubes in the structure shown. (Assume that the structure is solid.)

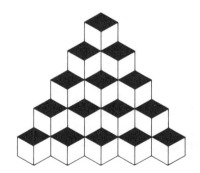

5. In how many ways can a group of 15 people elect a president, vice-president, and secretary?

6. How many two-member committees can be formed from a group of seven people?

When working these problems remember that the Fundamental Counting Principle will not work for some types of counting problems.

7. How many squares (of any size) are in the design shown?

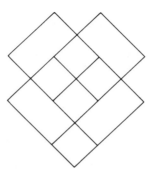

8. How many seven-digit telephone numbers are possible if the first two digits cannot be ones or zeros?

9. Foley's Village Inn offers the following menu in their restaurant:

Main Course	Dessert	Beverage
prime rib	ice cream	coffee
steak	sherbet	tea
chicken	cheesecake	milk
ham		Sanka
shrimp		

In how many different ways can someone order a meal consisting of one choice from each category?

HINT: All three flags must be used to convey a "message."

10. Boats often relay messages by using flags on a flagpole. How many messages can be made using three different flags?

Each arrangement of these three flags is a "code" for some message. Two different messages are shown here. Can you find all other possible messages?

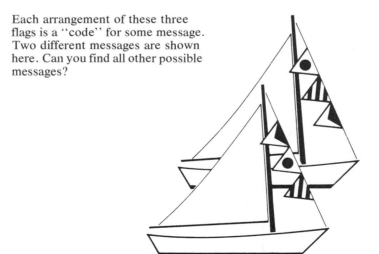

11. How many triangles are there in each figure?

a.    b.   c.

12. New York license plates consist of three letters followed by three numerals, and 245 letter arrangements are not allowed. How many plates can New York issue?

13. A typical Social Security number is 576-38-4459; how many Social Security numbers are possible if
a. the first digit cannot be zero?
b. neither of the first two digits can be a zero?

B Problems

14. A club consists of four men and three women. In how many ways can this club elect a president, vice-president, and secretary, in that order, if:
a. the president must be a woman and the other two officers must be men?
b. the president and the vice-president must be men and the secretary must be a woman?

15. How many triangles are there in the figure shown?

16. How many distinct arrangements are there of the letters in the word *math*? In the word *magnet*?

17. On page 38, we made the statement that 9^{9^9} bacteria would overflow the Milky Way, and the number of grains of sand on this and every other planet in our solar system would not be so large as this giant. How could we arrive at such a statement? We certainly could not obtain the result by direct counting. Give a convincing argument to support this statement.

18. The old lady jumping rope in the cartoon at the beginning of this section is counting one at a time. Assume that she jumps the rope 50 times per minute and that she jumps 8 hours a day, 5 days a week, 50 weeks a year. Estimate the length of time necessary for her to jump the rope the number of times indicated in the cartoon.

19. A die can be held so that one, two, or three of its faces can be seen at any one time. Is it possible to hold the die in different ways so that at different times the visible numbers add up to every number from 1 through 15?

An example of reaching a conclusion about the number of objects in a set is given by M. Cohen and E. Nagel in An Introduction to Logic *(Routledge & Kegan Paul Ltd., 1963).* They conclude that there are at least two people in New York City who have the same number of hairs on their heads. This conclusion was reached not through counting the hairs on the heads of 8 million inhabitants of the city, but through studies revealing that: (1) the maximum number of hairs on the human scalp could never be as many as 5000 per square centimeter, and (2) the maximum area of the human scalp could never reach 1000 centimeters. We can now conclude that no human head could ever contain $5000 \times 1000 = 5,000,000$ hairs. Since this number is less than the population of New York City, it follows that at least two New Yorkers must have the same number of hairs on their heads!

In the same vein, suppose you multiply the number of hairs on your head by the number of hairs on your neighbor's head. Continue this process for everyone in your town (multiplying the number of hairs on everyone's head). Next, multiply by the number of hairs on every person in the entire world. The exact *number of this product is* known by the author. Do you know what it is?

"I did it without counting. There are ---beans in the jar."

20. Outline a procedure that would allow the fellow in the cartoon to proclaim that he "did it without counting."

21. *Computer Problem*. The BASIC program that will print out the possibilities for Problem 5 is as follows:

```
5   PRINT
6   PRINT
10  PRINT "SUPPOSE THE PEOPLE IN THE GROUP ARE NUMBERED";
12  PRINT " FROM 1 TO 15."
14  PRINT
16  PRINT
20  PRINT "PRESIDENT      VICE-PRESIDENT      SECRETARY"
22  PRINT
30  FOR A = 1 TO 15
40  FOR B = 1 TO 15
50  FOR C = 1 TO 15
60  IF B = A GOTO 140
62  IF C = A GOTO 140
64  IF C = B GOTO 140
70  PRINT "      ";A;
80  PRINT "              ";B;
110 PRINT "                   ";C
140 NEXT C
150 NEXT B
160 NEXT A
170 END
```

This program will take a relatively long time to run. Modify it so that the computer will *count* the number of possibilities without actually printing them all out.

22. *Computer Problem*. Write a BASIC program showing all the arrangements for the word *math* in Problem 16.

23. *Instant Insanity*. There is a puzzle for sale in toy stores called Instant Insanity. It provides four cubes colored red, white, blue, and green. The puzzle is to assemble them into a 1 × 1 × 4 block so that all four colors appear on each side of the block.

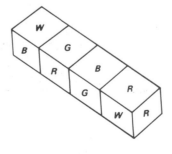

Instant Insanity is not a new puzzle; it has appeared under several patents, each in a different form. The present one is a trademark of Parker Brothers, Inc. Other versions involve numbers, dots, groceries, or card suits on the faces. All these forms are mathematically equivalent.

Estimate the number of possible arrangements for this puzzle. Hint: Use the Fundamental Counting Principle. (The answer is not 1296, since a single die can be arranged in more than six ways.)

Mind Bogglers

24. Consider the contest from Problem 20. Estimate the number of beans without actually counting them. Assume that the container is a cylinder 15 in. tall and 10 in. across and that it is filled with kidney beans.

25. *A Numerical Solution to Instant Insanity.* Let's associate numbers with the sides of the cubes of the Instant Insanity problem. Let

$$white = 1,$$
$$blue \ = 2,$$
$$green = 3,$$
$$red \ \ = 5.$$

We do not use 4, since 2 and 4 are not relatively prime.

Now, the product across the top must be 30, and the product across the bottom must also be 30 (why?). It follows that a solution must have a product of 900 for the faces on the top and bottom. Consider the four cubes:

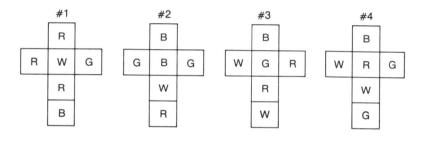

In Problem 23, we estimated the number of possible arrangements for the Instant Insanity puzzle, but we did not tell how to come up with a solution to the puzzle. If you are interested, this problem will tell you how to arrive at a solution.

These cubes do not show the solution.

Here are the possible products of top and bottom for cube #1:

			product
top:	red	= 5	25
bottom:	red	= 5	
top:	blue	= 2	2
bottom:	white	= 1	
top:	red	= 5	15
bottom:	green	= 3	

By a clever and systematic analysis of the products of top and bottom for cubes 2, 3, and 4, you will find that there are several ways of solving the top and bottom for a product of 900. Only one of these gives a product of 900 for front and back. Find the solution by using this method.

26. *Hexahexaflexagons.* Prepare a strip of paper as shown in Figure 8.2a.

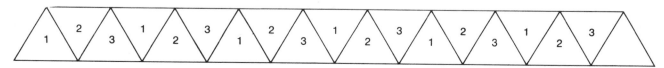

FIGURE 8.2a

Turn it over and mark the other side as shown in Figure 8.2b.

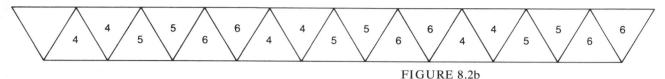

FIGURE 8.2b

It is important that all the triangles be equilateral.

Starting from the left of Figure 8.2b, fold the 4 onto the 4, the 5 onto the 5, 6 onto 6, 4 onto 4, and so on until your paper looks like the one shown in Figure 8.3.

𝕳istorical 𝕹ote

In 1939 Arthur Stone, a 23-year-old graduate student from England, trimmed an inch from his American notebook sheets to make them fit into his English binder. For lack of something else to do, he began to fold the trimmed-off strips of paper in various ways. By rearranging them he came up with several interesting flexagons.

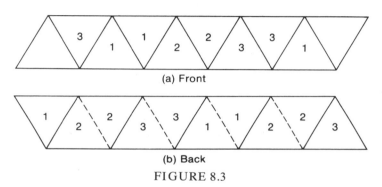

FIGURE 8.3

Continue by folding the 1 onto the 1 from the front, by folding the 1 onto the 1 from the back, and finally by bringing the 1 up from the bottom so that it rests on top of the 1 on the top. Your paper should now look like the one shown in Figure 8.4.

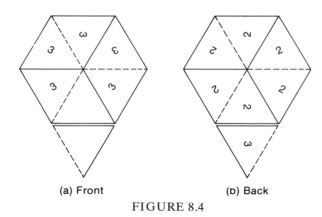

FIGURE 8.4

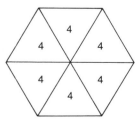

FIGURE 8.5 Folded hexahexaflexagon

Paste the blank onto the blank, and the result is called a *hexahexaflexagon*, as shown in Figure 8.5.

With a little practice you'll be able to "flex" your hexahexaflexagon so that you can obtain a side with all 1s, all 2s, . . . , all 6s. After you have become fairly proficient at "flexing," count the number of flexes you require to obtain all six "sides." What do you think is the fewest flexes necessary to obtain all six sides?

Problem for Individual Study

27. *Paper Folding.* How can all the constructions of Euclidean geometry be done by folding paper? What assumptions are made when paper folding is used to construct geometric figures? What polygons and polyhedrons can be formed by folding paper? What is a hexaflexagon? How can conic sections be formed by folding paper? What puzzles and tricks are based on paper folding?

Exhibit suggestions: models, geometric constructions, polyhedrons, conic sections, puzzles formed by paper folding, the five regular solids.
References: Gardner, Martin, *The Scientific American Book of Mathematical Puzzles and Diversions,* Chap.1 (New York: Simon and Schuster, 1959).
Johnson, Donovan, *Paper Folding for the Mathematics Class* (Washington, D.C.: National Council of Teachers of Mathematics, 1957).
Row, T. Saundara, *Geometrical Exercises in Paper Folding* (rev. ed.) (New York: Dover Publications, 1966).

8.2 PERMUTATIONS

Consider the election problem of the previous section. We wish to elect a president, secretary, and treasurer from among

{Frank, George, Happy, Iggy, Jerry, Ken}

We list some possibilities:

	President	Secretary	Treasurer
1.	Frank	George	Happy
2.	George	Frank	Happy
3.	George	Frank	Iggy

Certainly we see that direct listing would be too tedious. We seek an easier way. Is it possible to count the number of arrangements without actually counting them? We will look for some patterns in which the arrangement of the members of the set is important. That is, we will consider arrangements (1) and (2) as different; thus, the *order* in which we

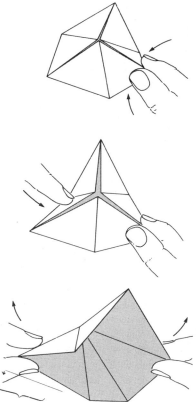

To "flex" your hexahexaflexagon, pinch together two of its triangles (top two figures). The inner edge may then be opened with the other hand (bottom). If the hexahexaflexagon cannot be opened, the adjacent pair of triangles is pinched. If it opens, turn it inside out, finding a side that was not visible before.

Definition of permutation

list the officers is important. Mathematicians call arrangements in which the order is important *permutations*.

We wish to choose a president, secretary, and treasurer. We say that this is a permutation of six objects (members of the club) taken three at a time (offices to be filled). We denote this by:

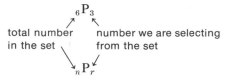

Notation for permutations

In general, $_nP_r$ is the number of permutations in which a total of n objects are under consideration and r objects are to be selected from the set. The problem, then, is to compute $_6P_3$ using the Fundamental Counting Principle.

President	Secretary	Treasurer
6	5	4

↑ number of ways of selecting a president

↑ number of ways of selecting a secretary

↑ number of ways of selecting a treasurer

Thus, $_6P_3 = 6 \cdot 5 \cdot 4 = 120$.

Let's look for some patterns. Suppose the set had seven members. Then how many selections could we make? We see that we have

$$_7P_3 = \underbrace{7 \cdot 6 \cdot \overset{\text{one more than } 7-3}{5}}_{\text{three factors}}$$

possibilities. What if the set contained ten members and we wished to elect four officers: a president, vice-president, secretary, and treasurer? Then we would have:

$$_{10}P_4 = \underbrace{10 \cdot 9 \cdot 8 \cdot \overset{\text{one more than } 10-4}{7}}_{\text{four factors}}$$

Do you see a pattern? In general, we say that the number of permutations of n objects taken r at a time is given by:

$$_nP_r = \underbrace{n \cdot (n-1) \cdot (n-2) \cdots \overset{\text{one more than } n-r}{(n-r+1)}}_{r \text{ factors}}$$

EXAMPLES: Evaluate the following.

1. $_5P_3 = 5 \cdot 4 \cdot 3 = 60$
2. $_6P_2 = 6 \cdot 5 = 30$

3. $_{10}P_3 = 10 \cdot 9 \cdot 8 = 720$
4. $_{10}P_2 = 10 \cdot 9 = 90$
5. $_{10}P_1 = 10$
6. $_{52}P_2 = 52 \cdot 51 = 2652$
7. Can r be larger than n?
   No, since n is the total number of objects.
8. Can $r = 0$?
   You must be careful. This example doesn't exactly fit our definition, so we must interpret it separately. Now, $_6P_0$ means the number of ways you can permute 6 objects by taking none of them. There is only one way to take no objects, and that is by not taking any. Thus, we define

$$_nP_0 = 1$$

Definition of $_nP_0$

9. Can $r = n$?
   Yes. Consider: $_5P_5 = 5 \cdot 4 \cdot 3 \cdot 2 \cdot 1 = 120$
   $_6P_6 = 6 \cdot 5 \cdot 4 \cdot 3 \cdot 2 \cdot 1 = 720$

In our work with permutations, we will frequently encounter products such as

$$6 \cdot 5 \cdot 4 \cdot 3 \cdot 2 \cdot 1$$
$$10 \cdot 9 \cdot 8 \cdot 7 \cdot 6 \cdot 5 \cdot 4 \cdot 3 \cdot 2 \cdot 1$$
$$52 \cdot 51 \cdot 50 \cdot 49 \cdots 4 \cdot 3 \cdot 2 \cdot 1$$

Since these are rather lengthy things to write down, we use *factorial notation*.

$$1! = 1$$
$$2! = 2 \cdot 1 = 2$$
$$3! = 3 \cdot 2 \cdot 1 = 6$$
$$4! = 4 \cdot 3 \cdot 2 \cdot 1 = 24$$
$$5! = 5 \cdot 4 \cdot 3 \cdot 2 \cdot 1 = 120$$
$$6! = 6 \cdot 5 \cdot 4 \cdot 3 \cdot 2 \cdot 1 = 720$$
$$7! = 7 \cdot 6 \cdot 5 \cdot 4 \cdot 3 \cdot 2 \cdot 1 = 5040$$
$$8! = 8 \cdot 7 \cdot 6 \cdot 5 \cdot 4 \cdot 3 \cdot 2 \cdot 1 = 40{,}320$$
$$9! = 9 \cdot 8 \cdot 7 \cdot 6 \cdot 5 \cdot 4 \cdot 3 \cdot 2 \cdot 1 = 362{,}880$$
$$10! = 10 \cdot 9 \cdot 8 \cdot 7 \cdot 6 \cdot 5 \cdot 4 \cdot 3 \cdot 2 \cdot 1 = 3{,}628{,}800$$

Historical Note

Factorial notation was first used by Christian Kramp in 1808.

Also, strange as it may seem, we define $0! = 1$. Notice that these numbers get big pretty fast. We need not wonder why the notation ! was chosen. For example,

$$52! \approx 8.065817512 \times 10^{67}$$

DEFINITION: For any counting number n,

$$n! = n(n - 1)(n - 2) \cdots 3 \cdot 2 \cdot 1$$

Also, $0! = 1$.

Definition of factorial

EXAMPLES:

1. find 5! − 4!

Solution: 5! = 5 · 4 · 3 · 2 · 1 = 120
4! = 4 · 3 · 2 · 1 = 24
Thus, 5! − 4! = 120 − 24 = 96.

2. Find (5 − 4)!.

Solution: Since 5 − 4 = 1, we have 1! = 1. Remember to do the parentheses first. Notice the difference between examples 1 and 2.

3. Find $\dfrac{7!}{5!}$

Solution: 7! = 7 · 6 · 5 · 4 · 3 · 2 · 1 = 5040

5! = 5 · 4 · 3 · 2 · 1 = 120

Thus, $\dfrac{7!}{5!} = \dfrac{5040}{120} = 42.$

A simpler solution arises by not multiplying the numbers out before dividing. That is, we could write:

$$\frac{7!}{5!} = \frac{7 \cdot 6 \cdot \cancel{5} \cdot \cancel{4} \cdot \cancel{3} \cdot \cancel{2} \cdot \cancel{1}}{\cancel{5} \cdot \cancel{4} \cdot \cancel{3} \cdot \cancel{2} \cdot \cancel{1}} = 7 \cdot 6 = 42$$

4. Find $\dfrac{10!}{7!}$.

Solution: $\dfrac{10!}{7!} = \dfrac{10 \cdot 9 \cdot 8 \cdot \cancel{7!}}{\cancel{7!}} = 10 \cdot 9 \cdot 8 = 720$

Notice that we could also write

$$\frac{10!}{7!} = 10 \cdot 9 \cdot 8$$

directly by noticing the pattern that 10!/7! means count down (multiplying all the while) from 10 to 7 (but don't include 7).

5. Find 20!/18!.

Solution: 20!/18! = 20 · 19 = 380

Using factorials, we can write some of our permutations a little more simply. For example, $_6P_6 = 6!$ and, in general, $_nP_n = n!$

By looking for patterns, we can define $_nP_r$ in terms of factorial notation. Notice that

We multiplied by 1 = $\dfrac{4!}{4!}$.

$$_7P_3 = 7 \cdot 6 \cdot 5 = \frac{7 \cdot 6 \cdot 5 \cdot 4!}{4!}$$

Thus,

$$_7P_3 = \frac{7!}{4!}$$

Also consider

$$_6P_3 = 6 \cdot 5 \cdot 4 = \frac{6!}{3!}$$

$$_{10}P_4 = 10 \cdot 9 \cdot 8 \cdot 7 = \frac{10!}{6!}$$

By studying this pattern, we see that it might be desirable to write these problems in terms of n and r.

$$_7P_3 = \frac{7!}{4!} = \frac{7!}{(7-3)!}$$

$$_6P_3 = \frac{6!}{(6-3)!}$$

$$_{10}P_4 = \frac{10!}{(10-4)!}$$

The general formula for this pattern is given below.

PERMUTATION FORMULA:

$$_nP_r = \frac{n!}{(n-r)!}$$

Notice that this formula works even for

$$_nP_0 = \frac{n!}{(n-0)!} = \frac{n!}{n!} = 1$$

EXAMPLES:

1. Find the number of license plates possible in a state using only three letters if none of the letters can be repeated. This is a permutation of 26 objects taken 3 at a time. Thus the solution is given by

$$_{26}P_3 = \underbrace{26 \cdot 25 \cdot 24}_{3 \text{ factors}} = 15,600$$

2. Find the number of ways a baseball coach can arrange the batting order. This is a permutation of 9 objects taken 9 at a time. Thus,

$$_9P_9 = 9! = 362,880$$

3. In Problem 16 of the previous problem set, we asked for the number of arrangements of the letters of the word *math*. The answer was found by using the Fundamental Counting Principle:

$$4 \cdot 3 \cdot 2 \cdot 1 = 24$$

We could also say that this is a permutation of 4 objects taken 4 at a time:

$$_4P_4 = 4! = 24$$

4. How many three-letter words can be formed by using the letters from the word *math*?

Solution: This is a permutation of 4 objects taken 3 at a time:

$$_4P_3 = 4 \cdot 3 \cdot 2 = 24$$

5. How many permutations are there of the letters in the word *soon*?

Solution: If we try to solve this problem as we did in Example 3, we would have

$$_4P_4 = 4! = 24$$

soon
sono
snoo
oson
osno
oosn
oons
onso
onos
noos
noso
nsoo

However, if we list the possibilities, we find only 12 different permutations. The difficulty here is that two of the letters in the word *soon* are *indistinguishable*. If we label them as so_1o_2n, we would find additional possibilities, such as

$$so_1o_2n$$
$$so_2o_1n$$
$$so_2no_1$$
$$\vdots$$

If we completed this list, we would find $4! = 24$ possibilities. This means that, since there are *two* indistinguishable letters, we divide the total, 4!, by 2 to find the result:

$$\frac{4!}{2} = \frac{24}{2} = 12$$

6. How many permutations are there of the letters in the word *assist*?

Solution: There are 6 letters, and, if we consider the letters as distinguishable, as in

$$as_1s_2is_3t$$

there are $_6P_6 = 6! = 720$ possibilities. However,

$$as_1s_2is_3t$$
$$as_1s_3is_2t$$
$$as_2s_1is_3t$$
$$as_2s_3is_1t$$
$$as_3s_1is_2t$$
$$as_3s_2is_1t$$

are all indistinguishable, so we must divide the total by $3! = 6$:

$$\frac{6!}{3!} = \frac{6 \cdot 5 \cdot 4 \cdot 3!}{3!} = 120$$

We see that there are 120 permutations of the letters in the word *assist*.

Examples 5 and 6 suggest a general result.

The number of permutations of n objects, of which a are alike, another b are alike, and another c are alike, is

$$\frac{n!}{a!b!c!}$$

This result extends to any number of objects that are alike.

PROBLEM SET 8.2

A Problems

Compute the results in Problems 1–20.

EXAMPLE: $(8 - 3)!$

Solution: $(8 - 3)! = 5!$
$= 120$

1. $7! - 5!$ 2. $(7 - 5)!$ 3. $\dfrac{10!}{8!}$ 4. $\dfrac{12!}{9!}$

5. $6! - 5!$ 6. $(10 - 4)!$ 7. $\dfrac{10!}{6!}$ 8. $\dfrac{10!}{4!6!}$

9. $\dfrac{52!}{3!(52 - 3)!}$ 10. $\dfrac{12!}{3!(12 - 3)!}$ 11. $6!$ 12. $6 \cdot 5!$

13. $\dfrac{8!}{4!}$ 14. $\left(\dfrac{8}{4}\right)!$ 15. $8! - 4!$ 16. $(8 - 4)!$

17. $\dfrac{8!}{5!3!}$ 18. $\dfrac{10!}{5!}$ 19. $\dfrac{9!}{6!3!}$ 20. $\dfrac{52!}{5!47!}$

Evaluate each of the numbers in Problems 21–40.

EXAMPLE: $_8P_3$

Solution: $_8P_3 = 8 \cdot 7 \cdot 6 = 336$

21. $_9P_1$ 22. $_9P_2$ 23. $_9P_3$ 24. $_9P_4$ 25. $_9P_0$

26. $_5P_4$ 27. $_{52}P_3$ 28. $_7P_2$ 29. $_4P_4$ 30. $_{100}P_1$

31. $_{12}P_5$ 32. $_5P_3$ 33. $_8P_4$ 34. $_8P_0$ 35. $_gP_h$

36. $_{92}P_0$ 37. $_{52}P_1$ 38. $_7P_5$ 39. $_{16}P_3$ 40. $_nP_4$

41. How many permutations are there of the letters of the word *holiday*?

42. How many permutations are there of the letters of the word *annex*?

43. How many permutations are there of the letters of the word *eschew*?

44. How many permutations are there of the letters of the word *obfuscation*?

45. In how many ways can a dot and a dash be arranged into groups of three? For example, $\cdot\,\cdot\,\cdot$, $\cdot\,\cdot\,-$, $\cdot\,-\,\cdot$, and so on.

46. A traveler must commute from Meridian to Centerville every day. There are four routes that he might take, as shown in Figure 8.6. This traveler would like to vary his trips as much as possible and always return on a different road. In how many different ways could he make the round trip?

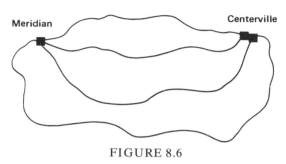

FIGURE 8.6

47. Write out a general formula for $_nP_r$ using only the numbers n, r, and the factorial. Show how you arrived at the answer.

B Problems

48. Give two original examples of arrangements of objects from everyday life that are permutations.

49. A certain mathematics test consists of ten questions. Peppermint Patty wishes to answer the questions without reading them. In how many ways can she fill in the answer sheet if:

   a. the possible answers are true and false?
   b. the possible answers are true, false, and maybe?

50. *Computer Problem.* Review the FOR command in BASIC. Write a BASIC program to compute $x!$.

51. *Computer Problem.* A BASIC program for finding permutations is as follows:

```
5    PRINT
6    PRINT
10   LET A = 1
20   PRINT "I WILL COMPUTE N/P/R = N!/(N − R)!"
25   PRINT "WHAT IS THE VALUE OF N";
30   INPUT N
35   PRINT "WHAT IS THE VALUE OF R";
40   INPUT R
45   IF R = 0 GOTO 58
47   FOR I = N − R + 1 TO N
```

```
50  LET A = A*I
55  NEXT I
57  GOTO 60
58  LET A = 1
60  PRINT "N/P/R = ";A
80  END
```

Use this program to compute:

a. $_7P_2$ b. $_{10}P_5$ c. $_{20}P_{10}$ d. $_{25}P_{25}$ e. $_{30}P_{20}$

52. *Computer Problem*. We know that, for $_nP_r$, n must be greater than or equal to r. Can you write this into the program for computing permutations? That is, can you add to the BASIC program given in Problem 51 so that the computer will tell you if $n < r$? A sample output for such a change is shown.

```
RUN

I WILL COMPUTE N/P/R=N!/(N-R)!
WHAT IS THE VALUE OF N? 3
WHAT IS THE VALUE OF R? 5
SHAME ON YOU. R CANNOT BE LARGER THAN N.
BUT I WILL GIVE YOU ANOTHER CHANCE. NOW, WHAT IS THE VALUE OF N? 5
WHAT IS THE VALUE OF R? 3
N/P/R= 60
```

Mind Bogglers

53. *Radar Puzzle*. Start on any one of the squares marked R, and, by consecutive moves to adjacent squares, spell out the word *RADAR*. In how many ways can this be done?

```
R  A  D  A  R
A  D  A  R  A
D  A  R  A  D
A  R  A  D  A
R  A  D  A  R
```

54. *Bridge Tournament*. Eight couples played bridge seven times. The people changed groupings every time they played, and, as it turned out, no man was ever paired with the same woman more than once as a partner or as an opponent. How were the couples distributed if no husband and wife were ever partners?

Problem for Individual Study

55. *Card-Shuffling Problem*. Have you ever watched someone shuffle a deck of cards and wondered how well the deck was being mixed? Suppose we define a *perfect* riffle shuffle of a deck with an even number of cards.

   (1) Divide the deck into two piles of equal size, putting the top half in your left hand and the bottom half in your right hand.
   (2) Alternate the cards from the two stacks by selecting the bottom card in

Hint for Problem 53:

```
      b
    o   o
      b
```

has $2 \times 1 = 2! \times 1! = 2$ ways of spelling bob.

```
        l
      e   e
    v   v   v
      e   e
        l
```

has $2 \times 3 \times 2 \times 1 = 3! \times 2! = 12$ ways of spelling level.

```
        r
      a   a
    t   t   t
  t   t   t   t
    l   l   l
      e   e
        r
```

has $2 \times 3 \times 4 \times 3 \times 2 = 4! \times 3!$ $= 144$ ways of spelling rattler.

the right pile, then the bottom card in the left pile, and so on. When you finish this, the first and last cards will not have changed positions (see examples below).

If we repeatedly perform a perfect shuffle, how many shuffles are necessary to return a deck of 52 cards to its original position? (*Hint:* Start with simpler cases and look for a pattern.)

EXAMPLE:

Two cards (numbered 1 and 2)
start: 1 2
shuffle: 1 2
Original position after *one* shuffle
Four cards (numbered 1 to 4)
start: 1 2 3 4
shuffle: 1 3 2 4
shuffle: 1 2 3 4
Original position after *two* shuffles
⋮

Ten cards (numbered 1 to 10)
start: 1 2 3 4 5 6 7 8 9 10
shuffle: 1 6 2 7 3 8 4 9 5 10
shuffle: 1 8 6 4 2 9 7 5 3 10
shuffle: 1 9 8 7 6 5 4 3 2 10
shuffle: 1 5 9 4 8 3 7 2 6 10
shuffle: 1 3 5 7 9 2 4 6 8 10
shuffle: 1 2 3 4 5 6 7 8 9 10
Original position after *six* shuffles

By "*original position*" *we mean that* all 52 cards *are in* exactly *the same order as they were to begin with.*

Election   Committee

{a,b}—①————————{a,b}
{b,a}—②

{a,c}—①————————{a,c}
{c,a}—②

{a,d}—①————————{a,d}
{d,a}—②

{a,e}—①————————{a,e}
{e,a}—②

{b,c}—①————————{b,c}
{c,b}—②

{b,d}—①————————{b,d}
{d,b}—②

{b,e}—①————————{b,e}
{e,b}—②

{c,d}—①————————{c,d}
{d,c}—②

{c,e}—①————————{c,e}
{e,c}—②

{d,e}—①————————{d,e}
{e,d}—②

20 possibilities   10 possibilities

FIGURE 8.7 Comparison of the election and committee problems (two members)

8.3 COMBINATIONS

Let's compare the committee and election problems of Section 8.1. We saw in the preceding section, while considering the election problem, that the following arrangements were *different*:

	President	Secretary	Treasurer
1.	Alfie	Bogie	Calvin
2.	Bogie	Alfie	Calvin

That is, the order was important. However, in the committee problem, the following three-member committees are the *same*:

1.	Alfie	Bogie	Calvin
2.	Bogie	Alfie	Calvin

In this case, the order is *not important*. When we list an arrangement of objects in which the order they are listed in is not important, we call the arrangement a *combination*.

Consider the example in Figure 8.7 comparing the election and committee problems. When we have a set of 5 elements taken 2 at a time, we see that we have twice as many permutations as combinations. That is,

$$_5C_2 = \frac{_5P_2}{2}$$

Notice that we used the notation $_5C_2$ to mean the number of combinations of 5 objects taken 2 at a time.

Consider a set of 5 elements taken 3 at a time, as shown in Figure 8.8. We see that we have 6 times as many permutations as combinations. That is,

$$_5C_3 = \frac{_5P_3}{6}$$

Finally, if we consider a set of 5 elements taken 4 at a time, we see that there are 24 times as many permutations as combinations, as shown in Figure 8.9.

Consider the following pattern.

$$_5C_2 = \frac{_5P_2}{2}$$

$$_5C_3 = \frac{_5P_3}{6}$$

$$_5C_4 = \frac{_5P_4}{24}$$

Do you see the pattern? It is rather difficult to see, but at least you can notice that

$$_nC_r = \frac{_nP_r}{\text{some number}}$$

What should this number be? The pattern 2, 6, 24 above suggests that it is $r!$. Using this pattern, the general rule can be stated.

COMBINATION FORMULA:

$$_nC_r = \frac{_nP_r}{r!}$$

That is, to find $_nC_r$, we proceed as follows:

1. Write $_nP_r$ in factored form.
2. Divide by $r!$.
3. Simplify the result.

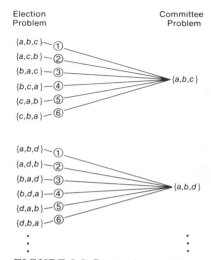

FIGURE 8.8 Comparison of the election and committee problems (three members)

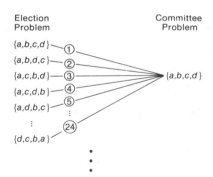

FIGURE 8.9 Comparison of the election and committee problems (four members)

EXAMPLES:

1. $_5C_3 = \dfrac{_5P_3}{3!} = \dfrac{5 \cdot 4 \cdot 3}{3 \cdot 2 \cdot 1} = 10$

2. $_6C_2 = \dfrac{_6P_2}{2!} = \dfrac{6 \cdot 5}{2} = 15$

3. $_{10}C_3 = \dfrac{_{10}P_3}{3!} = \dfrac{10 \cdot 9 \cdot 8}{3 \cdot 2 \cdot 1} = 120$

4. Find the number of ways different five-card hands can be drawn from an ordinary deck of cards.

$$_{52}C_5 = \dfrac{_{52}P_5}{5!} = \dfrac{52 \cdot 51 \cdot 50 \cdot 49 \cdot 48}{5 \cdot 4 \cdot 3 \cdot 2 \cdot 1} = 2{,}598{,}960$$

5. In how many ways can a diamond flush be drawn in poker? (A diamond flush is a hand of five diamonds.) This is a combination of 13 objects (diamonds) taken 5 at a time. Thus the solution is given by

$$_{13}C_5 = \dfrac{_{13}P_5}{5!} = \dfrac{13 \cdot 12 \cdot 11 \cdot 10 \cdot 9}{5 \cdot 4 \cdot 3 \cdot 2 \cdot 1} = 1287$$

PROBLEM SET 8.3

A Problems

Evaluate each of the expressions in Problems 1–20.

1. $_9C_1$
2. $_9C_2$
3. $_9C_3$
4. $_9C_4$
5. $_9C_0$
6. $_5C_4$ $=5$
7. $_{52}C_3$
8. $_7C_2$
9. $_4C_4$
10. $_{100}C_1$
11. $_7C_3$
12. $_5C_5$
13. $_{50}C_{48}$
14. $_{25}C_1$
15. $_9C_h$
16. $_7C_5$
17. $_8C_0$
18. $_{10}C_2$
19. $_{12}C_5$
20. $_nC_4$

21. A bag contains 12 pieces of candy. In how many ways can 5 pieces be selected?

22. If the Senate is to form a new committee of 5 members, in how many different ways can the committee be chosen if all 100 Senators are available to serve on this committee?

23. a. In how many ways can three aces be drawn from a deck of cards?
    b. In how many ways can two kings be drawn from a deck of cards?

B Problems

24. In how many ways can a heart flush be obtained? (A heart flush is a hand of five hearts.)

25. In how many ways can a full house of three aces and two kings be obtained?

Hint for Problem 25: Don't forget the Fundamental Counting Principle. Use it together with the results of Problem 24.

26. In how many ways can a flush be obtained? (Hint: See Problem 24.)

27. Recall that $_nP_r$ can be written as $_nP_r = n!/(n - r)!$. Find a formula for $_nC_r$ in terms of n and r only (along with the factorial symbol).

28. Compute:

$$_0C_0$$
$$_1C_0 \qquad _1C_1$$
$$_2C_0 \qquad _2C_1 \qquad _2C_2$$
$$_3C_0 \qquad _3C_1 \qquad _3C_2 \qquad _3C_3$$

See Figure 8.10 for Problems 28–30.

Do you see a relationship between combinations and Pascal's triangle, which was first introduced in Section 1.2?

29. Use Pascal's triangle to evaluate the following:

a. $_7C_3$ b. $_{11}C_9$ c. $_{13}C_6$

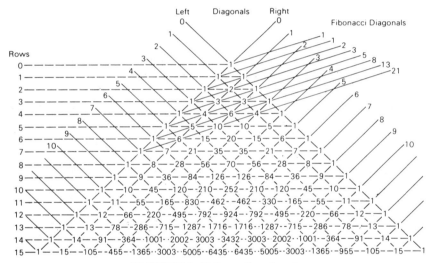

FIGURE 8.10 Pascal's triangle

30. *Computer Problem.* A BASIC program for computing Pascal's triangle is given. Modify this program so that the computer will accept only permissible values for the number of rows. That is, it doesn't make sense to say "Find the first 6.6 rows" or "Find the first ⁻7 rows." You might wish to use this program to help you answer some of the following problems.

```
 5  PRINT
 6  PRINT
10  PRINT "THIS PROGRAM WILL GENERATE THE N LINES OF PASCAL'S";
12  PRINT " TRIANGLE."
13  PRINT "WHAT IS N";
15  INPUT N
20  LET D = 2
```

```
22   LET P(1) = 1
24   LET P(2) = 1
26   PRINT
28   PRINT "PASCAL'S TRIANGLE FOR N = ";N;" IS:"
29   PRINT
30   PRINT " 1"
31   FOR I = 1 TO N
32   FOR J = 1 TO D
34   PRINT P(J);
35   NEXT J
36   LET D = D + 1
37   PRINT
38   FOR K = 2 TO D - 1
40   LET Q(K) = P(K) + P(K - 1)
41   NEXT K
42   LET P(D) = 1
44   FOR L = 2 TO D - 1
46   LET P(L) = Q(L)
47   NEXT L
50   NEXT I
100  END
```

31. Use Pascal's triangle to compute the following.

    a. In how many ways can we select a committee of 4 from a club with 10 members?
    b. In how many ways can we select a committee of 6 from a club with 8 members?
    c. In how many ways can we select a committee of 7 from a club with 11 members?

32. Examine Pascal's triangle carefully, look for patterns, and answer the following questions.

    a. What is the second number in the 100th row?
    b. What is the next-to-last number in the 200th row?
    c. Which rows contain only odd numbers?

33. Show that $_{n-1}C_{r-1} + {}_{n-1}C_r = {}_nC_r$.

34. *Computer Problem.* A BASIC program for finding combinations is given.

```
15   LET B = 1
20   PRINT "I WILL COMPUTE N/C/R = N!/R!(N - R)!"
25   PRINT "WHAT IS THE VALUE OF N";
30   INPUT N
35   PRINT "WHAT IS THE VALUE OF R";
40   INPUT R
42   IF R = 0 GOTO 85
45   FOR I = 1 TO R
48   LET D = N - I + 1
50   LET A = D/I
55   LET B = A*B
60   NEXT I
85   PRINT "N/C/R = ";B
90   END
```

Use this program to find:

a. $_7C_2$ b. $_{10}C_5$ c. $_{20}C_{10}$ d. $_{25}C_{25}$ e. $_{30}C_{20}$

35. *Computer Problem.* We know that, for $_nC_r$, n must be greater than or equal to r. Can you write this into the program for computing combinations? That is, can you add to the BASIC program given in Problem 34 so that the computer will tell you if $n < r$? A sample printout for such a change is shown.

```
RUN
I WILL COMPUTE N/C/R=N!/R!(N-R)!
WHAT IS THE VALUE OF N? -3
WHAT IS THE VALUE OF R? 5
THE VALUES FOR N AND R MUST BE POSITIVE INTEGERS.

WHAT IS THE VALUE OF N? 3
WHAT IS THE VALUE OF R? 5
SHAME ON YOU. R CANNOT BE LARGER THAN N. BUT I WILL GIVE
YOU ANOTHER CHANCE. NOW, WHAT IS THE VALUE OF N? 5
WHAT IS THE VALUE OF R? 3
N/C/R= 1Ø
```

Mind Bogglers

36. Prove that $_nC_r = {_nC_{n-r}}$.

37. *Calculator Problem.* Suppose the population of the United States is 220 million. In how many ways could we elect a President and a Vice-President? (Assume that everyone is eligible to be elected.) How many committees of two persons could we form?

Problems for Individual Study

38. How are the binomial coefficients related to Pascal's triangle?

39. Find a formula for the largest number in the nth row of Pascal's triangle.

40. *Continuation of the Card-Shuffling Problem.* If you tabulate the results of Problem 55 of Problem Set 8.2, you find: *See page 449*

Number of Cards	Number of Shuffles
4	2
6	4
8	3
10	6
12	10
14	12
16	4
⋮	⋮

The pattern is not so obvious as we had hoped. So let's approach the problem another way. Trace the position of the card we label "2." For 10 cards we begin:

$$1 \quad 2 \quad 3 \quad 4 \quad 5 \quad 6 \quad 7 \quad 8 \quad 9 \quad 10$$

The positions for the second card are:

$$2\text{-}3\text{-}5\text{-}9\text{-}8\text{-}6\text{-}2$$

which means that the second card moves to the third position after one shuffle, the fifth position after two shuffles, and so on. When it returns to the second position, all the cards will be in their original position, so we see that 10 cards require 6 shuffles. Find a pattern that predicts the position of the second card.

EXAMPLE: For 12 cards we would write: 2-3-5-9-6-11-10-8-4-7-2, which is 10 shuffles.

EXAMPLE: For 30 cards: 2-3-5-9-17-4-7-13-25-20-10-19-8-15-29-28-26-22-14-27-24-18-6-11-21-12-23-16-2, which is 28 shuffles.

8.4 COUNTING WITHOUT COUNTING

This chapter is concerned with methods of counting that are more efficient than the old "one, two, three, . . ." technique. There are many ways of counting. We saw that some counting problems can be solved by using the Fundamental Counting Principle, some by using permutations, and some by using combinations. Other problems required that we combine some of these ideas. However, it is important to keep in mind that not all counting problems fall into one of these neat categories.

Keep in mind, therefore, as we go through this section, that, although some problems may be permutation problems and some may be combination problems, there are many counting problems that are neither.

PERMUTATIONS AND COMBINATIONS

In practice, we are usually required to decide whether a given counting problem is a permutation or a combination before we can find a solution. For the sake of review, recall the difference between the election and committee problems of the previous sections. The election problem is a permutation problem, and the committee problem is a combination problem. We summarize the definitions of permutation and combination.

> DEFINITION: A *permutation* of a set of objects is an arrangement of certain of these objects in a *specified order*.
>
> DEFINITION: A *combination* of a set of objects is an arrangement of certain of these objects *without regard to their order*.

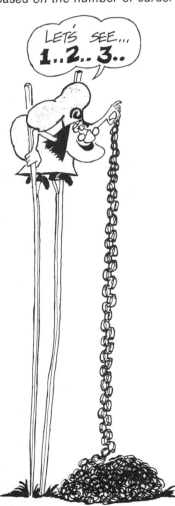

Hint: Try modular arithmetic based on the number of cards.

LET'S SEE...
1..2..3..

Remember: Permutation: order important.

Combination: order not important.

EXAMPLES: Classify the following as permutations or combinations.

1. The number of license plates possible in Florida.
2. The number of three-letter ''words'' that can be formed using the letters $\{m,a,t,h\}$.
3. The number of ways you can change a $1 bill with 5 nickels, 12 dimes, and 6 quarters.
4. The number of ways a five-card hand can be drawn from a deck of cards.
5. The number of different five-numeral combinations on a combination lock.

Solutions:

1. This is a permutation problem, since we count CWB072 and BCW072 as different plates. That is, the *order* in which the elements (in this case the letters) are arranged is important.
2. Permutation, since *mat* is different from *tam*. (In the examples that follow, any combination of letters is considered a ''word.'' Thus *atm* is also a ''word.'')
3. Combination, since ''2 quarters and 5 dimes'' is the same as ''5 dimes and 2 quarters.''
4. Combination, since the order in which you receive the cards is unimportant.
5. Permutation, since ''4 to the L, 6 to the R, 3 to the L, . . .'' is different from ''6 to the R, 4 to the L, 3 to the L, . . .'' We should not be misled by everyday usage of the word *combination*. We made a strict distinction between combination and permutation—one that is not made in everyday terminology. (The correct terminology would require that we call these ''permutation locks.'')

COUNTING BY COUNTING WHAT WE ARE NOT COUNTING

Sometimes it is easier to count what we are not counting than to enumerate all those items with which we are concerned. For example, suppose we wish to know the number of four-member committees that can be appointed from the club:

$$\{ \text{Alfie,Bogie,Calvin,Doug,Ernie} \}$$

A. We could count directly:

1. Alfie,Bogie,Calvin,Doug
2. Alfie,Bogie,Calvin,Ernie
3. Alfie,Bogie,Doug,Ernie
        .
        .
        .

B. We could use the formula:

$$_5C_4 = \frac{_5P_4}{4!} = \ldots$$

According to the 1974 Guinness Book of World Records, *the longest recorded paper-link chain was 6077 ft long and was made by the first- and second-grade children at Rose City Elementary School, Indiana.*

Combination lock or permutation lock?

C. We could count those we are not counting: For each four-member committee, there is one person left out. We can leave out Alfie, Bogie, Calvin, Doug, or Ernie. Therefore there are five four-member committees.

EXAMPLES:

1. Suppose you flip a coin three times and keep a record of the results. In how many ways can you obtain at least one head?

   Solution: Consider all possibilities:

$$
\begin{array}{ccc}
H & H & H \\
H & H & T \\
H & T & H \\
H & T & T \\
T & H & H \\
T & H & T \\
T & T & H \\
T & T & T \\
\end{array}
$$

We could count directly to obtain the answer, but we could also count what we are not counting by noticing that in only one out of eight possibilities do we obtain no heads. Thus there are $8 - 1 = 7$ possibilities in which we obtain at least one head.

2. How many California license plates have a repeating letter or numeral?

   Solution: The direct solution is *hard*. But it's easy to count how many don't:

$$26 \times 25 \times 24 \times 10 \times 10 \times 10 = 15,600,000.$$

From a previous example (on page 434), there are 17,576,000 plates possible, so the number of plates that have a repeating letter is $17,576,000 - 15,600,000 = 1,976,000$. This figure is almost $11\frac{1}{4}\%$ of the total number of license plates in the state!

3. The principle of counting without counting is particularly useful when the results become more complicated. For example, suppose we wish to know the number of ways of obtaining at least one diamond on drawing five cards from an ordinary deck of cards. This is very difficult if we proceed directly, but we can compute the number of ways of not drawing a diamond:

$$_{39}C_5 = \frac{39 \cdot 38 \cdot 37 \cdot 36 \cdot 35}{5 \cdot 4 \cdot 3 \cdot 2 \cdot 1} = 575,757$$

From Example 4, page 452, there is a total of 2,598,960 possibilities so the number of ways of drawing at least one diamond is $2,598,960 - 575,757 = 2,023,203$.

COUNTING BY COUNTING TOO MUCH

Suppose

$$U = \{1,2,3,4,5,6,7,8,9,10\}$$
$$A = \{2,4,6,8,10\}$$
$$B = \{1,3,5,7,9\}$$
$$C = \{1,2,3\}$$

If $|X|$ represents the number of elements in the set X, then $|U| = 10$, $|A| = 5$, $|B| = 5$, and $|C| = 3$.   Furthermore, we see that $|A \cap B| = 0$, $|A \cap C| = 1$, $|B \cap C| = 2$. We wish to know about the number of elements in the union of these sets.

First, we work the problem by counting:

$$A \cup B = \{1,2,3,4,5,6,7,8,9,10\}$$
$$A \cup C = \{1,2,3,4,6,8,10\}$$
$$B \cup C = \{1,2,3,5,7,9\}$$

Thus,

$$|A \cup B| = 10$$
$$|A \cup C| = 7$$
$$|B \cup C| = 6$$

If these sets were very large, the counting method would not be satisfactory. Can we work the problem without counting? Consider the general Venn diagram shown in Figure 8.11. Notice that, if we count X and then count Y, the intersection has been counted twice. Therefore we write:

$$|X \cup Y| = \quad \underbrace{|X| + |Y|} \quad - \quad \underbrace{|X \cap Y|}$$

The elements in the intersection are counted twice here.

This corrects for the "error" introduced by counting those elements in the intersection twice.

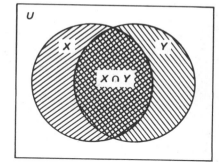

FIGURE 8.11 Venn diagram for the number of elements in the union of two sets

Thus,

$$|A \cup B| = |A| + |B| - |A \cap B|$$
$$= 5 + 5 - 0$$
$$= 10$$
$$|A \cup C| = |A| + |C| - |A \cap C|$$
$$= 5 + 3 - 1$$
$$= 7$$

$$|B \cup C| = |B| + |C| - |B \cap C|$$
$$= 5 + 3 - 2$$
$$= 6$$

EXAMPLE: What is the millionth positive integer that is not the square or cube of an integer?

Solution: We will solve this problem by counting what we are not counting; that is, we will count the squares and cubes and eliminate those from the rest of the numbers.

1̸,2,3,4̸,5,6,7,8̸,9̸,10,11,12, . . .

Squares	Cubes
$1 = 1^2$	$1 = 1^3$
$4 = 2^2$	$8 = 2^3$
$9 = 3^2$	$27 = 3^3$
$16 = 4^2$	$64 = 4^3$
$25 = 5^2$	$125 = 5^3$
$\vdots$	$\vdots$
$1,000,000 = (10^3)^2$	$1,000,000 = (10^2)^3$

If $S = \{squares\}$, then $|S| = 1000$, and if $C = \{cubes\}$, then $|C| = 100$. (Why?)
We find $S \cap C$ as follows:

$$1 = 1^6$$
$$64 = 2^6$$
$$729 = 3^6$$
$$\vdots$$
$$1,000,000 = 10^6$$

Thus, $|S \cap C| = 10$.
Now we can find $|S \cup C| = |S| + |C| - |S \cap C|$
$$= 1000 + 100 - 10$$
$$= 1090$$

Since the next square and cube ($1001^2 = 1,002,001$ and $101^3 = 1,030,301$, respectively) are both greater than 1,001,090, we see that there are no additional numbers to be excluded between 1,000,000 and 1,001,090. Thus the millionth number that is not a square or a cube is 1,001,090.

PROBLEM SET 8.4

A Problems

1. Explain the difference between a permutation and a combination.

2. Classify each of the following as permutations or combinations or neither.
   a. The number of arrangements of letters in the word *math*.

b. At Mr. Furry's Dance Studio, every man must dance the last dance. If there are five men and eight women, in how many ways can dance couples be formed for the last dance?

c. Martin's Ice Cream Store sells sundaes with chocolate, strawberry, butterscotch, or marshmallow toppings, nuts, and whipped cream. If you can choose exactly three of these, how many possible sundaes are there?

d. Five people are to dine together at a rectangular table, but the hostess cannot decide on a seating arrangement. In how many ways can the guests be seated?

e. In how many ways can three hearts be drawn from a deck of cards?

3. Classify each of the following as permutations or combinations or neither.

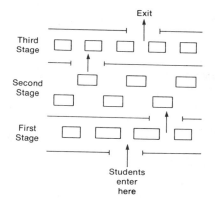

a. A shipment of 100 TV sets is received. Six sets are to be chosen at random and tested for defects. In how many ways can the six sets be chosen?

b. New students must register in person and be processed through three stages. At the first stage there are 4 tables (the student can go to any one of these), at the second stage there are 6 tables, and at the last stage there are 5 tables. In how many ways could the student be routed through the registration procedures?

c. How many subsets can be formed from a set of 100 elements?

d. A night watchman visits 15 offices every night. To prevent others from knowing when he will be at a particular office, he varies the order of his visits. In how many ways can this be done?

e. A certain manufacturing process calls for the mixing of six chemicals. One liquid is to be poured into the vat, and then the others are to be added in turn. All possible combinations must be tested to see which gives the best results. How many tests must be performed?

4. Classify each of the following as permutations or combinations or neither.

a. There are three boys and three girls at a party. In how many ways can they be seated if they can sit down four at a time?

b. If there are ten people in a club, in how many ways can they choose a dishwasher and a bouncer?

c. What is the number of three-digit numbers that can be formed using the numerals {3,5,6,7}?

d. In how many ways can you be dealt two cards from an ordinary deck of cards?

e. In how many ways can five taxi drivers be assigned to six cars?

5. Classify each of the following as permutations or combinations or neither.

a. How many arrangements are there of the letters in the word *gamble*?

b. A student is asked to answer 10 out of 12 questions on an exam. In how many ways can she select the questions to be answered?

c. In how many ways can a group of seven choose a committee of four?

d. In how many ways can seven books be arranged on a bookshelf?

e. In how many ways can we choose two books to read from a bookshelf containing seven books?

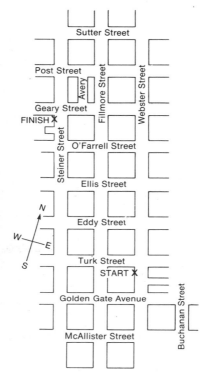

Sutter Street

Post Street

Avery

Fillmore Street

Webster Street

Geary Street

FINISH ✗

Steiner Street

O'Farrell Street

Ellis Street

N

W——E

S

Eddy Street

Turk Street

START ✗

Golden Gate Avenue

Buchanan Street

McAllister Street

Hint for Problem 12: Use Problem 9 and look for a pattern.

6. Use the notation $_nP_r$ or $_nC_r$ to write each of the parts of Problem 2, showing the appropriate choice for n and r. For example, Problem 2a is a permutation of 4 objects (letters, in this case) taken 4 at a time. Thus the appropriate notation is $_4P_4$.

7. If $U = \{1,2,3, \ldots ,100\}$, $A = \{\text{multiples of } 2\}$, and $B = \{\text{multiples of } 3\}$, find:
   a. $|A|$ b. $|B|$ c. $|A \cap B|$ d. $|A \cup B|$

8. A certain lock has five tumblers, and each tumbler can assume six positions. How many different possibilities are there?

9. Suppose you flip a coin and keep a record of the results. In how many ways could you obtain at least one head if you flip the coin:
   a. four times?   b. five times?   c. six times?

10. Tarot cards are used for telling fortunes, and, in a reading, the arrangement of the cards is as important as the cards themselves. How many different readings are possible if three cards are selected from a set of seven Tarot cards?

B Problems

11. In how many ways can a person in San Francisco walk from the corner of Webster and Turk Streets to the corner of Geary and Steiner Streets by traveling six blocks?

12. You flip a coin n times and keep a record of the results. In how many ways could you obtain at least one head?

13. Answer the questions posed in Problem 2.

14. Answer the questions posed in Problem 4.

15. Answer the questions posed in Problem 5.

16. a. In how many ways could a club of 15 members choose a president, vice-president, secretary, and treasurer?
    b. In how many ways could a committee of 4 be appointed by this club?

17. Using the variables n and r, discuss how $_nP_r$ and $_nC_r$ are related.

18. Compute each of the following (you may leave your answer in factored form):
    a. $_{52}P_5$ b. $_{52}C_5$ c. $_nP_n$ d. $_nC_n$ e. $_nC_2$

19. Assume that this class has 30 members.
    a. In how many ways could we select a president, vice-president, and secretary?
    b. How many committees of 3 persons can we choose?

Hint for Problem 20: Use the associative property to write $A \cup B \cup C$ as $(A \cup B) \cup C$. You will also need the distributive property: $(A \cup B) \cap C = (A \cap C) \cup (B \cap C)$.

Hint for Problem 21: Use the results of Problem 20.

20. In the text we found $|A \cup B| = |A| + |B| - |A \cap B|$. Show that $|A \cup B \cup C| = |A| + |B| + |C| - |A \cap B| - |A \cap C| - |B \cap C| + |A \cap B \cap C|$.

21. What is the millionth positive integer that is not a square, cube, or fifth power?

22. On January 24, 1971, The Santa Rosa *Press Democrat* began the contest shown in the clipping. If we assume that the contest was to be solved by chance (no skill involved), how many possibilities for matching the baby to the adult are there?

Mind Bogglers

23. Answer the problem posed in Peppermint Patty's textbook.

24. The advertisement reproduced here appeared in several national periodicals. It claims that the SEA graphic equalizer system can create 371,293 different sounds. Can this claim be substantiated from the advertisement and the knowledge of this chapter? If so, explain how; if not, tell why this number is impossible.

© 1974 United Feature Syndicate, Inc.

25. While the Dormouse lay sleeping in the middle of the table, the Mad Hatter and the March Hare sat down for tea. A number of places had been laid, and the Hatter and the Hare each moved alternately to a seat next to his old one. Each changed his direction of movement when, and only when, the other had just changed his or to avoid their both sitting on the same chair. They finished when both the Hatter and the Hare were again sitting in their initial positions. After how many moves was the tea party over?*

Problems for Individual Study

26. You have read about many mathematicians in the historical notes of this book. Write a news article about one of them as if you were a contemporary of the person you are writing about. You will need to do some additional research for this problem, but you can use one of the historical notes in this book as a starting point.

27. Prepare a list of black mathematicians from the history of mathematics.

28. Prepare a list of women mathematicians from the history of mathematics.

8.5 SUMMARY AND REVIEW

CHAPTER OUTLINE

I. **Fundamental Counting Principle.** If task A can be performed in m different ways, and, after task A is performed, a second task, B, can be performed in n different ways, then task A followed by task B can be performed in $m \cdot n$ different ways.

II. **Permutations: Order Important** (Election Problem)

$$_nP_r = \underbrace{n(n - 1)(n - 2) \cdots (n - r + 1)}_{r \text{ factors}}$$

III. **Combinations: Order Not Important** *(Committee Problem)*

$$_nC_r = \frac{_nP_r}{r!}$$

IV. Counting the Union of Two Sets: $|A \cup B| = |A| + |B| - |A \cap B|$.

*This problem was proposed by R. N. Lloyd in *The American Mathematical Monthly*, January 1970, and was solved by R. W. Sielaff in the November 1970 *Monthly*.

REVIEW PROBLEMS

Simplify each of the expressions in Problems 1–5.

1. $6! - 2!$ 2. $6 - 2!$ 3. $(6 - 2)!$ 4. $\left(\dfrac{6}{2}\right)!$ 5. $\dfrac{6!}{2!}$

Find the numerical value of each of the expressions in Problems 6–9.

6. $_5C_3$ 7. $_4P_2$ 8. $_2P_0$ 9. $_4C_2$

10. How many 3-member committees can be chosen from a group of 12 people?

11. How many permutations are there of the letters of the words *happy* and *college*?

12. Discuss and define permutations and combinations.

13. The Wednesday Luncheon Club, consisting of ten members, decided to celebrate its first anniversary by having lunch at a fancy restaurant. When the members arrived and were ready to take their seats, they could not decide where to sit.

    Just when they were ready to leave because of their embarrassment at being unable to decide how to sit around the rectangular table, the manager came to the rescue and told them to sit down just where they were standing. Then the secretary of the club was to write down where they were sitting, and they were to return the following day and sit in a different order. If they continued this until they had tried all arrangements, the manager would give them anything on the menu, free of charge.

    It sounded like a very good deal, so they agreed. However, the day of the free luncheon never came. Can you explain why?

14. An urn contains four red and six white balls. Three balls are drawn at random. In how many ways can at least one red ball be drawn?

If $U = \{1,2,3, \ldots ,49,50\}$, $A = \{odd\ numbers\}$, and $B = \{primes\}$, find the value requested in each of Problems 15–18.

15. $|A|$ 16. $|B|$ 17. $|A \cap B|$ 18. $|A \cup B|$

19. In how many ways can five people line up at a bank teller's window?

20. At Wendy's Old Fashioned Hamburgers, they advertise that you can have your hamburgers 256 ways. If they offer catsup, onion, mustard, pickles, lettuce, tomato, mayonnaise, and relish, is their claim correct? Explain why or why not.

CULTURAL

1753-1763: The Seven Year's War

1762: Rousseau's *Social Contract*
1767: Watt's steam engine

1776: Adam Smith's *Wealth of Nations*
1776: American Declaration of Independence

1789: French Revolution
1789: Washington elected President

Washington

1799: Metric system
1799: Napoleon rules France

1808: Beethoven's *Fifth Symphony*
1808: Goethe's *Faust*

1815: Battle of Waterloo

1821: Rosetta Stone deciphered

Napoleon

1830: Simon Bolivar liberates South America

1836: First telegraph

1843: Mendel's genetics research

1848: *Communist Manifesto* by Marx
1851: Melville's *Moby Dick*

1859: Darwin writes *Origin of the Species*
1861: U.S. Civil War
1862: Louis Pasteur's germ theory of infection

1867: Federation of Canada
1869: Tolstoy's *War and Peace*

1750
1775
1800
1825
1850

MATHEMATICAL

1750: Leonhard Euler— number theory, applied mathematics

1760: Compte de Buffon— connection between probability and π

1770: Johann Lambert— irrationality of π, non-Euclidean geometry, map projections

1780: Lagrange—calculus, number theory

Euler

Gauss— His work includes number theory, differential geometry, non-Euclidean geometry, and the Fundamental Theorem of Algebra

1805: Laplace—probability, differential equations, method of least squares, integrals

1820: Carl Gauss—considered one of the greatest mathematicians of all time
1822: Feuerbach—geometry of the triangle
1825: Bolyai and Lobachevski—non-Euclidean geometry
1830: Cauchy—calculus, complex variables

1837: Galois—groups, theory of equations
1837: Trisection of an angle and duplication of the cube proved impossible
1843: Hamilton—quaternions

1850: Cayley—invariants, hyperspace, matrices, and determinants
1854: Riemann—calculus
1854: Boole—logic, *Laws of Thought*
1855: Dirichlet—number theory

1872: Dedekind—irrational numbers
1873: Brocard—geometry of the triangle

9
THE NATURE OF PROBABILITY

There once was a breathy baboon
Who always breathed down a bassoon,
For he said, "It appears
That in billions of years
I shall certainly hit on a tune."

Sir Arthur Eddington

9.1 SOME EXPERIMENTS IN PROBABILITY

The nature of probability is best experienced by means of actual physical experiments. It is easy to say that the probability of obtaining a head on the toss of a coin is $1/2$, but what does that mean? We could define probability according to any scheme that is noncontradictory, but the definition would be "good" only insofar as it coincides with real-world events. In this section, we will perform experiments, make observations, and then form conjectures about the nature of probability. In the next section, we will formalize our conjectures and form mathematical models to fit the experiments.

EXPERIMENT 1: TOSSING A COIN

In this experiment, we flip a coin 50 times. Make sure that, each time the coin is flipped, it rotates several times in the air and lands on a table or on the floor. Keep a record similar to that in Table 9.1. The figures in each table of this section are the result of an actual experiment.

TABLE 9.1 Outcomes of Tossing a Single Coin

Number of Throw	Outcome	Number of Throw	Outcome
1	H	26	H
2	T	27	T
3	H	28	T
4	H	29	H
5	T	30	T
6	T	31	H
7	T	32	T
8	T	33	H
9	H	34	H
10	H	35	H
11	H	36	T
12	T	37	T
13	T	38	T
14	T	39	T
15	H	40	T
16	T	41	T
17	T	42	H
18	H	43	H
19	H	44	T
20	T	45	H
21	H	46	H
22	T	47	H
23	T	48	T
24	H	49	H
25	H	50	T

Totals: Heads, 24; Tails, 26

Repeat the experiment. The results are called *data*. We will study about the analysis of data in the next chapter. There are many ways to represent data, and you should choose the one most convenient for your purposes. For example, the findings in Table 9.1 could be recorded more conveniently by using tally marks, as in Table 9.2.

TABLE 9.2 Outcomes of Tossing a Single Coin—Tally Method

Outcome	Number of Occurrences	Total	Percentage
Heads	ＨＨＬ ＨＨＬ ＨＨＬ ＨＨＬ ‖‖	24	48%
Tails	ＨＨＬ ＨＨＬ ＨＨＬ ＨＨＬ ＨＨＬ ‖	26	52%

Can you formulate any conjectures concerning these data? Find the percentage of heads and tails for your data. For our example, we have:

$$\frac{24}{50} \text{ heads} \qquad \frac{26}{50} \text{ tails}$$

$$.48 \text{ heads} \qquad .52 \text{ tails}$$

$$48\% \text{ heads} \qquad 52\% \text{ tails}$$

From our experience, we see that this is as we would expect; that is, heads have about a "50–50" chance" of occurrence. If we write the percentage as a fraction, we say that the *probability* of heads occurring is about $1/2$. This means that, if we repeated the experiment a large number of times, we could expect heads to occur about $1/2$ the time.

EXPERIMENT 2: TOSSING THREE COINS

In this experiment, we flip three coins simultaneously 50 times. We are interested in the probability of obtaining 3 heads, 2 heads and 1 tail, 1 head and 2 tails, and 3 tails. The procedure is to flip the coins simultaneously and let them fall to the floor (it might be easier to flip them if you place them in a cup). Record the result in each case, as in Table 9.3.

In this section, no special knowledge is assumed. You are simply asked to perform some experiments and make some conjectures.

TABLE 9.3 Outcomes of Flipping Three Coins Simultaneously

Outcome	First Trial	Second Trial	Third Trial	Total	Percentage of Occurrence
Three heads	7	5	6	18	12%
Two heads and one tail	15	19	22	56	$37\frac{1}{3}\%$
Two tails and one head	24	22	13	59	$39\frac{1}{3}\%$
Three tails	4	4	9	17	$11\frac{1}{3}\%$
Total	50	50	50	150	100%

Repeat the experiment three times, and record the totals. Next, find the percentage of occurrence by dividing the number of times the event occurred by the total number of occurrences. Can you make any conjectures? It appears that these outcomes are *not* equally likely to occur. Can you make a conjecture about the probability of three heads occurring?

Notice some properties of probability. If we specify the probability of three heads as the percentage of occurrence (or as a fraction), we see that this percentage must be between 0% and 100% (why?). Therefore, if we express the probability as a fraction, the fraction will be between 0 and 1. The closer the fraction is to 0, the less likely the event is to occur; the closer it is to 1, the more likely the event is to occur. An event that can never occur has probability 0, and an event that is certain to occur has probability 1.

EXPERIMENT 3: ROLLING ONE DIE

In this experiment, we roll a single die* 50 times and record the results, as in Table 9.4.

TABLE 9.4 Outcomes of Rolling One Die

Outcome	First Trial	Second Trial	Third Trial	Total	Percentage of Occurrence
1	10	4	10	24	16%
2	8	5	6	19	$12^2/_3\%$
3	9	12	7	28	$18^2/_3\%$
4	6	11	11	28	$18^2/_3\%$
5	5	7	11	23	$15^1/_3\%$
6	12	11	5	28	$18^2/_3\%$
Total	50	50	50	150	100%

Repeat the experiment three times. Find the frequency of each outcome, and then compute the percentage of occurrence of each.

Can you make a conjecture about the probability of each? Notice that the average of each is close to $1/_6 = 16^2/_3\%$. That is, for one die, it appears that all outcomes are equally likely.

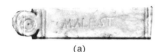

(a)

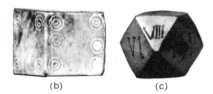

(b)   (c)

(d)

𝕳istorical 𝕹ote

Some ancient Greek and Roman dice: (a) a strip of bone marked with malest *(meaning ''bad luck'') on one side, (b) a bone die that is spotted very much like a modern die, (c) a stone with 14 faces marked with Roman numerals, and (d) a strip of ivory marked* victor *(''winner'') on one side.*

*The word *die* is the singular form of *dice*. All dice used in this text are fair; that is, they are not loaded unless so designated.

EXPERIMENT 4: ROLLING A PAIR OF DICE

In this experiment, we will roll two dice. Here the possible sums are 2, 3, 4, 5, 6, 7, 8, 9, 10, 11, and 12. In each of these examples, we have made a list of all possible outcomes. This list of outcomes is called the *sample space* of the experiment. Roll the pair of dice 50 times.

TABLE 9.5 Outcomes of Rolling a Pair of Dice

Outcome	First Trial	Second Trial	Third Trial	Total	Percentage of Occurrence
2	1	0	1	2	$1^1/_3\%$
3	3	2	2	7	$4^2/_3\%$
4	5	4	5	14	$9^1/_3\%$
5	2	7	6	15	10%
6	9	8	6	23	$15^1/_3\%$
7	8	11	7	26	$17^1/_3\%$
8	6	8	9	23	$15^1/_3\%$
9	6	3	7	16	$10^2/_3\%$
10	4	4	2	10	$6^2/_3\%$
11	5	2	5	12	8%
12	1	1	0	2	$1^1/_3\%$
Total	50	50	50	150	100%

Repeat the experiment three times. Now total to find the frequency of each outcome. Next, compute the percentage of occurrence for each.

Can you make a conjecture about the probability of each? In this problem, it is clear that there is no tendency for each outcome to be equally likely, as in the previous experiment. In fact, it seems as if there is a strong tendency for the middle outcomes to be more probable. Can you explain why?

EXPERIMENT 5: TOSSING A COIN AND ROLLING A DIE

In this experiment, we simultaneously toss a coin and roll a die. Repeat the experiment 50 times, and tabulate the results as shown in Table 9.6.

TABLE 9.6 Outcomes of Tossing a Coin and Rolling a Die

Outcome	First Trial	Second Trial	Third Trial	Total	Percentage of Occurrence
1H	5	4	3	12	8%
1T	6	3	3	12	8%
2H	1	8	2	11	$7^1/_3\%$
2T	1	3	7	11	$7^1/_3\%$
3H	4	5	4	13	$8^2/_3\%$
3T	3	2	7	12	8%
4H	8	5	1	14	$9^1/_3\%$
4T	7	3	1	11	$7^1/_3\%$
5H	1	1	9	11	$7^1/_3\%$
5T	8	5	3	16	$10^2/_3\%$
6H	4	5	6	15	10%
6T	2	6	4	12	8%
Total	50	50	50	150	100%

After performing this experiment, total and find the frequency and percentage of each outcome. Can you make a conjecture about the probabilities involved? We saw that the probability of heads should be $1/_2$, and the probability of rolling a 3 on a die is $1/_6$. How does the probability of obtaining 3H compare with the product of these probabilities?

(probability of 3) · (probability of H)

$$\frac{1}{6} \quad \cdot \quad \frac{1}{2} \quad = \quad \frac{1}{12}$$

This result follows when the events are *independent*—that is, when they are completely unrelated (the occurrence of one event does not affect the occurrence of the other event).

Suppose we reclassify this problem to determine the probability of obtaining an even number or a 3. This result can be tabulated as in Table 9.7. If we obtain an even number or a 3 in Table 9.6, we'll classify the result as a success; otherwise, we'll call it a failure. Do the same for

TABLE 9.7 Outcomes of Obtaining an Even Number or a 3 on Tossing a Coin and Rolling a Die

Outcome	First Trial	Second Trial	Third Trial	Total	Percentage of Occurrence
Success	30	37	32	99	66%
Failure	20	13	18	51	34%

your data. Compute the total percentage of each. Notice that the event of rolling an even number and the event of obtaining a 3 are *mutually exclusive*. That is, if either one occurs, the other one cannot occur. In our example, if a 3 appears, then an even number cannot appear on that *same* roll. Also, if an even number comes up, then a 3 can't come up. Compare the result you obtain if you compute the following:

$$\text{(probability of an even number)} + \text{(probability of 3)}$$

$$\frac{1}{2} \quad + \quad \frac{1}{6} \quad = \quad \frac{2}{3}$$

Do you think this will always be true? To find the probability of mutually exclusive occurrences, we *add* the probabilities of each.

TABLE 9.8 Outcomes of Obtaining an Even Number or a Head on Tossing a Coin and Rolling a Die

Outcome	First Trial	Second Trial	Third Trial	Total	Percentage of Occurrence
Success	33	40	37	110	$73\frac{1}{3}\%$
Failure	17	10	13	40	$26\frac{2}{3}\%$

On the other hand, suppose we reclassify the problem to determine the probability of obtaining an even number or a head. The result could be tabulated as in Table 9.8. Do the same for your data. If we obtain an even number or a head, we will call this result a success; otherwise, it will be a failure.

$$\text{(probability of even)} + \text{(probability of head)}$$

$$\frac{1}{2} \quad + \quad \frac{1}{2} \quad = \quad 1$$

Does this result fit the data obtained? Can you explain how the experiment of Table 9.7 differs from the experiment of Table 9.8? That is, for the probability of "an even or a 3," we can add the respective probabilities; but for the probability of "an even or a head," we apparently cannot. Can you make any conjectures concerning this experiment?

EXPERIMENT 6: THREE-CARD PROBLEM

In this experiment, you need to prepare three cards that are identical except for color. One card is black on both sides, one is white on both sides, and one is black on one side and white on the other. One card is selected at random and placed flat on the table. You will see either a black or a white card; record the color. This is not the probability with which we

Black on both sides

Black on one side, white on the other

White on both sides

are concerned; rather, we are interested in predicting the probability of the *other* side being black or white. Record the color of the second side, as in Table 9.9. Repeat the experiment 50 times, and find the percentage of occurrence of black or white with respect to the known color.

TABLE 9.9 Outcomes of Three-Card Experiment

Color of Chosen Card	Number of Times	Outcome (Color) of Second Side)	First Trial	Percentage of Occurrence[a]
White	26	White	17	65.38%
		Black	9	34.62%
Black	24	White	9	37.5%
		Black	15	62.5%

[a]This is with respect to the known color. For example, the first entry is found by dividing: $17 \div 26 \approx .6538$, or 65.38%.

Consider the following argument. The probability of a white on the underside of the shown card is $1/2$, since the chosen card must be one of two possibilities. Assume that the visible side is white. Then it cannot be the black-black card; it must be the white-white or the white-black. Thus the probability should be $1/2$. Does this argument match the data? If not, then either the data are biased, or the argument is not correct. You might wish to repeat the experiment.

The probabilities obtained in this chapter are *empirical probabilities*. In the next section, we will find *theoretical probabilities*. Our mathematical model will be good only insofar as these two probabilities are "about the same:" That is, if we perform the experiment enough times, the difference between the empirical and the theoretical probability can be made as small as we please.

PROBLEM SET 9.1

A Problems

1–6. Complete the six experiments as described in this section, and state any appropriate conclusions or conjectures.

7. *Computer Problem.* We can repeat Experiment 1 by *simulation*. That is, instead of actually flipping the coin, we can have a computer simulate this experiment for us. We will need to use the random-number generator,

RND. This command produces numbers between 0 and 1, and we must split these in half (to simulate heads or tails). To do this, we multiply the random number by 2 and then take the largest integer value contained in the result. For example, if the random number is .3276, then $2 \cdot .3276 = .6552$ and INT(.6552) = 0. If the original random number is less than .5, then the result is 0; we let this represent "tails." Similarly, if the original random number is greater than or equal to .5, then the result will be 1 (why?); we will let this represent "heads."

A BASIC program is given. First it asks how many times to flip the coin. Then it asks if we would like to see the results of the individual flips. Finally, it outputs the simulated results for the experment.

```
10    PRINT "THIS PROGRAM WILL SIMULATE THE EXPERIMENT OF FLIPPING ";
15    PRINT "A COIN N TIMES."
20    PRINT "HOW MANY TIMES WOULD YOU LIKE ME TO FLIP THE COIN";
30    INPUT N
35    IF N < 0 GOTO 340
36    IF INT(N) <> N GOTO 340
40    PRINT "WOULD YOU LIKE ME TO SHOW YOU THE RESULTS OF EACH FLIP ";
45    PRINT "(ANSWER 1 FOR"
47    PRINT "YES AND 2 FOR NO)";
50    INPUT A
60    IF A = 2 GOTO 210
70    IF A = 1 GOTO 100
80    PRINT "PLEASE ANSWER 1 OR 2"
90    GOTO 40
100   IF N > 500 GOTO 370
110   LET C = 0
120   FOR X = 1 TO N
125   RANDOMIZE
130   LET F = INT(2*RND(X))
140   IF F = 1 GOTO 170
150   PRINT "T";
160   GOTO 190
170   LET C = C + 1
180   PRINT "H";
190   NEXT X
200   GOTO 270
210   LET C = 0
220   FOR X = 1 TO N
225   RANDOMIZE
230   LET F = INT(2*RND(X))
240   IF F = 0 GOTO 260
250   LET C = C + 1
260   NEXT X
270   PRINT
280   PRINT C; "HEADS OUT OF";N;"FLIPS"
290   LET P = (C/N) * 100
300   PRINT P;"% HEADS, AND"; 100 - P;"% TAILS."
310   PRINT
320   PRINT
330   STOP
340   PRINT "N MUST BE A NATURAL NUMBER. PLEASE TRY AGAIN."
```

```
350   PRINT "NOW,";
360   GOTO 20
370   PRINT "ARE YOU SURE YOU WISH ME TO PRINT OUT THE RESULTS OF";
375   PRINT " EACH FLIP";
377   PRINT "(ANSWER "
378   PRINT "1 FOR YES AND 2 FOR NO)";
380   INPUT B
390   IF B = 1 GOTO 110
400   IF B = 2 GOTO 420
410   PRINT "PLEASE ANSWER 1 OR 2"
415   GOTO 370
420   PRINT "PLEASE MAKE UP YOUR MIND."
430   GOTO 40
440   END
```

Use the computer to complete Experiment 1.

8. *Computer Problem*. Use a computer to complete Experiment 2.
A BASIC simulation program is given.

```
10    PRINT "THIS PROGRAM WILL SIMULATE THE FLIPPING OF THREE COINS ";
15    PRINT "SIMULTANEOUSLY."
20    PRINT "HOW MANY TIMES WOULD YOU LIKE ME TO REPEAT THE EXPERIMENT";
30    INPUT N
35    IF INT(N) <> N GOTO 38
36    IF N < 0 GOTO 38
37    GOTO 40
38    PRINT "N MUST BE A NATURAL NUMBER. PLEASE TRY AGAIN."
39    GOTO 20
40    PRINT "THERE WILL BE A MOMENT'S WAIT WHILE I PERFORM THE ";
45    PRINT "EXPERIMENT. BE"
46    PRINT "PATIENT AND I WILL TABULATE THE RESULTS FOR YOU."
50    LET T0 = 0
52    LET T1 = 0
53    LET T2 = 0
54    LET T3 = 0
60    FOR X = 1 TO N
62    LET C = 0
64    LET D = 0
66    LET E = 0
70    RANDOMIZE
80    LET F = INT(2*RND(X))
90    LET G = INT(2*RND(X))
100   LET H = INT(2*RND(X))
110   IF F = 0 GOTO 130
120   LET C = C + 1
130   IF G = 0 GOTO 150
140   LET D = D + 1
150   IF H = 0 GOTO 170
160   LET E = E + 1
170   LET S = C + D + E
180   IF S = 3 GOTO 240
190   IF S = 2 GOTO 230
200   IF S = 1 GOTO 220
210   LET T3 = T3 + 1
```

```
215   GOTO 250
220   LET T2 = T2 + 1
225   GOTO 250
230   LET T1 = T1 + 1
235   GOTO 250
240   LET T0 = T0 + 1
250   NEXT X
252   PRINT
253   PRINT
255   PRINT      "OUTCOME        NUMBER      PERCENTAGE"
256   PRINT
260   PRINT "THREE HEADS";TAB(31);T0;TAB(46);T0/N*100
270   PRINT "TWO HEADS AND ONE TAIL";TAB(31);T1;TAB(46);T1/N*100
280   PRINT "TWO TAILS AND ONE HEAD";TAB(31);T2;TAB(46);T2/N*100
290   PRINT "THREE TAILS";TAB(31);T3;TAB(46);T3/N*100
300   END
```

9. *Computer Problem*. Use a computer to complete Experiment 3. A BASIC simulation program is given.

```
10    PRINT "THIS PROGRAM WILL SIMULATE THE ROLLING OF A SINGLE DIE."
20    PRINT "HOW MANY TIMES WOULD YOU LIKE ME TO ROLL THE DIE";
30    INPUT N
32    IF N < 0 GOTO 36
34    IF INT(N) <> N GOTO 36
35    GOTO 40
36    PRINT "N MUST BE A NATURAL NUMBER. PLEASE TRY AGAIN."
37    PRINT "NOW, ";
38    GOTO 20
40    PRINT "I WILL TABULATE THE RESULTS IN A MOMENT."
50    LET D1 = 0
52    LET D2 = 0
54    LET D3 = 0
56    LET D4 = 0
58    LET D5 = 0
60    LET D6 = 0
70    FOR X = 1 TO N
80    RANDOMIZE
90    LET F = INT(6*RND(Z))
100   IF F = 0 GOTO 160
102   IF F = 1 GOTO 150
104   IF F = 2 GOTO 140
106   IF F = 3 GOTO 130
108   IF F = 4 GOTO 120
110   LET D6 = D6 + 1
115   GOTO 170
120   LET D5 = D5 + 1
125   GOTO 170
130   LET D4 = D4 + 1
135   GOTO 170
140   LET D3 = D3 + 1
145   GOTO 170
150   LET D2 = D2 + 1
155   GOTO 170
160   LET D1 = D1 + 1
```

```
170   NEXT X
175   PRINT
176   PRINT
180   PRINT "    OUTCOME      NUMBER      PERCENTAGE"
181   PRINT
182   PRINT "    ONE";TAB(17);D1;TAB(30);D1/N*100
184   PRINT "    TWO";TAB(17);D2;TAB(30);D2/N*100
186   PRINT "    THREE";TAB(17);D3;TAB(30);D3/N*100
188   PRINT "    FOUR";TAB(17);D4;TAB(30);D4/N*100
190   PRINT "    FIVE";TAB(17);D5;TAB(30);D5/N*100
192   PRINT "    SIX";TAB(17);D6;TAB(30);D6/N*100
200   END
```

10. *Computer Problem*. Use a computer to complete Experiment 4. A BASIC simulation program is given.

```
 10   PRINT "THIS PROGRAM WILL SIMULATE THE ROLLING OF A PAIR OF DICE."
 20   PRINT "HOW MANY TIMES WOULD YOU LIKE ME TO ROLL THE DICE";
 30   INPUT N
 32   IF N < 0 GOTO 36
 34   IF INT(N) < > N GOTO 36
 35   GOTO 40
 36   PRINT "N MUST BE A NATURAL NUMBER. PLEASE TRY AGAIN."
 37   PRINT "NOW, ";
 38   GOTO 20
 40   PRINT "I WILL TABULATE THE RESULTS IN A MOMENT."
 50   LET D1 = 0
 52   LET D2 = 0
 54   LET D3 = 0
 56   LET D4 = 0
 57   LET D5 = 0
 58   LET D6 = 0
 60   LET E1 = 0
 62   LET E2 = 0
 64   LET E3 = 0
 66   LET E4 = 0
 68   LET E5 = 0
 70   FOR I = 1 TO N
 80   RANDOMIZE
 90   LET G = INT(6*RND(Z))
 95   LET H = INT(6*RND(Z))
 98   LET F = G + H
100   IF F = 0 GOTO 160
101   IF F = 10 GOTO 157
102   IF F = 1 GOTO 150
103   IF F = 9 GOTO 147
104   IF F = 2 GOTO 140
105   IF F = 8 GOTO 137
106   IF F = 3 GOTO 130
107   IF F = 7 GOTO 127
108   IF F = 4 GOTO 120
109   IF F = 6 GOTO 117
110   LET D6 = D6 + 1
115   GOTO 170
```

```
117   LET E1 = E1 + 1
118   GOTO 170
120   LET D5 = D5 + 1
125   GOTO 170
127   LET E2 = E2 + 1
128   GOTO 170
130   LET D4 = D4 + 1
135   GOTO 170
137   LET E3 = E3 + 1
138   GOTO 170
140   LET D3 = D3 + 1
145   GOTO 170
147   LET E4 = E4 + 1
148   GOTO 170
150   LET D2 = D2 + 1
155   GOTO 170
157   LET E5 = E5 + 1
158   GOTO 170
160   LET D1 = D1 + 1
170   NEXT I
175   PRINT
176   PRINT
180   PRINT "     OUTCOME      NUMBER     PERCENTAGE"
181   PRINT
182   PRINT "     TWO";TAB(17);D1;TAB(30);D1/N*100
184   PRINT "     THREE";TAB(17);D2;TAB(30);D2/N*100
186   PRINT "     FOUR";TAB(17);D3;TAB(30);D3/N*100
188   PRINT "     FIVE";TAB(17);D4;TAB(30);D4/N*100
190   PRINT "     SIX";TAB(17);D5;TAB(30);D5/N*100
192   PRINT "     SEVEN";TAB(17);D6;TAB(30);D6/N*100
194   PRINT "     EIGHT";TAB(17);E1;TAB(30);E1/N*100
196   PRINT "     NINE";TAB(17);E2;TAB(30);E2/N*100
198   PRINT "     TEN";TAB(17);E3;TAB(30);E3/N*100
200   PRINT "     ELEVEN";TAB(17);E4;TAB(30);E4/N*100
202   PRINT "     TWELVE";TAB(17);E5;TAB(30);E5/N*100
210   END
```

11. Explain why the probability of an event will be between 0 and 1.

B Problems

12. *Random Numbers*. We are all familiar with the idea of "picking a number at random." What do we mean by this? Can we really pick random numbers? In this experiment, we wish to choose 100 numbers (digits) between 0 and 9, inclusive, at random. How would you do this?

    Method 1: In your head. Try writing down 100 random digits between 0 and 9 inclusive. Remember to write down the numbers as you think of them. You may not go back and change the numbers once they have been put down.

    Method 2: Using cards. From an ordinary deck of cards, remove all jacks, queens, and kings. Consider the ten as 0, the ace as 1, and the other cards as their face value. Shuffle the cards thoroughly. Select one card, note the result, and return it to the deck. Shuffle, and repeat 100 times.

Method 3: Using a table of random numbers. Consult a table of random numbers (see Table 9.10). Go to any place on the table and pick 100 numbers.

TABLE 9.10 Random Numbers. This is a table of random digits, arranged into groups of 5 for ease of reading. There are books of random numbers, such as the Rand Corporation's book entitled *A Million Random Digits with 100,000 Normal Deviants* (New York: The Free Press, 1955).

76103	54833	70820	34913	98670	19279	09844	22971	92304	59293	35289	72050	01063	39217
48863	80541	05662	91234	68279	68524	38347	56276	46289	08490	30894	40350	38019	99324
45732	97777	95094	75664	28032	77949	29673	43353	12266	44999	17906	88336	39090	78957
06286	48453	55275	38581	05825	97943	33208	20674	15073	05128	14740	69565	73705	93054
08264	89736	27644	49229	53628	70723	50493	82657	86680	00443	05972	36225	93783	08026
55459	30688	81619	14591	32062	69629	68218	05069	63221	75975	21319	47239	98310	13552
82459	73179	00886	70194	31725	16658	98614	43381	30161	10846	22282	09708	24754	92896
67571	12653	88165	82561	07851	56660	71288	18642	72206	93517	42794	15872	19741	11757
15247	01518	55868	00166	76152	47436	69266	26621	05785	26612	41225	41310	87444	32098
58936	94717	62073	14288	42914	39416	23773	17905	09906	54660	12214	79899	50012	90279
24394	08154	42903	47442	78882	80560	28380	02828	56828	54497	61566	26977	24318	44348
18534	80284	07652	12150	77193	08640	91264	28554	09430	99046	20170	71345	50116	16327
11703	92464	94067	67479	93687	47464	77939	68513	74114	85589	89303	93201	20961	63648
41583	39061	12445	07282	69713	17034	02106	70092	63512	40342	90936	89590	17317	13310
77785	25025	94270	20441	94245	53364	40785	72264	45807	75318	35704	29122	86473	13771
32479	21037	67157	64614	00378	14772	30649	70212	02838	70342	27807	87898	45948	63632
69404	27495	56827	04026	43607	23164	99618	47578	33723	64703	32191	69755	73238	48101
60854	40778	20924	04486	12728	35928	17589	68749	97456	65785	25834	24803	99071	61100
52414	26291	55870	88153	99895	60996	97872	74404	10657	70924	11836	88313	51233	48771
29526	62514	00918	03119	31925	15446	93522	35609	26364	36632	90458	25664	42197	09952

Now, did these methods give truly random numbers? There are several tests for randomness. We will check only two of them.

a. Each digit should occur approximately the same number of times. This means that, given enough such digits, the percentage of occurrence for each should be close to 10%. Test the results of the three methods, and see if this holds true for each.

b. Are there any sequences of consecutive numbers? In naming 100 digits, we could expect a random selection to have a sequence of several digits, such as 6789. Does this hold true for each method?

In many applications of science, it is important to obtain random numbers. Computers have been a great help in selecting random numbers for us.

13. *Computer Problem.* Write a BASIC program that will output random numbers (see Table 9.10). Also, write the program so that, when finished, it will output the number of times each digit occurred.

14. *Computer Problem.* Write a BASIC program to simulate Experiment 5.

15. *Computer Problem.* Use a computer to complete Experiment 5 using the program you wrote for Problem 14.

16. *The Drunkard's Walk.* There is a drunk leaning on a lamppost, and he decides to take a walk, although, in his condition, he doesn't care where he is going.

   Suppose each step he takes is about 2 feet. Now, he takes a few steps this way and a few steps that way, and we would like to know how far the drunk is from the lamppost after, say, 100 steps.

   Even though this may seem like an impossible task, it has been found that the *most probable* distance the drunkard will be from the lamppost after x zigzags equals the average length of each step times $\sqrt{x}$. Thus, if he takes 100 steps, he should be about $2\sqrt{100} = 20$ feet away from the lamppost.

   Verify this result using the table of random numbers (Table 9.10). Assume that we arbitrarily divide the directions the drunk can go into 10 parts, as shown below.

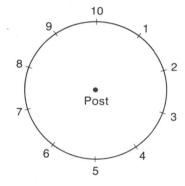

If the first random number is 1, then we will say the drunk moves 1 step in that direction. Then, if the second random number is 4, the drunk's next step will be in that direction.

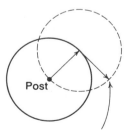

Drunk will be here after 2nd step.

Do this experiment on a full sheet of paper, and have the drunk take 100 steps (set some scale). Measure the (scaled) distance from the lamppost. Is he about 20 feet from it? Repeat the experiment.

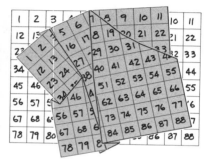

FIGURE 9.1 Crumpled-paper problem

17. On two full sheets of paper, construct 88 squares on each, as shown in Figure 9.1a. Crumple up one sheet of paper, and throw it on the other sheet at random, as shown in Figure 9.1b. What do you think the probability is that at least one number lies directly above its counterpart (1 above 1 or 2 above 2 or 3 above 3 . . .)? Perform the experiment several times. Is your original guess supported? Could you make a better conjecture?

18. *The Game of WIN.* Construct a set of nonstandard dice as shown.*

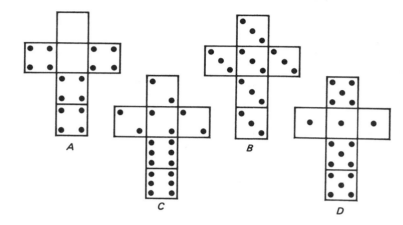

(Blanks are counted as zero.) Consider a game called WIN. The game is for 2 players. Each player is to choose one of the dice, and then the players roll the dice simultaneously. The player with the larger number wins the game. By experimenting with the dice, which one would you choose? If you were to play the game of WIN, would you like to choose your die first or second?

*These dice were designed by Bradley Efron of Stanford University. The game of WIN was suggested by Joseph Smyth of Santa Rosa Junior College.

Mind Bogglers

19. A tournament is organized between Shannon's Stompers and Melissa's Marvels. Suppose the teams each have an equal chance of winning and the Marvels are leading the series three games to two when play is interrupted. If the first team to win five games wins the tournament, use Table 9.10 to simulate the finish of the series and decide the winner. Record the name of the winner of the series. Repeat the experiment 25 times (or more if necessary), and record the frequency with which the Marvels win the series. Plot the relative frequencies after 5, 10, 15, 20, and 25 trials. Do these relative frequencies appear to be stabilizing around any particular number?

20. *Computer Problem.* Write a BASIC program to simulate Experiment 6. Complete the experiment on a computer.

21. *Computer Problem.* Write a BASIC program to simulate the Drunkard's Walk of Problem 16.

One way of using Table 9.10 to determine the winner is to select any digit in Table 9.10; if the digit is even then the game is won by the Marvels and if the digit is odd then the game is won by the Stompers. For subsequent games, move in any direction one digit at a time from the first digit.

Problem for Individual Study

22. *Probability and Chance.* What is probability? What is the risk involved in driving a car? What is Brownian motion? What are the odds of winning a lottery? What is the mathematical expectation in a carnival game? Does gambling pay? How is the law of disorder related to nuclear fission? How is probability used to derive new facts?

Exhibit suggestions: illustrate with toys or models the probability of accident, crop damage from storms, winning bridge hands, slot-machine odds, coin tosses, dice totals, contest winnings, lock combinations, multiple births.

References: Bakst, Aaron, *Mathematics, Its Magic and Mastery* (2nd ed.) (New York: Van Nostrand, 1952).

Bergamini, David, *Mathematics,* Chap. 6 (New York: Time, Inc., 1963).

Kac, Mark, "Probability," *Scientific American,* September 1964 (Vol. 211, No. 3).

Kasner, Edwin, and James Newman, *Mathematics and the Imagination* (New York: Simon and Schuster, 1940).

Kemeny, John, J. Laurie Snell, and Gerald L. Thompson, *Introduction to Finite Mathematics* (3rd ed.), Chap. 3 (Englewood Cliffs, N.J.: Prentice-Hall, 1974).

Kline, Morris, *Mathematics in Western Culture,* Chap. XXIII (New York: Oxford University Press, 1953).

Newman, James, *The World of Mathematics,* Vol. 2, Part VII (New York: Simon and Schuster, 1956).

9.2 DEFINITION OF PROBABILITY

Sample space

Event

Engraving by Darcis: *Le Trente-un*

We will now formalize the results of the previous section and develop a mathematical model giving *theoretical probabilities*. That is, we'll formulate definitions, assume some properties, and thus create a model that will serve as a predictor of the actual percentage of occurrence for a particular experiment. Our model will be good only insofar as it conforms to experimental data.

First we should settle on some terminology. We will consider experiments and events. The *sample space* of an experiment is the set of all possible outcomes. In each of the experiments of the previous section, we tabulated the sample space under the column headed "outcomes." Now, an *event* is simply a set of possible outcomes (a subset of the sample space). For example, in Experiment 5, we tossed a coin and rolled a die. The sample space is

$$\{1H,1T,2H,2T,3H,3T,4H,4T,5H,5T,6H,6T\}$$

and an event might be "obtaining an even number or a head." That is, the following subset of the sample space is considered to be this event:

$$\{1H,2H,2T,3H,4H,4T,5H,6H,6T\}$$

The event "obtaining a 5" is:

$$\{5H,5T\}$$

If the sample space can be divided into *mutually exclusive* and *equally likely* outcomes, we can define the probability of an event. Let's consider Experiment 1. We see here that a suitable sample space is:

$$S = \{ \text{Heads}, \text{Tails} \}$$

Suppose we wish to consider the event of obtaining heads; we'll call it event A.

$$A = \{ \text{heads} \}$$

We wish to define the probability of event A, which we denote by P(A). Notice that the outcomes in the sample space are *mutually exclusive*. That is, if one occurs, the others cannot occur. If we flip a coin, there are two possible outcomes, and *one and only one* outcome can occur on a toss. If each outcome in the sample space is *equally likely*, we define the probability of A as:

$$P(A) = \frac{\text{number of successful results}}{\text{number of possible results}} = \frac{|A|}{|S|}$$

A "successful" result is a result that corresponds to the probability we are seeking—in this case, {heads}. Since we can obtain a head in only one way (success), and the total number of possible outcomes is two (cardinality of the sample space), then the probability of heads is given by this definition as:

$$P(\text{heads}) = P(A) = \frac{1}{2}$$

This, of course, corresponds to our empirical results and to our experience of flipping coins. We now give a definition of probability.

DEFINITION: If an experiment can occur in any of n mutually exclusive and equally likely ways, and if s of these ways are considered favorable, then the probability of the event E, denoted by P(E), is:

$$P(E) = \frac{s}{n} = \frac{\text{number of outcomes favorable to } E}{\text{number of all possible outcomes}}$$

Definition of the probability of an event that can occur in any one of n mutually exclusive and equally likely ways

---

Notes: 1. Now, $s \geq 0$ (why?). Thus, P(E) ≥ 0.
2. Also, $s \leq n$ (since n represents the total of *all* possible outcomes), so we see that P(E) ≤ 1.
3. If $s = 0$, then P(E) = 0, which means that a success *cannot* occur.
4. If $s = n$, then P(E) = 1, which means that a success *must* occur.

Are these first four properties of probability consistent with the results of the previous section?

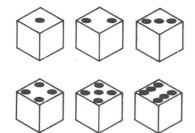

FIGURE 9.2 Sample space for tossing a single die.

In note 3 we said that, if an event *cannot* occur, then the probability is zero. The occurrence of Albino twins in cows had never before been recorded, and, according to *The Dairyman,* it more than likely will never happen again. This is an example of a highly unlikely event, and its probability is very close to zero. But this is not the same as saying that the probability *is* zero.

Outcomes	Theoretical Probability	Empirical Probability
2	.0909	.0133
3	.0909	.0467
4	.0909	.0933
5	.0909	.1000
6	.0909	.1533
7	.0909	.1733
8	.0909	.1533
9	.0909	.1067
10	.0909	.0667
11	.0909	.0800
12	.0909	.0133

The results obtained by applying this definition must be consistent with the results obtained by actually performing an experiment a large number of times. For example, consider Experiment 3 (rolling a single die). Since the sample space consists of six possibilities (see Figure 9.2), we might guess that the theoretical probability for each is $1/6 \approx .1667$. Let's compare this estimate with the results from Experiment 3:

Outcomes	Theoretical Probability	Empirical Probability*
1	.1667	.1600
2	.1667	.1267
3	.1667	.1867
4	.1667	.1867
5	.1667	.1533
6	.1667	.1867

*These figures vary according to the particular experiment and the number of trials. The figures used here are actual results of Experiment 3.

If we wish to find the probability of obtaining a 2, we would reason as follows: Let $A = \{2\}$. Then, $P(A) = 1/6$, since we have

$$\frac{\text{number of successful outcomes}}{\text{number of all possible outcomes}} = \frac{1}{6}$$

The empirical probabilites are ''close to'' our theoretical probabilities, and we get the feeling that, if we conducted Experiment 3 with a very large number of trials, the empirical probability would be very close to the theoretical probability. Thus our probability model is a good one for this example. Notice that, if the probabilities were *not* similar, we would be required to change our assumptions or theoretical model. We certainly cannot change the results of empirical evidence.

On the other hand, in Experiment 4, our sample space consisted of

$$\{2,3,4,5,6,7,8,9,10,11,12\}$$

If we wish to find the probability of obtaining a 2, we would *not* write $P(A) = 1/11$, even though

$$\frac{\text{number of successful outcomes}}{\text{number of all possible outcomes}} = \frac{1}{11} \approx 0.0909$$

Using the table in the margin, compare such a probability with the results of the experiment.

The empirical probabilities do *not* substantiate our theoretical probability. The problem here is our assumption that the outcomes are equally likely. Consider another sample space for a pair of dice (see Figure 9.3).

FIGURE 9.3 Sample space for tossing a pair of dice

There are 36 possibilities. Now we can compute realistic theoretical probabilities using this sample space of 36 equally likely possibilities:

$$P(2) = \frac{1}{36} \approx .0278$$

$$P(3) = \frac{2}{36} \approx .0556$$

$$P(4) = \frac{3}{36} \approx .0833$$

$$P(5) = \frac{4}{36} \approx .1111$$
.
.
.

We notice from the table in the margin that the empirical and theoretical probabilities are quite similar; thus we can assume that our sample space of 36 equally likely outcomes is correct.

How can we decide if several outcomes are equally likely? We can do it either empirically or by a careful examination of the problem. This is not always an easy task and must be considered for each problem. Keep in mind that the only time the definition of probability can be applied is when the outcomes in the sample space are mutually exclusive and equally likely.

Outcomes	Theoretical Probability	Empirical Probability
2	.0278	.0133
3	.0556	.0467
4	.0833	.0933
5	.1111	.1000
6	.1389	.1533
7	.1667	.1733
8	.1389	.1533
9	.1111	.1067
10	.0833	.0667
11	.0556	.0800
12	.0278	.0133

We can sometimes find probabilities by counting what we are not counting; that is, we can find the probabilities of *complementary events*. The *complement* of any event E is denoted by $\overline{E}$. Since any event either occurs or does not occur, we see that the sum of the probability of an event occurring and that of the same event not occurring is 1. In symbols,

$$P(E) + P(\overline{E}) = 1$$

This is usually written in a more useful form:

$$P(\overline{E}) = 1 - P(E)$$

A poker hand consists of 5 cards drawn from a deck of 52 cards.

For example, if we wish to find the probability that a poker hand has at least one ace in it, we could compute the probability directly *or* work the following simpler problem.

$$P(\text{no aces}) = \frac{_{48}C_5}{_{52}C_5} = \frac{\dfrac{48 \cdot 47 \cdot 46 \cdot 45 \cdot 44}{5 \cdot 4 \cdot 3 \cdot 2 \cdot 1}}{\dfrac{52 \cdot 51 \cdot 50 \cdot 49 \cdot 48}{5 \cdot 4 \cdot 3 \cdot 2 \cdot 1}}$$

$$\approx .6588$$

Thus,

$$P(\text{at least one ace}) \approx 1 - .6588$$
$$= .3412$$

Finding the probability of an event. (Remember that this scheme works only if the event occurs in n mutually exclusive and equally likely ways.)

> In summary, to find the probability of some event:
> 1. Describe and identify the sample space, and then count the number of elements (these should be equally likely). We call this number n.
> 2. Count the number of occurrences that interest us; we call this the number of successes and denote it by s.
> 3. Compute the probability of the event: $P(E) = s/n$.

The procedure just outlined will not always work. If it doesn't, you need a more complicated model, or else you must proceed experimentally, as in the last section. The procedure described will, however, be suitable for the applications in this text.

EXAMPLES:

1. A single card is selected from an ordinary deck of 52 cards.

    a. What is the probability that it is a heart?
    b. What is the probability that it is an ace?
    c. What is the probability that it is a 2 or a king?

Solutions:

a. There are 52 elements in the sample space, and 13 of these are hearts, or successes. Therefore, P(heart) = 13/52 = 1/4.

b. $P(ace) = \dfrac{4}{52} = \dfrac{1}{13}$.

c. $P(2 \text{ or king}) = \dfrac{8}{52} = \dfrac{2}{13}$.

2. An urn contains three red balls, two black balls, and five green balls. You draw one ball.

a. What is the probability of drawing a red ball?
b. What is the probability of drawing a black or a green ball?
c. What is the probability of drawing a black and a green ball?

Solutions:

a. A possible sample space is {red, green, black}, but these possibilities are not equally likely. (Why?) Consider the following sample space:

$$\{R_1, R_2, R_3, B_1, B_2, G_1, G_2, G_3, G_4, G_5\}$$

where R_1, R_2, R_3 represent the red balls, B_1, B_2, represent the black balls, and G_1, G_2, G_3, G_4, G_5 represent the green balls. Then the sample space, as now described, consists of equally likely outcomes, so we use this model. Thus,

$$P(red) = \frac{\text{number of favorable outcomes}}{\text{number of all possible outcomes}} = \frac{3}{10}$$

b. $P(\text{black or green}) = \dfrac{7}{10}$.

c. P(black and green) = 0/10 = 0. (The number of favorable outcomes is 0—you can't draw out a black and a green ball with only one draw.)

3. Suppose that in an assortment of 20 electronic calculators there are 5 with defective switches.

a. If a machine is selected at random, what is the probability that it has a defective switch?
b. If two are selected at random, what is the probability that they both have defective switches?
c. If three are selected at random, what is the probability that exactly one has a defective switch?
d. If three are selected at random, what is the probability that at least one has a defective switch?

Solutions:

a. The solution is given by

$$\frac{\text{number of ways of selecting 1 defective from the 5}}{\text{number of ways of selecting 1 machine from the 20}}$$

$$= \frac{{}_5C_1}{{}_{20}C_1} = \frac{5}{20} = \frac{1}{4}$$

b. The solution is given by

$$\frac{\text{number of ways of selecting 2 defectives from the 5}}{\text{number of ways of selecting 2 machines from the 20}}$$

$$= \frac{{}_5C_2}{{}_{20}C_2} = \frac{\dfrac{5 \cdot 4}{2}}{\dfrac{20 \cdot 19}{2}} = \frac{1}{19}$$

c. In this problem we use the Fundamental Counting Principle to determine the number of successes. Picking exactly one machine with a defective switch:

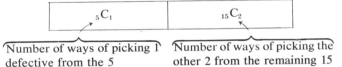

| ${}_5C_1$ | ${}_{15}C_2$ |

Number of ways of picking 1 defective from the 5   Number of ways of picking the other 2 from the remaining 15

The probability we are seeking is

$$\frac{{}_5C_1 \cdot {}_{15}C_2}{{}_{20}C_3} = \frac{\dfrac{5 \cdot 15 \cdot 14}{2}}{\dfrac{20 \cdot 19 \cdot 18}{3 \cdot 2 \cdot 1}} = \frac{5 \cdot 15 \cdot 7}{20 \cdot 19 \cdot 3} = \frac{35}{76}$$

d. Rather than calculate the probability of one defective or two defectives or three defectives directly, we will find the answer by counting what we are not counting. Let $A = \{$selecting at least one with a defective switch$\}$ and $\overline{A} = \{$selecting three with no defects$\}$.

$$P(\overline{A}) = \frac{{}_{15}C_3}{{}_{20}C_3} = \frac{\dfrac{15 \cdot 14 \cdot 13}{3 \cdot 2 \cdot 1}}{\dfrac{20 \cdot 19 \cdot 18}{3 \cdot 2 \cdot 1}} = \frac{15 \cdot 14 \cdot 13}{20 \cdot 19 \cdot 18} = \frac{91}{228}$$

$$P(A) = 1 - P(\overline{A}) = 1 - \frac{91}{228} = \frac{137}{228}$$

PROBLEM SET 9.2

A Problems

1. What is the difference between empirical and theoretical probabilities?

2. Define probability.

3. In the notes following the definition of probability, we said that $s \geq 0$. Why is this true?

4. A card is selected from an ordinary deck of cards. Find the probability that it is:
   a. a 2.   b. the ace of spades.   c. a king or queen.

5. An urn contains six red balls, four white balls, and three black balls. One ball is drawn at random. What is the probability of drawing:
   a. a red ball?   b. a black or a white ball?
   c. a black and a white ball?

6. A die is rolled. What is the probability that it is:
   a. a 5 or 6?   b. a 1, 2, 3, or 4?   c. odd or even?
   d. not a 5?   e. a 7?   f. a prime number?

7. A die is rolled. What is the probability that it is:
   a. a 4?   b. a 4 or 5?   c. even?
   d. less than 5?   e. a 4 and a 5?

8. A card is selected from an ordinary deck. Find the probability that it is:
   a. a king and a spade.   b. a heart or a club
   c. not a face card.   d. a king or a spade.

9. Suppose we randomly choose a natural number between 1 and 100, inclusive. Find the probability that it is:
   a. a 5.   b. less than 5.
   c. greater than 5.   d. less than 5 or greater than 5.
   e. a multiple of 5.   f. prime.

10. Suppose we toss a coin and roll a die. The sample space is shown.

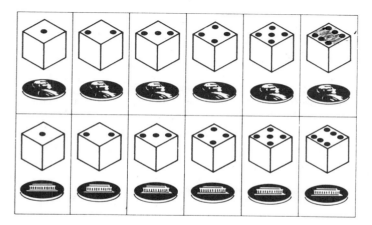

What is the probability that it is:
a. a tail?
b. a tail and a 5?
c. a tail or a 5?
d. a 1, 2, 3, 4, or a tail?
e. not a 6?

11. A coin is tossed and a die is rolled. What is the probability of obtaining:
   a. a head and a 2?   b. a head and an odd number?

12. A coin is tossed and a die is rolled. What is the probability of obtaining:
   a. a head and a 5?   b. a head and an even number?
   c. a head or a 5?   d. a head or an even number?
   e. a head or a number greater than 2?

B Problems

13. It is known that a company has 10 pieces of machinery wth no defects, 4 with minor defects, and 2 with major defects. Thus each of the 16 machines falls into one of these three categories. The inspector is about to come, and it is her policy to choose one machine and check it. Find the probability that the machine she checks:
   a. has no defects   b. has no major defects.
   c. has no defects or has major defects.

14. A pair of dice are rolled. Using the sample space shown in Figure 9.3 (page 487) find the probability that:
   a. a 4 is rolled (the sum of the two dice is 4).
   b. a 6 is rolled (the sum of the two dice is 6).
   c. a 4 or a 5 is on one of the dice.
   d. an even number is rolled (the sum is even).
   e. a number greater than or less than 5 (sum of the two dice) is rolled.

15. A pair of dice are rolled. Find the probability that:
   a. a 4 on the first die and a 5 on the second are rolled.
   b. a number greater than 7 is rolled (the sum of the two dice is over 7).

16. One integer is chosen at random from the numbers 1,2,...,50. What is the probability that the chosen number is:
   a. divisible by 2?   b. divisible by 3?   c. divisible by 2 and 3?

17. One integer is chosen at random from the numbers 1,2,3,...,50. What is the probability that the chosen number is:
   a. divisible by 5?   b. divisible by 6?   c. divisible by 1?
   d. divisible by 5 and 6? e. divisible by 5 or 6?

18. Dice is a popular game in gambling casinos. Two dice are tossed, and various amounts are paid according to the outcome. If a 7 or 11 occurs on the first roll, the player wins. What is the probability of winning on the first roll?

19. In dice, the player loses if the outcome of the first roll is a 2, 3, or 12. What is the probability of losing on the first roll?

20. In dice, a pair of ones is called snake eyes. What is the probability of losing a dice game by rolling snake eyes?

21. Three coins are tossed. Let A = {no heads}
    B = {at least 2 heads}
    C = {no heads or 3 heads}

    Find:

    a. P(A)   b. P(B)   c. P(C)

22. A news story reported that a man in Rochester, New York, took two chances at Russian roulette and killed himself on the second try. What is the probability that he would kill himself on the first try, assuming that there was only one bullet in his six-shooter? What is the probability that he would kill himself on the second try (assuming we know that he didn't kill himself on the first try)?

23. *Calculator Problem.* Find the probability of obtaining a royal flush (ace, king, queen, jack, and 10 of one suit).

24. *Calculator Problem.* Find the probability of obtaining a full house of three aces and a pair of twos.

25. Consider the *Buz Sawyer* cartoon, in which Stanley's secret has been blacked out. What is Stanley's secret? That is, how many socks must Rosco take out to be sure of getting a pair that match?

26. Consider the game of WIN described in Problem 18, page 482. Suppose your opponent picks die A. Suppose you pick die B. Then we can enumerate the sample space as shown. We see that the probability of A winning is 24/36, or 2/3. By enumeration of the sample space, find your probability of winning if you choose:
    a. die C.
    b. die D.
    c. If your opponent picks die A, which die would you pick?

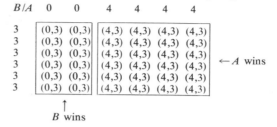

27. In WIN (see Problem 26), if your opponent picks die B, which would you pick?

28. In WIN (see Problem 26), if your opponent picks die C, which would you pick?

29. In WIN (see Problem 26), if your opponent picks die D, which would you pick?

Mind Bogglers

30. By considering Problems 26–29, see if you can write a strategy for playing WIN. If you follow this strategy, what is your probability of winning?

31. A man begins a stroll at point X. He walks north one block, at which time he flips a coin. If it's a head, he makes a right turn; if it's a tail, he makes a left turn. He walks four blocks, flipping the coin at each corner. What is the probability that he ends up at the starting point?

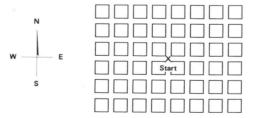

See the historical note on page 484.

32. *Calculator Problem*. Chevalier de Méré used to bet that he could get at least one 6 in four rolls of a die. He also bet that, in 24 tosses of a pair of dice, he would get at least one 12. He found that he won more often than he lost with the first bet, but not with the second. He did not know why, so he wrote to Pascal seeking the probabilities of these events. What are the probabilities for winning in these two games?

33. In a stunt on a TV variety show, a marksman fired three shots at a rapidly spinning disk that was divided into two semicircles. Amazingly, he always placed his three shots in the same semicircle. How much of this was skill, and how much was luck? That is, assuming he could always hit the target, what is the probability that the three shots would hit the same semicircle?

Problems for Individual Study

34. *Calculator Problem.* Verify as many entries of Table 9.11 as possible.

TABLE 9.11 Probabilities of Poker Hands

Hand		Number of Favorable Events	Probability
Royal flush		4	.00000153908
Other straight flush		36	.00001385169
Four of a kind		624	.00240096038
Full house		3744	.00144057623
Flush		5108	.00196540155
Straight		10,200	.00392464782
Three of a kind		54,912	.02112845138
Two pair		123,552	.04753901561
One pair		1,098,240	.42256902761
Other hands		1,302,540	.50117739403
Total		2,598,960	1.00000000000

35. *Friday-the-13th Problem.* Find the probability that the 13th day of a randomly chosen month will be a Friday. We might begin by saying that the sample space is {Sunday, Monday, Tuesday, Wednesday, Thursday, Friday, Saturday}. That is, the 13th must be one of these. Thus, P(Fri) = 1/7. This is *not correct*, since the outcomes listed are not equally likely. It turns out that the 13th of the month is more likely to be a Friday than any other day of the week. Show this.
   References: Bailey, William, "Friday-the-Thirteenth," *Mathematics Teacher,* 1969 (Vol. LXII, No. 5), pp. 363–364.
   Heuer, C. V., Solution to problem E1541 in *American Mathematical Monthly,* 1963 (Vol. 70, No. 7), p. 759.
   Ritter, G. L., et al. "An Aid to the Superstitious," *Mathematics Teacher,* 1977 (Vol. 70, No. 5), pp. 456–457.

36. What is the Monte Carlo method?

9.3 CONDITIONAL PROBABILITY

"If you were playing poker, your luck would be perfect."

At one time or another, we all become involved in a discussion of whether an expected child will be a boy or a girl, and many ideas of probability are used and misused in making these predictions. In this section, we will "keep it in the family" by considering applications of probability to family matters. We assume that each child is as likely to be a boy as a girl.

Let's begin by computing the probability of a family having two children of opposite sexes. The sample space might look this this:

> 2 boys
> a boy and a girl
> 2 girls

But this sample space presents problems in computing probabilities, since the outcomes are not equally likely. Consider the following tree diagram:

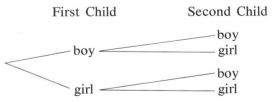

We see that there are four equally likely outcomes:

> boy——boy
> boy——girl
> girl——boy
> girl——girl

Thus the probability of having a boy and a girl with two children is:

$$\frac{\text{number of favorable outcomes}}{\text{total number of all possible outcomes}} = \frac{2}{4}$$

There is a probability of $1/2$ of having a boy and a girl family.

EXAMPLE: Suppose a family wanted to have four children. What is the probability of having two boys and two girls?

Solution: In this problem, there are 16 possibilities (can you list them?). Of these possibilities, there are 6 ways of obtaining 2 boys and 2 girls: BBGG, BGBG, BGGB, GBGB, GBBG, GGBB. Thus the probability is

$$\frac{6}{16} = \frac{3}{8}$$

A common fallacy is to say "$1/2$ boys and $1/2$ girls; thus the probability is $1/2$."

A couple with two boys wanted to have a girl. They knew a little about probability and reasoned that, since the chance of having three boys is only $1/8$, they really had a good chance of having a girl the next time. This is a common misconception. It is important to remember that the probability of a future event is unaffected by past occurrences. The probability of a girl on the next birth is the same as always—$1/2$—*regardless of the previous record*. The Emory Harrison family in Tennessee had 13 boys. The probability of this occurring is

$$\frac{1}{8192}$$

If the Harrisons were to have another baby, what would be the probability of this child being a boy? You *cannot* reason that the probability of 14

The Emory Harrison family

boys is 1/16,384 and thus their chances of another boy are very slight. Actually, the probability of their having another boy is still $1/2$.

Another example of this fallacy is given by Darrell Huff. On August 18, 1913, at a casino in Monte Carlo, black came up 26 times in a row on a roulette wheel. Had you bet $1 on black and continued for the entire run of blacks, you would have won $67,108,863. What actually happened confirms our statement that many people commit this fallacy. After about the 15th occurrence of black, there was a near-panic rush to bet on red. People kept doubling up on their bets in the belief that, after black came up the 20th time, there was not a chance in a million of another repeat. The casino came out ahead by several million francs. Remember: for each spin of the wheel, the probability of black remains the same, *regardless* of past performance.

Frequently, we wish to compute the probability of an event but have additional information that will alter the sample space. For example, suppose a family has two children. What is the probability that the family has two boys? P(2 boys) $= 1/4$ (why?). Let's complicate the problem a little. Suppose we know that the older child is a boy. We have altered the sample space.

Sample Space

$$\begin{array}{ll} B & B \\ B & G \end{array} \Big\} \quad \begin{array}{l} \text{This is the altered} \\ \text{sample space.} \end{array}$$

$\cancel{G \quad B}$
$\cancel{G \quad G}$

Gambling casinos like to take advantage of this fallacy in the thinking of their clients. There used to be a pair of dice displayed on a red velvet pillow behind glass in the lobby of the Desert Inn in Las Vegas. This pair of dice made 28 straight passes (winning tosses) in 1950. The man rolling them won only $750; if he had bet $1 and let his winning double for each pass, he would have won $268,435,455.

Use of an altered sample space

We see that the probability is now $1/2$.

If we know that at least one child is a boy, we alter the sample space as shown.

Sample Space

$$\begin{array}{ll} B & B \\ B & G \\ G & B \end{array}$$

$\cancel{G \quad G}$

Here we see that the probability is $1/3$!

Conditional probability

Notation for conditional probability

These are problems involving *conditional probability*. We speak about the *probability of an event E, given that another event F* has occurred. We denote this by $P(E \mid F)$. This means that, rather than simply compute $P(E)$, we re-evaluate $P(E)$ in light of the information that F has occurred. That is, we consider the altered sample space.

EXAMPLE: Suppose we toss three coins. What is the probability that two or more heads are obtained, given that at least one head is obtained?

Solution: We consider the altered sample space.

Sample Space
*HHH	*HHH
*HTH	HTT
*THH	THT
TTH	~~TTT~~

The probability is $^4/_7$.

We'll now look at one final example. We often hear about the "odds" of having a boy or a girl or the "odds of winning." Let's consider the odds of a family of four children having two boys. We say that the probability of two boys is $^3/_8$, so many people would incorrectly say "the odds of having two boys are 3 to 8." This is not correct. Since, for every 8 families, we would expect 3 of them to have two boys and 5 of them not to have two boys, we would say that the *odds are 3 to 5*.

DEFINITION: The *odds in favor* of an event E, where $P(E)$ is the probability that E will occur, and $P(\overline{E})$ is the probability that E will not occur, are given by:

$$\text{Odds in favor:} \quad \frac{P(E)}{P(\overline{E})} \quad \text{or} \quad \frac{P(E)}{1 - P(E)}$$

Similarly:

$$\text{Odds against:} \quad \frac{P(\overline{E})}{P(E)} \quad \text{or} \quad \frac{1 - P(E)}{P(E)}$$

Definition of odds

In terms of our example, we see that the odds in favor of two boys are:

$$\frac{3/8}{5/8} = \frac{3}{8} \cdot \frac{8}{5} = \frac{3}{5}$$

Thus the odds in favor of two boys are 3 to 5. The odds against two boys are:

$$\frac{5/8}{3/8} = \frac{5}{8} \cdot \frac{8}{3} = \frac{5}{3}$$

Thus the odds against two boys are 5 to 3.

EXAMPLES:

1. What are the odds in favor of drawing a heart from an ordinary deck of cards?

   Solution: Let H = {draw a heart}. Then $P(H) = ^1/_4$ and $P(\overline{H}) = 1 - P(H) = ^3/_4$. Thus the odds in favor of drawing a heart are given by

$$\frac{1/4}{3/4} = \frac{1}{4} \cdot \frac{4}{3} = \frac{1}{3}$$

The odds in favor of drawing a heart are 1 to 3.

2. If the odds in favor of an event are 1 to 3, what is the probability of the event occurring?

Solution:

Suppose we know $P(E) = s/n$ for some event E. Then the odds in favor are

$$\frac{\dfrac{s}{n}}{1 - \dfrac{s}{n}} = \frac{\dfrac{s}{n}}{\dfrac{n-s}{n}} = \frac{s}{n-s}$$

If the odds are given as s to b, then, from the above calculation, we see that $b = n - s$. Hence $n = b + s$, so the probability of E is

$$P(E) = \frac{s}{n} = \frac{s}{b+s}$$

In this example, $P(E) = \dfrac{1}{1+3} = \dfrac{1}{4}$.

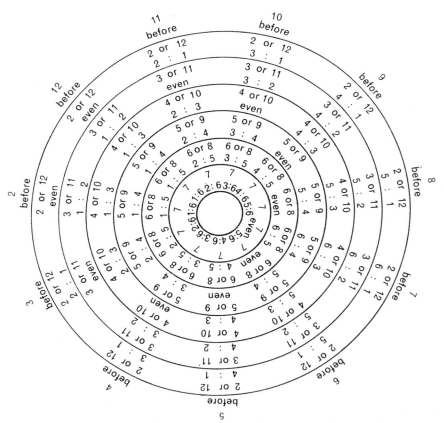

This chart shows the odds in favor of throwing any number before any other with two dice. Reading in from the outer circle, the odds in favor of throwing, say, a 9 before a 2 or a 12 are 4 to 1; before a 3 or an 11, 4 to 2; before a 4 or a 10, 4 to 3; and so on.

3. If the odds of an event E are 3 to 5, what is the probability of E?
   Solution:

$$P(E) = \frac{3}{3 + 5} = \frac{3}{8}$$

PROBLEM SET 9.3

A Problems

1. Suppose a family wants to have four children.
   a. What is the sample space?
   b. What is the probability of 4 girls? 4 boys?
   c. What is the probability of 1 girl and 3 boys? 1 boy and 3 girls?
   d. What is the probability of 2 boys and 2 girls?
   e. What is the sum of your answers in parts b through d?

2. In your own words, explain what we mean by "conditional probability."

3. What is the probability of a family of three children containing exactly two boys, given that at least one of them is a boy?

4. What is the probability of flipping a coin four times and obtaining exactly three heads, given that at least two are heads?

5. What are the odds in favor of drawing an ace from an ordinary deck of cards?

6. What are the odds in favor of drawing a heart from an ordinary deck of cards?

7. What are the odds against a family of four children containing four boys?

8. Suppose the odds that a man will be bald by the time he is 60 are 9 to 1. State this as a probability.

9. Suppose the odds are 33 to 1 that someone will lie to you at least once in the next seven days. State this as a probability.

10. Racetracks quote the approximate odds for a particular race on a large display board called a tote board. Some examples from a particular race are:

Horse Number	Odds
1	2 to 1
2	15 to 1
3	3 to 2
4	7 to 5
5	1 to 1

What would be the probability of winning for each of these horses?

EXAMPLE: The odds stated are for the horse losing. Thus,

$$P(\text{horse 1 losing}) = \frac{2}{2 + 1} = \frac{2}{3}. \quad P(\text{horse 1 winning}) = 1 - \frac{2}{3} = \frac{1}{3}.$$

Historical Note

Girolamo Cardano (1501–1576) was a doctor, mathematician, and astrologer. He investigated many interesting problems in probability and wrote a gambler's manual, probably because he was a compulsive gambler. He was the illegitimate son of a jurist and was at one time imprisoned for heresy for publishing a controversial horoscope of Christ's life. The reason we used the date of Cardano's death rather than the date of his birth in our list in Problem 15 is that it seems more significant. He had a firm belief in astrology and predicted the date of his death. However, when the day came, he was alive and well—so he drank poison toward the end of the day to make his prediction come true!

B Problems

11. On a single roll of a pair of dice, what is the probability that the sum is 7, given that at least one die came up 2?

12. On a single roll of a pair of dice, suppose we are given that at least one die came up 2. What is the probability that the sum is

    a. 3?   b. 4?   c. 2?

13. What are the odds in favor of rolling a 7 or 11 on a single roll of a pair of dice?

14. Problem 1 can be solved using Pascal's triangle. Can you show how?

See the historical note on page 501.

15. *Experiment*. Consider the birthdates of some of the mathematicians we've met in this book:

Abel	August 5, 1802
Cardano	September 21, 1576 (died)
Descartes	March 31, 1596
Euler	April 15, 1707
Fermat	August 20, 1601 (baptized)
Galois	October 25, 1811
Gauss	April 30, 1777
Newton	December 25, 1642
Pascal	June 19, 1623
Riemann	September 31, 1815

    Add to this list the birthdates of the members of your class. What is your *guess* as to the probability that at least two persons in this group will have exactly the same birthdate (not counting the year)? Be sure to make your guess *before* finding out the birthdates of your classmates. The answer, of course, depends on the number of persons on the list. We've listed ten mathematicians, and you may have 20 persons in your class, giving 30 names on the list.

We do not know Fermat's birthdate, so we will use his baptismal date instead.

Mind Bogglers

16. Bill says to Pat, "You roll your dice and I'll roll mine. If I roll a 7 before you roll an 8, you win." Pat refuses because he knows the probability of rolling a 7 is 1/6, while the probability of rolling an 8 is 5/36. Bill then says, "OK, then I'll roll a 7 before you roll a 6." Again, Pat refuses because the probability of rolling a 6 is the same as that of rolling an 8. "Here's my final offer," continues Bill. "I'll bet you that I can roll both a 6 and an 8 in fewer rolls than you can roll two 7s." "Ha, that bet I'll take," says Pat. Was he wise? Discuss the probability of these events.

17. Harry the Hustler, a world-renowned pool player, never went anywhere without his pool cue. One day while he was riding a cable car in San Francisco, his cue was knocked out of his hand, and it broke into three neat pieces. What is the probability that the three pieces can be placed together to form a triangle?

18. There is a game of solitaire called Tournament. The game involves two ordinary decks of cards shuffled together. Eight cards are dealt face up. In the game, it is desirable to have an ace or a king turned up. What is the probability that you will turn up at least one ace or king in the first eight cards?

Problems for Individual Study

19. *Birthday Problem.* Let's actually perform the experiment suggested by the *B.C.* cartoon on page 502 (except that we are concerned only with birthdates, not with mentalities). This problem is related to Problem 15. Instead of standing on one spot, let's consult an almanac and look up the birthdates of the Presidents of the United States. What would you guess is the probability that two of them were born on the same day (not counting the year)? As it turns out, Polk and Harding were both born on November 2.

TABLE 9.12 Probabilities for the Birthday Problem

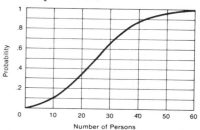

If 23 or more people are selected at random, the probability that two or more of them will have the same birthday is greater than 50%! This is a seemingly paradoxical situation that will fool most people. You might try it sometime when you are in a group of more than 23 persons. For example, if the class has 35 persons, the probability that two will have a birthday on the same day is almost 80%. For 2 persons, the probability is $1/365$; for more than 60 persons, it is almost 1 (after 60 the curve is indistinguishable from a straight line). The probability of 1 is not reached, of course, until there are 366 persons (see Table 9.12).

a. Pick 23 names at random from a biological dictionary or a *Who's Who* and verify some of the probabilities of the table. The more you go past 23, the greater the probability of the same birthday.

b. Find the probabilities for part (a). You can check almost any probability text for a discussion of this problem. One possibility is Emanuel Parzen's *Modern Probability Theory and Its Applications* (New York: Wiley, 1964), pp. 46–47.

20. *Computer Problem.* Use the following BASIC program to help you answer Problem 19.

```
 5   PRINT
 6   PRINT
10   PRINT TAB(10);"BIRTHDAY PROBLEM"
15   PRINT
16   PRINT
20   PRINT "NUMBER OF      PROBABILITY OF TWO"
25   PRINT "PERSONS      HAVING SAME BIRTHDAY"
26   PRINT
27   FOR I = 5 TO 100 STEP 5
28   LET M = 1
29   LET N = 1
30   FOR J = 1 TO N − 1
35   LET T = 365 − J
40   LET M = M * T/365
50   NEXT J
60   PRINT "     ";J;TAB(15);1 − M
65   NEXT I
70   END
```

Birthday Problem. If we select 23 persons out of a crowd, the chances are about even that two of them will have the same birthday.

9.4 MATHEMATICAL EXPECTATION

Smiles toothpaste is giving away $10,000. All you must do to have a chance to win is send in a postcard with your name on it (the fine print says you do not need to buy a tube of toothpaste). Is it worthwhile to enter?

Suppose the contest receives one million postcards (a conservative estimate). We wish to compute your *expected value* (or your expectation) in entering this contest. We denote the expectation by E and compute:

$$E = \text{(amount to win)} \cdot \text{(probability of winning)}$$
$$= \$10{,}000 \cdot \frac{1}{1{,}000{,}000}$$
$$= \$.01$$

A game is *fair* if the expected value equals the cost of playing the game. Is this game fair? If the toothpaste company charges you 1¢ to play the game, it is fair. But how much does it really cost you to enter the contest? How much does the postcard cost? We see that it is not a fair game.

EXAMPLE: Suppose you are going to roll two dice. You will be paid $3.60 if you roll two 1s. You do not receive anything for any other outcome. How much should you be willing to pay for the privilege of rolling the dice?

Solution: You should be willing to pay the mathematical expectation of the game, which is computed as follows.

$$E = \text{(amount of winnings)} \cdot \text{(probability of winning)}$$
$$= \quad \$3.60 \quad \cdot \quad \frac{1}{36}$$
$$= \$.10$$

You should be willing to pay 10¢ to play the game.

When we say that the expectation is 10¢, we certainly do *not* mean that you will win 10¢ every time you play. (Indeed, you will *never* win 10¢; it will be either $3.60 or nothing.) We mean that, if you were to play this game a very large number of times, you could expect to win an *average* of 10¢ per game.

Most games of chance will not be fair (the "house" must make a living). The question often is "Which games come closest to being fair?" We will consider this question in some of the problems.

Sometimes there is more than one payoff. A recent contest offered one grand prize worth $10,000, two second prizes worth $5000 each, and ten

According to the Encyclopedia Britannica, *each game gives the casino a mathematical expectancy of winning. Over the course of a year the odds of a casino ending up without a profit have been calculated at about 1 chance in 50 billion!*

third prizes worth $1000 each. The expectation can be computed by multiplying the amounts to be won by their respective probabilities and then adding the result. That is,

E = (amount of 1st prize) · P(1st prize) + (amount of 2nd prize) · P(2nd prize) + (amount of 3rd prize) · P(3rd prize).

Now, if we assume that there were one million entries and that winners' names were replaced in the pool after being drawn, we see that:

$$P(\text{1st prize}) = \frac{1}{1,000,000}$$

$$P(\text{2nd prize}) = \frac{2}{1,000,000}$$

$$P(\text{3rd prize}) = \frac{10}{1,000,000}$$

Hence,

$$E = \$10,000 \cdot \frac{1}{1,000,000} + \$5,000 \cdot \frac{2}{1,000,000} + \$1,000 \cdot \frac{10}{1,000,000}$$

$$= \$.01 + \$.01 + \$.01$$

$$= \$.03$$

Now we will give a formal definition of expectation.

DEFINITION: *Mathematical expectation:* If an event has several possible outcomes with probabilities $p_1, p_2, p_3, \ldots$, and for each of these outcomes the amount that can be won is $a_1, a_2, a_3, \ldots$, the *mathematical expectation,* E, of the event is

$$E = a_1 \cdot p_1 + a_2 \cdot p_2 + a_3 \cdot p_3 + \cdots$$

Definition of mathematical expectation

EXAMPLE: Suppose you roll one die. You are paid $5 if you roll a 1, and you pay $1 otherwise. What is the expectation?

Solution:

$$E = \begin{pmatrix} \text{amount} \\ \text{to win} \end{pmatrix} \begin{pmatrix} \text{probability} \\ \text{of winning} \end{pmatrix} + \begin{pmatrix} \text{amount} \\ \text{to lose} \end{pmatrix} \begin{pmatrix} \text{probability} \\ \text{of losing} \end{pmatrix}$$

$$= (5)\left(\frac{1}{6}\right) + (^-1)\left(\frac{5}{6}\right)$$

$$= 0$$

Notice that the amount to lose is a negative amount because it is something you must pay (instead of receive) if this event occurs.

We say that a game is fair if the expectation is 0. If the expectation is positive, the game is favorable to the player; if it is negative, the game is unfavorable to the player.

EXAMPLE: The contest by Kraft Miracle Whip offered the following prizes (as indicated in the fine print):

Prize	Value	Probability of Winning
Grand prize trip	$1500 $= x_1$	.000026 $= p_1$
Weber Kettle	110 $= x_2$	.000032 $= p_2$
Magic Chef Range	279 $= x_3$	.000016 $= p_3$
Murray Bicycle	191 $= x_4$	.000021 $= p_4$
Lawn Boy Mower	140 $= x_5$	.000026 $= p_5$
Samsonite Luggage	183 $= x_6$	.000016 $= p_6$

What is the expected value for this contest?

Solution:

$$E = x_1p_1 + x_2p_2 + x_3p_3 + x_4p_4 + x_5p_5 + x_6p_6$$
$$= 1500(.000026) + 110(.000032) + 279(.000016)$$
$$+ 191(.000021) + 140(.000026) + 183(.000016)$$
$$= .057563$$

The expected value is a little less than $6^¢$. Suppose we mail in our entry as specified in the rules. What is the cost of the stamp and the envelope?

EXAMPLE: Suppose you turned 21 years old today and wish to take out a $10,000 insurance policy for one year. How much should you be willing to pay for the policy?

Solution: This is also an expectation problem, in which you must find the probability that a 21-year-old person will die during his or her 21st year. Consulting a mortality table (see Table 9.13), we see that, for 96,478 persons of age 21, we could expect 177 of them to die within a year. Thus,

P(death during 21st year) = .0018

The expected value is $E = \$10,000 \cdot .0018$
$= \$18.00$

This means that, if the insurance company charged $18.00, their gain would be zero. The actual premium, then, would be $18.00 plus administrative costs and profit.

TABLE 9.13 Mortality Table. Based on 100,000 living at age 0, giving: ℓ_x, number of living; d_x, number of deaths; p_x, probability of living; q_x, probability of dying for age x from 0 to 99.

x	ℓ_x	d_x	p_x	q_x	x	ℓ_x	d_x	p_x	q_x
0	100000	708	.9929	.0071	50	87624	729	.9917	.0083
1	99292	175	.9982	.0018	51	86895	792	.9909	.0091
2	99117	151	.9985	.0015	52	86103	858	.9900	.0100
3	98966	144	.9986	.0015	53	85245	928	.9891	.0109
4	98822	138	.9986	.0014	54	84317	1003	.9881	.0119
5	98684	133	.9987	.0014	55	83314	1083	.9870	.0130
6	98551	128	.9987	.0013	56	82231	1168	.9858	.0142
7	98423	124	.9987	.0013	57	81063	1260	.9845	.0156
8	98299	121	.9988	.0012	58	79803	1357	.9830	.0170
9	98178	119	.9988	.0012	59	78446	1458	.9814	.0186
10	98059	119	.9988	.0012	60	76988	1566	.9797	.0204
11	97940	120	.9988	.0012	61	75422	1677	.9778	.0222
12	97820	123	.9988	.0013	62	73745	1793	.9757	.0243
13	97697	129	.9987	.0013	63	71952	1912	.9734	.0266
14	97568	136	.9986	.0014	64	70040	2034	.9710	.0291
15	97432	142	.9986	.0015	65	68006	2159	.9683	.0318
16	97290	150	.9985	.0016	66	65847	2287	.9653	.0347
17	97140	157	.9984	.0016	67	63560	2418	.9620	.0381
18	96983	164	.9983	.0017	68	61142	2548	.9583	.0417
19	96819	168	.9983	.0017	69	58594	2672	.9544	.0456
20	96651	173'	.9982	.0018	70	55922	2784	.9502	.0498
21	96478	177	.9982	.0018	71	53138	2877	.9459	.0542
22	96301	179	.9982	.0019	72	50261	2948	.9414	.0587
23	96122	182	.9981	.0019	73	47313	2993	.9368	.0633
24	95940	183	.9981	.0019	74	44320	3019	.9319	.0681
25	95757	185	.9981	.0019	75	41301	3030	.9266	.0734
26	95572	187	.9981	.0020	76	38271	3030	.9208	.0792
27	95385	190	.9980	.0020	77	35241	3020	.9143	.0857
28	95195	193	.9980	.0020	78	32221	2998	.9070	.0931
29	95002	198	.9979	.0021	79	29223	2957	.8988	.1012
30	94804	202	.9979	.0021	80	26266	2888	.8901	.1100
31	94602	207	.9978	.0022	81	23378	2790	.8807	.1194
32	94395	212	.9978	.0023	82	20588	2659	.8709	.1292
33	94183	218	.9977	.0023	83	17929	2499	.8606	.1394
34	93965	226	.9976	.0024	84	15430	2314	.8500	.1500
35	93739	235	.9975	.0025	85	13116	2113	.8389	.1611
36	93504	247	.9974	.0027	86	11003	1901	.8272	.1728
37	93257	261	.9972	.0028	87	9102	1685	.8149	.1851
38	92996	280	.9970	.0030	88	7417	1470	.8018	.1982
39	92716	301	.9968	.0033	89	5947	1263	.7876	.2124
40	92415	326	.9965	.0035	90	4684	1068	.7720	.2280
41	92089	354	.9962	.0039	91	3616	888	.7544	.2456
42	91735	383	.9958	.0042	92	2728	725	.7342	.2658
43	91352	414	.9955	.0045	93	2003	579	.7109	.2891
44	90938	447	.9951	.0049	94	1424	450	.6840	.3160
45	90491	484	.9947	.0054	95	974	341	.6499	.3501
46	90007	525	.9942	.0058	96	633	253	.6003	.3997
47	89482	569	.9937	.0064	97	380	185	.5132	.4869
48	88913	618	.9931	.0070	98	195	129	.3385	.6615
49	88295	671	.9924	.0076	99	66	66	.0000	1.0000

Based on the 1958 CSO Mortality Table prepared in cooperation with the National Association of Insurance Commissioners. Courtesy of the Society of Actuaries, Chicago, Illinois.

EXAMPLE: Suppose your "friend" George shows you three cards. One is white on both sides, one is black on both sides, and the last is black on one side and white on the other. He mixes the cards and lets you select one at random and place it on the table. Suppose the upper side turns out to be black. It is not the white-white card; it must be the black-black or the black-white card. "Thus," says George, "I'll bet you $1 that the other side is black." Would you play?

Perhaps you hesitate. Now George says he feels generous. You need to pay him 75¢ if you lose, and he will pay you $1 if he loses. Would you play now?

Compare this with Experiment 6 of Section 9.1.

Solution: Both these propositions can be answered by finding the expectation for the problem.

$$E = (\text{amount of black}) \cdot P(\text{black}) + (\text{amount if white}) \cdot P(\text{white})$$

Notice: An *incorrect* assumption is that $P(\text{black}) = P(\text{white}) = 1/2$. Recall that the empirical results of Experiment 6 should that $P(\text{black}) \neq 1/2$. What is it then? Consider the sample space. Let's distinguish between the front (side 1) and back (side 2) of each card. The sample space of *equally likely* events is:

	Card 1		Card 2		Card 3	
Side showing:	B_1	B_2	B	W	W_1	W_2
Side not showing:	B_2	B_1	W	B	W_2	W_1

We see that

$$P(\text{black face down}) = \frac{3}{6} = \frac{1}{2}$$

But we have additional information. We wish to compute the *conditional probability* of black, given that a black card is face up. Alter the sample space to take into account the additional information.

	Card 1		Card 2		Card 3	
Side showing:	B_1	B_2	B	~~W~~	~~W_1~~	~~W_2~~
Side not showing:	B_2	B_1	W	~~B~~	~~W_2~~	~~W_1~~

The conditional probability is:

$$P(\text{a black is on the bottom given that a black is on top}) = \frac{2}{3}$$

Thus the desired expectation is:

$$E = \binom{\text{amount}}{\text{to lose}} \binom{\text{probability}}{\text{of losing}} + \binom{\text{amount}}{\text{to win}} \binom{\text{probability}}{\text{of winning}}$$

$$= (^-.75)\left(\frac{2}{3}\right) + (1)\left(\frac{1}{3}\right)$$

$$\approx {}^-.17$$

Since the expectation is negative, you should not play.

PROBLEM SET 9.4

A Problems

1. Suppose you roll two dice. You will be paid $5 if you roll a double. You will not receive anything for any other outcomes. How much should you be willing to pay for the privilege of rolling the dice?

2. A magazine subscription service is having a contest in which the prize is $80,000. If the company receives one million entries, what is the expectation of the contest?

3. Suppose you have 5 quarters, 5 dimes, 10 nickels, and 5 pennies in your pocket. You reach in and choose a coin at random so you can tip your barber. What is the barber's expectation? What tip is the barber most likely to receive?

4. A game involves tossing two coins and receiving 50¢ if they are both heads. What is a fair price to pay for the privilege of playing?

5. Krinkles potato chips is having a "Lucky Seven Sweepstakes." The one grand prize is $70,000; 7 second prizes each pay $7000; 77 third prizes each pay $700; and 777 fourth prizes each pay $70. How much is the expectation of this contest if there are 10 million entries?

6. A punch-out card contains 100 spaces. One space pays $100, 5 spaces pay $10, and the others pay nothing. How much should you pay to punch out one space?

7. Use the mortality table (Table 9.13, page 507) to answer the following questions.
   a. What is the expected value for a 1-year $10,000 policy issued at age 10?
   b. What is the expected value for a 1-year $10,000 policy issued at age 38?
   c. What is the expected value for a 1-year $10,000 policy issued at age 65?

8. *Calculator Problem.* A company with 210 franchises conducted the contest described in the table at the right. What is the expectation for this game?

9. *Calculator Problem.* A company held a bingo contest for which the following chances of winning were given:

Prize	Number of Prizes Available	Approximate Probability of Winning
$1000	13	.000005
100	52	.00002
10	520	.0002
1	28,900	.010989
Total	29,485	.011111

Playing one card, your chances of winning are at least:

	One Card 1 Time	One Card 7 Times	One Card 13 Times
$25 prize	1 in 21,252	1 in 3036	1 in 1630
$3 prize	1 in 2125	1 in 304	1 in 163
$1 prize	1 in 886	1 in 127	1 in 68
Any prize	1 in 609	1 in 87	1 in 47

What is the expectation for playing one card 13 times?

State and national lotteries are often held to increase government income. The prizes are large, but, because of the large number of tickets sold, the expectation is low. The percentage returned to the gamblers is much smaller than with other forms of legal gambling.

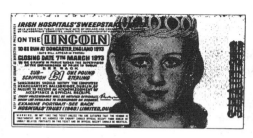

B Problems

10. A game involves drawing a single card from an ordinary deck. If an ace is drawn, you receive 50¢; if a heart is drawn, you receive 25¢; if the queen of spades is drawn, you receive $1. If the cost of playing is 10¢, should you play?

11. A realtor who takes the listing on a house to be sold knows that she will spend $800 trying to sell the house. If she sells it herself, she will earn 6% of the selling price. If another realtor sells a house from her list, our realtor will earn only 3% of the price. If the house is unsold in 6 months, she will lose the listing. Suppose the probabilities are as follows:

	Probability
Sell by herself	.50
Sell by another realtor	.30
Not sell in 6 months	.20

What is the expected profit for listing a $85,000 house?

12. Consider the following game where a player rolls a single die. If a prime (2, 3, or 5) is rolled, the player wins $2. If a square (1 or 4) is rolled, the player

wins \$1. However, if the player rolls a perfect number (6), then it costs the player \$11. Is this a good deal for the player or not?

13. *Slot Machine Problem.* Suppose a slot machine has three identical independent wheels, each with 13 symbols as follows:

1 bar	2 lemons
2 bells	3 plums
2 cherries	3 oranges

a. What is the probability of a jackpot (three bars)?
b. What is the probability of a cherry on the first wheel?
c. What is the probability of cherries on the first two wheels?
d. What is the probability of cherries on the first two wheels and a bar on the third wheel?
e. What is the probability of three cherries?
f. What is the probability of three oranges?
g. What is the probability of three plums?
h. What is the probability of three bells?

14. *Calculator Problem.* Suppose a slot machine has three independent wheels, as shown in Figure 9.4. Answer the questions posed in Problem 13.

Historical Note

This is a late-19th-century version of Charles Fey's first slot machine.

First Wheel	Second Wheel	Third Wheel
Orange	Cherry	Cherry
Cherry	Bar	Plum
Lemon	Plum	Cherry
Bell	Bar	Plum
Plum	Orange	Cherry
Orange	Bell	Cherry
Bar	Cherry	Plum
Plum	Bell	Bar
Orange	Cherry	Orange
Lemon	Plum	Bell
Orange	Orange	Cherry
Bar	Bell	Plum
Lemon	Cherry	Bar
Orange	Bell	Orange
Lemon	Bar	Plum
Plum	Bell	Cherry
Lemon	Bar	Orange
Orange	Bell	Cherry
Cherry	Cherry	Plum
Lemon	Lemon	Cherry

FIGURE 9.4

15. *Calculator Problem. Slot Machine Expectations.* Using Problem 14, compute the expectation for playing a slot machine that gives the following payoffs:

First one cherry	2 coins
First two cherries	5 coins
First two cherries and a bar	10 coins
Three cherries	10 coins
Three oranges	14 coins
Three plums	14 coins
Three bells	20 coins
Three bars (jackpot)	50 coins

16. *Calculator Problem.* Calculate the expectation for the Reader's Digest Sweepstakes described below.

You have six chances to win

$100.00 A MONTH FOR LIFE.

Or, if you prefer, $24,000.00 cash or $2,000.00 a month for a year.
Your six Sweepstakes Numbers will give you six chances to win it --
plus any of 2,815 other cash prizes.

IMPORTANT! Mail this Official Sweepstakes Record Card
not later than October 20, 1972...to have six chances to win any prize.

----------Tear off here—You may keep top portion for your records if you wish—Mail Official Record Card today!-----------

OFFICIAL PRIZE LIST

1 Prize of
$20,000.00 CASH

1 Prize of
$10,000.00 CASH

3 Prizes of
$5,000.00 CASH

10 Prizes of
$1,000.00 CASH

2,800 Prizes of
$25.00 CASH

© 1972, The Reader's Digest Association, Inc.

How the Sweepstakes Works:

1. All prizes will be awarded. Any prize set aside for a winning number that is not returned will be awarded in a drawing conducted by The Reuben H. Donnelley Corporation from all eligible entries submitted.
2. The approximate numerical odds of winning each prize are: Grand Prize – one in 6,851,000; $20,000 – one in 15,970,000; $10,000—one in 15,970,000; $5,000—one in 5,323,000; $1,000—one in 1,597,000; $25—one in 5,704.
3. Names, cities and states of residence of major prize winners will be published in Reader's Digest; all winners will be notified by mail; a printed list of winners will be sent upon request if you enclose a self-addressed, stamped envelope.
4. No purchase is required.
5. This Sweepstakes will begin when the first entry is received and the final closing date is December 26, 1972. (Postmarks determine dates.) Closing date for the Grand Prize is October 20, 1972.
6. Winners must verify that they are not employees of Reader's Digest or The Reuben H. Donnelley Corporation or members of their immediate families.
7. This is a national Sweepstakes.

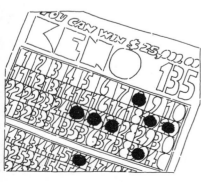

Problems 17–19 are concerned with the game of Keno.

In Keno, a player tries to guess in advance which numbers will be selected from a pot containing 80 numbers. Twenty numbers are selected at random. The player may choose from 1 to 15 spots and is paid according to the amounts shown in Table 9.14.

TABLE 9.14 Keno Payoffs

MARK 1 SPOT
70¢ Minimum

Catch	Play .70	Play 1.40
1	2.10	4.20

MARK 2 SPOTS
70¢ Minimum

Catch	Play .70	Play 1.40
2	8.50	17.00

MARK 3 SPOTS
70¢ Minimum

Catch	Play .70	Play 1.40
2	.70	1.40
3	30.00	60.00

MARK 4 SPOTS
70¢ Minimum

Catch	Play .70	Play 1.40
2	.70	1.40
3	3.00	6.00
4	75.00	150.00

MARK 5 SPOTS
70¢ Minimum

Catch	Play .70	Play 1.40
3	.70	1.40
4	6.50	13.00
5	580.00	1160.00

MARK 6 SPOTS
70¢ Minimum

Catch	Play .70	Play 1.40
3	.40	.80
4	2.50	5.00
5	70.00	140.00
6	1300.00	2600.00

MARK 7 SPOTS
70¢ Minimum

Catch	Play .70	Play 1.40
4	.70	1.40
5	16.00	32.00
6	260.00	520.00
7	6000.00	12000.00

MARK 8 SPOTS
70¢ Minimum

Catch	Play .70	Play 1.40
5	6.30	12.60
6	63.00	126.00
7	1155.00	2310.00
8	12600.00	**25000.00**

MARK 9 SPOTS
70¢ Minimum

Catch	Play .70	Play 1.40
5	2.10	4.20
6	31.50	63.00
7	234.50	469.00
8	3290.00	6580.00
9	12950.00	**25000.00**

MARK 10 SPOTS
70¢ Minimum

Catch	Play .70	Play 1.40
5	1.40	2.80
6	14.00	28.00
7	99.40	198.80
8	700.00	1400.00
9	3150.00	6300.00
10	13300.00	**25000.00**

MARK 11 SPOTS
70¢ Minimum

Catch	Play .70	Play 1.40
6	7.00	14.00
7	52.50	105.00
8	266.00	532.00
9	1400.00	2800.00
10	8750.00	17500.00
11	13650.00	**25000.00**

MARK 12 SPOTS
70¢ Minimum

Catch	Play .70	Play 1.40
6	4.00	8.00
7	20.00	40.00
8	160.00	320.00
9	700.00	1400.00
10	1800.00	3600.00
11	12500.00	**25000.00**
12	**25000.00**	**25000.00**

MARK 13 SPOTS
70¢ Minimum

Catch	Play .70	Play 1.40
6	1.50	3.00
7	11.00	22.00
8	55.00	110.00
9	500.00	1000.00
10	2500.00	5000.00
11	5000.00	10000.00
12	15000.00	**25000.00**
13	**25000.00**	**25000.00**

MARK 14 SPOTS
70¢ Minimum

Catch	Play .70	Play 1.40
6	2.00	4.00
7	6.00	12.00
8	23.00	46.00
9	200.00	400.00
10	600.00	1200.00
11	1800.00	3600.00
12	8000.00	16000.00
13	**25000.00**	**25000.00**
14	**25000.00**	**25000.00**

MARK 15 SPOTS
70¢ Minimum

Catch	Play .70	Play 1.40
6	1.00	2.00
7	6.00	12.00
8	16.00	32.00
9	60.00	120.00
10	160.00	320.00
11	2000.00	4000.00
12	8000.00	16000.00
13	20000.00	**25000.00**
14	**25000.00**	**25000.00**
15	**25000.00**	**25000.00**

EXAMPLE: Suppose a person picks one number and is paid $2.10 if he wins. For the privilege of playing this game, he pays $.70. Is this a fair game?

First we compute the probability of picking a winning number.

$$P(\text{one number}) = \frac{\text{number of ways of choosing one number from the 20}}{\text{number of ways of choosing one number from the 80}}$$

$$= \frac{_{20}C_1}{_{80}C_1} = \frac{20}{80} = \frac{1}{4}$$

Now we compute the expectation:

$$E = \$2.10 \cdot \frac{1}{4} = \$.53$$

The game becomes more complicated for picking more numbers. We see that on a three-spot ticket (the player chooses three numbers) there are four payoffs as follows:

Pick 0 numbers, win $0.
Pick 1 number, win $0.
Pick 2 numbers, win $.70.
Pick 3 numbers, win $30.00.

We compute the expected value as follows:

$$P(\text{picking 3}) = \frac{_{20}C_3}{_{80}C_3} = \frac{20 \cdot 19 \cdot 18}{80 \cdot 79 \cdot 78} \approx .0139$$

$$P(\text{picking 2}) = \frac{_{20}C_2 \cdot {_{60}C_1}}{_{80}C_3} = \frac{20 \cdot 19 \cdot 60 \cdot 3}{80 \cdot 79 \cdot 78} \approx .1388$$

We see that the expected value of one 70¢ play is as follows:

$$E \approx \$30.00 \cdot .0139 + \$.70 \cdot .1388$$

$$\approx \$.51$$

Thus we can expect to win only about 51¢ while paying 70¢ to play the game.

17. Using Table 9.14, what is the expectation in playing Keno of picking one number and paying $1.40 to play?

18. What is the expectation for playing Keno by picking two numbers and paying 70¢ to play?

19. What is the expectation for playing a three-spot ticket and paying $1.40 to play?

Mind Bogglers

20. *Computer Problem.* Since, as we can see from Problems 17–19, the arithmetic in computing the expectations in Keno soon becomes unmanageable, let's turn to the computer. The BASIC program for computing the amount you can expect to receive by playing an "X-SPOT" Keno ticket is shown. This program is rather difficult to follow because of the various FOR

commands, and it should be examined only for individual study. After receiving the RUN command, the computer asks for the value of X (a positive integer) and then asks for the payoffs (these must be positive). The first value is the payoff for obtaining 0 of the numbers (this payoff is normally $0); the next value is the payoff for obtaining 1 number; and so on. Study the sample runs of this program.

```
RUN


THIS PROGRAM WILL COMPUTE THE AMOUNT YOU CAN EXPECT TO RECEIVE
BY PLAYING AN X-SPOT KENO TICKET.

WHAT IS THE VALUE OF X? 4

WHAT IS THE PAYOFF FOR CATCHING Ø NUMBERS? Ø
WHAT IS THE PAYOFF FOR CATCHING 1 NUMBERS? Ø
WHAT IS THE PAYOFF FOR CATCHING 2 NUMBERS? .7Ø
WHAT IS THE PAYOFF FOR CATCHING 3 NUMBERS? 3.ØØ
WHAT IS THE PAYOFF FOR CATCHING 4 NUMBERS? 75.ØØ


THE EXPECTATION FOR AN 4 -SPOT KENO TICKET IS .5Ø83429


READY


RUN


THIS PROGRAM WILL COMPUTE THE AMOUNT YOU CAN EXPECT TO RECEIVE
BY PLAYING AN X-SPOT KENO TICKET.

WHAT IS THE VALUE OF X? 5

WHAT IS THE PAYOFF FOR CATCHING Ø NUMBERS? Ø
WHAT IS THE PAYOFF FOR CATCHING 1 NUMBERS? Ø
WHAT IS THE PAYOFF FOR CATCHING 2 NUMBERS? Ø
WHAT IS THE PAYOFF FOR CATCHING 3 NUMBERS? .7Ø
WHAT IS THE PAYOFF FOR CATCHING 4 NUMBERS? 6.5Ø
WHAT IS THE PAYOFF FOR CATCHING 5 NUMBERS? 58Ø.ØØ


THE EXPECTATION FOR AN 5 -SPOT KENO TICKET IS .5114111


READY
```

```
 5   PRINT
 6   PRINT
10   PRINT "THIS PROGRAM WILL COMPUTE THE AMOUNT YOU CAN EXPECT TO ";
15   PRINT "RECEIVE"
20   PRINT "BY PLAYING AN X-SPOT KENO TICKET."
22   PRINT
25   PRINT "WHAT IS THE VALUE OF X";
30   INPUT R
```

```
 33   PRINT
 35   FOR H = 0 TO R
 40   PRINT "WHAT IS THE PAYOFF FOR CATCHING";H;"NUMBERS";
 45   INPUT P(H)
 50   NEXT H
 55   PRINT
 56   PRINT
 60   LET A1 = 0
110   LET C8 = 1
113   IF R = 0 GOTO 145
115   FOR I = 1 TO R
120   LET A = (81 − I)/I
125   LET C8 = A * C8
130   NEXT I
135   GOTO 165
145   LET C8 = 1
165   FOR X = 0 TO R
170   GOSUB 195
175   NEXT X
177   PRINT "THE EXPECTATION FOR AN";r;"-SPOT KENO TICKET IS";
180   PRINT A1
182   PRINT
185    GOTO 350
195   IF X = 0 GOTO 225
198   LET C6 = 1
200   FOR I = 1 TO X
205   LET A = (61 − I)/I
210   LET C6 = A * C6
215   NEXT I
220   GOTO 260
225   LET C6 = 1
260   IF R − X = 0 GOTO 300
263   LET C2 = 1
265   FOR I = 1 TO R − X
270   LET A = (21 − I)/I
275   LET C2 = A * C2
280   NEXT I
285   GOTO 330
300   LET C2 = 1
330   LET A1 = A1 + ((C2 * C6/C8) * P(R − X))
340   RETURN
350   END
```

Using this program, determine which Keno ticket gives the highest expectation for a 70¢ ticket.

21. *St. Petersburg Paradox.* Suppose you toss a coin and will win $1 if it comes up heads. If it comes up tails, you toss again. This time you will receive $2 if it comes up heads. If it comes up tails, toss again. This time you will receive $14 if it is heads. You continue in this fashion until you finally toss a head.

Would you pay $100 for the privilege of playing this game? What is the mathematical expectation for this game?

22. *Calculator Problem.* In Problem 24, Section 9.2, we found the probability of a particular full house. Find the probability of obtaining *any* full house.

Problem for Individual Study

23. *Computer Problem: Gambler's Financial Ruin.* In all casinos, the payoffs are adjusted so that the "house" has the advantage. This means that your expectation is always less than the amount you must pay to play the game. Thus, if you continue to play long enough, regardless of your original bankroll, you will meet your financial ruin. The BASIC program for finding the time to a gambler's ruin is shown.

All that is necessary is a beginning bankroll and the computed expectation for the game (both the bankroll and expectation must be greater than 0). Notice that the program makes the assumption that a $1 bet is made every minute continuously—with no time off. If you have ever been in a gambling casino, you will know that some people try to do precisely that—bet continuously, with no time off for anything!

"How did you do with your new blackjack system?"

```
  5   PRINT
  6   PRINT
  7   PRINT
 10   PRINT "THIS PROGRAM WILL FIND THE LENGTH OF TIME TO A GAMBLER'S";
 13   PRINT " RUIN."
 15   PRINT "THE PROGRAM ASSUMES A $1 BET PER MINUTE."
 17   PRINT
 20   PRINT "WHAT IS THE GAMBLER'S ORIGINAL BANKROLL";
 22   INPUT B
 24   PRINT "WHAT IS THE EXPECTATION FOR THE GAME";
 26   INPUT E
 30   PRINT
 35   IF E > 1 GOTO 200
 40   IF E = 1 GOTO 250
 45   LET M = B/(1 − E)
 50   LET H = M/60
 55   LET D = H/24
100   LET A1 = D − INT(D)
105   LET A2 = 24*A1
110   LET A3 = A2 − INT(A2)
115   LET A4 = 60*A3
120   PRINT "FINANCIAL RUIN IN";
125   PRINT INT(D);
130   PRINT "DAYS,";
135   PRINT INT(A2);
140   PRINT "HOURS, AND ";
145   PRINT INT(A4);
150   PRINT "MINUTES."
155   PRINT "GOOD LUCK. BUT AS YOU SEE, FINANCIAL RUIN IS INEVITABLE."
157   PRINT
158   PRINT
159   PRINT
160   GOTO 300
200   PRINT "YOU HAVE FOUND THE PERFECT GAME! IF YOU PLAY LONG ENOUGH";
203   PRINT " YOU WILL"
205   PRINT "BE A WEALTHY PERSON."
206   PRINT
207   PRINT
208   PRINT
210   GOTO 300
```

```
250   PRINT "THIS IS A FAIR GAME. YOU CAN BREAK EVEN IF YOU PLAY";
255   PRINT "LONG ENOUGH."
260   PRINT
270   PRINT
280   PRINT
290   PRINT
300   END
```

Some sample runs of this program are shown.

```
RUN


THIS PROGRAM WILL FIND THE LENGTH OF TIME TO A GAMBLER'S RUIN.
THE PROGRAM ASSUMES A $1 BET PER MINUTE.

WHAT IS THE GAMBLER'S ORIGINAL BANKROLL? 100
WHAT IS THE EXPECTATION FOR THE GAME? .9474

FINANCIAL RUIN IN 1 DAYS, 7 HOURS, AND  41 MINUTES.
GOOD LUCK. BUT AS YOU SEE, FINANCIAL RUIN IS INEVITABLE.


READY

RUN


THIS PROGRAM WILL FIND THE LENGTH OF TIME TO A GAMBLER'S RUIN.
THE PROGRAM ASSUMES A $1 BET PER MINUTE.

WHAT IS THE GAMBLER'S ORIGINAL BANKROLL? 100
WHAT IS THE EXPECTATION FOR THE GAME? 1

THIS IS A FAIR GAME. YOU CAN BREAK EVEN IF YOU PLAY LONG ENOUGH.


READY

RUN


THIS PROGRAM WILL FIND THE LENGTH OF TIME TO A GAMBLER'S RUIN.
THE PROGRAM ASSUMES A $1 BET PER MINUTE.

WHAT IS THE GAMBLER'S ORIGINAL BANKROLL? 100
WHAT IS THE EXPECTATION FOR THE GAME? 1.25

YOU HAVE FOUND THE PERFECT GAME! IF YOU PLAY LONG ENOUGH YOU WILL
BE A WEALTHY PERSON.



READY
```

Using a computer, verify as many entries of Table 9.15 as possible.

TABLE 9.15 Time to Gambler's Ruin

Game	Expectation per Game ($1 bet)	Time Required to Lose $100 by Betting $1 per Minute		
		Days	Hours	Minutes
Roulette (with 0 and 00)				
1. Black or Red bet	$.9474	1	7	41
2. Odd or Even bet	.9474	1	7	41
3. Single Number bet	.9211	0	21	7
4. Two number bet	.8947	0	15	49
5. Three Number bet	.8684	0	12	39
6. Four Number bet	.8421	0	10	33
7. Five Number bet	.7895	0	7	55
8. Six Number bet	.7895	0	7	55
9. Twelve Number bet	.6316	0	4	31
Dice				
1. Eleven (or Three)	.8333	0	9	59
2. Twelve (or two)	.8333	0	9	59
3. Craps	.7778	0	7	30
4. Seven	.6667	0	5	0
5. Any Doubles	.6667	0	5	0
6. Field	.4444	0	3	0
7. Under Seven (or over)	.4167	0	2	51
Slot Machine (our hypothetical machine)	.6410	0	4	38
Keno				
1. One Spot	.7500	0	6	40
2. Two Spot	.7515	0	6	42
3. Three Spot	.7412	0	6	26
4. Four Spot	.7502	0	6	40
5. Five Spot	.7458	0	6	33
6. Six Spot	.7492	0	6	38
7. Seven Spot	.7477	0	6	36
8. Eight Spot	.7345	0	6	16
9. Nine Spot	.7453	0	6	32
10. Ten Spot	.7498	0	6	39
11. Eleven Spot	.7445	0	6	31
12. Twelve Spot	.7448	0	6	31
13. Thirteen Spot	.7490	0	6	38
14. Fourteen Spot	.7468	0	6	34
15. Fifteen Spot	.7444	0	6	31

9.5 SUMMARY AND REVIEW

CHAPTER OUTLINE

I. **Empirical Probability** is the result of experimentation.
Theoretical Probability is the result obtained by using a mathematical model. The result of the theoretical probability should be predictive of the empirical probability for a large number of trials.

II. **Definition of Probability:** If an experiment can occur in any of n mutually exclusive and equally likely ways, and if s of those ways are considered favorable, then the *probability* of the event E is given by

$$P(E) \;=\; \frac{s}{n} = \frac{\text{number of outcomes favorable to } E}{\text{number of all possible outcomes}}$$

III. Probability of the **Complement of an Event:**
$P(\overline{E}) = 1 - P(E)$

IV. The **Conditional Probability** of an event E, given that the event F has occurred, denoted by $P(E \mid F)$, is the probability of E *given* the knowledge that F has occurred. We consider the altered sample space and then compute the probability of E.

V. The **Odds in Favor** of an event E:

$$\frac{P(E)}{1 - P(E)}$$

The **Odds Against** an event E:

$$\frac{1 - P(E)}{P(E)}$$

VI. **Definition of Mathematical Expectation:** If an event has several possible outcomes with probabilities $p_1, p_2, p_3, \ldots$, and for each of these outcomes the amount that can be won is $a_1, a_2, a_3, \ldots$, the mathematical expectation of the event is $E = a_1 \cdot p_1 + a_2 \cdot p_2 + a_3 \cdot p_3 + \cdots$.

REVIEW PROBLEMS

1. A die is rolled. What is the probability that it comes up:
   a. 3?   b. 3 or 4?   c. 3 and 4?

2. A card is selected from an ordinary deck of cards. What is the probability that it is:
   a. a diamond?   b. a diamond and a 2?   c. a diamond or a 2?

3. Define probability.

4. What is the probability that Clumsy Carp will take a fall, according to the *B.C.* cartoon?

5. Two dice are tossed. What is the probability that the sum is 6 or at least one of the two faces is odd?

6. An urn contains three orange and two purple balls, and two are chosen at random. What is the probability of obtaining two orange balls if we know that the first drawn was orange and we draw the second ball:
   a. after replacing the first ball?
   b. without replacing the first ball?

7. Two dice are thrown. What is the probability of at least one being a 6? What is the probability of both being a 6?

8. If $P(A) = {}^5/_{11}$, what is $P(\overline{A})$?

9. If Ferdinand the Frog has twice as much chance of winning the jumping contest as the other two champion frogs:
   a. What is the probability that Ferdinand will lose?
   b. What are the odds in favor of Ferdinand's winning?

10. A lottery offers a prize of a color TV (value $500), and 800 tickets are sold. What is the expectation if you buy three tickets? If the tickets cost $1 each, should you buy them? How much should you be willing to pay for the three tickets?

CULTURAL		MATHEMATICAL
1876: Bell invents telephone	**1875**	
1879: Edison's first light bulb		1880: Georg Cantor—irrational numbers, transcend-ental numbers, transfinite numbers
		1882: Lindemann—π a transcendental number, squaring the circle proved impossible
		1890: Peano—axioms for natural numbers
		1895: Poincaré—analysis
		1896: Hadamard and Pousson—proof of prime-number theorem
1903: First powered aircraft	**1900**	1899: Hilbert—calculus; poses 23 famous problems
		1900: Russell and Whitehead—*Principia Mathematica*, logic
1908: Henry Ford's first Model T		1906: Fréchet—abstract spaces
1914: World War I begins		1916: Einstein—general theory of relativity
1917: Russian Revolution		1917: Hardy and Ramanujan—analytic number theory
1920: U.S. women gain right to vote		
1924: Lenin dies, Stalin succeeds him		
1927: Lindbergh's solo transatlantic flight	**1925**	
1928: Fleming discovers penicillin		1931: Gödel's theorem
1929: Stock-market crash		
1933: Hitler becomes Chancellor of Germany		
1939: World War II begins		1939: Bourbaki—*Elements*
1941: Japan bombs Pearl Harbor		
1942: First controlled nuclear chain reaction		
1945: United Nations formed		
1946: First electronic computer		
1950: Korean War begins	**1950**	
1953: Watson and Crick discover double helix structure of DNA		1955: Homological algebra
1955: Salk polio vaccine		
1957: Sputnik I		
1963: Vietnam War begins		1963: Cohen—continuum hypothesis
1967: First human heart transplant		
1969: First man on the moon		Albert Einstein
1976: Physicists discover the "Charmed Quark"	**1975**	1976: Appel and Haken solve four-color problem

10
THE NATURE OF
STATISTICS

There are three kinds of lies:
lies, damned lies, and statistics.

Benjamin Disraeli

10.1 FREQUENCY DISTRIBUTIONS

Undoubtedly, you have some idea about what is meant by the term *statistics*. For example, we hear about:

1. the latest statistics on the cost of living;
2. statistics on population growth;
3. the Gallup Poll's use of statistics to predict election outcomes;
4. the Nielsen ratings, which show that one show has 30% more viewers than another;
5. baseball statistics.

We could go on and on, but we see that there are two main uses for the word *statistics*. First, we use the term to mean a mass of data, including charts and tables. Second, the word refers to *statistical methods,* which are techniques used in the collection of numerical facts, called *data,* the analysis and interpretation of these data, and, finally, the presentation of these data. In this chapter we'll introduce some of these statistical methods.

In the last chapter, we performed experiments and gathered data. For example, in Experiment 4, we rolled a pair of dice 50 times. The outcomes were:

3	2	6	5	3	8	8	7	10	9
7	5	12	9	6	11	8	11	11	8
7	7	7	10	11	6	4	8	8	7
6	4	10	7	9	7	9	6	6	9
4	4	6	3	4	10	6	9	6	11

The next step was to organize the data in a convenient way. We used a *frequency distribution,* as shown in Table 10.1.

TABLE 10.1 Frequency Distribution for an Experiment of Rolling a Pair of Dice 50 Times

Outcome	Tally	Frequency
2	I	1
3	III	3
4	IIII	5
5	II	2
6	IIII IIII	9
7	IIII III	8
8	IIII I	6
9	IIII I	6
10	IIII	4
11	IIII	5
12	I	1

EXAMPLE: The waiting time in 1974 for a marriage license by state was as follows:

Alabama—None　　　　　Alaska—3 days
Arizona—2 days　　　　　Arkansas—3 days
California—None　　　　　Colorado—None
Connecticut—4 days　　　Delaware—1 day
Florida—3 days　　　　　Georgia—3 days
Hawaii—None　　　　　　Idaho—3 days
Illinois—None　　　　　　Indiana—3 days
Iowa—3 days　　　　　　Kansas—3 days
Kentucky—3 days　　　　Louisiana—3 days
Maine—5 days　　　　　　Maryland—48 hours
Massachusetts—3 days　　Michigan—3 days
Minnesota—5 days　　　　Mississippi—3 days

Missouri—3 days
Nebraska— None
New Hampshire—5 days
New Mexico—3 days
North Carolina— None
Ohio—5 days
Oregon—7 days
Rhode Island— None
South Dakota— None
Texas— None
Vermont—5 days
Washington—3 days
Wisconsin—5 days

Montana—5 days
Nevada— None
New Jersey—72 hours
New York—1 day
North Dakota— None
Oklahoma—3 days
Pennsylvania—3 days
South Carolina—24 hours
Tennessee—3 days
Utah— None
Virginia— None
West Virginia—3 days
Wyoming— None

Prepare a frequency distribution for these data.

Solution: we make three columns. The first column will contain the number of days' wait (notice that some times are given in hours; these must be converted to days), the second will have the tally, and the third, the frequency.

Days; Wait for Marriage License	Tally	Frequency (Number of States)
0	⊬⊬⊬ ⊬⊬⊬ ⊬⊬⊬	15
1					3
2				2
3	⊬⊬⊬ ⊬⊬⊬ ⊬⊬⊬ ⊬⊬⊬		21
4			1
5	⊬⊬⊬ ||	7
6		0
7			1
Total		50

A disadvantage of a frequency distribution is that it lacks visual appeal. Data classified into groups are often represented by means of bar graphs. To construct a bar graph, draw a horizontal axis and a vertical axis, as shown. We have labeled the vertical axis ''Frequency'' and the horizontal axis ''Outcomes of Experiment of Rolling a Pair of Dice.'' The data are from Table 10.1

EXAMPLE: The numbers of years several leading batters played in the major leagues are as follows:

Ty Cobb	24	George Sisler	16
Rogers Hornsby	23	Nap Lajoie	21
Ed Delehanty	16	Cap Anson	22

𝕳istorical 𝕹ote

Even though the origins of statistics are ancient, the development of modern statistical techniques began in the 16th century. Adolph Quetelet (1796–1874) was the first person to apply statistical methods to accumulated data. He correctly predicted (to his own surprise) the crime and mortality rates from year to year. Florence Nightingale (1820–1910), Gregor Mendel (1822–1884), and Sir Francis Galton (1822–1911) were all proponents of the use of statistics in their work. Since the advent of the computer, statistical methods have come within the reach of almost everyone. Today, statistical methods are strongly influencing agriculture, biology, business, chemistry, economics, education, electronics, medicine, physics, psychology, and sociology.

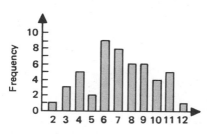

Outcomes of Experiment of Rolling a Pair of Dice

Dan Brouthers	19	Sam Thompson	15
Willie Keeler	19	Al Simmons	20
Ted Williams	19	Eddie Collins	25
Tris Speaker	22	Paul Waner	20
Billy Hamilton	14	Stan Musial	22
Harry Heilmann	17	Henie Manush	17
Babe Ruth	22	Honus Wagner	21
Jesse Burkett	16	Joe Dimaggio	13
Bill Terry	14	Jimmy Foxx	20
Lou Gehrig	17		

Give the frequency distribution and draw the bar graph for these data.

Solution:

Number of Years	Tally	Frequency	Number of Years	Tally	Frequency
13	|	1	20	|||	3
14	||	2	21	||	2
15	|	1	22	||||	4
16	|||	3	23	|	1
17	|||	3	24	|	1
18		0	25	|	1
19	|||	3			

The bar graph for this distribution is given in Figure 10.1

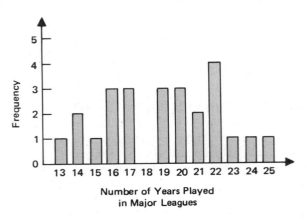

FIGURE 10.1 Bar graph

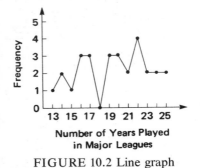

FIGURE 10.2 Line graph

Another type of graph is the *line graph,* which uses dots that are marked and then connected in order. The line graph for Figure 10.1 is shown in Figure 10.2.

EXAMPLES: The monthly normal rainfall for Honolulu and New Orleans is given in Figure 10.3.

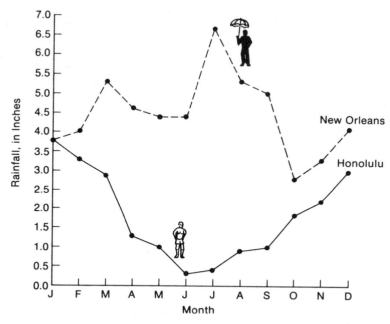

FIGURE 10.3 Monthly normal rainfall for two U.S. cities

1. During which month is there the most rain in New Orleans? In Honolulu?
2. During which month is there the least rain in New Orleans? In Honolulu?

Solutions:

1. It rains most in New Orleans in July and in Honolulu in January.
2. The least rain falls in New Orleans in October and in Honolulu in June.

There are other kinds of graphs, some of which are discussed in the problem set.

PROBLEM SET 10.1

A Problems

1. The heights of 30 students are as follows (figures rounded to the nearest inch):

66	68	64	70	67	67	68	64	65	66
64	70	72	71	69	64	63	70	71	63
68	67	67	65	69	65	67	66	69	69

Give the frequency distribution, and draw a bar graph to represent these data.

Hank Aaron hitting his 715th home run

2. The wages of the employees of a small accounting firm as as follows:

$10,000	$15,000	$15,000	$15,000
$20,000	$20,000	$25,000	$40,000
$60,000	$8000	$8000	$8000
$8000	$8000	$6000	$4000

Give the frequency distribution, and draw a bar graph to represent these data.

3. Hank Aaron's home-run record for 1954–1974 is as follows:

1954	13	1955	27	1956	26
1957	44	1958	30	1959	39
1960	40	1961	34	1962	45
1963	44	1964	24	1965	32
1966	44	1967	39	1968	29
1969	44	1970	38	1971	47
1972	34	1973	40	1974	20

Since there are 21 years, the graph would be quite large if we handled each year separately. Instead, we group together some of the years as shown in Table 10.2. Show these figures by using a line graph.

TABLE 10.2

Years	Number of Home Runs Hit by Aaron
1954–1956	66
1957–1959	113
1960–1962	119
1963–1965	100
1966–1968	113
1969–1971	129
1972–1974	94

The purchasing power of the dollar in 1940 was $2.38. In 1950 it was $1.39.

4. The purchasing power of the dollar (1967 = $1.00) is shown (to the nearest cent):

1960	$1.13	1962	$1.10	1964	$1.07
1966	$1.03	1968	$ .96	1970	$ .86
1972	$ .80	1974	$ .68	1976	$ .59
1978	$ .49				

Show these figures by using a line graph.

5. The amount of time it takes for three leading pain relievers to reach your bloodstream is as follows:

Brand A   480 seconds
Brand B   490 seconds
Brand C   500 seconds

a. Make a bar graph using the scale shown in Figure 10.4a.
b. Make a bar graph using the scale shown in Figure 10.4b.

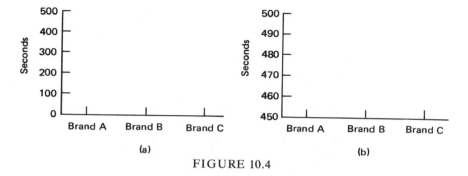

(a)

(b)

FIGURE 10.4

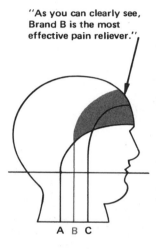

"As you can clearly see, Brand B is the most effective pain reliever."

Brands of Pain Relievers

c. If you were an advertiser working on a promotion campaign for Brand A, which graph would seem to give your product a more distinct advantage?
d. Consider the graph at the right. Does it tell you anything at all about the effectiveness of the three pain relievers? Discuss.

6. The percentage of claims paid by five leading insurance companies is as follows:

Company A   96.2%
Company B   94.1%
Company C   97.6%
Company D   96.1%
Company E   95.9%

a. Make a bar graph using the scale shown in Figure 10.5a.
b. Make a bar graph using the scale shown in Figure 10.5b.

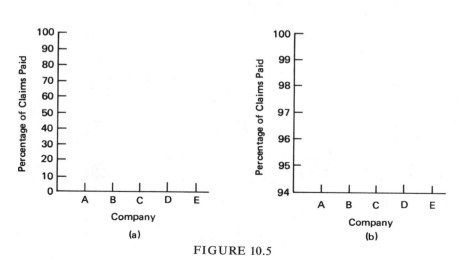

(a)

(b)

FIGURE 10.5

   c. If you were an advertiser working on a promotion campaign for Company C, which graph would seem to give your client a more distinct advantage?

7. Comment on the following graph and advertising statement.

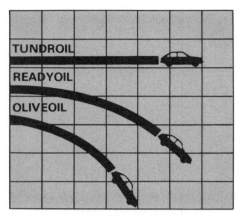

To prove that Tundroil is better suited for smaller, hotter, higher-revving engines we tested Tundroil against Readyoil and Oliveoil. As the graph above plainly shows, only Tundroil didn't break down.

8. Use the graphs in Figure 10.6 to answer the following questions.
   a. How many deaths occurred in 1975?
   b. What was the death rate per 100 million miles of travel in 1975?
   c. How many miles were traveled in 1976?
   d. Which year had the fewest number of deaths, and which year had the lowest death rate?

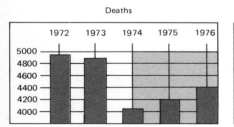

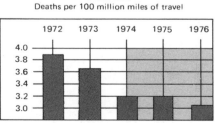

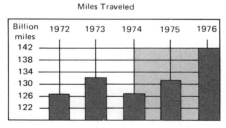

Shaded area reflects the speed limit change in 1974 to 55 mph.

FIGURE 10.6

B Problems

9. Following is a list of holidays in 1974. Find the frequency distribution for the number of holidays celebrated on each day of the week, and draw a bar graph to represent the data.

January

1 (Thu) New Year's Day
8 (Thu) Battle of New Orleans (Lousiana)
16 (Fri) Arbor Day (Florida)
19 (Mon) Robert E. Lee's Birthday

February

12 (Thu) Lincoln's Birthday
14 (Sat) Admission Day (Arizona)
16 (Mon) Washington's Birthday

March

2 (Tue) Texas Independence Day (Texas)
         Town Meeting Day (Vermont)
25 (Thu) Maryland Day (Maryland)
26 (Fri) Kuhio Day (Hawaii)
29 (Mon) Seward's Day (Alaska)

April

2 (Fri) Pascua Florida Day (Florida)
13 (Tue) Jefferson's Birthday (Missouri)
16 (Fri) Good Friday
19 (Mon) Patriot's Day (Maine &
         Massachusetts)
21 (Wed) San Jacinto Day (Texas)
22 (Thu) Arbor Day (Nebraska)
26 (Mon) Fast Day (New Hampshire)
         Confederate Memorial Day
         (Alabama & Mississippi)
30 (Fri) Arbor Day (Utah)

May

4 (Tue) Rhode Island Independence
         (Rhode Island)
10 (Mon) Confederate Memorial Day
         (North Carolina & South Carolina)
20 (Thu) Mecklenburg Day (North Carolina)
31 (Mon) Memorial Day

June

3 (Thu) Confederate Memorial Day
14 (Mon) Flag Day (Pennsylvania)
20 (Sun) West Virginia Day (West Virginia)

July

4 (Sun) Independence Day
24 (Sat) Pioneer Day (Utah)

August

2 (Mon) Colorado Day (Colorado)
9 (Mon) Victory Day (Rhode Island)
16 (Mon) Bennington Battle Day (Vermont)
20 (Fri) Admission Day (Hawaii)
27 (Fri) Lyndon Johnson's Birthday (Texas)
30 (Mon) Huey B. Long's Birthday
         (Louisiana)

September

6 (Mon) Labor Day
9 (Thu) Admission Day (California)
12 (Sun) Defenders Day (Maryland)

October

11 (Mon) Columbus Day
25 (Mon) Veterans Day
31 (Sun) Nevada Day (Nevada)

November

1 (Mon) All Saints Day (Louisiana)
2 (Tue) Election Day
25 (Thu) Thanksgiving

December

10 (Fri) Wyoming Day (Wyoming)
25 (Sat) Christmas Day

10. Listed are the home-run champions for the American and National Leagues for every year from 1901 to 1978. Divide the data into five even categories, and construct a line graph for the number of players in each category. Do the National League and the American League separately, but put the graphs on the same coordinate system.

National League

Year	Player	No. of Home Runs	Year	Player	No. of Home Runs	Year	Player	No. of Home Runs	Year	Player	No. of Home Runs
1901	Crawford	16	1921	Kelly	23	1939	Mize	28	1959	Mathews	46
1902	Leach	6	1922	Hornsby	42	1940	Mize	43	1960	Banks	41
1903	Sheckard	9	1923	Williams	41	1941	Camilli	34	1961	Cepeda	46
1904	Lumley	9	1924	Fournier	27	1942	Ott	30	1962	Mays	49
1905	Odwell	9	1925	Hornsby	39	1943	Nicholson	29	1963	Aaron	44
1906	Jordan	12	1926	Wilson	21	1944	Nicholson	33		McCovey	
1907	Brain	10	1927	Wilson	30	1945	Holmes	28	1964	Mays	47
1908	Jordan	12		Williams		1946	Kiner	23	1965	Mays	52
1909	Murray	7	1928	Wilson	31	1947	Kiner	51	1966	Aaron	44
1910	Beck	10		Bottomley			Mize		1967	Aaron	39
	Schulte		1929	Klein	43	1948	Kiner	40	1968	McCovey	36
1911	Schulte	21	1930	Wilson	56		Mize		1969	McCovey	45
1912	Zimmerman	14	1931	Klein	31	1949	Kiner	54	1970	Bench	45
1913	Cravath	19	1932	Klein	38	1950	Kiner	47	1971	Stargell	48
1914	Cravath	19		Ott		1951	Kiner	42	1972	Bench	40
1915	Cravath	24	1933	Klein	28	1952	Kiner	37	1973	Stargell	44
1916	Robertson	12	1934	Ott	35		Sauer		1974	Schmidt	36
	Williams			Collins		1953	Mathews	47	1975	Schmidt	38
1917	Robertson	12	1935	Berger	34	1954	Kluszewski	49	1976	Schmidt	38
	Cravath		1936	Ott	33	1955	Mays	51	1977	Foster	52
1918	Cravath	8	1937	Ott	31	1956	Snider	43	1978	Foster	40
1919	Cravath	12		Medwick		1957	Aaron	44			
1920	Williams	15	1938	Ott	36	1958	Banks	47			

American League

Year	Player	No. of Home Runs	Year	Player	No. of Home Runs	Year	Player	No. of Home Runs	Year	Player	No. of Home Runs
1901	Lajoie	13	1921	Ruth	59	1941	Williams	37	1962	Killebrew	48
1902	Seybold	16	1922	Williams	39	1942	Williams	36	1963	Killebrew	45
1903	Freeman	13	1923	Ruth	41	1943	York	36	1964	Killebrew	49
1904	Davis	10	1924	Ruth	46	1944	Etten	22	1965	Conigliaro	32
1905	Davis	8	1925	Meusel	33	1945	Stephens	24	1966	Robinson	49
1906	Davis	12	1926	Ruth	47	1946	Greenberg	44	1967	Yastrzemski	44
1907	Davis	8	1927	Ruth	60	1947	Williams	32		Killebrew	
1908	Crawford	7	1928	Ruth	54	1948	DiMaggio	39	1968	Howard	44
1909	Cobb	9	1929	Ruth	46	1949	Williams	43	1969	Killebrew	49
1910	Stahl	10	1930	Ruth	49	1950	Rosen	37	1970	Howard	44
1911	Baker	9	1931	Gehrig	46	1951	Zernial	33	1971	Melton	33
1912	Baker	10		Ruth		1952	Doby	32	1972	Allen	37
1913	Baker	12	1932	Foxx	58	1953	Rosen	43	1973	Jackson	32
1914	Baker	8	1933	Foxx	48	1954	Doby	32	1974	Allen	32
	Crawford		1934	Gehrig	49	1955	Mantle	37	1975	Jackson	36
1915	Roth	7	1935	Foxx	36	1956	Mantle	52		Scott	
1916	Pipp	12		Greenberg		1957	Sievers	42	1976	Nettles	32
1917	Pipp	9	1936	Gehrig	49	1958	Mantle	42	1977	Rice	39
1918	Ruth	11	1937	DiMaggio	46	1959	Colavito	42	1978	Rice	46
	Walker		1938	Greenberg	58		Killebrew				
1919	Ruth	29	1939	Foxx	35	1960	Mantle	40			
1920	Ruth	54	1940	Greenberg	41	1961	Maris	61			

11. Refer to Table 9.10, on page 480.
    a. Select 100 random digits. Make a frequency distribution for the digits 0,1,2, . . . ,8,9. Draw a line graph.
    b. Repeat part (a) for a different 100 random digits. Draw a line graph.
    c. Combine the frequencies for the digits from parts (a) and (b), and draw a line graph.

12. Another type of graph is the *circle graph*, which is useful in displaying percentages or parts of a whole. For example, the amount of electricity used in a typical "all-electric" home is shown by the graph in Figure 10.7. If in a certain month a home used 1100 kwh (kilowatt-hours), the number of kwh used by the water heater is found by looking at the graph and finding that the water heater consumes about 11% of the electricity:

$$1100 \text{ kwh} \times .11 = 121 \text{ kwh}$$

    a. Find the amount of electricity used by the clothes dryer in this home.
    b. Find the amount of electricity used by the refrigerator in this home.

13. *Calculator Problem.* According to the UCLA General Catalog, a student's principal expenses for a year are:

Registration fee	$ 300	Books & supplies	200
Educational fee	300	Room & board	1325
Miscellaneous fees	25	Miscellaneous	500
		Total	$2650

    Make a circle graph showing these data.

14. The typical percentage of income allotted to various items is shown in Figure 10.8. Use this figure to answer the following questions.
    a. If a family's monthly income increases from $850 to $900, which items remain about the same in percentage?
    b. If a family's income increases from $840 to $920, what is the percentage of change in the amount spent for food?
    c. Does a family with a monthly income of $940 spend more for food than a family with a monthly income of $830? How much more or less?

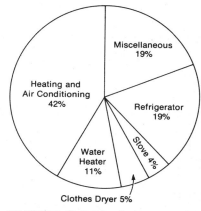

FIGURE 10.7 Distribution of the electricity used by a typical home during one month

Hint for Problem 13: When constructing a circle graph, you must first find the central angle for each of the given categories. For example, the registration fee represents

$$\frac{300}{2650} \approx .1132$$

of the total. Thus the angle must be .1132 of one revolution (360°), or

$$.1132(360) \approx 40.75$$

This means that the central angle for the registration fee is about 41°. Find the angles for the other categories in the same way.

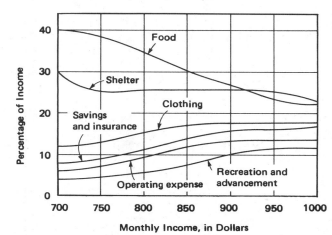

FIGURE 10.8 Family income and expenditures

15. Answer the following questions by referring to Figure 10.9.
   a. In which year was the proportion of national defense expenditures to total expenditures the smallest?
   b. In which year was the proportion of national defense expenditures to total income the smallest?
   c. In which year was the proportion of national defense expenditures to total expenditures the greatest?
   d. In which year was the expenditures over income the greatest?

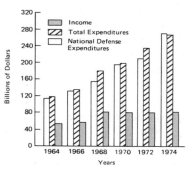

FIGURE 10.9 Government income and expenditures for selected years

Pictogram

16. Another graphical technique, which is sometimes misused, is the *pictogram*. Consider the marital status of persons age 65 and older in 1978. The raw data are shown in Table 10.3 and are illustrated by a bar graph in Figure 10.10. However, this information might also be shown as a pictogram, as in Figure 10.11.

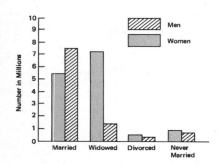

FIGURE 10.10 Bar graph showing the marital status of persons age 65 and older in 1978

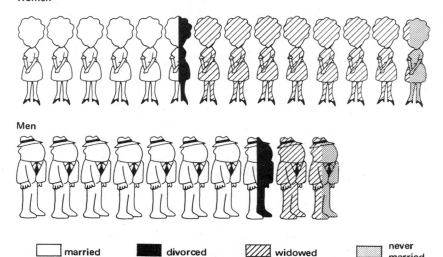

FIGURE 10.11 Pictogram showing the marital status in 1978 of persons 65 and older (each figure represents approximately one million persons).

TABLE 10.3 1978 Marital Status of Persons Age 65 and Older (in Millions)

	Married	Widowed	Divorced	Never Married
Women	5.4	7.2	.4	.8
Men	7.5	1.4	.3	.6

Figures rounded to the nearest 100,000.

Some writers wish to exaggerate the differences, so they use the size of the objects as the measuring point in drawing the pictogram, as shown in Figure 10.12. This is very misleading, since the figures in Figures 10.11 and 10.12 are viewed as three dimensional, and the differences seem much larger than they actually are. Let's look at a simpler case. Consider the following "pictogram":

FIGURE 10.12 Pictogram showing the number of widowed persons age 65 and older in 1978

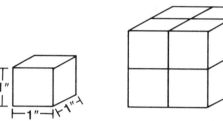

a. How do the heights of the objects compare?
b. How do the areas of the faces compare?
c. How do the volumes compare?

Mind Bogglers

17. In Chapter 6, we worked on some modular codes. Frequency distributions are also used to break codes. In ordinary English the expected frequency of each letter of the alphabet is known and is given in Table 10.4. This means that, if a message contained 200 letters, we would expect the letter *E* to occur about 26 times and the *R* to occur about 13 times. The longer the message, the more accuracy we could expect from these precentages. Consider the following coded message:

VKU.OUY.VX.VKU.JPXJUP.QUEBJKUPBFC.XI.DUEPUV.
NUDDTCUD.BD.JTVBUFEU.TFQ.JUPDUPLUPTFEU,.BI.
YXR.HPUTO.VKU.EXQU.QX.FXV.HU.T.VTVVWUVTWU,

a. Prepare a frequency distribution for the letters.
b. The letter to occur most often is probably an *E* or a space. Which character do you think represents a space in this coded message?
c. The second most frequent letter is probably the *T*. Look for combinations of three letters that might be the word *the*. Now you probably know the letters, *T, H,* and *E*, as well as the space.
d. Next, find the coded letters for the *A* and the *O*. There appear to be three possibilities. Suppose you know that the words *secret messages* appear in the message. Finish the problem by breaking the code.

TABLE 10.4 Decoding Table

Letter	Expected Frequency
E	13%
T	9%
A, O	8%
N	7%
I, R	6.5%
S, H	6%
D	4%
L	3.5%
C, U, M	3%
F, P, Y	2%
W, G, B	1.5%
V	1%
K, X, J	.5%
Q, Z	.2%
SPACE	Frequency varies greatly depending on word size but will be about 10–20%.

18. Read the SAAB advertisement carefully. There is a fallacy in the statistics given about the car. See if you can find it.

There are two cars built in Sweden. Before you buy theirs, drive ours.

When people who know cars think about Swedish cars, they think of them as being strong and durable. And conquering some of the toughest driving conditions in the world.

But, unfortunately, when most people think about buying a Swedish car, the one they think about usually isn't ours. (Even though ours doesn't cost any more.)

Ours is the SAAB 99E. It's strong and durable. But it's also a lot different from their car.

Our car has Front-Wheel Drive for better traction, stability and handling.

It has a 1.85 liter, fuel-injected, 4-cylinder, overhead cam engine as standard in every car. 4-speed transmission is standard too. Or you can get a 3-speed automatic (optional).

Our car has four-wheel disc brakes and a dual-diagonal braking system so you stop straight and fast every time.

It has a wide stance. (About 55 inches.) So it rides and handles like a sports car.

Outside, our car is smaller than a lot of "small" cars. 172″ overall length, 57″ overall width.

Inside, our car has bucket seats up front and a full five feet across in the back so you can easily accommodate five adults.

It also has more headroom than a Rolls Royce and more room from the brake pedal to the back seat than a Mercedes 280. And it has factory air conditioning as an option.

There are a lot of other things that make our car different from their car. Like roll cage construction and a special "hot seat" for cold winter days.

So before you buy their car, stop by your nearest SAAB dealer and drive our car. The SAAB 99E. We think you'll buy it instead of theirs. **SAAB 99E**

Phone 800-243-6000 toll-free for the name and location of the SAAB dealer nearest you. In Connecticut, call 1-800-942-0655.

Problems for Individual Study

19. Collect examples of good statistical graphs and examples of misleading graphs. Use some of the leading newspapers and national magazines.

20. Read *How to Lie with Statistics,* by Darrell Huff (New York: Norton, 1954), and make a presentation to the class. Illustrate your presentation with graphs and examples from the book.

10.2 DESCRIPTIVE STATISTICS

In the last section, we organized data into a frequency distribution and then discussed their representation as a graph. However, there are some properties of data that can help us to interpret masses of information.

Lucy is using one of these properties in the cartoon. Do you suppose her dad necessarily bowled better on Monday nights than on Thursday nights? Don't be too hasty to say "yes" unless you first look at the scores making up these averages.

	Monday Night	*Thursday Night*
Game 1:	175	180
Game 2:	150	130
Game 3:	160	161
Game 4:	180	185
Game 5:	160	163
Game 6:	183	185
Game 7:	287	186
Total:	1295	1190

To find the averages used by Lucy, we divide the total scores by the number of games:

$$\text{Monday Night} \qquad \text{Thursday Night}$$
$$\frac{1295}{7} = 185 \qquad \frac{1190}{7} = 170$$

However, if we consider the games separately, we see that Lucy's dad did better on Thursday in five out of seven games. Are there any other properties of the bowling scores that would tell us this fact?

The average used by Lucy is only one kind of statistical measure that can be used. It is the measure that most of us think of when we hear someone use the word "average." It is called the *mean.* Other statistical measures, called *measures of central tendency,* are given by the following definition:

Measures of central tendency

Mean

Median

Mode

DEFINITION: Given a set of data, three common *measures of central tendency* are:

1. *The mean:* the number found by adding the data and then dividing by the number of data. The mean is usually denoted by $\bar{x}$.
2. *The median:* the middle number when the numbers are arranged in order of size. If there are two middle numbers (in the case of an even number of data), the median is the mean of these two middle numbers.
3. *The mode:* the value that occurs most frequently. If there is no number that occurs more than once, there is no mode. It is possible to have more than one mode.

If we consider these other measures for the bowling scores, we find:

(a) The median

Monday Night		Thursday Night
150		130
160		161
160	The middle	163
175	←number is the→	180
180	median.	185
183		185
287		186

For the median, we arrange the data in order from lowest to highest.

(b) The mode

Monday Night		Thursday Night
150		130
160 ⎱	The mode is	161
160 ⎰	the number that	163
175	occurs	180
180	most frequently.	⎱185
183		⎰185
287		186

There is a rather nice physical example that illustrates the idea of the mean. Consider a seesaw that consists of a plank and a movable support (called a fulcrum).

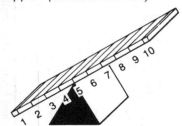

We assume that the plank has no weight and is marked off into units as shown. Now let's place some one-pound weights in the

When we compare the three measures of the bowling scores, we find the following:

	Monday Night	Thursday Night
Mean	185	170
Median	175	180
Mode	160	185

We are no longer convinced that Lucy's dad "did better" on Monday nights than on Thursday nights.

EXAMPLES: Find the mean, median, and mode for the following sets of numbers.
1. 3,5,5,8,9
2. 4,10,9,8,9,4,5
3. 6,5,4,7,1,9

Solutions:

1. Mean: $\dfrac{\text{sum of the terms}}{\text{number of terms}} = \dfrac{3 + 5 + 5 + 8 + 9}{5}$

$= \dfrac{30}{5}$

$= 6$

Median: Arrange in order: 3,5,5,8,9.
The middle term, 5, is the median.
Mode: The term that occurs most frequently is the mode, which is 5.

2. Mean: $\dfrac{4 + 10 + 9 + 8 + 9 + 4 + 5}{7} = \dfrac{49}{7}$

$= 7$

Median: 4,4,5,8,9,9,10
The median is 8.
Mode: This set of data is *bimodal*, with modes 4 and 9.

3. Mean: $\dfrac{6 + 5 + 4 + 7 + 1 + 9}{6} = \dfrac{32}{6}$

≈ 5.33

Median: 1,4,5,6,7,9

$\dfrac{5 + 6}{2} = \dfrac{11}{2} = 5.5$

Mode: There is no mode.

Suppose our data are presented in a frequency distribution. For example, consider the days one must wait for a marriage license, as shown in Table 10.5. Suppose we're interested in the average wait for a marriage license in the United States.

1. *Mean*
To find the mean, we could, of course, add all 50 individual numbers. But, instead, notice that

0 occurs 15 times, so we write $0 \cdot 15$;
1 occurs 3 times, so we write $1 \cdot 3$;
2 occurs 2 times, so we write $2 \cdot 2$;
3 occurs 21 times, so we write $3 \cdot 21$;
⋮

positions of the numbers in our distribution. The balance point is the mean. In Example 1, we place weights at 3, 5 (two weights), 8, and 9.

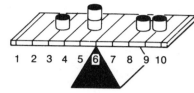

If we place the fulcrum at the mean, 6, the seesaw will balance.

Bimodal means that the set of data has two modes.

TABLE 10.5

Days' Wait for a Marriage License	Frequency (Number of States)
0	15
1	3
2	2
3	21
4	1
5	7
6	0
7	1
Total	50

Thus the mean is

$$\bar{x} = \frac{0 \cdot 15 + 1 \cdot 3 + 2 \cdot 2 + 3 \cdot 21 + 4 \cdot 1 + 5 \cdot 7 + 6 \cdot 0 + 7 \cdot 1}{50}$$

$$= \frac{0 + 3 + 4 + 63 + 4 + 35 + 0 + 7}{50}$$

$$= \frac{116}{50}$$

$$= 2.32$$

2. *Median*

Since there are 50 values, the mean of the 25th and 26th largest values is the median. From Table 10.5, we see that the 25th term is 3 and the 26th term is 3, so the median is

$$\frac{3 + 3}{2} = \frac{6}{2} = 3$$

3. *Mode*

The mode is the value that occurs most frequently, which is 3.

DISPERSION

The measures we've been discussing can help us interpret information, but they do not give the whole story. For example, consider the two sets of data given in Figure 10.13.

	First Example	Second Example
	8	2
	9	9
	9	9
	9	12
	10	13
Mean:	9	9
Median:	9	9
Mode:	9	9

First Example (don't forget that the plank has no weight):

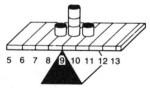

5 6 7 8 9 10 11 12 13

Second Example:

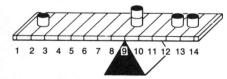

1 2 3 4 5 6 7 8 9 10 11 12 13 14

FIGURE 10.13 Comparison of three statistical measures for two sets of data

Notice that the two examples have the same mean, median, and mode, but the second set of data is more spread out than the first set. There are three measures of dispersion that we'll consider: the *range,* the *variance,* and the *standard deviation.*

The simplest measure of dispersion is the *range.*

Definition of range

DEFINITION: The *range* in a set of data is the difference between the largest and the smallest numbers in the set.

The ranges for the examples given in Figure 10.13 are:

First example: $10 - 8 = 2$
Second example: $13 - 2 = 11$

Notice that the range is determined only by the largest and the smallest numbers in the set; it does not give us any information about the other numbers. It thus seems reasonable to invent another measure of dispersion that takes into account all the numbers in the data.

Suppose we use the example of Figure 10.13. First we subtract each number from the mean:

First Example

Data	Difference from the Mean
8	1
9	0
9	0
9	0
10	$^-1$

Second Example

Data	Difference from the Mean
2	7
9	0
9	0
12	$^-3$
13	$^-4$

Remember that the mean is 9.

Note that some differences are positive and others are negative. Remember, we wish to find a measure of total dispersion. But if we add all these differences, we do not obtain the total variability. Indeed, if we add the differences for either example, the sum is zero. But we don't wish to say there is no dispersion. To resolve this difficulty with positive and negative differences, we square each difference.

Note: After we square each difference, the result is always nonnegative.

First Example

Data	Difference from the Mean	Square of the Difference
8	1	1
9	0	0
9	0	0
9	0	0
10	$^-1$	1

Second Example

Data	Difference from the Mean	Square of the Difference
2	7	49
9	0	0
9	0	0
12	$^-3$	9
13	$^-4$	16

Next, we sum the squares of these differences and divide by one less than the number of pieces of data in our problem. This number is called the *variance* (abbreviated *var*). In our examples:

	First Example	Second Example
Sum of squares of differences:	2	74
Variance:	$\dfrac{2}{4} = .5$	$\dfrac{74}{4} = 18.5$

It may seem strange to you that you divide by one less than the number of pieces of data rather than simply find the mean of the squares of the differences. The proof of this formula is beyond the scope of this course, but an empirical justification is given in Problem Set 10.2.

The larger the variance, the more dispersion there is in the original data.

A third measure of dispersion, and by far the most commonly used, is called the *standard deviation* and is found by taking the square root of the variance.

Definition of standard deviation

Finding a square root is the opposite operation of squaring a number. It is not assumed that you know how to find square roots. Table 10.6 presents a list of square roots, or you can use a calculator.

> DEFINITION: The *standard deviation*, denoted by σ, is the square root of the variance. That is,
>
> $$\sigma = \sqrt{\mathrm{var}}$$
>
> To find the standard deviation of a set of n numbers:
> 1. Determine the mean of the numbers.
> 2. Subtract each number from the mean.
> 3. Square each of these differences.
> 4. Find the sum of the squares of these differences.
> 5. Divide this sum by $n - 1$; this is the variance.
> 6. Take the square root of the variance.

The standard deviation for our examples is:
$\sqrt{.5} \approx .71$ for the first example;
$\sqrt{18.5} \approx 4.30$ for the second example.

EXAMPLE: G. Thumb, the leading salesman for the Moe D. Lawn Landscaping Company, turned in the following summary of his sales for the week of October 23–28. Find the mean, range, variance, and standard deviation for the number of clients contacted.

Date	Number of Clients Contacted by G. Thumb
Oct. 23	12
Oct. 24	9
Oct. 25	10
Oct. 26	16
Oct. 27	10
Oct. 28	21

Solution:

Mean: $\dfrac{12 + 9 + 10 + 16 + 10 + 21}{6} = \dfrac{78}{6}$

$= 13$

The range is $21 - 9 = 12$.

Historical Note

The Wright brothers wrote to the United States Weather Bureau seeking a location that offered privacy and a consistent wind so that they could conduct experiments. Kitty Hawk, North Carolina, was suggested. Yet the reported winds represented an annual average of calms and gales. Many days were spent in waiting and frustration because the winds were unsuitable. If the Weather Bureau's report had included information about dispersion, the Wright brothers might have recognized that the acceptable average was produced by unacceptable extremes.

TABLE 10.6 Squares and Square Roots

n	n^2	$\sqrt{n}$	$\sqrt{10n}$	n	n^2	$\sqrt{n}$	$\sqrt{10n}$
1	1	1.000	3.162	51	2601	7.141	22.583
2	4	1.414	4.472	52	2704	7.211	22.804
3	9	1.732	5.477	53	2809	7.280	23.022
4	16	2.000	6.325	54	2916	7.348	23.238
5	25	2.236	7.071	55	3025	7.416	23.452
6	36	2.449	7.746	56	3136	7.483	23.664
7	49	2.646	8.367	57	3249	7.550	23.875
8	64	2.828	8.944	58	3364	7.616	24.083
9	81	3.000	9.487	59	3481	7.681	24.290
10	100	3.162	10.000	60	3600	7.746	24.495
11	121	3.317	10.488	61	3721	7.810	24.698
12	144	3.464	10.954	62	3844	7.874	24.900
13	169	3.606	11.402	63	3969	7.937	25.100
14	196	3.742	11.832	64	4096	8.000	25.298
15	225	3.873	12.247	65	4225	8.062	25.495
16	256	4.000	12.649	66	4356	8.124	25.690
17	289	4.123	13.038	67	4489	8.185	25.884
18	324	4.243	13.416	68	4624	8.246	26.077
19	361	4.359	13.784	69	4761	8.307	26.268
20	400	4.472	14.142	70	4900	8.367	26.458
21	441	4.583	14.491	71	5041	8.426	26.646
22	484	4.690	14.832	72	5184	8.485	26.833
23	529	4.796	15.166	73	5329	8.544	27.019
24	576	4.899	15.492	74	5476	8.602	27.203
25	625	5.000	15.811	75	5625	8.660	27.386
26	676	5.099	16.125	76	5776	8.718	27.568
27	729	5.196	16.432	77	5929	8.775	27.749
28	784	5.292	16.733	78	6084	8.832	27.928
29	841	5.385	17.029	79	6241	8.888	28.107
30	900	5.477	17.321	80	6400	8.944	28.284
31	961	5.568	17.607	81	6561	9.000	28.460
32	1024	5.657	17.889	82	6724	9.055	28.636
33	1089	5.745	18.166	83	6889	9.110	28.810
34	1156	5.831	18.439	84	7056	9.165	28.983
35	1225	5.916	18.708	85	7225	9.220	29.155
36	1296	6.000	18.974	86	7396	9.274	29.326
37	1369	6.083	19.235	87	7569	9.327	29.496
38	1444	6.164	19.494	88	7744	9.381	29.665
39	1521	6.245	19.748	89	7921	9.434	29.833
40	1600	6.325	20.000	90	8100	9.487	30.000
41	1681	6.403	20.248	91	8281	9.539	30.166
42	1764	6.481	20.494	92	8464	9.592	30.332
43	1849	6.557	20.736	93	8649	9.644	30.496
44	1936	6.633	20.976	94	8836	9.695	30.659
45	2025	6.708	21.213	95	9025	9.747	30.822
46	2116	6.782	21.448	96	9216	9.798	30.984
47	2209	6.856	21.679	97	9409	9.849	31.145
48	2304	6.928	21.909	98	9604	9.899	31.305
49	2401	7.000	22.136	99	9801	9.950	31.464
50	2500	7.071	22.361	100	10000	10.000	31.623

On a calculator:

Algebraic

Press	Display
1	1.
+	1.
16	16.
+	17.
9	9.
+	26.
9	9.
+	35.
9	9.
+	44.
64	64.
÷	108.
5	5.
=	21.6
√	4.6475800 15

RPN

Press	Display
1	1.
ENTER	1.
16	16.
+	17.
9	9.
+	26.
9	9.
+	35.
9	9.
+	44.
64	64.
+	108.
5	5.
÷	21.6
√x	4.6475800 15

For the variance and standard deviation, we find the mean and then construct the following chart.

Number	Difference from Mean	Square of the Difference
12	$13 - 12 = 1$	1
9	$13 - 9 = 4$	16
10	$13 - 10 = 3$	9
16	$13 - 16 = {}^-3$	9
10	$13 - 10 = 3$	9
21	$13 - 21 = {}^-8$	64

$$\text{Var} = \frac{1 + 16 + 9 + 9 + 9 + 64}{5}$$

This is one less than the number of pieces of data we are considering.

$$= \frac{108}{5}$$

$$= 21.6$$

$$\sigma = \sqrt{\text{var}} = \sqrt{21.6}$$

$$\approx 4.65$$

The range is 9, the variance is 21.6, and the standard deviation is 4.65 clients contacted per day by G. Thumb.

PROBLEM SET 10.2

A Problems

Find the mean, median, mode, range, and standard deviation for each of the sets of values in Problems 1–10.

1. 1,2,3,4,5
2. 17,18,19,20,21
3. 103,104,105,106,107
4. 765,766,767,768,769
5. 4,7,10,7,5,2,7
6. 15,13,10,7,6,9,10
7. 3,5,8,13,21
8. 1,4,9,16,25
9. 79,90,95,95,96
10. 70,81,95,79,85

11. Compare Problems 1–4. What do you notice about the mean and standard deviation?

12. By looking at Problems 1–4 and discovering a pattern, find the mean and standard deviation of the set of numbers being juggled in the illustration in the margin at the top of page 545.

13. Find the mean, median, and mode of the following salaries of the Moe D. Lawn Landscaping Company:

Salary	Frequency
$ 5000	4
8000	3
10,000	2
15,000	1

14. Find the mean, median, and mode of the following test scores:

Test Score	Frequency
90	1
80	3
70	10
60	5
50	2

15. Suppose a variance is zero. What can you say about the data?

B Problems

16. A class obtained the following scores on a test:

Score	Frequency
90	1
80	6
70	10
60	4
50	3
40	1

Find the mean, median, mode, and range for the class.

See Problem 26 for the standard deviation.

17. A class obtained the following scores on a test:

Score	Frequency
90	2
80	4
70	9
60	5
50	3
40	1
30	2
0	4

Find the mean, median, mode, and range for the class.

See Problem 27 for the standard deviation.

18. A professor gives five exams. Two students' scores have the same mean, although one student seemed to do better on all the tests except one. Give an example of such scores.

Position	Salary
President	$80,000
1st VP	30,000
2nd VP	30,000
Supervising Manager	24,000
Accounting Manager	20,000
Personnel Manager	20,000

See Problem 28 for the standard deviation.

See Problem 29 for the standard deviation.

Many calculators have keys for finding the mean, variance, and standard deviation.

19. A professor gives six exams. Two students' scores have the same mean, although one student's scores have a small standard deviation and the other student's scores have a large standard deviation. Give an example of such scores.

20. The salaries for the executives of a certain company are shown at the left. Find the mean, median, and mode. Which measure seems to best describe the average executive salary for the company?

21. The numbers of miles driven on five tires were 17,000, 19,000, 19,000, 20,000, and 21,000 miles. Find the mean, range, and standard deviation for these tires.

22. What is the average length of the words in the first paragraph of this section? Find the mean, median, and mode. Which average seems more descriptive of the data? Find the range and standard deviation of the word size in this problem.

23. Repeat Problem 22 for the historical note on page 542.

24. Roll a single die until all 6 numbers occur at least once. Repeat the experiment 20 times. Find the mean, median, mode, and range of the number of tosses.

25. Roll of a pair of dice until all 11 numbers occur at least once. Repeat the experiment 20 times. Find the mean, median, mode, and range of the number of tosses.

26. *Calculator Problem.* Compute the standard deviation for the data of Problem 16.

27. *Calculator Problem.* Compute the standard deviation for the data of Problem 17.

28. *Calculator Problem.* Compute the standard deviation for the data of Problem 24.

29. *Calculator Problem.* Compute the standard deviation for the data of Problem 25.

30. *Computer Problem.* Below is a BASIC program for finding the mean, variance, and standard deviation. A sample output (using the data from Problem 5) is also shown.

```
102   REM: COMPUTES THE MEAN, VARIANCE, AND STANDARD DEVIATION
104   LET T = 0
105   LET A = 0
106   LET B = 0
107   PRINT "INPUT DATA (TYPE 9999 TO INDICATE END OF DATA)"
112   INPUT N
114   IF N = 9999 THEN 120
115   LET T = T + 1
116   LET A = A + N
118   B = B + N*N
119   GOTO 112
120   LET M = A/T
122   LET X = B*T − A*A
124   LET D = SQR(X)/SQR(T*T − T)
126   PRINT "THE MEAN IS "M
```

```
130   PRINT
140   PRINT "THE VARIANCE IS "X/(T*T − T)
141   PRINT
145   PRINT "THE ST. DEVIATION IS "D
998   DAT 9999
999   END
```

```
RUN

THIS PROGRAM COMPUTES THE MEAN, VARIANCE, AND STANDARD DEVIATION.

INDICATE THE END OF YOUR DATA BY TYPING 9999.
WHAT IS YOUR DATA? 4
? 7
? 1Ø
? 7
? 5
? 2
? 7
? 9999

THE MEAN IS  6

THE VARIANCE IS  6.666667

THE STANDARD DEVIATION IS  2.581989

READY
```

The computer, of course, is most useful for problems that have a large number of data. Use this program to find the mean, variance, and standard deviation of the heights given in Problem 1 of Problem Set 10.1 (page 527).

31. *Computer Problem.* Find the mean, variance, and standard deviation of the number of home runs hit by the yearly champions in the National League, as given in Problem 10 of Problem Set 10.1 (pages 531–532).

You can also use a calculator for Problem 31.

Mind Bogglers

32. If you roll a die 36 times, the expected number of times for rolling each of the numbers is summarized at the right. A graph of this table is shown in Figure 10.14. Find the mean, variance, and standard deviation for this model.

Experiment: Rolling a Pair of Dice 36 Times

Outcome	Expected Frequency
2	1
3	2
4	3
5	4
6	5
7	6
8	5
9	4
10	3
11	2
12	1

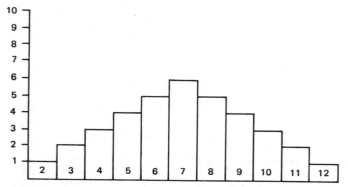

FIGURE 10.14 Expected frequency from rolling a pair of dice 36 times

A calculator may be helpful for Problems 32–35.

33. *Continuation of Problem 32*. Roll a pair of dice 36 times. Construct a table and a graph similar to the ones shown for Problem 32. Find the mean, variance, and standard deviation.

34. *Continuation of Problems 32 and 33*. Compare the results of Problems 32 and 33. If this is a class problem, you might wish to pool the entire class's results for Problem 33 before making the comparison.

35. *Continuation of Problems 32–34*. In the text, we noted that it was beyond the scope of this course to prove that you should divide the sum of the squares of the differences by $n - 1$ instead of by n when finding the standard deviation. However, we can verify this result empirically. Repeat Problems 32–34 using n instead of $n - 1$. How does this change the results?

Problems for Individual Study

36. Prepare a report or exhibit showing how statistics are used in baseball.

37. Prepare a report or exhibit showing how statistics are used in educational testing.

38. Prepare a report or exhibit showing how statistics are used in psychology.

39. Prepare a report or exhibit showing how statistics are used in business. Use a daily report of the New York Stock Exchange's transactions. What inferences can you make from the information reported?

40. Investigate the work of Quetelet, Galton, Pearson, Fisher, and Nightingale. Prepare a report or an exhibit of their work in statistics.

10.3 THE NORMAL CURVE

The cartoon suggests that there are more children with above-normal intelligence than with normal intelligence. But what do we mean by *normal* or *normal intelligence*?

Suppose we survey the results of 20 children's scores on an IQ test. The scores (rounded to the nearest 5 points) are: 115, 90, 100, 95, 105, 105, 95, 105, 105, 95, 125, 120, 110, 100, 100, 90, 110, 100, 115, and 80. The mean is 103, the standard deviation is 10.93, and the frequency is shown in Figure 10.15.

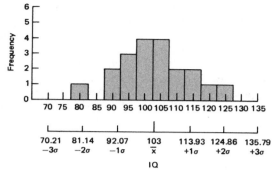

FIGURE 10.15 Frequencies of IQs for a sample of 20 persons

If we consider 10,000 IQ scores instead of only 20, we might obtain the frequency distribution shown in Figure 10.16. As we can see, the frequency distribution approximates a curve. If we connect the endpoints of the bars in Figure 10.16, we obtain a curve that is very close to something called a *normal frequency curve,* or simply a *normal curve,* as shown in Figure 10.17.

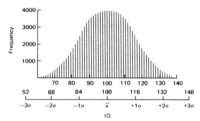

FIGURE 10.16 Frequencies of IQs for a sample of 10,000 persons

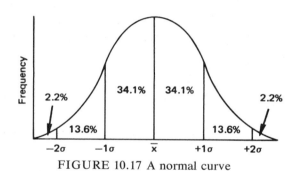

FIGURE 10.17 A normal curve

The distribution represented by the normal curve is very common and occurs in a wide variety of circumstances. For example, the heights of horses, the nodes on a jellyfish, baseball averages, the size of peas, the density of stars, the life span of light bulbs, chest size, and intelligence quotients all are normally distributed. That is, if we obtain the frequency distribution of a large number of measurements (as with the IQ example), the corresponding graph tends to look like a normal curve.

The interesting and useful property of the normal curve is that roughly 68% of all values lie within one standard deviation above and below the mean. About 96% lie within two standard deviations, and virtually all (99.8%) values lie within three standard deviations of the mean. These percentages are the same regardless of the particular mean or standard deviation. Thus, for 1000 people taking an IQ test, about 34%, or 340 persons, would have scores between 100 and 116. These results are summarized in Table 10.7.

TABLE 10.7 Sample of 1000 IQ Scores

Range	Percentage of Scores	Expected Number of Scores
Below 52	.1	1
52–68	2.2	22
69–84	13.6	136
85–100	34.1	341
101–116	34.1	341
117–132	13.6	136
133–148	2.2	22
Above 148	.1	1
Total		1000

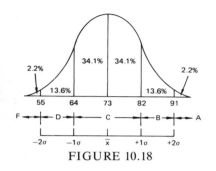

FIGURE 10.18

EXAMPLE: A teacher claims to grade "on a curve." This means the teacher believes that the scores on a given test are normally distributed. If 200 students take the exam, with mean 73 and standard deviation 9, what are the grades the teacher would give?

Solution: First draw a normal curve with mean 73 and standard deviation 9, as shown in Figure 10.18. The range of 73 to 82 will contain about 34% of the class, and the range of 82 to 91 will contain about 14% of the class. Finally, about 2% of the class will score higher than 91. The teacher would therefore give grades according to the following table:

Grade on Final	Letter Grade	Number Receiving Grade	Percentage of Class
Above 91	A	4	2%
83–91	B	28	14%
65–82	C	136	68%
55–64	D	28	14%
Below 55	F	4	2%

EXAMPLE: The Eureka Light Bulb Company tested a new line of light bulbs and found them to be normally distributed, with a mean life of 98 hours and a standard deviation of 13.
a. What percentage of bulbs will last fewer than 72 hours?
b. What is the probability that a bulb selected at random will last more than 111 hours?

Solution: Draw a normal curve with mean 98 and standard deviation 13, as shown in Figure 10.19.
a. About 2% (2.2%) will last under 72 hours.
b. We see that about 16% (15.8%) of the bulbs last longer than 111 hours, so

P(bulb will last longer than 111 hours) ≈ .16.

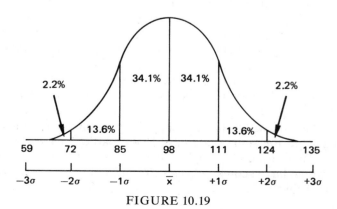

FIGURE 10.19

PROBLEM SET 10.3

A Problems

In Problems 1–5, suppose that people's heights (in centimeters) are normally distributed, with a mean of 170 and a standard deviation of 5. We take a sample of 50 persons.

1. How many would you expect to be between 165 and 175 cm tall?

2. How many would you expect to be taller than 160 cm?

3. How many would you expect to be taller than 175 cm?

4. If a person is selected at random, what is the probability that he or she is taller than 165 cm?

5. What is the variance for this sample?

In Problems 6–10, suppose that, for a certain exam, a teacher grades on a curve. It is known that the mean is 50 and the standard deviation is 5. There are 45 students in the class.

6. How many students should receive a C?

7. How many students should receive an A?

8. What score would be necessary to obtain an A?

9. If an exam paper is selected at random, what is the probability that it will be a failing paper?

10. What is the variance for this exam?

11. Suppose the breaking strength of a rope (in pounds) is normally distributed, with a mean of 100 pounds and a standard deviation of 16. What is the probability that a certain rope will break with a force of 132 pounds?

12. The diameter of an electric cable is normally distributed, with a mean of .9 inch and a standard deviation of .01. What is the probability that the diameter will exceed .91 inch?

13. Suppose the annual rainfall in Ferndale, California, is known to be normally distributed, with a mean of 35.5 inches and a standard deviation of 2.5. About 2.2% of the time, the rainfall will exceed how many inches?

14. In Problem 13, what is the probability that the rainfall will exceed 30.5 inches in Ferndale?

B Problems

15. The diameter of a pipe is normally distributed, with a mean of .4 inch and a variation of .0004. What is the probability that the diameter will exceed .44 inch?

16. The breaking strength (in pounds) of a certain new synthetic is normally distributed, with a mean of 165 and a variance of 9. The material is considered defective if the breaking strength is under 159 pounds. What is the probability that a sample chosen at random will be defective?

17. Suppose the neck size of men is normally distributed, with a mean of 15.5 inches and a standard deviation of .5. A shirt manufacturer is going to introduce a new line of shirts. How many of each of the following sizes should be included in a batch of 1000 shirts?
    a. 14   b. 14.5   c. 15   d. 15.5
    e. 16   f. 16.5   g. 17

18. A package of Toys Galore Cereal is marked "Net Wt. 12 oz." The actual weight is normally distributed, with a mean of 12 oz and a variance of .04.
    a. What percentage of the packages will weigh under 12 oz?
    b. What weight will be exceeded by 2.2% of the packages?

19. Instant Dinner comes in packages that are normally distributed, with a standard deviation of .3 oz. If 2.2% of the dinners weigh more than 13.5 oz, what is the mean weight?

Mind Boggler

FIGURE 10.20 Floorboards

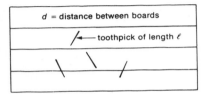

Equipment needed: *A box of toothpicks of uniform length and a large sheet of paper with equidistant parallel lines. A hardwood floor works as well as a sheet of paper. The length of the toothpicks should be less than the perpendicular distance between the parallel lines.*

To find p, you will divide the number crossing a line by the number of toothpicks tossed (1000 in this case).

20. *Buffon's Needle Problem.* Toss a toothpick onto a hardwood floor 1000 times (see description in the margin), or toss 1000 toothpicks, one at a time, onto the floor. Let ℓ be the length of the toothpick and d be the distance between the parallel lines determined by the floorboards (see Figure 10.20.)
    a. Make a guess as to the probability, p, that a toothpick will cross a line. Do this before you begin the experiment.
    b. Perform the experiment and find p empirically.
    c. By direct measurement, find ℓ and d.
    d. Calculate the numbers 2ℓ and pd.
    e. Divide 2ℓ by pd. What is the result?

Problems for Individual Study

21. Select something that you think might be normally distributed (for example, the ring size of students at your college). Next, select a sample of 100 people, and make the appropriate measurements (in this example, ring size). Calculate the mean and standard deviation. Are your data normally distributed? Make a presentation of your findings to the class.

22. *Statistics.* How can statistics be summarized? What are the measures of central tendency? What are measures of dispersion? How is the relationship between sets of measures determined? What is a normal curve? What is quality control? How is the accuracy of a sample measured? How are statistics used to draw conclusions?
    Exhibit suggestions: sample distributions of original data with analysis and graphs; examples of statistics found in advertisements and newspapers; examples of misuses of statistics; models of random-sampling devices and Gauss' probability board.
    References: Hogben, Lancelot, *Mathematics for the Millions* (New York: Norton, 1937).
        Huff, Darrell, and Irving Geis, *How to Lie with Statistics* (New York: Norton, 1954).

Kline, Morris, *Mathematics in Western Culture,* Chap. XII (New York: Oxford University Press, 1953).

Newman, James, *The World of Mathematics,* Vol. 3, Part VIII (New York: Simon and Schuster, 1956).

Weaver, Warren, "Statistics," *Scientific American,* January 1952, (Vol. 186, No. 1).

10.4 SAMPLING

The first sections of this chapter dealt with the accumulation of data, measures of central tendency, and dispersion. However, a more important part of statistics is its ability to help us make predictions about a population based on a sample from that population. Sampling necessarily involves some error, and therefore statistics is also concerned with estimating the error involved in predictions based on samples.

A sample is a small group of items chosen to represent a larger group. The larger group is called a population.

In national polls, not everyone interviewed; rather, the predictions are based on the responses of a small sample. The inferences drawn from a poll can, of course, be wrong. In 1936, the *Literary Digest* predicted that Alfred Landon would defeat Franklin D. Roosevelt—who was subsequently re-elected by a landslide. (The magazine ceased publication the following year.) Another classic example of drawing an incorrect conclusion from the data gathered in a poll occurred in the 1948 Presidential election, when it was predicted that Thomas E. Dewey would defeat Harry S. Truman. The *Chicago Daily Tribune* readied its morning edition proclaiming Dewey the victor, but, when the votes were counted, Truman, not Dewey, became the 33rd President of the United States.

In an attempt to minimize error in their predictions, statisticians follow very careful procedures:

Step 1: Propose some hypothesis about a population.
Step 2: Gather a sample from the population.
Step 3: Accept or reject the hypothesis. (It is also important to be able to estimate the error involved with making this decision.)

Suppose a friend hands you a coin to flip and wants to bet on the outcome. Now, this friend is known to be of dubious integrity, and you suspect that the coin is "rigged." You decide to test this hypothesis by taking a sample. You flip the coin twice, and it is heads both times. You say "Aha, I knew it was rigged." Your friend says "Don't be silly. Any coin can come up heads twice in a row."

You decide to go along with your friend, for the time being, and accept the hypothesis that the coin is fair. But you decide to perform an experiment of flipping the coin 100 times. The result is:

Heads: 55   Tails: 45

Do you accept or reject the hypothesis "The coin is fair"? The expected number of heads is 50, but certainly a fair coin might well produce the results above.

As you can readily see, there are two types of error possible:

Type I: Rejection of the hypothesis when it is true.
Type II: Acceptance of the hypothesis when it is false.

How, then, can we proceed to minimize the possibility of making either error?

Let's carry our example further and repeat the experiment of flipping the coin 100 times:

Trial Number	Number of Heads
1	55
2	52
3	54
4	57
⋮	

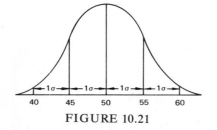

FIGURE 10.21

If the coin is fair and we repeat the experiment a large number of times, the distribution would be normal, with a mean of 50 and a standard deviation of 5, as shown in Figure 10.21. Suppose you are willing to accept the coin as fair only if the number of heads falls between 45 and 55 ($\pm 1\sigma$). You know you will be correct 68% of the time. But your friend says "Wait, you're rejecting a lot of fair coins! If you accept coins between 40 and 60 ($\pm 2\sigma$), you'll be correct 96% of the time."

You say "But suppose the coin really favors heads, so that the mean is 60 with a standard deviation of 5. I'd be accepting all those coins in the shaded region of Figure 10.22."

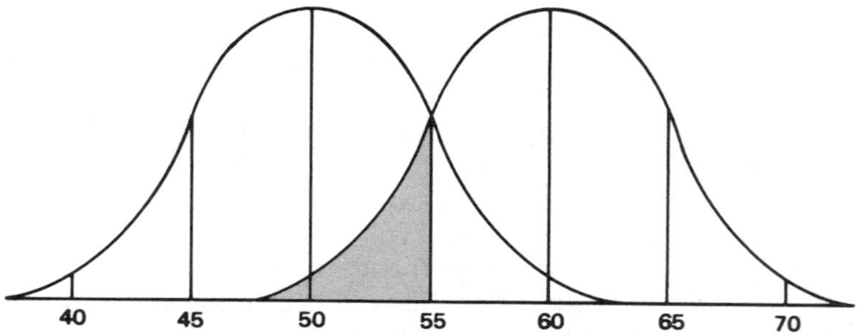

FIGURE 10.22

As you can see, a decrease in Type I error increases the Type II error, and vice versa. Deciding which type of error to minimize depends on the stakes involved as well as on some statistical calculations that are beyond the scope of this course. Consider a company that produces two types of valves. The first type is used in jet aircraft, and the failure of this valve might cause many deaths. A sample of the valves is taken, and the company must accept or reject the entire shipment. Under these circumstances, they would rather reject many good valves than accept a bad one.

On the other hand, the second valve is used in toy airplanes; the failure of this valve would cause the crash of the model. In this case, the company would not care to reject too many good valves, so they would minimize the probability of rejecting the good valves.

PROBLEM SET 10.4

Problems for Individual Study

1. You are interested in knowing the number and ages of children (0–18 years) in a part (or all) of your community. You will need to sample 50 families and find out the number of children in each family and the age of each child. It is important that you select the 50 families at random.

   Step 1: Determine the geographical boundaries of the area with which you are concerned.
   Step 2: Consider various methods for selecting the families at random. For example, could you: (a) select the first 50 homes at which someone is at home when you call? (b) select 50 numbers from the phone book?
   Step 3: Consider different ways of asking the question. Can the way the family is approached affect the response?
   Step 4: Gather your data.
   Step 5: Organize your data. Construct a frequency distribution for the children, with integral values from 0 to 18.
   Step 6: Find out the number of families actually living in the area you've selected. If you can't do this, assume that the area has 1000 families.

   a. What is the average number of children per family?
   b. What percentage of the children are in the first grade (age 6)?
   c. If all the children of ages 12–15 are in junior high, how many are in junior high for the geographical area you are considering?
   d. See if you can actually find out the answers to parts (b) and (c), and compare these answers with your projections.
   e. What other inferences can you make from your data?

2. Five identical boxes must be prepared for this problem, with contents as follows:

Examples of how to prepare for Problem 2:

1. *Use shoe boxes and marbles, beads, or poker chips.*
2. *Use paper cups and beads. Cut a window out of each cup, and use a clear plastic wrap to make a cover for the cups.*
3. *Use cardboard salt shakers and some small beads (a little larger than the openings at the top).*

Box	Contents (Marbles, Beads or Poker Chips)
#1	15 red, 15 white
#2	30 red, 0 white
#3	25 red, 5 white
#4	20 red, 10 white
#5	10 red, 20 white

Select one of the boxes at random so that you do not know its contents.

Step 1: Shake the box.
Step 2: Select one marker, note the result, and return it to the box.
Step 3: Repeat the first two steps 20 times.

a. What do you think is inside the box you've sampled?
b. Could you have guessed the contents by repeating the experiment five times? Ten times? Do you think you should have more than 20 observations? Discuss.

10.5 SUMMARY AND REVIEW

CHAPTER OUTLINE

The mean is the most sensitive average. It reflects the entire distribution and is probably the most important of the averages.

The median gives the middle value. It is especially useful when there are a few extraordinary values that distort the mean.

The mode is the average that measures "popularity." It is

I. Frequency Distributions
   A. Constructing tables
   B. Bar graphs
   C. Line graphs
   D. Circle graphs
   E. Pictograms

II. Measures of Central Tendency
   A. The **mean** is the number found by adding the values and then dividing by the number of values.
   B. The **median** is the middle number when the numbers are arranged in order of size.
   C. The **mode** is the value that occurs most frequently.

III. Measures of Dispersion
   A. The **range** in a set of data is the difference between the largest and the smallest numbers in the set.
   B. The **variance** is the sum of the squares of the differences of each number from the mean, divided by one less than the number of values.
   C. The **standard deviation** is the square root of the variance.

IV. The Normal Curve
  A. The normal curve is shown in Figure 10.23.
  B. About 68% of all values lie within one standard deviation of the mean.
  C. About 96% of all values lie within two standard deviations of the mean.
  D. About 99.8% of all values lie within three standard deviations of the mean.

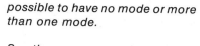

possible to have no mode or more than one mode.

See the summary on how to find the variance and standard deviation on page 542.

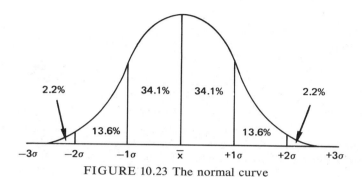

FIGURE 10.23 The normal curve

V. Sampling
  A. Statistical predictions
    1. Propose some hypothesis about a population.
    2. Gather a sample from the population.
    3. Accept or reject the hypothesis.
  B. The accuracy of the predictions depends on the way the sample is chosen as well as on the sample size.
  C. Two types of sampling error
    1. Reject the hypothesis when it is true.
    2. Accept the hypothesis when it is false.

REVIEW PROBLEMS

1. Find the mean, median, and mode of the following: 23,24,25,26,27.

2. Make a frequency table and draw a bar graph for the following data:7,8,8,7,8,9,6,9,8,6,8,9,10,7,7.

3. Find the range, variance, and standard deviation for the data given in Problem 1.

4. Find the mean, median, mode, and range for the data given in Problem 2.

5. Comment on a sales manager who tells the staff that no one should have sales below the average for the week. Is it possible for the staff to comply?

Size	Number of Cases Sold
6 oz	5
10 oz	10
12 oz	35
16 oz	50

6. A small grocery store stocked several sizes of Copycat Cola last year. The sales figures were as shown at the left. Find the mean, median, and mode for these data. If the store manager decides to cut back the variety and stock only one size, which measure of central tendency will be most useful in making this decision?

7. A student's scores in a certain math class are 72, 73, 74, 85, and 91. Find the mean, median, and mode. Which measure of central tendency is most representative of the student's scores?

8. Find the range, variance, and standard deviation for the student's scores in Problem 7.

9. The earnings of the employees of a small real-estate company for the past year were (in thousands of dollars) 12, 18, 15, 9, 11, 22, 18, 17, and 10. Find the mean, median, and mode. Which measure of central tendency is most representative of the employees' earnings?

10. The heights of 10,000 women students at an Eastern college are known to be normally distributed, with a mean of 67 inches and a standard deviation of 3. How many women would you expect to be between 64 and 70 inches tall?

11. For the data of Problem 10, how many women would you expect to be shorter than 5'1"?

12. How would you obtain a sample of one-syllable words used in this text? How would you apply those results to the whole book?

13. Two instructors gave the same Math 1A test in their classes. Both classes had the same mean, but one class had a standard deviation twice as large as that of the other class. Which class do you think would be easier to teach, and why?

14. Determine the upper and lower scores for the middle 68% of a normally distributed test in which

$$\bar{x} = 73 \quad \text{and} \quad \sigma = 8$$

15. A student received a score of 91 on a history test for which the mean was 86 and the standard deviation 5. She also received a score of 79 on a mathematics test for which the mean was 75 and the standard deviation 2. On which test did she have the better score in comparison to her classmates?

16. An advertisement claims "Four out of five dentists surveyed recommend Brightdent Sugarless Gum for their patients who chew gum." State why you would or would not accept this as a valid claim to product superiority.

17. In a random sample of 1000 chilren, IQ scores were normally distributed, with a mean of 100 and a standard deviation of 16. How many children could you expect to have an IQ greater than 132?

18. A writer, wishing to emphasize the tremendous shrinking of the dollar over the years, selects 1930 as his base year and constructs a graph for 1930 to

1980 (see, for example, *Building Your Fortune with Silver,* by Robert Preston, Salt Lake City: Hawkes Publications, 1973). Another writer, wishing to minimize the effects, selects 1973 as his base year and depicts the years 1970 to 1973. A third writer selects for his base a "neutral year"—say 1967—as is done in most current almanacs (see, for example, Problem 4 on page 528). Discuss how the selection of different base years might support different viewpoints.

19. In 1974, the U.S. government spent 47% of its budget on human resources, 10% on physical resources, 30% on national defense, 7% on interest, and 6% on other items. Construct a circle graph depicting these percentages.

20. Consider the figures at the right, which show the U.S. government's expenditures for space research and technology.
    a. *Viewpoint 1: Space and technology expenditures rose only .074 billion dollars from 1973 to 1974.* Draw a graph showing very little increase in the expenditures.
    b. *Viewpoint 2: Space and technology expenditures rose $74,000,000 from 1973 to 1974.* Draw a graph showing a large growth in expenditures.

Year	Expenditure (in Millions of Dollars)
1970	3749
1971	3381
1972	3422
1973	3061
1974	3135

FINAL REVIEW A

It is always difficult to summarize a course and review for a final exam. The following set of problems, although lengthy, will help you to prepare for the final examination. The answers to all of these problems are given at the end of Appendix B. If you miss any of the problems, you can refer to the section in parentheses for review.

1. (1.1) Compute $15^2 - 14^2$. Do not obtain your answer by direct multiplication. Show all of your work.

   Hint: Compute $1^2 - 0^2; 2^2 - 1^2; \ldots$

2. (1.2) Fill in the missing terms.
   a. 19,50,81,____
   b. 99,33,11,_____
   c. 1,1,2,2,2,3,3,3,____

3. (1.2) Classify the sequences in Problem 2 as arithmetic, geometric, or neither.

4. (1.3) a. Standing with your arms at your sides, measure the distance from your elbow joint to the floor. Divide this number by your height. What is the result?
   b. Measure the distance from your shoulder to your elbow joint. Divide this number by the distance from your elbow joint to your fingertips of the same arm. What is the result?
   c. Measure the distance between your shoulders when your arms are held at your sides. Divide this number by the distance from your shoulder to your fingertips of the same arm. What is the result?
   d. Compare your answers to parts a, b, and c. All of them should be about the same number. What is this number?

   The answer to Problem 4d should look familiar—does it?

5. (1.5) Give the metric prefix meaning
   a. one hundredth.

561

    b. one hundred.

    c. one thousand.

    d. one tenth.

6. (1.5) Name the basic metric unit for measuring
   a. capacity.
   b. length.
   c. mass.
   d. temperature.

7. (2.1–2.3) Define conjunction, disjunction, negation, conditional, and biconditional.

8. (2.1) According to our definition, which of the following sentences are statements? If possible, use an appropriate quantifier to change the sentence into a statement.
   a. Ms. Laura Enertia did not close the deal.
   b. Did Ms. Mat E. Matic find out about the contract?
   c. Ms. Laura Enertia and Ms. Mat E. Matic are friends.
   d. She didn't know the amount of the contract.
   e. All the employees knew about the deal.

9. (2.2) Construct a truth table for $(p \wedge q) \to (p \vee \sim q)$.

10. (2.2) Construct a truth table for $(p \vee q) \vee (\sim r)$.

11. (2.1–2.3) Translate each of the following statements into words. Let p: $\triangle ABC$ is isosceles; q: The base angles of $\triangle ABC$ are congruent.

    a. $p \wedge q$ b. $p \vee q$ c. $p \to q$ d. $p \leftrightarrow q$
    e. $\sim q \to \sim p$

12. (2.4) Draw an Euler circle diagram for each part of Problem 11.

13. (2.1–2.3) Assume that p and q are T (true). Which of the statements in Problem 11 are true?

14. (2.3) Give an example of a statement that is true and its converse, which is false.

15. (2.3) Give an example of a statement that is true and its converse, which is also true.

16. (2.3) Use the biconditional connective in the example you wrote for Problem 15.

17. (2.3) Which of the following statements are true?
   a. If $111 + 1 = 1000$, then I'm a monkey's uncle.
   b. If $6 + 2 = 10$ and $7 + 2 = 9$, then $5 + 2 = 7$.
   c. If $6 + 2 = 10$ or $5 + 2 = 7$, then $7 + 2 = 9$.
   d. If $5 + 2 = 7$ and $7 + 2 \neq 9$, then $6 + 2 = 10$.
   e. If the moon is made of green cheese, then $5 + 7 = 57$.

18. (2.3) Write the following pair of statements as one statement by using the appropriate connective.

    i. If $ad = bc$, then $\dfrac{a}{b} = \dfrac{c}{d}$.

    ii. If $\dfrac{a}{b} = \dfrac{c}{d}$, then $ad = bc$.

19. (2.5) Let $U = \{1,2,3,4,5,6,7,8,9,10\}$
$A = \{5,7,9,10\}$ $B = \{2,3,5\}$

    a. Find $A \cap B$.   b. Find $\overline{A}$.   c. Find $A \cup B$.

20. (2.5) For the sets in Problem 19, find $A \cup (\overline{B \cap A})$.

21. (2.5) Prove or disprove $A \cap (B \cup C) = (A \cap B) \cup (A \cap C)$.

22. (2.6) Form a valid conclusion using all the following statements.
If we go to the concert, then we will be broke.
We are not broke.

23. (2.6) Form a valid conclusion using all the following statements.
If you are not careful, you will fall.
If you obey your parents, you will not fall.
You obey your parents.

24. (2.6) Form a valid conclusion using all the following statements.
If San Francisco loses, then Dallas wins.
If Dallas wins, then they will go to the Super Bowl.
San Francisco loses.

25. (2.6) Form a valid conclusion using all the following statements.
If your parents both have type O blood, then you have type O blood.
You don't have type O blood.

26. (3.1) Instructions for solving a problem are followed in the part of a computer called _____.

27. (3.1) Name the five basic components of a computer.

28. (3.2) *Calculator Problem.* What do you do when the stock market falls and you seek advice, only to receive the following calculation?
$.05(961 + 769 \cdot 2 \cdot 10^2 - 60)$

For the advice, turn your calculator over after doing the calculation.

29. (3.3) Write 5.9×10^{25}
a. in fixed-point notation.   b. in floating-point notation.

30. (3.3) Write 578,000,000
a. in scientific notation.   b. in floating-point notation.

31. (3.3) Write 7.5E − 8
a. in scientific notation.   b. in fixed-point notation.

32. (3.4) Write a BASIC program to find the average of two numbers. Include a flow chart.

33. (3.5) Write a BASIC program to output the first n Fibonacci numbers. Recall that Fibonacci numbers are numbers in the pattern

$$1,1,2,3,5,8,13,21, \ldots$$

34. (3.5) What will the computer type if you RUN the following program?

```
10   GOSUB 200
20   PRINT "TO ";
30   GOTO 60
40   PRINT "TOO ";
50   GOTO 140
60   PRINT "THE SEVENTH ";
70   PRINT "POWER IS "
80   FOR I = 1 TO 6
90   GOSUB 200
100  PRINT "TIMES";
110  NEXT I
115  PRINT "TWO"
120  PRINT "AND THIS IS ";
130  GOTO 40
140  PRINT "MUCH.";
150  GOTO 300
200  PRINT "TWO ";
210  RETURN
300  END
```

35. (4.1) In your own words, explain what is meant by a positional numeration system.

36. (4.2) Write 4,000,056.43 in expanded notation.

37. (4.1) Explain the difference between a number and a numeral.

38. (4.3; 6.1) In your own words, and without using the text as a reference, discuss each of the following properties: closure, associative, identity, inverse, commutative, and distributive.

39. (4.4) Simplify.

  a. $6 - {}^-14$ b. $6 - 14$ c. $8(2 - 6)$

  d. $\dfrac{14 + 3(^-2)}{2}$ e. $(6 - 5) - (3 - 12)$

40. (4.4) Solve for x.

  a. $(4 - 9) - (6 - 2) = x$ b. $\lfloor(^-3) \cdot (^-8)\rfloor \cdot (^-4) = x$

41. (4.5) Simplify.

  a. $\dfrac{2}{3} + \dfrac{1}{2}$ b. $\dfrac{2}{3} \cdot \dfrac{4}{5} + \dfrac{1}{3} \cdot \dfrac{4}{5}$

  c. $3 - \dfrac{3}{4}$ d. $15(\dfrac{2}{3} + \dfrac{1}{5})$

  e. $\dfrac{1/2}{2/3}$

42. (4.5) Solve for x.

  a. $\dfrac{6}{18} \cdot \dfrac{10}{7} = x$ b. $\dfrac{3}{5} \div \dfrac{^-11}{7} = x$

43. (2.5; 4.6) Let $A = \{6, {}^-9, 5.3, .\overline{345}, \sqrt{3}, 0, 3.1415, \sqrt{25}, {}^-3.2\}$

$$R = \{\text{reals}\} \qquad N = \{\text{naturals}\}$$
$$Z = \{\text{integers}\} \qquad Q = \{\text{rationals}\}$$

    a. Locate the elements of A on a real number line.
    b. Find $A \cap Z$.   c. Find $A \cap N$.
    d. Find $A \cap Q$.   e. Find $Z \cap N$.

44. (4.6) Express $.\overline{54}$ as a quotient of two integers.

45. (4.6) Is the answer given in Problem 44 rational or irrational?

46. (4.6) Express .3333 as a quotient of two integers.

47. (5.1) Explain the difference between simple and compound interest.

48. (5.1) If you assume an 8% inflation rate, use Table 5.1 on page 281 or a calculator to project the price in the year 2001 of a cup of coffee selling for 25¢ in 1976.

49. (5.2) What is APR?

50. (5.2) What is the difference between open-end and closed-end credit?

51. (5.3) If a $4995.00 automobile is advertised at $148.11 per month for 48 months with no down payment,

1978 MONARCH 2 DR.
$4995⁰⁰
NO DOWN PAYMENT

$148 11 per month

    a. find the total interest.
    b. use Table 5.2 on page 289 to find the APR.

52. (5.3) Suppose the Monarch advertised in Problem 51 has a list price of $4854 with the following options added:

automatic transmission:	$193
power steering:	$148
power brakes:	$ 63
air conditioning:	$494

If the cost factor for car and options is 0.85, make a 5% offer.

53. (5.4) If your gross monthly income is $2100 with current monthly payments of $285, estimate the maximum monthly payment you can make for a home?

54. (5.4) If the interest rate is 12.5% + 2 pts + $200 for a $60,000 loan, what is the comparable interest rate?

55. (5.5) Discuss, compare, and contrast *term, whole life,* and *endowment insurance.*

56. (6.1) Which of the field properties are satisfied for the integers and the operations of addition and multiplication?

57. (6.1) Repeat Problem 56 for the rationals.

58. (6.1) Find the additive inverse of each of the following, if possible.

   a. $\dfrac{3}{4}$ b. $^-17$ c. 6.2   d. 0

59. (6.1) Find the multiplicative inverse of each of the numbers in Problem 58, if possible.

60. (6.2) Solve for x.

   a. $x + 10 = 115$
   b. $3x + 4 = x - 7$
   c. $2(x + 3) = 5(x - 3) - 6$
   d. $\dfrac{2x}{3} + 1 = 7$

61. (6.3) What is a prime number?

62. (6.4) A perfect number is a natural number such that the sum of its proper divisors is equal to the original number. Show that 28 is a perfect number.

63. (6.4) What is the least common multiple of two relatively prime numbers?

64. (6.5) Are 1890 and 1960 relatively prime?

65. (6.5) Simplify:

$$\frac{\dfrac{2}{3} \cdot \dfrac{1}{5} + \dfrac{1}{4}}{2}$$

66. (6.6) Solve for x.
   a. $6 - 10 \equiv x \pmod{12}$ b. $5x \equiv 4 \pmod 7$
   c. $x/5 \equiv 8 \pmod{10}$ d. $5x^2 + 3x + 1 \equiv 0 \pmod 3$

67. (7.2) What is a Saccheri quadrilateral?

68. (7.3) Which of the following figures are topologically equivalent?

a.   b.    c.

d.   e.

69. (7.3) Count the number of regions, arcs, and vertices for the figures in Problem 68, and verify Euler's formula $V + R = A + 2$.

70. (7.3) Is a doughnut topologically equivalent to
   a. a sphere?
   b. a funnel?

71. (7.3) Draw a map with exactly six countries that requires:
   a. no more than two colors.
   b. three, but no more than three, colors.
   c. four colors.

72. (7.3) Suppose a square is drawn on a "completely flexible" inner tube, with a point X inside the square and a point Y outside the square. The line segment XY is also drawn. Can the square be distorted so that
   a. it becomes a circle?
   b. the points X and Y are 1/1000 in. apart?
   c. the points X and Y coincide?
   d. the point X is outside the square?

73. (7.3) Draw a three-ring figure with the topological property that no two rings are joined, but all three rings taken together are joined. That is, if any one ring is removed, the other two will also be free.

74. (7.4) Are the following networks transversible?

a.    b.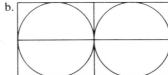

75. (7.4) Color the networks of Problem 74 with as few colors as possible.

76. (7.3) Insert doors in each of the figures in Problem 74 so that each room is accessible and also so that:
   a. each floor-plan problem has a solution.
   b. each floor-plan problem does not have a solution.

77. (8.1) In your own words, define *permutation* and *combination*. Tell how to find each, and discuss how they are related. Give some original examples of each.

78. (8.2–8.4) Evaluate each of the following:
   a. ${}_{17}C_4$ b. ${}_{16}P_3$ c. ${}_xP_y$ d. ${}_wC_z$

79. (8.2–8.4) How many five-man committees can be formed from a club consisting of six members?

80. (8.2–8.4) If a breeder of hybrid peas is experimenting with seven different varieties, in how many ways can a cross-pollination be obtained by combining two of the varieties?

81. (8.2–8.4) If Happy Harry is giving a party and has invited seven people, in how many ways can he seat the guests at a rectangular table?

82. (8.4) If you flip a coin five times, in how many ways can you obtain at least one tail?

83. (9.2) Suppose we randomly choose a natural number between 1 and 25, inclusive. What is the probability that it is:
a. even?   b. prime?   c. a multiple of 4?
d. a multiple of 5?   e. a multiple of 4 and 5?

84. (9.2) A pair of dice, one red and one green, are rolled. What is the probability that the sum of the dice is:
a. 8?
b. 8, if one of the dice is a 3?
c. 8, if the red die is a 3?

85. (9.2) If three coins are tossed simultaneously, what is the probability of obtaining at least one tail?

86. (9.2) A coin is flipped two consecutive times. What is the probability of two tails if we know that at least one was a tail?

87. (9.2) What are the odds against winning in roulette if you bet on a single number? (Assume that there are 36 positive numbers, 0, and 00 on the roulette wheel.)

88. (9.3) Five cards are drawn at random from a deck of cards that have been numbered consecutively from 1 to 97. What is the probability that the cards drawn are in order of increasing magnitude?

89. (9.3) What is the expectation for playing the game described in Problem 87 if you will win $36 by picking the number that comes up next?

90. (9.4) A contest offered at $10,000 first prize, 10 second prizes worth $1000 each, and 1000 third prizes worth $100 each. What is the expectation for this contest if we assume that there are one million entries?

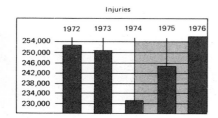

91. (10.1) According to the graph, what was the approximate number of automobile injuries in 1973? In 1974?

92. (10.2) What is the mean, median, and mode for the data given in Problem 91?

93. (10.2) Discuss what we mean by "measures of central tendency." Illustrate with examples.

94. (10.2) Discuss what is meant by "measures of dispersion." Illustrate with examples.

95. (10.2) You are shopping for a thermometer, but you notice that they do not all register the same temperature. How would you decide which one to select?

96. (10.2) In 1973, the U.S. government's receipts totaled approximately $233 million. The sources (in millions of dollars) were: individual

income taxes, $100; corporate income taxes, $67; social insurance taxes and contributions, $37; excise taxes, $14; other sources, $15. Depict these receipts in a circle graph.

Name	Number of Minutes
Linda	120
Melissa	70
Shannon	80
Jack	110
Rosamond	100
Niels	90
Rigmor	45
Ron	70
Loraine	65
Todd	110
Tim	40

97. (10.2) The time spent by several students doing Math 10 homework is summarized in the margin. Find the mean, median, and mode. Which measure of central tendency is most representative?

98. (10.3) Find the range, variance, and standard deviation for the students' homework time in Problem 97.

99. (10.4) An advertisement claims "ninety-five percent of the chief engineers polled said they preferred solid-state television." State why you would or would not accept this as a valid claim to product superiority.

100. (1.1) We have begun each of the chapter reviews with a figure as shown.

Chapter 1

Chapter 2

Chapter 3

Chapter 4

Chapter 5

Chapter 6

Chapter 7

Chapter 8

Chapter 9

Chapter 10

If this textbook had another chapter, Chapter 11, predict the logo that would probably be used. How about the one for Chapter 12?

ANSWERS TO THE ODD-NUMBERED PROBLEMS B

PROBLEM SET 1.1, PAGES 6–11

1. 11,111,111   **3.** a. 142,857   b. 285,714   c. 428,571   d. 571,428   e. 714,285   f. 857,142   g. The answers all exhibit the same sequence, only with a different starting place.   **5.** a. 4444444404   b. 5555555505   c. 8888888808   d. 9999999909   **7.** Answers vary; the next figure would be a 5×5 square with the new rows of squares unshaded.   **9.** a. 4,8,3,7,2,6,1,5,9,4,8,3,7, . . .   b. 3,6,9,3,6,9,3, . . .   c. 2,4,6,8,1,3,5,7,9,2,4, . . .   d. 1,2,3,4,5,6,7,8,9,1,2,3, . . .   **11.** a. 1.444 . . .   b. 1.555 . . .   c. 1.666 . . .   **13.** a. $25^2 = 625$ b. $50^2 = 2500$ c. $100^2 = 10,000$ **15.** a. It should be approximately 90°.   b. approximately the same   c. They are always 90°.   **17.** a. 61   b. 80   c. 98   d. (add 2 and divide by 13) 118

PROBLEM SET 1.2, PAGES 18–23

1.   1   9   36   84   126   126   84   36   9   1
  1   10   45   120   210   252   210   120   45   10   1
3. a. They are the numbers in the first diagonal.   b. They are the numbers in the second diagonal.   c. They are the numbers in the third diagonal.   **5.** They are all arithmetic.   **7.** a. 96   b. ⁻1215   c. 85   d. pq^5
9. a. geometric   b. geometric   c. neither   d. geometric   **11.** a. neither   b. geometric; common ratio 2   c. geometric; common ratio 1/2   d. neither   **13.** 7 (The numbers are arranged alphabetically.)
15. Fibonacci numbers are numbers that appear in the following progression: 1,1,2,3,5,8,13, . . .   **17.** answers vary   **19.** 1   **21.** answers vary; for example, 5,16,8,4,2,1 or 9,28,14,7, . . .,4,2,1.

PROBLEM SET 1.3, PAGES 29–32

1. a. 1   b. 3   c. 8   d. 21   **3.** a. 2   b. 4   c. 7   d. 12   **5.** a. 1   b. 1   c. 4   d. 9
7. 5×5 square   **9.** 8×13 rectangle   **11.** $F_5 \times F_6$ **13.** 1,3,4,7,11,18,29,47,76,123
15. $(F_1)^2 + (F_2)^2 + \cdots + (F_n)^2 = F_n \times F_{n+1}$

17. $\dfrac{3}{1} = 3$; $\dfrac{4}{3} \approx 1.333$; $\dfrac{7}{4} = 1.75$; $\dfrac{11}{7} \approx 1.5714$; $\dfrac{18}{11} \approx 1.6364$; $\dfrac{29}{18} \approx 1.611$; $\dfrac{47}{29} \approx 1.6207$; $\dfrac{76}{47} \approx 1.6170$;

$\dfrac{123}{76} \approx 1.618$ **19.** answers vary

PROBLEM SET 1.4, PAGES 39–43

1. a. one million   b. 10   c. 6   d. 1,000,000 or $10 \cdot 10 \cdot 10 \cdot 10 \cdot 10 \cdot 10$ **3.** $3.2 \cdot 10^3$ **5.** $5.629 \cdot 10^3$
7. $3.5 \cdot 10^{10}$ **9.** $6.3 \cdot 10^7$ **11.** 10^{-5} **13.** $2.2 \cdot 10^9$; $2.5 \cdot 10^{-6}$ **15.** $3 \cdot 10^{10}$ **17.** $5.9 \cdot 10^7$ **19.** $6 \cdot 10^{11}$
21. 64   **23.** .0021   **25.** 216   **27.** .00000041   **29.** 907   **31.** 333,000   **33.** .00000003   **35.** 1,280,000
37. $1973 = 1.973 \cdot 10^3$; 1.6 million $= 1.6 \cdot 10^6$; half million $= 5 \cdot 10^5$; half billion $= 5 \cdot 10^8$; $63{,}000 = 6.3 \cdot 10^4$
39. $5.157 \cdot 10^{15}$ **41.** There are approximately 17 pennies per inch. $1{,}000{,}000/17 \approx 58{,}823$ in. or 4902 ft, or .92 mi. Thus, a million pennies would be about a mile high.
43. 186,282 mi/sec $\approx 1.9 \cdot 10^5$ mi/sec

$\begin{aligned}
&= 1.9 \cdot 10^5 \cdot 60 \cdot 60 \cdot 24 \cdot 365 \cdot 25 \text{ mi/year} \\
&= (1.9 \cdot 10^5)(6 \cdot 10)(6 \cdot 10)(2.4 \cdot 10)(3.6 \cdot 10^2) \\
&= (1.9)(6)(6)(2.4)(3.6) \cdot 10^5 \cdot 10 \cdot 10 \cdot 10 \cdot 10^2 \\
&\approx 590 \cdot 10^{10} \\
&\approx 5.9 \cdot 10^{12}
\end{aligned}$

45. a. 3 ft = 36 in., and on the scale shown, 1 in. is about 26 light years, so 3 ft is about 936 light-years away.
b. $(9.4 \cdot 10^2)(5.9 \cdot 10^{12}) = (9.4)(5.9) \cdot 10^2 \cdot 10^{12} \approx 55.46 \cdot 10^{14} \approx 5.5 \cdot 10^{15}$ **47.** a. $.5 \cdot 10^6$ or $5 \cdot 10^5$ persons per sq mi
b. $.7 \cdot 10^4$ or 7,000 sq mi (This is less than the area of New Jersey.)   c. $1.5 \cdot 10^{-2}$ or .015 sq mi per person. There are
$.015 \cdot 640 \approx 9.6$ acres per person.   **49.** 5,888,896 digits; about 68 days

PROBLEM SET 1.5, PAGES 52–56

1. a. cm (or m)   b. mm   c. km   d. m (or cm)   e. cm   **3.** a. g   b. g   c. kg   d. kg (or g)
e. kg   **5.** B   **7.** C   **9.** C   **11.** a. 6230   b. 450   c. 60   d. 4.8   e. 3.3   **13.** a. 35,250
b. 23   c. .75   d. 1000   e. .458   **15.** a. about 2 m (actually, about 205 cm)   b. 2 mm   c. 5 cm
d. 12 cm   e. 30 cm (or 3 dm)   **17.** a. 180 mi   b. 540 mi   c. 160 mi (or 162 mi)   d. 380 mi (or 384 mi)
e. 2100 mi   **19.** a. 320 km   b. 140 km (or 136 km)   c. 1440 km   d. 19 km (or 19.2 km)   e. 4640 km
21. 37°C
23. 6 to 8 apples, sliced (about 1000 mℓ)   Peel and slice apples thinly; put them in a baking dish. Add the water.
60 mℓ water   Combine the sugar, flour, cinnamon, and salt; blend in the butter until
180 mℓ sugar   crumbly in consistency. Pour over the apples and press down. Bake
120 mℓ cake flour   uncovered in 190°C oven for about 1 hour.
5 mℓ cinnamon
90 mℓ butter
2.5 mℓ salt

PROBLEM SET 1.6, PAGES 58–59

1. 12345678987654321   **2.** It is the reasoning from particular facts or individual cases to a general conjecture.
3. a. 10   b. 12   c. 65   **4.** a. 3a is a geometric progression   b. 1/10   **5.** a. .000579   b. 401,000

6. a. .1   b. 43,210,000   **7.** a. $3.4 \cdot 10^{-3}$ b. $4.0003 \cdot 10^6$ **8.** a. $1.74 \cdot 10^4$ b. 5   **9.** meter, liter, gram   **10.** a. 1/100   b. 10   c. 1000   d. 1/1000   e. 1/10   **11.** a. mass   b. length   c. capacity   d. capacity   e. length   **12.** a. km   b. cm   c. $m\ell$ d. m^3 e. g   **13.** a. 4.3   b. .001   c. 48   d. 2880   e. 1.6   **14.** C   **15.** C   **16.** E   **17.** C   **18.** E   **19.** B   **20.** C

PROBLEM SET 2.1, PAGES 68–72

1. a, b, d, and e are statements.   **3.** a, c, and e are statements.   **5.** a. T   b. T   c. T   d. F   e. T
7. a. T   b. F   c. F   d. T   e. F   **9.** a. F   b. F   c. T   d. T   e. T

11. answers may vary

p	q	$p \wedge q$
T	T	T
T	F	F
F	T	F
F	F	F

13. inductive reasoning   **15.** east; deductive   **17.** We do not know the starting point, but the time of departure was probably sometime between the time she goes to bed and the time she wakes up—a period of probably less than eight hours. This means that the distance traveled is probably less than 400 miles, if by car. Therefore, the probable starting point is Las Vegas. inductive reasoning   **19.** eight persons; speaker, brother, mama, Billy Joe, papa, Tom, preacher, and Becky; deductive reasoning   **21.** Choctaw Ridge in Carroll County (probably Mississippi); inductive reasoning   **23.** You can raise many questions about what she and Billy Joe were throwing off the bridge, but this question asked what she was throwing off the bridge, and the answer is clearly stated in the last two lines of the song. She was throwing flowers off the Tallahatchee Bridge; deductive reasoning.   **25.** $\~s \wedge \~a$ where s: Sam will seek the nomination; a: Sam will accept the nomination.   **27.** $e \wedge \~l$ where e: Fat Albert eats to live; l: Fat Albert lives to eat.   **29.** F
31. T   **33.** T

PROBLEM SET 2.2, PAGES 79–80

1.

p	q	$\~p$	$\~p \vee q$
T	T	F	T
T	F	F	F
F	T	T	T
F	F	T	T

3.

p	q	$p \wedge q$	$\~(p \wedge q)$
T	T	T	F
T	F	F	T
F	T	F	T
F	F	F	T

5.

r	$\~r$	$\~(\~r)$
T	F	T
F	T	F

7.

p	q	$\~q$	$p \wedge \~q$
T	T	F	F
T	F	T	T
F	T	F	F
F	F	T	F

9.

p	q	$\~p$	$\~p \wedge q$	$\~q$	$(\~p \wedge q) \vee \~q$
T	T	F	F	F	F
T	F	F	F	T	T
F	T	T	T	F	T
F	F	T	F	T	T

11.

p	q	$\tilde{p}$	$\tilde{p} \vee q$	$q \wedge p$	$(\tilde{p} \vee q) \wedge (q \wedge p)$
T	T	F	T	T	T
T	F	F	F	F	F
F	T	T	T	F	F
F	F	T	T	F	F

13.

p	q	r	$p \vee q$	$\tilde{r}$	$(p \vee q) \wedge (\tilde{r})$	$[(p \vee q) \wedge \tilde{r}] \wedge r$
T	T	T	T	F	F	F
T	T	F	T	T	T	F
T	F	T	T	F	F	F
T	F	F	T	T	T	F
F	T	T	T	F	F	F
F	T	F	T	T	T	F
F	F	T	F	F	F	F
F	F	F	F	T	F	F

15. p is T and q is T. Using truth tables, we see that (1) is true, (2) is false, and (3) is false.

17. 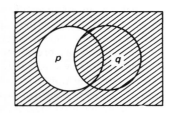   **19.** 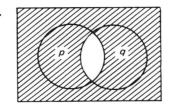   **21.**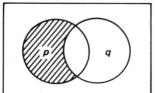

23. 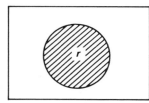   **25.**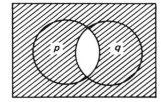

27. a. 2   b. 4   c. 8   d. 16   e. 2^n

PROBLEM SET 2.3, PAGES 86–89

1. a. If there is time enough, then everything happens to everybody sooner or later.   b. If we make a proper use of those means which the God of Nature has placed in our power, then we are not weak.   c. If it is a useless life, then there is an early death.   d. If it is work, then it is noble.   e. If only you can find it, then everything has got a moral.   **3.** a. T   b. T   c. T   d. T   e. F

5.

					a.		b.
p	q	$p \wedge q$	$p \vee q$	$\sim(p \wedge q)$	$(p \vee q) \wedge \sim(p \wedge q)$	$q \rightarrow p$	$(p \wedge q) \wedge (q \rightarrow p)$
T	T	T	T	F	F	T	T
T	F	F	T	T	T	T	F
F	T	F	T	T	T	F	F
F	F	F	F	T	F	T	F

7. a.

p	q	$p \rightarrow q$	$p \vee (p \rightarrow q)$
T	T	T	T
T	F	F	T
F	T	T	T
F	F	T	T

b.

p	q	$\sim p$	$\sim p \rightarrow q$	$p \rightarrow (\sim p \rightarrow q)$
T	T	F	T	T
T	F	F	T	T
F	T	T	T	T
F	F	T	F	T

9. a.

p	q	$p \wedge q$	$(p \wedge q) \rightarrow p$
T	T	T	T
T	F	F	T
F	T	F	T
F	F	F	T

b.

p	q	$p \rightarrow p$	$\sim q$	$q \rightarrow \sim q$	$(p \rightarrow p) \rightarrow (q \rightarrow \sim q)$
T	T	T	F	F	F
T	F	T	T	T	T
F	T	T	F	F	F
F	F	T	T	T	T

11. a.

p	q	$p \wedge q$	$p \vee q$	$(p \wedge q) \leftrightarrow (p \vee q)$
T	T	T	T	T
T	F	F	T	F
F	T	F	T	F
F	F	F	F	T

b.

p	q	$p \rightarrow q$	$\sim p$	$\sim p \vee q$	$(p \rightarrow q) \leftrightarrow (\sim p \vee q)$
T	T	T	F	T	T
T	F	F	F	F	T
F	T	T	T	T	T
F	F	T	T	T	T

13. *Statement:* If p, then q. *Converse:* If q, then p. *Inverse:* If not p, then not q. *Contrapositive:* If not q, then not p.
15. *Statement:* If your car is air-conditioned, then I will go with you. *Converse:* If I go with you, then your car is air-conditioned. *Inverse:* If your car is not air-conditioned, then I will not go with you. *Contrapositive:* If I do not go with you, then your car is not air-conditioned.   **17.** *Statement:* If I get paid, then I will go on Saturday, then I will get paid. *Inverse:* If I do not get paid, then I will not go on Saturday. *Contrapositive:* If I do not go on Saturday, then I do not get paid.   **19.** *Statement:* If you brush your teeth with Smiles toothpase, then you will have fewer cavities. *Converse:* If you have fewer cavities, then you brush your teeth with Smiles toothpaste. *Inverse:* If you do not brush your teeth with Smiles toothpaste, then you will not have fewer cavities. *Contrapositive:* If you do not have fewer cavities, then you do not brush your teeth with Smiles toothpaste.
21. *Statement:* $r \rightarrow t$. *Converse:* $t \rightarrow \sim r$. *Inverse:* $r \rightarrow \sim t$. *Contrapositive:* $\sim t \rightarrow r$.   **23.** *Statement:* $\sim t \rightarrow \sim s$. *Converse:* $\sim s \rightarrow \sim t$. *Inverse:* $t \rightarrow s$. *Contrapositive:* $s \rightarrow t$.   **25.** Let j: Jacob will be late. k: Ken will be late. *Solution:* $j \wedge k$.
27. Let c: Harry has a headache. h: Harry is reasonable. *Solution:* $c \rightarrow \sim h$.   **29.** Let e: It is an elephant. b: It is big. *Solution:* $e \rightarrow b$.   **31.** Let t: Tom was at the scene of the crime. u: I was at the scene of the crime. *Solution:* $(t \vee u) \wedge [\sim(t \wedge u)]$.   **33.** Let p: He is a mathematician. o: He is an ogre. *Solution:* $p \rightarrow o$.   **35.** Let h: I will buy a new house. p: All provisions of the sale are clearly understood. $\sim p \rightarrow \sim h$ **37.** Let m: It is a man. i: It is an island. $m \rightarrow \sim i$ **39.** Let n: Be nice to people on your way up. m: You will meet people on your way down. $(n \wedge m) \wedge (m \rightarrow n)$ **41.** Let f: The majority, by mere force of numbers, deprives a minority of a clearly written constitutional right. r: Revolution is justified. $f \rightarrow r$

PROBLEM SET 2.4, PAGES 96–98

1. a. T   b. F   c. T   d. T   e. F (zero is not divisible by itself)   **3.** a. ∃   b. ∀   c. ∃   d. ∃
e. ∀

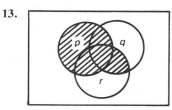

5.   **7.**   **9.**

11.   **13.**

15.

p	$\~p$	$\~(\~p)$	$p \leftrightarrow \~(\~p)$
T	F	T	T
F	T	F	T

Since all possibilities are true, $p \Longleftrightarrow \~(\~p)$.

17.

p	q	$p \rightarrow q$	$\~q$	$\~p$	$\~q \rightarrow \~p$	$(p \rightarrow q) \leftrightarrow (\~q \rightarrow \~p)$
T	T	T	F	F	T	T
T	F	F	T	F	F	T
F	T	T	F	T	T	T
F	F	T	T	T	T	T

Since all possibilities are true, $(p \rightarrow q) \Longleftrightarrow (\~q \rightarrow \~p)$.
19. regions 2, 3, and 4   **21.** region 6   **23.** a. T   b. T   c, d, and e are not necessarily true.   **25.** valid
27. valid   **29.** not valid   **31.** not valid   **33.** valid

PROBLEM SET 2.5, PAGES 106–110

1. answers vary   **3.** *Subset:* A is a subset of B if every element of A is in B. *Proper subset:* A is a proper subset of B if
every element in A is in B and $A \neq B$.   **5.** $A \cup B$ is the set consisting of elements of A or B or both. $A \cap B$ is the set
consisting of elements of both A and B. $\overline{A}$ is the set consisting of elements not in A.

7. a. {5}

b.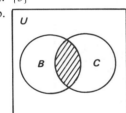

9. a. {1,2,4,5,6,7}

b.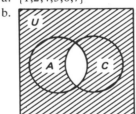

11. a. {1,2,3}

b.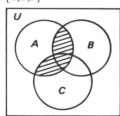

13. a. {1,2,4,5,6,7}

b.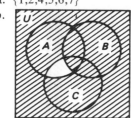

15. a. $(A \cap B) \cap C = \varnothing \cap \{2,5,8,9,10\} = \varnothing$

b.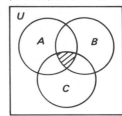

17. a. $(A \cup B) \cap (A \cap C) = \{2,4,5,6,8,9\} \cap \{2,4,5,6,8,9,10\}$
$= \{2,4,5,6,8,9\}$

b.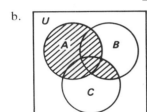

19. a. $A \cap (\overline{B \cup C}) = \{2,4,6,8,\} \cap \overline{\{2,5,8,9,10\}}$
$= \{2,4,6,8\} \cap \{1,3,4,6,7\}$
$= \{4,6\}$

b.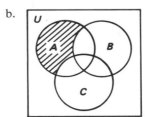

21. a. $\overline{A} \cup (B \cap C) = \{1,3,5,7,9,10\} \cup \{5,9\}$
$= \{1,3,5,7,9,10\}$ or $\overline{A}$

b.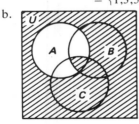

23. a. $A \cap (\overline{B} \cup C) = \{2,4,6,8\} \cap \{1,2,3,4,5,6,7,8,9,10\}$
$= \{2,4,6,8\}$ or A

b.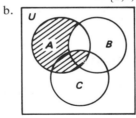

25.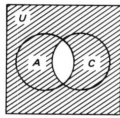

27.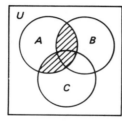

29.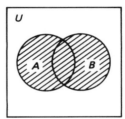

31.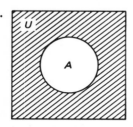

33.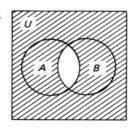

35.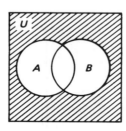

37.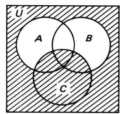

39. {s,e,c,o,n,d,r}　　**41.** Yes

43. No because there are 102 responses instead of 100.　　**45.** F　　**47.** T

49. F

51. a.

p	q	$p \vee q$	$\sim(p \vee q)$	$\sim p$	$\sim q$	$(\sim p) \wedge (\sim q)$	$\sim(p \vee q) \leftrightarrow (\sim p) \wedge (\sim q)$
T	T	T	F	F	F	F	T
T	F	T	F	F	T	F	T
F	T	T	F	T	F	F	T
F	F	F	T	T	T	T	T

b.

p	q	$p \wedge q$	$\sim(p \wedge q)$	$\sim p$	$\sim q$	$(\sim p) \vee (\sim q)$	$\sim(p \wedge q) \leftrightarrow (\sim p) \vee (\sim q)$
T	T	T	F	F	F	F	T
T	F	F	T	F	T	T	T
F	T	F	T	T	F	T	T
F	F	F	T	T	T	T	T

53. a. B　　**b.** A　　**c.** U　　**d.** $\emptyset$　　**55. a.** {5,6,7}　　**b.** {3,4}　　**c.** {5,6}　　**57.** The difference $A - B$ of two sets A and B is the set of all elements that belong to A and not to B.

PROBLEM SET 2.6, PAGES 115–120

1. a. direct　　**b.** indirect　　**c.** transitive　　**3.** If you can learn mathematics, then you understand human nature.

5. All trebbles are expensive.   **7.** If a nail is lost, then a kingdom is lost.   **9.** b ≠ 0   **11.** I will not eat that piece of pie.   **13.**

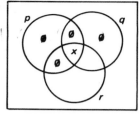

15. valid; indirect   **17.** valid; indirect   **19.** not valid   **21.** not valid   **23.** not valid   **25.** I will participate in student demonstrations.   **27.** You will obey the law.   **29.** Babies cannot manage crocodiles.   **31.** My poultry are not officers.   **33.** yes

PROBLEM SET 2.7, PAGES 123–125

1. inductive reasoning; answers vary

2.

p	q	$p \wedge q$	$p \vee q$	$\tilde{p}$	$\tilde{q}$	$p \rightarrow q$	$p \leftrightarrow q$
T	T	T	T	F	F	T	T
T	F	F	T	F	T	F	F
F	T	F	T	T	F	T	F
F	F	F	F	T	T	T	T

3. a.

p	q	$p \wedge q$	$\tilde{}(p \wedge q)$
T	T	T	F
T	F	F	T
F	T	F	T
F	F	F	T

b.

p	q	$\tilde{}(p \wedge q)$	$\tilde{p}$	$\tilde{q}$	$(\tilde{p}) \vee (\tilde{q})$	$[\tilde{}(p \wedge q)] \leftrightarrow [(\tilde{p}) \vee (\tilde{q})]$
T	T	F	F	F	F	T
T	F	T	F	T	T	T
F	T	T	T	F	T	T
F	F	T	T	T	T	T

4. Answers may vary.   a. If C. D. Money is trustworthy, then he inherits a fortune.   b. C. D. Money is trustworthy unless he inherits a fortune.   **5.** a. Let c: It is a computer. s: It is capable of self-direction. *Translation*: $c \rightarrow \tilde{s}$.   *b. *Contrapositive*: $s \rightarrow \tilde{c}$. Or, if it is capable of self-direction, then it is not a computer.   **6.** *Converse*: If I don't make a lot of money, then I didn't go to college. *Inverse*: If I go to college, then I will make a lot of money. *Contrapositive*: If I make a lot of money, then I went to college.

7. Direct reasoning $p \rightarrow q$

$$\frac{p}{\therefore q}$$

Proof:

p	q	$p \rightarrow q$	$(p \rightarrow q) \wedge p$	$[(p \rightarrow q) \wedge p] \rightarrow q$
T	T	T	T	T
T	F	F	F	T
F	T	T	F	T
F	F	T	F	T

It is always true and is, therefore, proved.

8. Answers vary but should fit the pattern: $p \rightarrow q$

$$\frac{\tilde{q}}{\therefore \tilde{p}}$$

9. Yes; it illustrates indirect reasoning.   **10.** a. $\exists$ b. $\exists$ c. $\forall$ d. $\exists$ e. $\forall a \forall b$
11. a. $\{1,2,3,4,5,6,7,9,10\}$ b. $\{9\}$ c. $\{1,3,5,7,8\}$ **12.** a. $\{1,3,5,7,8,9\}$ b. $\{2,4,6,10\}$ **13.** $\emptyset$
14. a. F   b. T   c. T   **15.** a. T   b. F   c. F   **16.** a. not true   b. true   **17.** Mr. Gent obtained a tan.   **18.** No prune is an artificial sweetener.   **19.** I do not attend to my duties.   **20.** If we go on a picnic, then we must go to the store.

PROBLEM SET 3.1, PAGES 138–144

1. a. 27   b. 63   c. 243   d. 432   e. 504   **3.** It introduced the idea of the punched card and a program to complete a task.   **5.** 10^{-9} sec   **7, 9,** and **11** answers vary, but you should support your opinions whenever possible.   **13.** a. 13   b. 35   c. 64   d. 21   e. 34

PROBLEM SET 3.2, PAGES 151–153

1. arithmetic, algebraic, and RPN; answers vary   **3.** 57334 (HEELS)   **5.** 57334.4614 (HIGH·HEELS)
7. 53045 (SHOES)   **9.** 4914 (HIGH)   **11.** 40 (OH)   **13.** 510714 (HILOIS)   **15.** $2.814749767 \cdot 10^{14}$ or 281,474,976,710,656   **17.** 30,814 minutes or 21 days, 9 hours, and 34 minutes   **19.** .618   **21.** The result is the original number.   **23.** .1666 . . .   **25.** .142857142857 . . .   **27.** .0036900369 . . .
29. a. .142857142857 . . .   b. .285714285714 . . .   c. .428571428571 . . .   d. .571428571428 . . .
e. .714285714285 . . .   f. .857142857142 . . .

PROBLEM SET 3.3, PAGES 161–165

1. Analog: continuous measurement of data and information. Digital: discrete measurement.   **3.** Paper tape; magnetic tape; punched cards   **5.** Floating-point is a representation of a number that is written as a decimal between 1 and 10 and a power of 10.   **7.** a. 5.37E − 5   b. 3.14159   **9.** a. 3.78596E + 5   b. 5.002E + 2   **11.** a. 4560
b. .000179   **13.** a. 30,500   b. 705.6   **15.** a. 5,932,600   b. .00000 3217   **17.** answers vary   **19.** answers vary

PROBLEM SET 3.4, PAGES 172–176

1. A flow chart is a device that lists the steps to be performed in sequence when writing a program.   **3.** *Step 1:* Analyze the problem to understand what is being done. *Step 2:* Prepare a flow chart. *Step 3:* Put the flow chart into a language that the machine can "understand." *Step 4:* "Debug" the program.

5.

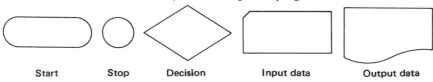

| Start | Stop | Decision | Input data | Output data |

7. GOOD MORNING.
CAN YOU DO THIS PROBLEM CORRECTLY?
5

READY

9. a. $35x^2 - 13x + 2$ b. $6x - 7$ **11.** a. $6.29^{14} - 7$ b. $(x - 5)(2x + 4)^2$

13. $\dfrac{16.34}{12.5}(42.1^2 - 64)$ **15.** a. $(2/3)*X\uparrow2$ b. $3*X\uparrow2 - 17$ **17.** a. $12*(X\uparrow2 + 4)$ b. $(15*X + 7)/2$

19. $(X + 1)*(2*X - 3)*(X\uparrow2 + 4)$ **21.** Yes. $2(x + 6) = 2x + 12; 2x + 12 - 10 = 2x + 2;$
$\dfrac{2x + 2}{2} = x + 1; (x + 1) - x = 1$

23. 10 PRINT "I'M A HAPPY COMPUTER."   **25.** 10 PRINT $(1 + .08)\uparrow12$ **27.** 10 PRINT $(23.5\uparrow2 - 5*61.1)/2$
    20 END   20 END   20 END

29.   **31.**

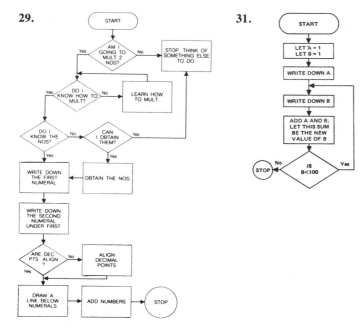

33. answers vary

PROBLEM SET 3.5, PAGES 184–187

1. (1) to do calculations, as in 10 PRINT 4 + 5: (2) to type out text, as in 10 PRINT "HELLO"; (3) to generate a line feed, as in 10 PRINT.   **3.** INPUT, READ, or LET

5. 9   **7.** 19

    HELLO, I LIKE YOU.   —   5   84
    4 + 5 = 9

9. There is no operation symbol between the parentheses on line 20.

11. I WILL CALCULATE IQ.
    WHAT IS MENTAL AGE? [Input 12 here]
    WHAT IS CHRONOLOGICAL AGE? [Input 10 here]

    IQ IS 120

13.
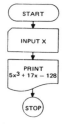

```
10   PRINT "WHAT IS X";
20   INPUT X
30   PRINT 5*X↑3 + 17*X − 128
40   END
```

15.
```
10   PRINT "I WILL COMPUTE 24*X↑3 + 3*X↑2 + 6. WHAT IS X";
20   INPUT X
30   PRINT "24*X↑3 + 3*X↑2 + 6 = ",24*X↑3 + 3*X↑2 + 6
40   PRINT "WHEN X =",X
50   END
```

$x = 1, 32$; $x = 10, 24,296$; $x = 4.63, 2447.7490$

17.
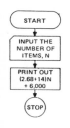

```
10   PRINT "WHAT IS THE NUMBER OF ITEMS";
20   INPUT N
30   PRINT "THE MONTHLY COST IS";
40   PRINT (2.68 + 14)↑N + 6000
50   END
```

PROBLEM SET 3.6, PAGES 193–197

1. A loop is a repetition of one or more lines in a program.   **3.** HI FRIEND.

5. 10   PRINT "HI FRIEND."
20   END

7. A IS BETWEEN 1 AND 100 INCLUSIVE.   **9.** A IS LESS THAN 1.   **11.** 15   **13.** 5   **15.** 720   **17.** 6

19. 1
2
3

21. 10   FOR N = 1 to 99
20   PRINT 101 − N;
30   PRINT "BOTTLES OF BEER ON THE WALL."
40   NEXT N
50   PRINT " 1   BOTTLE OF BEER ON THE WALL."
60   PRINT "THAT'S ALL FOLKS!"
70   END

23. answers vary
10   PRINT "I WILL OUTPUT NUMBERS IN THE FIBONACCI SEQUENCE."
20   PRINT "HOW MANY CONSECUTIVE FIBONACCI NUMBERS WOULD YOU"   } Lines 10–25 are not necessary.
25   PRINT "LIKE TO SEE";
30   INPUT N
40   IF N < 0 GOTO 210
50   IF INT(N) <> N GOTO 210   Lines 40, 50, 80, 90, 120, 130, 200, 210 and 215
60   LET A = 1   are all making sure that N is a permissible value;
70   PRINT A;   this checking is not essential for this problem.
80   IF N > 1 GOTO 100
90   STOP
100 LET B = 1
110 PRINT B;
120 IF N > 2 GOTO 140
130 STOP
140 FOR I = 3 TO N
150 LET C = A + B
160 PRINT C;   Lines 140–190 are where the requested problem
170 LET A = B   is being done.
180 LET B = C
190 NEXT I
200 STOP
210 PRINT "N MUST BE A POSITIVE INTEGER. NOW, ";
215 GOTO 20
220 END

25. N^2

PROBLEM SET 3.7, PAGES 199–201

1. answers vary; see Section 3.1   **2.** a. 44.1   b. 3.009375   **3.** 7735.05 (SO'SELL)   **4.** 3,857,620   **5.** It refers to a computer incorporating the circuit board and microminiaturization into its design.   **6.** *Input:* means of putting information into the computer. *Memory:* the part of the computer that saves or stores information. *Accumulator:*

that part of the computer in which the arithmetic operations are carried on. *Control:* that part of the computer that directs the flow of data to and from the various components of the computer. It is also the part in which the steps of the program are followed. *Output:* means of getting results from the computer.   **7.** A floating-point number is a representation of a number as a decimal between 1 and 10 times an appropriate power of 10. For example, 230,000 may be written as $2.3E + 5$. Floating-point numbers are needed whenever the magnitude of the number exceeds the capacity of the computer for a single number.   **8.** a. $5.76E + 11$ b. $2E - 6$ **9.** a. 380,000,000   b. .00000 00574   **10.** Computer programming refers to the process by which we give a computer a series of step-by-step instructions to complete a particular task or problem. The level of a program and the type of instructions the computer can "understand" may vary greatly.   **11.** *Machine language:* the native language of the computer. It consists of giving the computer instructions in terms of a binary "code." *Symbolic language:* this language uses words to represent certain sequences of zeros and ones in machine language. *Algorithmic language:* this is a more conversational language than the previous two. Here an entire list of instructions is generally replaced by a single command. BASIC is an example of an algorithmic language.

12.

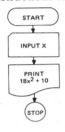

```
10   INPUT X
20   PRINT 18*X↑2 + 10
30   END
```

13. $5(x + 2)^2$ **14.** $3*(X + 4)*(2X - 7)↑2$ **15.**

```
       START
         │
      INPUT X
         │
      PRINT
   15x³ − 2x² + 7
         │
       STOP
```

16.
```
10   PRINT "WHAT IS X";
20   INPUT X
30   PRINT 15*X↑3 − 2*X↑2 + 7
40   END
```
17. THE MISSISSIPPI IS WET.
18.
```
10   PRINT "THE MISSISSIPPI IS WET."
20   END
```

19. NOT IGNORANCE, BUT THE IGNORANCE OF IGNORANCE, IS THE DEATH OF KNOWLEDGE.
20.
```
10   PRINT "WHAT IS THE RADIUS";
20   INPUT R
30   PRINT "THE AREA IS";
40   PRINT 3.141592*R↑2
50   END
```

PROBLEM SET 4.1, PAGES 210–213

1. Answers vary; a number is an idea or a concept, while a numeral is a symbol for a number.   **3.** a. yes, since it is not a positional system   b. no   **5.** a. ∩∩|||||| / ∩∩

b. ∩∩∩ / ∩∩∩∩ ||||| c. 99 / 99 ∩∩∩| d. 9999 / 9999 ∩∩∩∩ / ∩∩∩ |||||| e. 99 ∩∩∩∩ / 99 ∩∩∩∩ / ∩ ||

7. answers vary   **9.** a. 1256   b. 1/200   **11.** a. 24   b. 261   **13.** a. 123 1/2   b. 1/100
15. a. 152   b. 4881
17. 𝄡999∩IIIIIIII   **19.** 99III   **21.** ⊲⊲⊲⊲⊲   **23.** ▾▾⊲▾▾
                                99IIII
25. answers vary   **27.** one

PROBLEM SET 4.2, PAGES 217–219

1. a. $b^n = \underbrace{b \cdot b \cdot b \cdots b}_{n \text{ factors}}$ b. $b^0 = 1$ c. $b^{-n} = 1/b^n$ **3.** a. 1000   b. .0008   **5.** a. 521,658

b. 60,004,001   **7.** 3028.5462   **9.** a. $(6 \times 10^{-2}) + (4 \times 10^{-3}) + (2 \times 10^{-4}) + (1 \times 10^{-5})$
b. $(2 \times 10^1) + (7 \times 10^0) + (5 \times 10^{-1}) + (7 \times 10^{-2}) + (2 \times 10^{-3})$
11. a. $(4 \times 10^2) + (2 \times 10^1) + (8 \times 10^0) + (3 \times 10^{-1}) + (1 \times 10^{-2})$
b. $(2 \times 10^6) + (3 \times 10^5) + (4 \times 10^4) + (5 \times 10^3) + (6 \times 10^2) + (8 \times 10^1) + (1 \times 10^0)$
13. a. $(5 \times 10^{-6}) + (2 \times 10^{-7}) + (7 \times 10^{-8})$
b. $(5 \times 10^3) + (2 \times 10^2) + (4 \times 10^1) + (5 \times 10^0) + (5 \times 10^{-1})$ **15.** a. 5×10^0 b. 5×10^2 c. 5×10^{-3}
17. answers vary   **19.** Let the given two digit number have tens digit a and unit digit b. Then the number can be
represented by $a \cdot 10 + b$. Since $11 = 10 + 1$, the product of the given number and 11 can be written as follows:

$$(a \cdot 10 + b)(10 + 1) = (a \cdot 10 + b) \cdot 10 + (a \cdot 10 + b) \cdot 1$$
$$= a \cdot 10^2 + b \cdot 10 + a \cdot 10 + b$$
$$= \underbrace{a \cdot 10^2}_{} + \underbrace{(b + a) \cdot 10}_{} + \underbrace{b}_{}$$

Hundreds   Tens digit is   Units digit
digit is   the sum of   is the same
the same   the digits.   as the units
as the   digit of the
tens digits   given number.
of the
given
number.

PROBLEM SET 4.3, PAGES 225–231

1. commutative   **3.** commutative   **5.** both   **7.** commutative   **9.** commutative   **11.** no   **13.** both
15. answers vary   **17.** a. ^-i b. $^-1$ c. i d. $^-1$ **19.** reasons vary   a. yes   b. yes
c. yes   **21.** reasons vary   a. yes   b. yes   c. no   **23.** yes
25. (a)

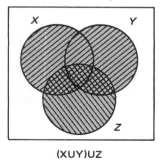

(XUY)UZ   XU(YUZ)

Since the shaded parts are the same, $(X \cup Y) \cup Z = X \cup (Y \cup Z)$

27. Yes; proof is similar to that shown in the answer to Problem 25.

29. No; yes, the set of odd numbers is closed for multiplication.   **31.** a. A   b. F
c. A   d. F   e. R   **33.** It is closed for both ↑ and ↓.

35. It is the same as the program in the text but with the following changes:

```
30   IF A – B = B – A GOTO 80
100  PRINT "THE SET IS COMMUTATIVE FOR SUBTRACTION."
```

37.
```
10   FOR A = 0 TO 1
20   FOR B = 0 TO 1
30   FOR C = 0 TO 1
40   IF (A*B)*C = A*(B*C) GOTO 100
50   PRINT "NOT ASSOCIATIVE, SINCE I HAVE FOUND A";
55   PRINT " COUNTER-EXAMPLE WHEN"
60   PRINT "A = ";A;
70   PRINT ", B = ";B;
80   PRINT ", AND C = C";C
90   STOP
100  PRINT "CHECKING A = ";A;", B = ";B;", AND C = ";C
120  PRINT "(A*B)*C = ";(A*B)*C;", AND A*(B*C) = ";A*(B*C)
130  PRINT
140  PRINT
150  NEXT C
160  NEXT B
170  NEXT A
180  PRINT "THE SET IS ASSOCIATIVE FOR MULTIPLICATION."
190  END
```

PROBLEM SET 4.4, PAGES 241–244

1. a. You moved to the *right* 7 units.   b. $^+6$ c. $^-8$ **3.** a. $^+4$ b. $^-4$ c. $^-14$ d. $^+135$ e. $^-19$
f. $^-1$ **5.** a. $^+3$ b. $^-3$ c. $^+10$ d. $^+5$ e. $^+10$ f. $^+10$ **7.** a. $^-47$ b. $^-35$ c. $^-4$
d. $^+7$ e. $^+1$ f. 0   **9.** a. $^-18$ b. $^-20$ c. $^+2$ d. $^+2$ e. $^-26$ f. $^+2$ **11.** a. $^-3$
b. $^+42$ c. $^+132$ d. $^+170$ e. $^-1$ f. $^+40$ **13.** a. 0   b. $^-13$ c. $^+6$ d. $^+7$ e. $^+44$
f. $^+15$ **15.** a. $^-56$ b. $^-75$ c. $^+30$ d. $^-14$ e. $^+2$ f. $^+18$

17.

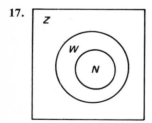

19. a. An operation * is commutative for a set S if $a*b = b*a$ for all elements a and b in S.   b. Yes for addition; no for
subtraction   c. Yes for multiplication; no for division   **21.** An operation * is associative for a set S if $(a*b)*c$
$= a*(b*c)$ for all elements a, b, and c in S.   b. Yes for addition; no for subtraction   c. Yes for multiplication; no for
division   **23.** a. $^+16$ b. $^-1$ c. $^+1$ d. $^+1$ **25.** Yes. Consider the set $\{^-1,0,^+1\}$.

PROBLEM SET 4.5, PAGES 250–252

1. Equal   **3.** Equal   **5.** Equal   **7.** 13/9   **9.** $^-92/105$ **11.** 63/25   **13.** 1/9   **15.** 1/15

17. 239/1000   **19.** 10/21   **21.** 30/7   **23.** 2/(⁻7) or ⁻2/7   **25.** 2/3   **27.** 21   **29.** ⁻625/686

31.

Sets	5	⁻7	0	4/5	⁻1/2
Natural numbers	√	X	X	X	X
Whole numbers	√	X	√	X	X
Integers	√	√	√	X	X
Rational numbers	√	√	√	√	√

33. Given any two rationals a/b and c/d where $c \neq 0$. To show $a/b \div c/d$ is rational. Now, $a/b \div c/d = ad/bc$ by the definition of division. ad is an integer because a and d are integers; also, bc is a nonzero integer because b and c are nonzero integers. Therefore, ad/bc is a rational number, by the definition of a rational number.   **35.** closure, associative, and commutative   **37.** Since $b \neq 0$ and $d \neq 0$, $1/b$ and $1/d$ are rationals. Then

$$ad = bc$$

$$ad\left(\frac{1}{b}\right)\left(\frac{1}{d}\right) = bc\left(\frac{1}{b}\right)\left(\frac{1}{d}\right)$$

$$a\left(\frac{1}{b}\right)\left(d \cdot \frac{1}{d}\right) = c\left(\frac{1}{d}\right)\left(b \cdot \frac{1}{b}\right)$$

$$a\left(\frac{1}{b}\right)(1) = c\left(\frac{1}{d}\right)(1)$$

$$\frac{a}{b} = \frac{c}{d}$$

PROBLEM SET 4.6, PAGES 265–271

1.

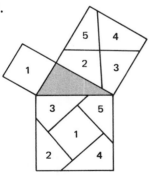

3. answers vary; about 1.41

5.

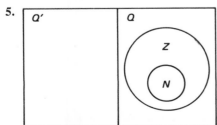

R = {reals}
Q = {rationals}
Q' = {irrationals}
Z = {integers}
N = {naturals}

7. a. .6   b. 1.8   **9.** a. .875   b. $.\overline{428571}$ **11.** a. $14.\overline{18}$ b. $52.\overline{4}$ **13.** a. 5   b. $1.\overline{09}$

15. a. 5/11   b. 26/111   **17.** a. 889/9   b. 7/11   **19.** a. 111/200   b. 5/9   **21.** a. 4567/9999
b. 484/125   **23.** a. 21,28,36   b. 35,56,84   c. Any number can be written as the sum of three triangular
numbers.   **25.** Let $2m$ and $2n$ be any 2 even integers. Then $2m \cdot 2n = 2(2mn)$, which is even since it is of the form $2k$ for
some number k.   **27.** Let $2m + 1$ be any odd integer. Then $(2m + 1)^2 = 4m^2 + 4m + 1 = 2(2m^2 + 2m) + 1$, which is
odd since it is of the form $2k + 1$.   **29.** answers vary   **31.** answers vary
33.

```
 5 PRINT
 6 PRINT
10 PRINT "THIS PROGRAM WILL OUTPUT THE PARTIAL ";
12 PRINT "SUMS FOR THE SERIES"
14 PRINT "S/(2↑N). THE PROGRAM WILL CONTINUE ";
16 PRINT "UNTIL YOU TYPE 'STOP'."
20 PRINT "FOR WHAT VALUE OF N WOULD YOU LIKE ME TO BEGIN";
25 INPUT N
30 IF N < 0 GOTO 36
32 IF INT(N) <> N GOTO 36
34 GOTO 45
36 PRINT "N MUST BE A WHOLE NUMBER. PLEASE TRY AGAIN."
37 PRINT
38 PRINT "NOW, ";
40 GOTO 20
45 PRINT
46 PRINT
50 PRINT "  NUMBER OF TERMS          PARTIAL SUM"
52 PRINT
55 LET M = 1
57 LET S = 1/2↑N
60 PRINT TAB(8);M;TAB(24);S
70 LET M = M + 1
75 LET N = N + 1
80 LET S = S + 1/2↑N
85 GOTO 60
90 END
```

A sample output is given below.
THIS PROGRAM WILL OUTPUT THE PARTIAL SUMS FOR THE SERIES
S/(2↑N). THE PROGRAM WILL CONTINUE UNTIL YOU TYPE 'STOP'.
FOR WHAT VALUE OF N WOULD YOU LIKE ME TO BEGIN? 1

NUMBER OF TERMS	PARTIAL SUM
1	.5
2	.75
3	.875
4	.9375
5	.96875
6	.984375
7	.9921875
8	.9960938
9	.9980469
10	.9990234

STOP
READY
35. answers vary

PROBLEM SET 4.7, PAGES 274–275

1. The position in which the individual digits are listed is relevant. Examples will vary.   **2.** Addition is easier in a
simple grouping system. Examples will vary.   **3.** It uses ten symbols, it is positional and additive, it has a placeholder
symbol (0), and it uses 10 as its basic unit for grouping.

4. a. $b^n = \underbrace{b \cdot b \cdots b}_{n \text{ factors}}$ **b.** $b^{-n} = 1/b^n$ **c.** $b^0 = 1$

5. $(4 \times 10^2) + (3 \times 10^1) + (6 \times 10^0) + (2 \times 10^{-1}) + (1 \times 10^{-5})$ **6.** 4,020,005.62   **7.** answers vary

8.

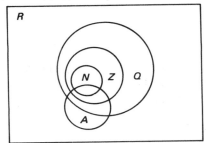

The universe is the set of reals; Q' is the set of numbers outside of Q within the universe.

9. a. A **b.** $\{5,0,{}^-6,17,\sqrt{16}\}$ **c.** $\{5,.7,0,{}^-6,5/7,192.1,17,\sqrt{16}\}$ **10. a.** $\{5,17,\sqrt{16}\}$ **b.** Z **c.** $\{\sqrt{2}\}$
11. answers vary   **12. a.** $^-7$ **b.** $^+23$ **c.** $^-17$ **d.** $^-3$ **e.** $^-8$ **f.** $^-20$ **13. a.** $^+9$ **b.** $^+15$
c. $^-14$ **d.** $^-4$ **e.** $^-11$ **f.** $^{36}/_5$ **14. a.** $^{71}/_{63}$ **b.** $^{-1}/_{36}$ **c.** $^{35}/_3$ **d.** 2 (use the commutative and associative properties to simplify first)   **e.** $^7/_4$ **15. a.** $^{12}/_5$ **b.** $^1/_{15}$ **c.** $^{31}/_{30}$ **d.** $^{15}/_8$ **e.** 7 (use the commutative and associative properties to simplify first)   **16.** If triangle ABC is a right triangle with legs a and b and hypotenuse c, then $c^2 = a^2 + b^2$.   **17.** .625   **18.** $^8/_{11}$ **19.** $^{333}/_{500}$

20. $\dfrac{3}{5}\left(\dfrac{25}{97} + \dfrac{72}{97}\right) = \dfrac{3}{5} \cdot 1 = \dfrac{3}{5}$

PROBLEM SET 5.1, PAGES 284–286

1. $27.50   **3.** $288   **5.** $1152   **7.** $2400   **9.** $3200   **11.** $27.50   **13.** $311.65   **15.** $1619.51
17. $2934.37   **19.** $20,724.52   **21.** $1.80   **23.** $631   **25.** $14,414   **27.** $180,177   **29.** $29,960
31. 6.17%   **33.** Effective yield of 5¼% compounded daily is 5.39%. Therefore, the 5½% compounded annually is the better interest rate.   **35.** The difference is almost $13 million   **37.** $77.51 (no difference, to the nearest cent)

PROBLEM SET 5.2, PAGES 291–294

1. Answers vary; open-end allows purchases up to a specific amount with a flexible repayment schedule while closed-end is an installment loan to pay off a specific amount over a certain specified length of time.   **3.** annual percentage rate; answers vary   **5.** about 14½%   **7.** about 12%   **9.** about 15½%   **11.** about 16%   **13.** $176.14
15. $735.28

17.

Date					
2/1/79	$600.00	----	$ 50	$550.00	Total payments $654.70
3/1/79	$558.25	$8.25	$ 50	$508.25	APR: (54.70/600)100 ≈ 9.12
4/1/79	$515.87	$7.62	$ 50	$465.87	From Table 5.2, about 16½%.
5/1/79	$472.86	$6.99	$ 50	$422.86	
6/1/79	$429.20	$6.34	$ 50	$379.20	
7/1/79	$384.89	$5.69	$ 50	$334.89	
8/1/79	$339.91	$5.02	$ 50	$289.91	
9/1/79	$294.26	$4.35	$ 50	$244.26	
10/1/79	$247.92	$3.66	$ 50	$197.92	
11/1/79	$200.89	$2.97	$ 50	$150.89	
12/1/79	$153.15	$2.26	$ 50	$103.15	
1/1/80	$104.70	$1.55	$104.70	0	

19. a. 18% (given)   **b.** (244.72/800)100 = 30.59; about 20% (from Table 5.2)   **c.** (128.56/800)100 = 16.07; about 14½% (from Table 5.2)   **21.** 6.38   **23.** $106.96

PROBLEM SET 5.3, PAGES 297–299

1. $2870.13   **3.** $3639.09   **5.** $4944.24   **7.** $4710.40   **9.** $10,368.82   **11.** $720.44   **13.** $136.50
15. $938.60   **17.** $666.05   **19.** $585.20   **21.** $3895 to $4075   **23.** $4089 to $4278   **25.** $6377 to $6671
27. $5845 to $6114   **29.** $11,702 to $12,249   **31.** 16½%   **33.** 14½%   **35.** $197.96   **37.** $375.94

PROBLEM SET 5.4, PAGES 306–307

1. a loan contract   **3.** an agreement between a buyer and a seller stating the terms of the sale   **5.** a one-time charge to cover a lender's administrative costs in providing the loan   **7.** about $210   **9.** about $240   **11.** $2425
13. $17,000   **15.** 11.63%   **17.** 11.00%   **19.** 11.38%   **21.** $507.29; $75,701.22   **23.** $474.11;
$124,586.80   **25.** $568.15; $114,580.08   **27.** $701.76; $100,436   **29.** $647.36; $165,036   **31.** $740.10;
$143,707.20

PROBLEM SET 5.5, PAGES 313–314

1. Deductible refers to the amount that the insured must pay before the company makes any payment for collision coverage.   **3.** The cash value of a whole life policy is a forced savings that is paid along with the premium.
5. Comprehensive coverage on an automobile policy pays for loss or damage to the insured's car due to something other than collision.   **7.** Extended coverage applies to home insurance and is protection against certain specific damages in addition to fire protection.   **9.** answers vary   **11. a.** 40/80/10 means the company will pay a maximum of $40,000 for bodily injury or death to one person, $80,000 if more than one person is injured, and $10,000 for property damage.
b. 100/200/20 means the company will pay a maximum of $100,000 for bodily injury or death to one person, $200,000 if more than one person is injured, and $20,000 for property damage.   **13.** $69.75   **15.** $43.75   **17.** $888.20
19. $55.65   **21.** $555   **23.** $35,130   **25.** $7724   **27.** $10,204   **29.** $5726

PROBLEM SET 5.6, PAGES 316–317

1. $1600   **2.** $4660.96   **3.** 8.24%   **4.** closed-end is an installment loan, and open-end is a credit card or revolving credit purchase   **5.** annual percentage rate; answers vary   **6.** about 16½% (16¼%)   **7.** $1633.16   **8.** about 17%   **9.** 5% offer is $3943 and a 10% offer is $4130; since the price is above a 10% offer, it is not a particularly good sale price   **10. a.** $5289   **b.** $4811   **c.** $5040   **11. a.** $7.75   **b.** $7.50   **c.** $8.25   **12. a.** $3.25
b. $0.75   **b.** $8.25   **13.** about $1400   **14. a.** $6,500   **b.** $13,000   **c.** $16,250   **15. a.** $601.97
b. $495.04   **c.** $464.10   **16.** $366.25; $26,530   **17.** Lender A is 12.25%; Lender B is 12.38%; Lender A is the better offer by about ⅛%.   **18. a.** $52,780   **b.** $69,160   **c.** $86,359   **19.** $19,310   **20.** $44,410

PROBLEM SET 6.1, PAGES 324–332

See text for **1, 3,** and **5.**   **7.** 7   **9.** commutative   **11.** commutative   **13.** inverse   **15.** commutative
17. inverse

19.

Set	Natural Numbers N			
Operation	+	×	−	÷
Closure	yes	yes	no	no
Associative	yes	yes	no	no
Identity	no	yes	no	no
Inverse	—	no	—	—
Commutative	yes	yes	no	no
Distributive	× over + yes		× over − yes	

21.

Set	Integers Z			
Operation	+	×	−	÷
Closure	yes	yes	yes	no
Associative	yes	yes	no	no
Identity	yes	yes	no	no
Inverse	yes	no	—	—
Commutative	yes	yes	no	no
Distributive	× over + yes		× over − yes	

23.

Set	Reals R			
Operation	+	×	−	÷
Closure	yes	yes	yes	yes
Associative	yes	yes	no	no
Identity	yes	yes	no	no
Inverse	yes	yes	—	—
Commutative	yes	yes	no	no
Distributive	× over + yes		× over − yes	

25. a.

#	1	2	3	4
1	1	2	3	4
2	2	4	6	8
3	3	6	9	12
4	4	8	12	16

b.

*	1	2	3	4
1	2	2	2	2
2	4	4	4	4
3	6	6	6	6
4	8	8	8	8

c.

Property	#	*
Closure	no	no
Associative	yes	no
Identity	yes	no
Inverse	no	no
Commutative	yes	no

Distributive property of # over *: no.
Distributive property of * over #: no.
(Other distributive properties could
be checked.)

27. a. yes   b. yes   c. no, $A \star E \neq E \star A$

29.

Ω	A	B	C	D	E	F
A	A	B	C	D	E	F
B	B	A	D	C	F	E
C	C	E	A	F	B	D
D	D	F	B	E	A	C
E	E	C	F	A	D	B
F	F	D	E	B	C	A

31. a. yes; it is A

b. yes; inverse of A is A;
of B is B;
of C is C;
of D is E;
of E is D;
of F is F.

PROBLEM SET 6.2, PAGES 335–337

1. $A = 5$ **3.** $C = 10$ **5.** $E = 1/3$ **7.** $G = 6$ **9.** $I = 7$ **11.** $K = 9$ **13.** $M = ^-9$ **15.** $\emptyset = ^-10$
17. $Q = ^-2$ **19.** $S = 8$ **21.** $U = 3$ **23.** $W = ^-15$ **25.** $Y = ^-28$

PROBLEM SET 6.3, PAGES 343–347

1. A prime number is a counting number that has exactly two divisors. **3.** a. not prime b. prime
c. prime d. prime e. not prime **5.** a. T b. T c. F d. T e. T **7.** a. no b. yes
9. 2,3,5,7,11,13,17,19,23,29,31,37,41,43,47,53,59,61,67,71,73,79,83,89,97,101,103,107,109,113,127,131,
137,139,149,151,157,163,167,173,179,181,191,193,197,199,211,223,227,229,233,239,241,251,257,263,269,
271,277,281,283,and 293 **11.** answers vary **13.** See solution to Problem 9 for some possible pairs. **15.** yes
17. See answer to Problem 9 for primes less than 300; 307,311,313,317,331,337,347,349,353,359,367,373,379,383,389,
397,401,409,419,421,431,433,439,443,449,457,461,463,467,479,487,491,499.

PROBLEM SET 6.4, PAGES 351–354

1. a. $2^3 \cdot 3$ b. $2 \cdot 3 \cdot 5$ **3.** a. $2^2 \cdot 7$ b. $2^2 \cdot 19$ **5.** a. $2^2 \cdot 3^3$ b. $2^2 \cdot 5 \cdot 37$ **7.** a. $2^3 \cdot 3 \cdot 5$ b. $2 \cdot 3^2 \cdot 5$
9. a. $2 \cdot 5 \cdot 7^2$ b. $2^4 \cdot 3^3 \cdot 11$ **11.** prime **13.** prime **15.** $13 \cdot 29$ **17.** $3 \cdot 5 \cdot 7$ **19.** prime
21. $3^2 \cdot 5 \cdot 7$ **23.** answers vary **25.** $2^5 \cdot 3^3 \cdot 5 \cdot 7 \cdot 17$ **27.** $3^4 \cdot 7$ **29.** $19 \cdot 151$
31.
```
10   PRINT "WHAT IS N";
20   INPUT N
25   LET N = ABS(N)
30   PRINT "POSITIVE FACTORS OF N ARE:"
40   FOR I = 1 TO N
50   GOSUB 200
60   NEXT I
70   GOTO 300
200  IF INT(N/I) <> N/I GOTO 220
210  PRINT I
220  RETURN
300  END
```
33. a. RUN   b. RUN   c. RUN   d. RUN
WHAT IS N? 1763   WHAT IS N? 15049   WHAT IS N? 128137   WHAT IS N? 614777
41   101   97   701
43   149   1321   877
READY   READY   READY   READY
35. Suppose that the nth prime is the largest prime. Denote these primes by $p_1, p_2, p_3, \ldots, p_n$. Consider $M = p_1 \cdot p_2 \cdot p_3$
$\cdots p_n + 1$. Obviously $M > p_n$. M is prime or composite. If M is prime, then we have found a larger prime. If M is

composite, it has a prime divisor by the Fundamental Theorem of Arithmetic. Now, $p_1, p_2, \ldots, p_n$ do not divide M, so there is some prime larger than p_n that divides M. In either case we have found a prime larger than p_n. Therefore, there is no largest prime.

PROBLEM SET 6.5, PAGES 359–364

1. The greatest common factor is 1.   **3.** a. $2^2 \cdot 3$ b. $2 \cdot 3^3$ c. $3^2 \cdot 19$ d. prime   **5.** 95   **7.** 1
9. 2   **11.** 3   **13.** 360   **15.** 2052   **17.** 180   **19.** 252   **21.** 1800   **23.** The product of the g.c.f. and l.c.m. for any two numbers equals the product of those numbers.   **25.** yes   **27.** no   **29.** a. 6/35
b. 3/20   c. 5/14   d. 7/15   **31.** 11/15   **33.** $^{-}3/4$ **35.** $^{-}53/48$ **37.** $^{-}15/56$ **39.** 131/25
41. $^{-}92/29$ **43.** $^{-}5\ 23/40$ **45.** 18/35   **47.** 4/5   **49.** 22/105   **51.** $^{-}21/40$ **53.** 8/3   **55.** a. 73
b. 47   c. 457
57.
```
 10   PRINT "WHAT ARE THE TWO NUMBERS";
 20   INPUT A,B
 23   LET N = A*B
 25   IF A < = 0 GOTO 98
 26   IF B < = 0 GOTO 98
 27   PRINT
 28   IF INT(B) < > B GOTO 98
 30   IF A > = B GOTO 60
 40   LET C = A
 45   LET A = B
 50   LET B = C
 60   LET D = A − B*INT(A/B)
 65   LET A = B
 70   LET B = D
 80   IF D < >0 GOTO 60
 90   PRINT "THE GREATEST COMMON FACTOR IS ";
 91   PRINT A
 93   PRINT "THE LEAST COMMON MULTIPLE IS ";
 95   PRINT N/A
 96   GOTO 105
 98   PRINT "PLEASE INPUT POSITIVE INTEGERS ONLY."
 99   PRINT
100   GOTO 10
105   END
```

PROBLEM SET 6.6, PAGES 373–382

1. a. 3   b. 10   c. 3   d. 2   **3.** a. 12   b. 10   c. 8   d. 3   **5.** a. no solution   b. 12
7. a. T   b. F   c. T   d. F   **9.** a. T   b. T   c. F   d. F   **11.** a. 3 (mod 4)   b. 3 (mod 5)
c. 4 (mod 8)   d. 2 (mod 9)   **13.** a. 3 (mod 5)   b. 9 (mod 11)   c. 0 (mod 2)   d. 6 (mod 13)   **15.** a. 6 (mod 9)   b. no solution   c. 1 and 4 (mod 5)   d. 5 (mod 13)   **17.** a. 5 (mod 6)   b. 3 (mod 7)   c. 0 (mod 4)   d. 3 (mod 4)   **19.** On a 24-hour clock, $8 + 8k \equiv 0, 8, 16 \pmod{24}$ for any integral value of k; thus you will never have to take your medication between midnight and 8 A.M. The medication would be taken at 8:00 A.M, 4:00 P.M., and 12 midnight.   **21.** You must buy $12x$ inches of rope, where x is the number of multiples you will buy. Also, you need $7k + 80$ inches, where k is the number of pieces you will obtain. This means that $12x \equiv 80 \pmod{7}$. Solving, $x \equiv 2 \pmod{7}$. Thus, $x = 2, 9, 16, 23, \ldots$. If $x = 16$, you will have one piece 80 inches long and 16 pieces 7 inches long with no waste. Thus, you must buy 192 inches of rope (or **16** ft of rope).   **23.** April and July
25. a. yes   b. yes, 12   c. yes

27.

×	0	1	2	3	4	5
0	0	0	0	0	0	0
1	0	1	2	3	4	5
2	0	2	4	0	2	4
3	0	3	0	3	0	3
4	0	4	2	0	4	2
5	0	5	4	3	2	1

a. *Closure*
   Yes, since the table has no new entries.
b. *Associative*
   Yes, try several examples.
c. *Identity*
   Yes, the identity is 1.
d. *Inverse*
   No
   $0 \cdot ? = 1$; 0 has no inverse.
   $1 \cdot 1 = 1$; inverse of 1 is 1.
   2 has no inverse.
   3 has no inverse.
   4 has no inverse.
   $5 \cdot 5 = 1$; inverse of 5 is 5.

29. The only change necessary is on line 105; change + to∗.   **31.** a. yes   b. yes   c. yes   **33.** This program assumes that a positive integer will be the input value. Note: those numbers not listed in each modulus do not have multiplicative inverses.

a.
Number	Multiplicative Inverse
1	1
2	3
3	2
4	4

b.
Number	Multiplicative Inverse
1	1
5	5

c.
Number	Multiplicative Inverse
1	1
2	9
3	6
4	13
5	7
6	3
7	5
8·	15
9	2
10	12
11	14
12	10
13	4
14	11
15	8
16	16

35.
Number	Multiplicative Inverse
1	1
3	7
7	3
9	9

37. Humpty Dumpty is a fall guy.   **39.** Midas had a gilt complex.
41. D UPEOTEHEQAUXUJGEWAUOEHEQHGHEXAUNTDDCRZEDN   UULEHRQECEPC LLED UPEVUJEHERCRNUOXJGTAB

PROBLEM SET 6.7, PAGES 385–387

1. Answers vary; see text for statements of properties.   **2.** $a \mid b$ means that there exists a natural number k so that b

$= ak.$ **3.** a. True, since $714 = 6 \cdot 119.$ b. False, since $8 \backslash 812.$ c. False, since $11 \backslash 095$ (095 was found by the scratch method).   d. False, $9 \backslash 35$ (35 was found by adding the digits).

4.

+	0	1	2	3
0	0	1	2	3
1	1	2	3	0
2	2	3	0	1
3	3	0	1	2

×	0	1	2	3
0	0	0	0	0
1	0	1	2	3
2	0	2	0	2
3	0	3	2	1

5. It satisfies the closure, associative, identity, inverse, and commutative properties for addition. It satisfies the closure, associative, identity, and commutative properties for multiplication. It does not satisfy the inverse property for multiplication. Therefore, it is not a field.   **6.** $3 \cdot 5^2 \cdot 7 \cdot 13$ **7.** a. prime   b. not prime   c. prime
8. answers vary   **9.** The greatest common factor of a set of numbers is the largest number that divides evenly into all members of the set.   **10.** It is the prime factorization of a number that is written in exponential notation with the factors written in order of increasing magnitude.   **11.** a. 3   b. 7   **12.** a. $2 \cdot 3^2 \cdot 5 \cdot 7 \cdot 11 = 6930$ b. $7^4 \cdot 11 \cdot 13$ $= 343,343$ **13.** 17/19   **14.** a. 7/72   b. $^-7/9$ **15.** a. 1396/3465   b. 708/7007 or $708/7^2 \cdot 11 \cdot 13$
16. a. 2   b. 2   c. 2   d. 4   e. 2   **17.** a. T   b. F   c. F   d. T   e. F   **18.** a. $x \equiv 2 \,(\text{mod}$ 8)   b. $x \equiv 1 \,(\text{mod } 2)$ c. $x \equiv 5 \,(\text{mod } 7)$ d. $x \equiv 3 \,(\text{mod } 6)$ e. $x \equiv 4 \,(\text{mod } 7)$ **19.** a. $x = 3$
b. $x = 4$ c. $x = 2$ d. $x = 1/4$ e. $x = 9$ **20.** a. $x = {}^-7/3$ b. $x = 3$ c.
d. $x = 11/12$ e. $x = 17/3$

PROBLEM SET 7.1, PAGES 394–400

1. An axiom is accepted without proof; a theorem is a proved result.   **3.** You should be able to see both a young and an old woman.   **5.** a. From the photograph of this apple, it appears symmetric.   b. yes   c. no   d. yes
e. no   **7.** answers vary   **9.** g. Answer should be about .618.   **11.** c. Answer should be about .6   d. They should be about the same.

PROBLEM SET 7.2, PAGES 406–411

1. Saccheri quadrilateral   **3.** not a Saccheri quadrilateral   **5.** one line   **7.** a. They never intersect   b. Only great circles are considered to be lines.   **9.** The sum of the angles is greater than 180°.   **11.** North Pole
13. answers vary   **15.** answers vary

PROBLEM SET 7.3, PAGES 415–419

Problems 1, 8, and 12; 2, 9, and 13; 3 and 10; 5, 6, and 7; and 4, 14, and 15 are topologically equivalent. Problems 17, 18, 20, 23, 24, 25, 26, 29, 30, and 31; 19, 21, and 27; and 22, and 28 are topologically equivalent.   **33.** B and C are inside; A is outside.   **35.** A and B are outside; C is inside.   **37.** It will cross an odd number of times for $\overline{AX}$ and $\overline{CX}$ and an even number of times for $\overline{BX}$.   **39.** It will cross an odd number of times for $\overline{AX}$ and $\overline{CX}$ and an even number of times for $\overline{BX}$.   **41.** C, I, L, M, N, S, U, V, W, and Z; D and O; E, F, G, T, and Y; K and X; and Q and R are topologically equivalent.

PROBLEM SET 7.4, PAGES 422–426

1. traversible   **3.** not traversible   **5.** traversible   **7.** not traversible   **9.** traversible   **11.** not traversible, since there are odd vertices at $C, S, L,$ and H **13.** Transform the problem into the following network:

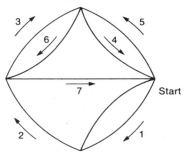

It is now traversible, as shown by the arrows.

15. possible   **17.** possible   **19.** If all the rooms have an even number of doors, or if there are exactly two rooms with an odd number of doors, then the problem can be done.   **21.** $V + R - 2 = A$ **23.** The result is a twisted band twice as long and half as wide as the band you began with.   **25.** Two interlocking pieces, one twice as long as the other   **27.** One edge; one side; after cutting, you will have a single loop with a knot in it.

PROBLEM SET 7.5, PAGES 428–429

1. A Saccheri quadrilateral is a quadrilateral $ABCD$ with base angles A and B right angles and with sides AC and BD with equal lengths. If the summit angles C and D are right angles, then the result is Euclidean geometry. If they are acute, then the result is hyperbolic geometry. If they are obtuse, the result is elliptic geometry.   **2.** Answers vary; see Table 7.1, page 405   **3.** a, c, and e; b and d   **4.** a and d   **5.** a. 2   b. 1   c. 0   d. 1   e. more than 3
6. a. inside   b. inside   **7.** not traversible   **8.** traversible   **9.** possible   **10.** possible   **11.** answers vary   **12.** answers vary   **13.** 2 edges, 2 sides; the result is two interlocking loops.
14.

	F	E	V
tetrahedron	4	6	4
hexahedron	6	12	8
octahedron	8	12	8
dodecahedron	12	30	20
icosahedron	20	30	12

15. $V + F = E + 2$

PROBLEM SET 8.1, PAGES 435–441

1. If task A can be performed in m ways, and, after task A is performed, a second task, B, can be performed in n ways, then task A followed by task B can be performed in $m \cdot n$ ways.   **3.** 786,240   **5.** 2730   **7.** 11 squares   **9.** 60
11. a. 10   b. 8   c. 8   **13.** a. 9×10^8 b. 8.1×10^8 **15.** 48   **17.** answers vary   **19.** No; all numbers but 13 are possible.
21.
```
10   Let I = 0
20   FOR A = 1 TO 15
30   FOR B = 1 TO 15
40   FOR C = 1 TO 15
60   IF B = A GOTO 140
```

```
 62   IF C = A GOTO 140
 64   IF C = B GOTO 140
 70   LET I = I + 1
140   NEXT C
150   NEXT B
160   NEXT A
170   PRINT I
180   END
RUN
2730
READY
```
23. 41,472 possibilities

PROBLEM SET 8.2, PAGES 447–450

1. 4920   **3.** 90   **5.** 600   **7.** 5040   **9.** 22,100   **11.** 720   **13.** 1680   **15.** 40,296   **17.** 56
19. 84   **21.** 9   **23.** 504   **25.** 1   **27.** 132,600   **29.** 24   **31.** 95,040   **33.** 1680   **35.** $g!/(g - h)!$
37. 52   **39.** 3360   **41.** 5040   **43.** 360   **45.** 8   **47.** $n!/(n - r)!$ **49.** a. 1024   b. 59,049
51. a. 42   b. 30,240   c. 6.704426E + 11   d. 1.551121E + 25   e. 7.309658E + 25

PROBLEM SET 8.3, PAGES 452–456

1. 9   **3.** 84   **5.** 1   **7.** 22,100   **9.** 1   **11.** 35   **13.** 1225   **15.** $g!/h!(g - h)!$ **17.** 1   **19.** 792
21. 792   **23.** a. 4   b. 6   **25.** 24   **27.** $_nC_r = {_nP_r}/r!$ or $n!/r!(n - r)!$ **29.** a. 35   b. 55   c. 1716
31. a. 210   b. 28   c. 330
35.
```
 15   LET B = 1
 20   PRINT "I WILL COMPUTE N/C/R = N!/R!(N − R)!"
 25   PRINT "WHAT IS THE VALUE OF N";
 30   INPUT N
 35   PRINT "WHAT IS THE VALUE OF R";
 40   INPUT R
 41   IF INT(R − N)<>R − N GOTO 100
 42   IF N <0 GOTO 100
 43   IF R <0 GOTO 100
 44   IF N − R <0 GOTO 110
 45   IF R =0 GOTO 95
 46   FOR I = 1 TO R
 48   LET D = N − I + 1
 50   LET A = D/I
 55   LET B = A*B
 60   NEXT I
 85   PRINT "N/C/R = ";B
 90   GOTO 130
 95   LET B = 1
 97   GOTO 85
100   PRINT "THE VALUES FOR N AND R MUST BE   ";
102   PRINT "POSITIVE INTEGERS."
104   PRINT
105   GOTO 25
110   PRINT "SHAME ON YOU. R CANNOT BE   ";
112   PRINT "LARGER THAN N. BUT I WILL GIVE";
115   PRINT "YOU ANOTHER CHANCE. NOW,   ";
120   GOTO 25
130   END
```

PROBLEM SET 8.4, PAGES 460–464

1. In a permutation the order is important; in a combination the order of arrangement of the objects is not important.
3. a. combination   **b.** neither   **c.** neither   **d.** permutation   **e.** permutation   **5. a.** permutation
b. combination   **c.** combination   **d.** permutation   **e.** combination   **7. a.** 50   **b.** 33   **c.** 16
d. 67   **9. a.** 15   **b.** 31   **c.** 63   **11.** 15   **13. a.** 24   **b.** 6720   **c.** 120   **d.** 120   **e.** 286
15. a. 720   **b.** 66   **c.** 35   **d.** 5040   **e.** 21   **17.** answers vary   **19. a.** 24,360   **b.** 4060
21. 1,001,101

PROBLEM SET 8.5, PAGE 465

1. 718   **2.** 4   **3.** 24   **4.** 6   **5.** 360   **6.** 10   **7.** 12   **8.** 1   **9.** 6   **10.** 220   **11.** Happy: 60
College: 1260   **12.** Answers may vary. In a permutation the order of the arrangement is important, but in a combination the order of arrangement is not relevant. Permutations can be found in the following way:

$$\underbrace{{}_nP_r = n(n - 1)(n - 2) \cdots (n - r + 1)}_{r \text{ factors}}$$

Combinations are related to permutations rather closely, as we see by writing the formula for finding combinations:

$$_nC_r = \frac{_nP_r}{r!}$$

13. The total number of ways possible is $_{10}P_{10} = 10! = 3,628,800 \approx 10,000$ years!
14. The number of ways three balls can be drawn is

$$_{10}C_3 = \frac{10 \cdot 9 \cdot 8}{3 \cdot 2 \cdot 1} = 120.$$

The number of ways of drawing no red balls (all white balls) is

$$_6C_3 = \frac{6 \cdot 5 \cdot 4}{3 \cdot 2 \cdot 1} = 20.$$

Thus, the number of ways of drawing at least one red is
$$120 - 20 = 100.$$

15. 25   **16.** 15   **17.** 14   **18.** 26   **19.** 120   **20.** Their claim is correct. There are eight ingredients, so the number of hamburgers is the number of subsets of a set with eight elements or $2^8 = 256$.

PROBLEM SET 9.1, PAGES 474–483

11. If we specify the probability as a percent, the probability must be between 0% (never occurs) and 100% (always occurs). Thus the probability of an event will always be between 0 and 1 (writing the percent as a fraction).
13.
```
 5  PRINT
 6  PRINT
 7  DIM A(5), C(10)
 8  LET C = -8
10  GOSUB 200
20  FOR J = 1 TO 100
25  FOR K = 1 TO 7
30  FOR I = 1 TO 5
40  RANDOMIZE
50  LET X = INT(RND(Z)*10)
```

```
 60   LET A(I) = X
 65   GOSUB 300
 70   NEXT I
 80   LET A = A(1) + 10*A(2) + 100*A(3) + 1000*A(4) + 10000*A(5)
 85   LET C = C + 8
 90   IF A(5) = 0 GOTO 125
 95   PRINT TAB(C);A;
100   NEXT K
102   Let C = −8
104   PRINT
105   NEXT J
110   PRINT
111   PRINT
112   PRINT C(1); C(2); C(3); C(4); C(5); C(6); C(7); C(8); C(9); C(0)
120   STOP
125   IF A(4) = 0 GOTO 140
130   PRINT TAB(C + 1); A;
135   GOTO 100
140   IF A(3) = 0 GOTO 160
145   PRINT TAB(C + 2); A;
150   GOTO 100
160   PRINT TAB(C + 3); A;
165   GOTO 100
200   FOR K = 0 TO 9
210   LET C(K) = 0
220   NEXT K
230   RETURN
300   IF A(I) = 0 GOTO 320
302   IF A(I) = 1 GOTO 322
304   IF A(I) = 2 GOTO 324
306   IF A(I) = 3 GOTO 326
308   IF A(I) = 4 GOTO 328
310   IF A(I) = 5 GOTO 330
312   IF A(I) = 6 GOTO 332
314   IF A(I) = 7 GOTO 334
316   IF A(I) = 8 GOTO 336
318   IF A(I) = 9 GOTO 338
320   LET C(0) = C(0) + 1
321   RETURN
322   LET C(1) = C(1) + 1
323   RETURN
324   LET C(2) = C(2) + 1
325   RETURN
326   LET C(3) = C(3) + 1
327   RETURN
328   LET C(4) = C(4) + 1
329   RETURN
330   LET C(5) = C(5) + 1
331   RETURN
332   LET C(6) = C(6) + 1
333   RETURN
334   LET C(7) = C(7) + 1
335   RETURN
336   LET C(8) = C(8) + 1
337   RETURN
338   LET C(9) = C(9) + 1
339   RETURN
400   END
```

The above program assumes that N is a positive integer.

17. The probability is 1; one number will always be directly above its counterpart.

PROBLEM SET 9.2, PAGES 491–496

1. Empirical probability is the result of experimentation and could vary from experiment to experiment; it is dependent on the number of times the event is repeated. The theoretical probability is the numerical value obtained by using a mathematical model. The theoretical probability should be predictive of the empirical probability for a large number of trials.   **3.** Answers vary. If s is the number of ways an event can favorably occur, then it must occur at least 0 times.   **5.** a. 6/13   b. 7/13   c. 0   **7.** a. 1/6   b. 1/3   c. 1/2   d. 2/3   e. 0   **9.** a. 1/100   b. 1/25   c. 19/20   d. 99/100   e. 1/5   f. 1/4   **11.** a. 1/12   b. 1/4   **13.** a. 5/8   b. 7/8   c. 3/4   **15.** a. 1/36   b. 5/12   **17.** a. 1/5   b. 4/25   c. 1   d. 1/50   e. 17/50   **19.** 1/9   **21.** a. 1/8   b. 1/2   c. 1/4   **23.** .00000 15390 77   **25.** 3   **27.** A; you would win 2/3 of the time.   **29.** C; you would win 2/3 of the time.

PROBLEM SET 9.3, PAGES 501–503

1. a. BBBB; BBBG; BBGB; BBGG; BGBB; BGBG; BGGB; BGGG; GBBB; GBBG; GBGB; GBGG; GGBB; GGBG; GGGB; GGGG   b. 1/16   c. 1/4   d. 3/8   e. 1   **3.** 3/7   **5.** 1 to 12   **7.** 15 to 1   **9.** 33/34   **11.** 2/11   **13.** 2 to 7

PROBLEM SET 9.4, PAGES 509–519

1. $.83   **3.** The barber's expectation is about 9¢, but he is most likely to receive a nickel.   **5.** $.02   **7.** a. $12   b. $30   c. $318   **9.** About 5¢   **11.** $2515   **13.** a. 1/2197   b. 2/13   c. 4/169   d. 4/2197   e. 8/2197   f. 27/2197   g. 27/2197   h. 8/2197   **15.** .691   **17.** $1.05   **19.** $1.03

PROBLEM SET 9.5, PAGES 520–521

1. a. 1/16   b. 1/3   c. 0   **2.** a. 1/4   b. 1/52   c. 4/13   **3.** If an event E can occur in any one of n mutually exclusive and equally likely ways, and if s of those ways are considered successful, then the probability of the event is given by

$$P(E) = \frac{s}{n}.$$

4. 10/11   **5.** 29/36   **6.** a. 3/5   b. 1/2   **7.** 11/36; 1/36   **8.** 6/11   **9.** a. 1/2   b. 1 to 1   **10.** No; he should be willing to pay about 62¢ each for the tickets.

PROBLEM SET 10.1, PAGES 527–537

1.

Outcome	Tally	Frequency
72	|	1
71	||	2
70	|||	3
69	||||	4
68	|||	3
67	⊮	5
66	|||	3
65	|||	3
64	||||	4
63	||	2

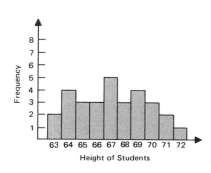

3.

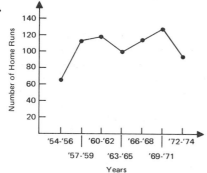

5. a.

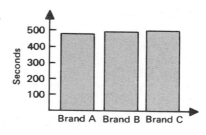

b.

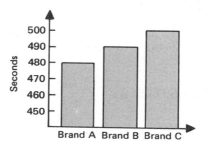

c. graph *b*

d. no; answers vary

7. Answers vary; the graph is meaningless without labeling the units.

9.

Outcome	Tally	Frequency
Sun	||||	4
Mon	Ⅲ Ⅲ Ⅲ |	16
Tue	||||	4
Wed	|	1
Thu	Ⅲ ||||	9
Fri	Ⅲ |||	8
Sat	|||	3

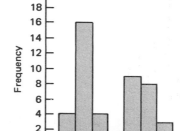

11. answers vary

13.

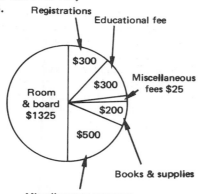

15. a. 1974   b. 1974   c. 1968   d. 1972

PROBLEM SET 10.2, PAGES 544–548

1. Mean = 3; median = 3; no mode; range = 4; σ = 1.58   **3.** Mean = 105; median = 105; no mode; range = 4; σ = 1.58   **5.** Mean = 6; median = 7; mode = 7; range = 8; σ = 2.58   **7.** mean = 10; median = 8; no mode; range = 18, σ = 7.21   **9.** Mean = 91; median = 95; mode = 95; range = 17; σ = 7.11   **11.** Same range and same standard deviation   **13.** Mean = $7900; median = $8000; mode = $5000   **15.** All data are the same.   **17.** Mean = 56; median = 65; mode = 70; range = 90   **21.** Mean = 19,200; range = 4000; standard deviation = 1483.24   **23.** Mean = 5.55; median = 5; mode = 3; range = 11; σ = 2.68; median (or mean) is most descriptive.   **25.** answers vary   **27.** 26.73   **31.** Mean = 32.62820513; variance = 194.8859474; standard deviation = 13.96015571

PROBLEM SET 10.3, PAGES 551–553

1. 34   **3.** 8   **5.** 25   **7.** 1   **9.** .022   **11.** .978   **13.** It will exceed 40.5 inches about 2.2% of the time.   **15.** .022   **17.** a. 1   b. 22   c. 136   d. 682   e. 136   f. 22   g. 1   **19.** 12.9 oz mean weight

PROBLEM SET 10.5, PAGES 557–559

1. mean = 25; median = 25; no mode
2. Tally   Frequency
　6　　　2
　7　　　4
　8　　　5
　9　　　3
　10　　　1

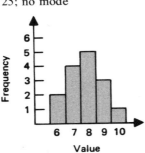

3. range = 4; variance = 2.5; std. dev. = 1.581139   **4.** mean = 7.8; median = 8; mode = 8; range = 4   **5.** Not possible, unless every person had *exactly* the same sales.   **6.** mean = 13.5; median = 14; mode = 16; mode is the most useful.   **7.** mean = 79; median = 74; no mode; mean is most representative.   **8.** range = 19; variance = 72.5; std. dev. 85.15   **9.** mean = $14,666; median = $15,000; mode = $18,000; median is most representative.
10. 68.2%   **11.** About 220 women   **12.** answers vary   **13.** The class with the smaller standard deviation would be more homogeneous and consequently easier to teach.   **14.** Lower score is 65; upper score is 81.   **15.** She did better on her math test since she was 2 std. dev. above the mean whereas in history she was only 1 std. dev. above the mean.   **16.** We don't know how many were surveyed; we don't know how the sample was chosen; we don't know what percentage of dentists didn't respond; we don't know how the question was asked.   **17.** About .022(1000) = 22 children.
19. .47(360°) ≈ 169°
　.10(360°) ≈ 36°
　.30(360°) ≈ 108°
　.07(360°) ≈ 25°
　.06(360°) ≈ 22°

20. a.

b.

(Note: These figures could be even more exaggerated if we wished.)

FINAL REVIEW ANSWERS

1. 29   **2.** a. 112   b. 11/3   c. 3   **3.** a. arithmetic   b. geometric   c. neither   **4.** a, b, and c
answers vary   d. all should be about .6   **5.** a. centi   b. hecto   c. kilo   d. deci   **6.** a. liter
b. meter   c. gram   d. Celsius

7.

		conjunction	disjunction	negation	conditional	biconditional
p	q	$p \wedge q$	$p \vee q$	$\sim p$	$p \rightarrow q$	$p \leftrightarrow q$
T	T	T	T	F	T	T
T	F	F	T	F	F	F
F	T	F	T	T	T	F
F	F	F	F	T	T	T

8. a. statement   b. not a statement   c. statement (provided "friends" is well defined)   d. statement (there
exists a person who . . .)   e. statement

9.

p	q	$p \wedge q$	$\sim q$	$p \vee \sim q$	$(p \wedge q) \rightarrow (p \vee \sim q)$
T	T	T	F	T	T
T	F	F	T	T	T
F	T	F	F	F	T
F	F	F	T	T	T

10.

p	q	r	$p \vee q$	$\sim r$	$(p \vee q) \vee (\sim r)$
T	T	T	T	F	T
T	T	F	T	T	T
T	F	T	T	F	T
T	F	F	T	T	T
F	T	T	T	F	T
F	T	F	T	T	T
F	F	T	F	F	F
F	F	F	F	T	T

11. Answers may vary.   a. $\triangle ABC$ is isosceles and the base angles are congruent.   b. $\triangle ABC$ is isosceles or its base angles are congruent.   c. If $\triangle ABC$ is isosceles, then the base angles are congruent.   d. $\triangle ABC$ is isosceles if and only if its base angles are congruent.   e. If the base angles of $\triangle ABC$ are not congruent, then it is not an isosceles triangle.

12. a.

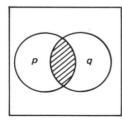

b.

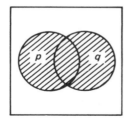

c.

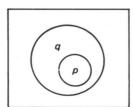

d.

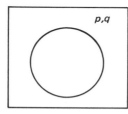

e.

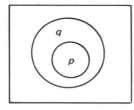

13. a. T   b. T   c. T   d. T   e. T   **14.** answers vary   **15.** answers vary   **16.** answers vary

17. They are all true.

18. $\dfrac{a}{b} = \dfrac{c}{d}$ if an only if $ad = bc$ **19.** a. $\{5\}$ b. $\{1,2,3,4,5,6,8\}$ c. $\{2,3,5,7,9,10\}$

20. $\{1,2,3,4,5,6,7,8,9,10\}$ or U **21.**

$A \cap (B \cup C)$

=

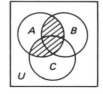

$(A \cap B) \cup (A \cap C)$

22. We did not go to the concert.   **23.** You are careful.   **24.** Dallas goes to the Super Bowl.   **25.** Your parents don't both have type O blood.   **26.** control   **27.** input, control, memory, accumulator, and output   **28.** 7735.05 (SO·SELL)   **29.** a. 59,000,000,000,000,000,000,000,000   b. 5.9E + 25   **30.** a. $5.78 \cdot 10^8$ b. 5.78E + 8

31. a. $7.5 \cdot 10^{-8}$ b. .000000075

32.
```
10  PRINT "WHAT IS THE FIRST NUMBER";
20  INPUT N
30  PRINT "WHAT IS THE SECOND NUMBER";
40  INPUT M
50  PRINT "THE AVERAGE IS";
60  PRINT (N + M)/2
70  END
```

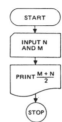

33.
```
10  LET A = 1
20  PRINT A;
30  LET B = 1
```

```
40   PRINT B;
50   LET C = A + B
60   PRINT C
70   LET A = B
80   LET B = C
90   GOTO 50
100  END
```

34. TWO TO THE SEVENTH POWER IS
TWO TIMES TWO TIMES TWO TIMES TWO TIMES TWO TIMES TWO TIMES TWO
AND THIS IS TOO MUCH.

35. Answers vary; a positional system is one in which the position of a numeral relative to the other numerals determines the value of that numeral.   **36.** $4 \times 10^6 + 5 \times 10 + 6 + 4 \times 10^{-1} + 3 \times 10^{-2}$ **37.** Answers vary; a number is the concept and a numeral is the symbol used to represent that number.   **38.** Answers vary; see text for precise statements.   **39.** a. 20   b. $^-8$ c. $^-32$ d. 4   e. 10   **40.** a. $^-9 = x$ b. $^-96 = x$ **41.** a. 7/6   b. 4/5   c. 9/4   d. 13   e. 3/4   **42.** a. 10/21   b. $^-21/55$

43. a.

b. $\{6, ^-9, 0, \sqrt{25}\}$ c. $\{6, \sqrt{25}\}$ d. $\{6, ^-9, 5.3, .\overline{345}, 0, 3.1415, \sqrt{25}, ^-3.2\}$ e. N **44.** 6/11   **45.** rational
46. 3333/10,000   **47.** Simple interest earns interest only on the principal even after the first compounding period; compound interest earns interest on the interest as well as on the principal after the first compounding period.
48. $1.71   **49.** annual percentage rate   **50.** You must apply for credit each time you use closed-end credit, and each purchase is paid for over a fixed amount of time with fixed payments; it is an installment purchase. Open-end credit approves a credit limit, and you can purchase any amounts as long as the cumulative total doesn't exceed the credit limit; it is a credit-card-type purchase.
51. a. $2114.28   b. 18%   **52.** 5133.66   **53.** $454   **54.** 12.75%   **55.** Answers vary; term is life insurance only and rates go up as your age increases; whole life gives uniform rates based on your age when you take out the policy, and you also build up some cash value; an endowment is a form of forced savings along with your insurance, which guarantees you a cash sum at some specific time in the future.   **56.** Addition: closure, associative, commutative, identity, and inverse. Multiplication: closure, associative, commutative, and identity.   **57.** Inverse for multiplication in addition to those mentioned in Problem 56.   **58.** a. $^-3/4$ b. 17   c. $^-6.2$ d. 0   **59.** a. 4/3   b. $^-1/17$ c. 5/31   d. none   **60.** a. $x = 105$ b. $x = ^-11/2$ c. $x = 9$ d. $x = 9$ **61.** A prime is a counting number with exactly two divisors.   **62.** Proper divisors of 28 are 1,2,4,7, and 14: $1 + 2 + 4 + 7 + 14 = 28$
63. their product   **64.** yes   **65.** 23/120   **66.** a. $x \equiv 8 \pmod{12}$ b. $x \equiv 5 \pmod 7$ c. $x \equiv 0 \pmod{10}$ d. $x \equiv 1, 2, \pmod 3$ **67.** A Saccheri quadrilateral is a quadrilateral $ABCD$ with base angles A and B right angles and sides AC and BD congruent (same length).   **68.** a and b; c, d, and e

69.

	R	A	V	V + R	A + 2	
a.	5	8	5	10	=	10
b.	5	8	5	10	=	10
c.	3	5	4	7	=	7
d.	3	3	2	5	=	5
e.	3	3	2	5	=	5

70. a. no   b. yes   **71.** answers vary   **72.** a. yes   b. yes   c. no   d. no

73.

74. a. no   b. no   **75.** two colors for each   **76.** a.

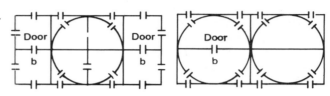

b. Add two doors as shown (labeled b).

77. Answers vary; see text for precise statement.   **78.** a. 2380   b. 3360

c. $\dfrac{x!}{(x-y)!}$ d. $\dfrac{w!}{z!(w-z)!}$ **79.** 6   **80.** 21   **81.** 5040   **82.** 31 ways   **83.** a. 12/25   b. 9/25

c. 6/25   d. 1/5   e. 1/25   **84.** a. 5/36   b. 2/11   c. 1/6   **85.** 7/8   **86.** 1/3   **87.** 37 to 1

88. $\dfrac{92}{_{97}P_5}$ or $\dfrac{92}{97 \cdot 96 \cdot 95 \cdot 94 \cdot 93}$ or $1.19 \cdot 10^{-8}$ **89.** $\dfrac{36}{38} \approx .9474$ or about 95¢   **90.** $.12   **91.** 251,000; 233,000

92. 1972: 252,000   mean $\approx$ 247,000; median $\approx$ 251,000; no mode
      1973: 251,000
      1974: 233,000
      1975: 243,000
      1976: 256,000

93. mean, median, and mode; examples will vary.   **94.** range, variance, and standard deviation; examples will vary.   **95.** Select the one that registers closest to the mode (that is, if several recorded the same temperature, select one of those).

96.

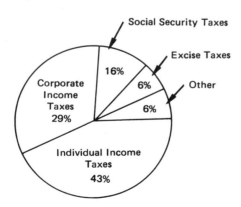

Social Security Taxes

Excise Taxes

16%

6%   Other

Corporate
Income
Taxes
29%

6%

Individual Income
Taxes
43%

97. mean $\approx$ 81.82 minutes; median $\approx$ 80 minutes; bimodal at 70 and 110 minutes; the median is most representative.

98. range = 70 minutes; variance ≈ 711.36 minutes; standard deviation ≈ 26.67 minutes.   **99.** Not acceptable, since we do not know enough about the poll.

100. Chapter 11   Chapter 12

SUBJECT INDEX

INDEX OF
SPECIAL PROBLEMS

INDEX OF MATHEMATICS PROJECTS

Ideas for mathematics projects and reports are listed throughout this book. References, sources, and specific ideas are provided. This list summarizes those ideas and gives a few additional suggestions. (If there is no page reference, consult an encyclopedia or history of mathematics.)